Skin and Arthropod Vectors

Skin and Anthropod vectors

Skin and Arthropod Vectors

Edited by

Nathalie Boulanger
Université de Strasbourg, Strasbourg, France

Academic Press is an imprint of Elsevier
125 London Wall, London EC2Y 5AS, United Kingdom
525 B Street, Suite 1800, San Diego, CA 92101-4495, United States
50 Hampshire Street, 5th Floor, Cambridge, MA 02139, United States
The Boulevard, Langford Lane, Kidlington, Oxford OX5 1GB, United Kingdom

Notices
Knowledge and best practice in this field are constantly changing. As new research and experience
broaden our understanding, changes in research methods, professional practices, or medical
treatment may become necessary.

Practitioners and researchers must always rely on their own experience and knowledge in
evaluating and using any information, methods, compounds, or experiments described herein. In
using such information or methods they should be mindful of their own safety and the safety of
others, including parties for whom they have a professional responsibility.

To the fullest extent of the law, neither the Publisher nor the authors, contributors, or editors,
assume any liability for any injury and/or damage to persons or property as a matter of products
liability, negligence or otherwise, or from any use or operation of any methods, products,
instructions, or ideas contained in the material herein.

Library of Congress Cataloging-in-Publication Data
A catalog record for this book is available from the Library of Congress

British Library Cataloguing-in-Publication Data
A catalogue record for this book is available from the British Library

ISBN: 978-0-12-811436-0

For information on all Academic Press publications visit our website at
https://www.elsevier.com/books-and-journals

  Working together
to grow libraries in
developing countries

www.elsevier.com • www.bookaid.org

Publisher: Andre G. Wolff
Acquisition Editor: Anna Valutkevich
Editorial Project Manager: Pat Gonzalez
Production Project Manager: Mohanapriyan Rajendran
Cover Designer: Matthew Limbert

Typeset by TNQ Books and Journals

Contents

Contributors

Chetan Aditya, Institut Pasteur, Paris, France

Rogerio Amino, Institut Pasteur, Paris, France

Pavlína Bartíková, Biomedical Research Centre, Institute of Virology, Slovak Academy of Sciences, Bratislava, Slovakia

Quentin Bernard, University of Maryland, College Park, MD, United States

Sarah Bonnet, UMR BIPAR 956 INRA-ANSES-ENVA, Maisons-Alfort, France

Nathalie Boulanger, Université de Strasbourg, Strasbourg, France

Cláudia Ida Brodskyn, Instituto Gonçalo Moniz, FIOCRUZ-Bahia, Salvador, Brazil

Cláudia Brodskyn, Gonçalo Moniz Institute- Osvaldo cruz Foundation (IG-FIOCRUZ-Ba), Salvador, Brazil

Guy Caljon, University of Antwerp, Wilrijk, Belgium

Van-Mai Cao-Lormeau, Pôle de recherche et de veille sur les maladies infectieuses émergentes, Institut Louis Malardé, Papeete, French Polynesia

Léa Castellucci, Serviço de Imunologia do Hospital Universitário Edgar Santos-Federal University of Bahia, Salvador, Brazil

Iliano V. Coutinho-Abreu, National Institute of Allergy and Infectious Diseases, National Institutes of Health

Claudia Demarta-Gatsi, Institut Pasteur, Unité de Biologie des Interactions Hôte Parasites, Paris, France; Centre National de la Recherche Scientifique ERL9195, Paris, France; INSERM U1201, Paris F-75015, France

Gérard Duvallet, Univ. Paul Valéry Montpellier 3, Univ. Montpellier, EPHE, CNRS, IRD, CEFE UMR 5175, Montpellier, France

Pauline Formaglio, Otto-von-Guericke University, Magdeburg, Germany

Ema Helezen, Université de Strasbourg, Strasbourg, France

Joppe W. Hovius, University of Amsterdam, Amsterdam, The Netherlands

Camila Indiani de Oliveira, Gonçalo Moniz Institute- Osvaldo cruz Foundation (IG-FIOCRUZ-Ba), Salvador, Brazil

Shaden Kamhawi, National Institute of Allergy and Infectious Diseases, National Institutes of Health

Mária Kazimírová, Institute of Zoology, Slovak Academy of Sciences, Bratislava, Slovakia

Cédric Lenormand, Université de Strasbourg, Strasbourg, France

Laura Mac-Daniel, Institut Pasteur, Paris, France; Loyola University Chicago, Maywood, IL, United States

Dorien Mabille, University of Antwerp, Wilrijk, Belgium

Louis Maes, University of Antwerp, Wilrijk, Belgium

Lauren M.K. Mason, University of Amsterdam, Amsterdam, The Netherlands

Salaheddine Mécheri, Institut Pasteur, Unité de Biologie des Interactions Hôte Parasites, Paris, France; Centre National de la Recherche Scientifique ERL9195, Paris, France; INSERM U1201, Paris F-75015, France

Robert Ménard, Institut Pasteur, Paris, France

Christopher G. Mueller, CNRS UPR3572, Université de Strasbourg, Strasbourg, France

Richard E. Paul, Institut Pasteur, Functional Genetics of Infectious Diseases Unit, Paris, France; Centre National de la Recherche Scientifique, Unité de Recherche Associée 3012, Paris, France

Jennifer Richardson, UMR 1161 Virologie INRA-ANSES-ENVA, Maisons-Alfort, France

Vincent Robert, MIVEGEC, IRD, CNRS, Université de Montpellier, Montpellier, France

Jeffrey G. Shannon, National Institute of Allergy and Infectious Diseases (NIAID), National Institutes of Health (NIH), Hamilton, MT, United States

Ladislav Šimo, UMR BIPAR 956 INRA-ANSES-ENVA, Maisons-Alfort, France

Jeroen Spitzen, Wageningen University & Research, Wageningen, The Netherlands

Iveta Štibrániová, Biomedical Research Centre, Institute of Virology, Slovak Academy of Sciences, Bratislava, Slovakia

Joana Tavares, Universidade do Porto, Porto, Portugal

Natalia Tavares, Gonçalo Moniz Institute- Osvaldo cruz Foundation (IG-FIOCRUZ-Ba), Salvador, Brazil

Jesus G. Valenzuela, National Institute of Allergy and Infectious Diseases, National Institutes of Health

Niels O. Verhulst, Wageningen University & Research, Wageningen, The Netherlands; University of Zurich, Zurich, Switzerland

Preface

The World Health Organization reported in 2016 that arthropod-borne diseases (ABDs) account for more than 17% of all infectious disease cases, causing more than 1 million deaths annually. Parasitic diseases such as malaria are responsible for around half of these annual deaths. Viral infections, such as dengue, chikungunya, yellow fever, and Zika, transmitted by the *Aedes* mosquitoes have extended their areas of threat not only due to globalization of travel and trade but also due to climatic changes. Although less in the spotlight, tick-borne diseases such as Lyme borreliosis, relapsing fevers, and rickettsioses are expanding as well, as a result of anthropogenic changes in our environment.

Pharmacological treatments are available for many of these diseases affecting humans and animals worldwide, but they are often impaired by drug resistances. Only few efficient vaccines are available as alternatives so far, which target mainly viral infections such as yellow fever, Japanese encephalitis, or tick-borne encephalitis. Regarding the arthropod vectors, remarkable results have been achieved over the last years to better control their spread by insecticide/acaricide intervention, unfortunately at the price of growing resistances. Impressive and innovative research is focusing on the development of transgenic arthropods, but there are still gaps to overcome on the way toward an efficient control of ABDs.

Nevertheless, we have made progress. Tremendous multidisciplinary and translational approaches have been undertaken to improve our understanding of ABDs. Thanks to these efforts, ABDs are now much better perceived as interplay of three actors: the vector, the pathogen, and the host, always considering the role of each other and their mutual interactions. The host has been too long for the only concern of research. Control of pathogen invasion and of the clinical manifestations seemed to be the single approach to fight these diseases, until the arthropod vector received the appropriate attention as a key player in the development of ABDs. The worldwide spread of the *Aedes* mosquito has promoted this paradigm shift and emphasized the need of new control strategies.

Arthropod vectors, which were originally described as simple needles inoculating pathogens, are now seen as complex and sophisticated machineries used by pathogens to efficiently ensure their transmission. They control the hosted pathogen population and the development of their antigenicity by their innate immunity. Insects and ticks acting as vectors for ABDs are bloodsucking arthropods. They inoculate infectious microorganisms to the vertebrate host at the skin interface, thus actively transmitting pathogens between humans and from animals to humans. Not only mosquitoes are by far the best known and studied disease vectors but also other arthropods, including ticks, tsetse flies, sand flies, fleas, triatomine bugs, are widely investigated.

One aspect, which has attracted increasing interest during the last years, is the role of arthropod saliva. It has been clearly shown in different models that pathogens injected via a syringe in absence of saliva were less infectious than those inoculated through an arthropod bite. Arthropod saliva plays thus a major role in pathogen transmission. It modulates the pharmacology of the vertebrate host, inhibiting hemostasis, itch, pain, and vasoconstriction. It also influences the immunology by inducing an immunosuppressive effect. Two models, leishmaniasis and Lyme borreliosis, have been particularly well studied on this aspect. Transcriptomics and proteomics have also been very helpful to identify a large catalog of bioactive molecules. The discovery of the invertebrate microbiome has not only enlarged but also complicated the understanding of the transmission process of arthropod-borne pathogens. To perfect the efficacy of pathogen transmission, infected arthropods modify their behavior toward the host. For *Yersinia*-infected flea, *Leishmania*-infected sand flies, and malaria-infected mosquitoes, it is well documented that the infection increases their probing time. In the case of ticks, an interesting study on *Ixodes–Borrelia* revealed that *Borrelia* infection increases the tick survival (more fat and more resistance to desiccation) and the questing period (less need to move to the litter zone to rehydrate). The bacterial infection thus enhances chances for a tick to find a host and to subsequently transmit the pathogens.

These examples may give a flavor on the smart strategies, which pathogens are able to develop to make their transmission process a real success.

The fulminant spread of certain **pathogens** is a real concern. Although some have been circulating for decades, their impact has strongly amplified during the last years, mainly because of major economic and ecological changes. All types of pathogens, viruses, parasites, bacteria, except fungi are actively transmitted by arthropods vectors. Pathogen virulence is amazing in the context of ABDs. Only a few hundreds of pathogens are generally inoculated, emphasizing the efficacy of arthropod saliva in the transmission process. More and more of these pathogens are described as potential threats for domestic animals and humans. However, the system is not so simple. Although the DNA from different microorganisms can be easily detected by sophisticated high sequencing techniques in arthropod vectors, it does not necessarily mean that the presence of these microorganisms will lead to a disease in vertebrate hosts. What indeed is often forgotten is the essential concept of vector competence: it says that arthropods interact in a specific manner with the hosted microorganisms that will be transmitted later to the host. Once there, the organism needs to further develop to become a pathogen causing clinical manifestations. This concept has been well described in the past but is too often neglected. In addition, the immune status of the vertebrate host, in case it is not a reservoir but an accidental host, seems to be critical for the development of clinical manifestations. Not all hosts will develop a pathological infection.

Arthropod-borne viruses have become a real burden for human and animal health. Dengue, tick-borne encephalitis, and the new Zika virus constitute major threats for humans, especially driven by the accelerated spread of infection. Blue tongue and African swine fever are viral infections, which represent a major concern in veterinary medicine. Arthropod-transmitted parasitic diseases remain as health problems of utmost importance, the most prominent always being malaria. Although the promotion and use of impregnated bed nets has substantially reduced the number of deaths from malaria worldwide, and especially in the sub-Saharan region, this disease still devastates more than half a million lives every year, especially children. In veterinary medicine, babesiosis and theileriosis transmitted by ticks have a major impact on cattle in tropical countries with important economic losses. Arthropod-borne bacterial infections such as Lyme disease, rickettsioses, and relapsing fever occupy an important position in human health, whereas anaplasmosis and ehrlichiosis affect significantly wild and domestic animals.

The **vertebrate host** affected by these ABDs has to face all these challenges. The need for new and more powerful strategies is unquestioned. Progress in immunology, especially the discovery of innate immunity in the 1990s opened new avenues for research. Epithelia are no longer seen not only as a physical barrier but also as a potent and sophisticated immune tissue. The skin was long considered as a simple hurdle derived from ectodermic cells producing skin appendages (hair, nails…) and armed with secretory functions. The discovery of innate immunity with the Toll receptors in humans and in various animal models has unraveled the powerful role of the skin and its keratinocytes in the recognition of pathogens and danger signals. In recent years, our knowledge on skin immunity has greatly improved with important studies, which described the roles of skin cells consisting of immune cells and resident cells, by characterizing the skin as a very complex network with intense trafficking of cells from the blood to the skin tissue and from the skin to peripheral lymph nodes. This picture became even more complex with the discovery of the skin microbiome.

It became obvious for a number of researchers that the role of the skin as the very first contact between the arthropod and the vertebrate host in ABDs deserved further attention and investigations. New techniques, such as intravital microscopy, produce amazing images of pathogens within the skin network. The processes in the skin are recognized as essential events during pathogen transmission and for the development of an efficient immune response. A better understanding of these early events should help to identify the mechanism of skin immunity against these invaders and to characterize the involved key molecules of the host and the pathogen. This could become a fundament for the development of better vaccines and diagnostic tools, a high need in times of rapidly increasing drug and insecticide resistances.

The purpose of this book *Skin and arthropod vectors* is to provide a comprehensive update and overview on the latest research on the role of the skin in arthropod-borne diseases, with special attention to the interplay of the three key actors: the host and its skin immunity, the arthropod and its saliva, and, last but not least, the pathogen with its virulence factors. I am grateful to all the authors for their contributions and the stimulating discussions, which made the preparation of this book a fascinating scientific adventure.

Nathalie Boulanger

Chapter 1

Skin Immunity and Microbiome

Nathalie Boulanger, Cédric Lenormand
Université de Strasbourg, Strasbourg, France

INTRODUCTION

The skin is one of the largest organs of human body. While its main function is to provide a physical barrier against the external environment, it also contributes to numerous additional critical physiological functions among which protection against water and electrolyte loss, protection against ultraviolet radiation damage, thermoregulation, and synthesis of metabolic products such as vitamin D. The immune functions of the skin have been formally recognized only in 1978, when Streilein used for the first time the term "skin-associated lymphoid tissue" to describe the continuous traffic of leukocytes between the skin, draining lymph nodes, and the blood (Streilein, 1978). Since then, an accumulating body of evidence has contributed to refine our perception of the complex network of interactions between epithelial, stromal, and resident immune cells that ensure host defense against pathogens while preserving tissue homeostasis (Di Meglio et al., 2011; Pasparakis et al., 2014). Moreover, recent insights into the previously unexpected influence of skin microbiota on these immune functions in health and disease have revealed an additional level of complexity, paving the way to new strategies to modulate skin immunity (Belkaid and Segre, 2014).

Although the skin acts as a highly efficient barrier against surface pathogenic microbes, it is still the major portal of entry for most arthropod-borne pathogens. Indeed, taking advantage from the breakage of the host's epidermis by the arthropod's mouthparts during the blood meal, these organisms are directly inoculated into the dermis, where they have developed elaborated strategies to successfully circumvent the local immune defenses. Unraveling these mechanisms of immune escape is a prerequisite to identify and elaborate efficient strategies to better fight these major threats to global public health in the future, e.g., via optimization of vaccine design.

This chapter does not pretend to exhaustively describe the full picture of knowledge on skin immunity and microbiome, but to rather provide a comprehensive insight into the organization of the human skin and its interactions with arthropod vectors. It describes as well the current, albeit preliminary knowledge

Skin and Arthropod Vectors. https://doi.org/10.1016/B978-0-12-811436-0.00001-0

1

about the resident microbiota in the skin and their interaction with the various arthropods in the context of the diseases they transmit: from a distance (attractiveness) and also directly at the skin interface during pathogen transmission.

Skin Immunity

Overview of the Skin Immune System Organization

The skin can be divided in three major components: a stratified epithelium, the *epidermis*, which represents the main physical barrier with the external environment; a connective tissue, the *dermis*, which confers its mechanical properties of pliability, elasticity, and trauma resistance to the skin; and an adipose tissue, the *hypodermis*, dedicated to thermal insulation and energy supply (Fig. 1.1). Epidermis is separated from the dermis by a basal membrane that provides resistance against external shearing forces. The dermis is richly irrigated by both blood and lymphatic vessels, which represent entry and exit portals for circulating immune cells. The lymphatics channels conduct interstitial fluid enriched with cells, proteins (including free antigenic peptides), lipids, bacteria, and degraded substances to the draining lymph nodes, where immune cells such as naïve T and B cells are waiting to meet appropriate stimulation to give rise either to potent adaptive cellular and humoral-mediated immune responses or to promote tolerance against exogenous or endogenous antigens from harmless sources.

The skin immune system can be schematically subdivided in innate and adaptive functional compartments, which are in fact strongly intricate. The innate immune system is composed of various cell types (i.e., keratinocytes, fibroblasts, mast cells, dendritic cells [DCs], macrophages, innate lymphoid cells

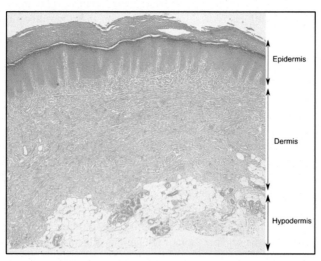

FIGURE 1.1 Low magnification of human skin (hand). Magnification × 40, hematoxylin eosin.

[ILCs]) present in the different anatomical layers of the skin and which share strong abilities to sense the presence of pathogens and to trigger an inflammatory response in reaction to skin injuries. The adaptive immune system relies on various specialized subpopulations of skin-resident T cells located both in the epidermis and in the dermis, which are responsible for the immune memory of the skin, allowing strong and rapid coordinated response in the case of rechallenge with an already encountered pathogen.

It is important to note that major anatomical and immunological differences exist between human and murine skin, hampering the interpretation of results obtained in murine models of skin infection or inflammation (Di Meglio et al., 2011). Mouse skin is much thinner than human skin, covered by an abundant waterproofing fur, and contains a muscle layer (*panniculus carnosus*) that allows rapid wound healing by contraction, limiting the scarring usually observed in human skin where granulation tissue formation is the main skin repair mechanism. Moreover, mice possess specific populations of immune resident cells such as dendritic epidermal T cells and dermal γδ T cells, which are deemed to play important roles in response to skin injury but do not have any known human counterpart. Finally, unlike mice, most of the human epidermis is interfollicular epidermis, an important point to consider as hair follicles have been described as areas of relative immune privilege (Ito et al., 2008). Thus, unless otherwise specified, the data presented here are mainly focused on human skin immune system.

Keratinocytes and the Cells of the Epidermis

Keratinocytes

They are the main cellular component of the epidermis, the outer compartment of the skin, which is composed of four layers (Fig. 1.2) (Nestle et al., 2009). Basal keratinocytes are columnar undifferentiated cells that form a one-cell-thick layer and renew constantly in the lower part of the epidermis, the *stratum basale* or basal layer, generating cells for the more superficial layers in a clonal way, also known as the epidermal proliferative unit. While detaching from the basement membrane, basal keratinocytes differentiate into polyhedral cells, which constitute the next layer, the *stratum spinosum* or prickle cell layer, interconnected by abundant desmosomes. These intercellular bridges bring an important contribution to the efficient resistance of the epidermis to mechanical stress. Immediately above is the *stratum granulosum*, where keratinocytes are characterized by numerous intracellular granules of keratohyalin. At this level, most of the physical cohesion between cells is provided by tight junctions, which also represent a first highly regulated functional barrier, which seals the intercellular spaces. This granular layer is dedicated to the synthesis of a number of structural components of the epidermal barrier. The outermost layer of the epidermis, the *stratum corneum*, is made of stacked, flatted keratinocytes that have lost their nuclei and cytoplasmic organelles, called the corneocytes.

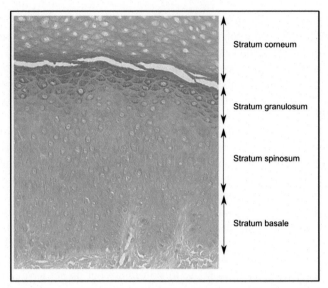

FIGURE 1.2   High magnification of human epidermis (hand). Magnification × 200, hematoxylin eosin.

These terminally differentiated cells form bricks embedded in a lipid matrix mortar composed of ceramides, cholesterol, and free fatty acids, constituting a physical wall that confers both mechanic protection and impermeability to the epidermis (Chu, 2012).

Besides keratinocytes, which constitute at least 80% of the epidermal cells, the epidermis also contains different other specialized cell types that interact with keratinocytes and contribute to the epidermal barrier.

Melanocytes are neural crest-derived cells primarily located in the basal layer, which are dedicated to the synthesis of the main skin pigment, melanin. Following synthesis, melanin is packed into cytoplasmic granular organelles, the melanosomes, which are then transferred to the adjacent keratinocytes where they accumulate above the nucleus forming supranuclear "caps" that shield DNA from ultraviolet radiation.

Merkel cells are mechanoreceptors enriched in sites dedicated to tactile sensitivity, also predominantly located in the basal layer of the epidermis. They are anchored to adjacent keratinocytes by the mean of desmosomes and are believed to release neuropeptides in response to mechanical stimuli, thus modulating the activity of low-threshold sensory neurons innervating the epidermis.

Langerhans cells (LCs) are the resident antigen-presenting cells of the epidermis, accounting for 2%–8% of the total epidermal cell population. Their main features and crucial role in skin immune system are developed below, as well as those of *resident intraepidermal lymphocytes,* a minor but specialized subpopulation of skin lymphocytes.

Immune Function of Keratinocytes

Keratinocytes were historically thought to be passive bystanders during the skin immune and inflammatory reactions. It is now well established that these cells are integral components of the skin innate immune system, playing a pivotal role in the first steps following the injury of the epidermis, either by physical stress, irritant chemicals, nonionizing radiation or microbial pathogens (Di Meglio et al., 2011). At steady state, only a few cytokines are constitutively produced by keratinocytes, among which interleukin (IL)-1, IL-7, and transforming growth factor-β (TGF-β), a key regulator of LCs development. However, keratinocytes are also abundant producers of proinflammatory cytokines that are stored in an inactive state (pro-IL-1β and pro-IL-18), waiting the action of the enzyme caspase 1 to be processed in their active form (IL-1β and IL-18) when the inflammasome is activated by danger signals. Interestingly, keratinocytes have the ability to sense the presence of various microbial pathogens by recognizing conserved molecular structures known as pathogen-associated molecular patterns via different pattern recognition receptors (PRRs) such as Toll-like receptor (TLR) 1, 2, 3, 4, 5, 6, and 9 (Lebre et al., 2007). The response of keratinocytes to activation by danger signals appears indeed highly diversified, including initiation of the inflammation (IL-1, TNF-α, IL-6, attractant chemokines such as CCL3, CCL20, CCL27, CXCL9, or CXCL10), T cell activation (IL-15, IL-18), and polarization (type 1 interferons and IL-12 for Th1 induction, thymic stromal lymphopoietin for Th2 induction, IL-23 for Th17 induction), or even inhibition (IL-10 and TGF-β for Treg induction), according to the context (Di Meglio et al., 2011).

Keratinocytes are not only danger sensors and alarm signal generators but they also play a direct effector role in the defense against pathogens by the way of numerous antimicrobial peptides (AMPs), of which they are the main providers in the steady-state epidermis (Pivarcsi et al., 2005). AMPs are a family of small peptides with antibiotic-like properties that target a broad spectrum of bacteria, fungi, and viruses. In the epidermis, they are mainly produced by the keratinocytes of the basal and suprabasal layers (Clausen and Agner, 2016). Some AMPs can also modulate both cytokine/chemokine production and immune cell attraction, providing a link between innate and adaptive immunity. Following their biosynthesis in basal keratinocytes, they progressively reach the stratum corneum where they accumulate and constitute an authentic antimicrobial chemical barrier (Gallo and Hooper, 2012; Nakatsuji and Gallo, 2012; Schauber and Gallo, 2009). *Human β-defensins* (HBD) are small cysteine-rich cationic AMPs endowed with various antibacterial activities. HBD1 is constitutively expressed, noninducible, and has shown moderate activity against gram-negative bacteria. HBD2 expression is induced by the presence of different pathogens including *Pseudomonas aeruginosa* or *Staphylococcus aureus* and by proinflammatory cytokines (TNF-α, IL-1). It has shown good bactericidal activity against gram-negative bacteria (*Escherichia coli, P. aeruginosa*) but has only a bacteriostatic effect on *S. aureus*. HBD3 is a highly potent anti-*S. aureus* AMP, with a broad spectrum of activity against other gram-positive (*Streptococcus pyogenes*) and -negative (*E. coli*, *P. aeruginosa*) bacteria and even fungi (*Candida albicans*). *Cathelicidin* (including LL37 in human and CRAMP in mouse) is cationic amphipathic AMP, which needs to be processed by proteases and activated by vitamin D to gain full antimicrobial activity (Morizane et al., 2010). Cathelicidin demonstrated potent activity not only against various bacteria including *S. aureus* but also against viruses such as *Herpes simplex virus* (HSV). It plays, like all AMPs, an interesting role of "alarmin," recruiting neutrophils, T cells, mast cells, and monocytes to the site of infection (Peric et al., 2009; Yang et al., 2009). *RNase7* is another highly potent AMP, demonstrating strong activity against multiple bacterial pathogens, including *S. aureus*, *E. coli*, and *P. aeruginosa*, even at low doses. Calcium-binding *Psoriasin* (also known as S100A7) is not only a potent antibacterial AMP but also acts as a strong enhancer of neutrophil host defense functions (Clausen and Agner, 2016). When the keratinocyte's basal secretion of AMPs fails to clear an infection, upregulation of AMPs synthesis and afflux of innate immune cells such as neutrophils or mast cells recruited by danger signals will occur, thus initiating the inflammatory response.

Additional innate immune functions of keratinocytes include: secretion of various immunomodulatory molecules of the eicosanoid family (Rosenbach et al., 1990), production of antimicrobial reactive oxygen species (ROS) (Bickers and Athar, 2006), expression of complement-regulating proteins such as C3b, C3d, and C5a receptors (CR1/CD35, CR2/CD21, and CD88, respectively), membrane cofactor protein (CD46), decay-accelerating factor (CD55), and complement protectin (CD59) (Dovezenski et al., 1992; Modlin et al., 2012). Interestingly, keratinocytes may also be endowed with adaptive immune properties: under the action of interferon gamma, they express major compatibility class II (MHC II) molecules and may efficiently present antigen peptides to $CD4^+$ T cells (Albanesi et al., 1998; Meister et al., 2015).

Skin Mononuclear Phagocyte System

The mononuclear phagocyte system is a family of functionally related cells of hematopoietic cell lineage, composed of monocytes, macrophages, and DCs. These are highly specialized cells of the immune system that share endocytic, phagocytic, and antigen-presentation properties but differ, for instance, by their ability to prime the adaptive immune response. Their main distinctive features are summarized in Table 1.1.

Dendritic Cells

DCs are classically subdivided in conventional DCs (cDCs), plasmacytoïd DCs (pDCs), and monocyte-derived DCs (MoDCs). During the steady state, pDCs are absent from the skin and therefore not discussed further here. In injured skin, they populate the dermis and participate to wound healing via the recognition of nucleic acids by TLR7 and TLR9 leading to type I interferon production (Gregorio et al., 2010).

TABLE 1.1 Types of Resident Dendritic Cells and Macrophages in Human Skin

Type	Site	Specific Markers	Distinctive Immune Properties
LCs	Epidermis	CD1a⁺ CD207⁺ BG⁺	Induction of tolerance at the steady state
CD141⁺ cDCs	Dermis	CD141⁺ XCR1⁺ CD209⁺	Cross-presentation
CD1c⁺ cDCs	Dermis	CD1c⁺ CD1a⁺ CD207⁻	Defense against *Mycobacteria*
CD14⁺ DCs	Dermis	CD14⁺ AF^low	Poor migrating capacities
Mac	Dermis	CD14⁺ FXIIIa⁺ AF^hi	Scavenging and phagocytosis

AF, autofluorescence; *BG*, Birbeck's granules; *cDCs*, conventional dendritic cells; *LCs*, Langerhans cells; *Mac*, resident macrophages.

Langerhans Cells Discovered at the end of the 19th century by a German medical student, Paul Langerhans, LCs are the resident DCs of the epidermis, accounting to approximately 8% of the epidermal cell population. They have been traditionally viewed as a prototype of tissue-resident immunogenic DCs, playing the role of sentinels at the interface between host and microbes and endowed with powerful abilities to prime immune responses in reaction to invading pathogens. However, recent findings in knockout mice models as well as in humans argue instead for a more balanced role between immunostimulatory and tolerogenic functions (Romani et al., 2012).

LCs have a unique ontogeny among the family of DCs. They are supposed to originate from a yolk sac-derived myeloid precursor during the early phase of embryogenesis and to self-renew at the steady state during the rest of the life, while the other cDCs populations have a shorter life span and need to be continuously replenished by blood-borne precursors (Malissen et al., 2014). However, under inflammatory conditions, such blood-borne precursors can differentiate in LCs (inflammatory LCs) that repopulate the damaged epidermis. LCs are characterized by highly specific intracellular organelles, the Birbeck's granules, which are subdomains of the endosomal recycling compartment. The genesis of these rod- or racket-shaped pentalamellar structures is strictly dependent on the expression of langerin (CD207), a trimeric C-type lectin, which is specifically expressed in LCs and is therefore widely used as a specific marker to identify these cells. Langerin has been shown to be a PRR implicated in recognition and uptake of various pathogens such as *Mycobacterium* spp., *C. albicans*, the human immunodeficiency virus type 1 (HIV-1), or the influenza A virus

(Ng et al., 2015). LCs are primarily located in the suprabasal layers of the epidermis, where they are anchored to the neighboring keratinocytes by adhesion molecules such as e-cadherin. Those interactions are crucial to maintain LCs in an immature state (i.e., with low expression of MHC II and costimulatory molecules). LCs extend their dendrites through the tight junctions of the granulous layer, which allows them to sense the external environment and capture antigens that have not penetrated the stratum corneum, a feature that may be useful to develop preemptive immunity against potentially pathogenic surface microbes (Ouchi et al., 2011). A schematic view of the fine-tuned functions of LCs is presented as follows. Under the steady state, LCs continuously leave the epidermis and migrate across the dermis to reach the regional lymph nodes via the lymph vessels, where they arrive in a fully matured state to present the collected antigens to the naïve T cells in the T cell-rich area of the skin-draining lymph node. In the absence of danger signal in the epidermis, LCs arrive in a full mature but inactivated state, and stimulation of T cells will not lead to a strong activation but instead induce a tolerogenic state. In the case of epidermal injury, on the presence of danger signals LCs strongly upregulate their costimulatory molecules and thus reach the lymph nodes in a fully mature and activated state. The stimulation of cognate naïve T cells will then lead to rapid clonal expansion, and blood circulation of antigen-specific effector T cells targeted to the site of injury (Romani et al., 2012).

It is of note that most of the published work on the immune functions of LCs have only examined their ability to stimulate naïve T cells in the skin-draining lymph nodes, while the interaction of LCs with skin-resident T cells has been only very recently investigated. Interestingly, LCs were shown to selectively induce the activation of skin-resident regulatory T cells (Treg) if in a resting state, while inducing a strong activation and proliferation of skin-resident effector memory T cells if previously activated by the presence of a pathogen, *C. albicans* (Seneschal et al., 2012).

Before the discovery of langerin/CD207, LCs were identified by their expression of another specific (albeit slightly less specific than langerin) cell surface marker, CD1a (formerly known as OKT-6). CD1a is a member of the lipid and glycolipid-presenting molecule family CD1, also expressed on some dermal DCs and macrophages, which has increasingly known implications in the immune functions of LCs. It has been shown, for example, that numerous autoreactive αβ T cells are restricted by CD1a and react to LCs expressing CD1a-self-lipid complexes. These lipid-reactive T cells are mainly producing IL-22 (Th22 cells), a cytokine with central role in the homeostasis of the epidermis including epithelial repair functions (de Jong et al., 2014). This suggests that beyond their role in the immune response against pathogens, LCs may also play a crucial role in the epidermis reparation following injury.

Dermal Dendritic Cells The diversity and functional features of dermal DCs have been consistently studied in the murine model, but less data are known about their human counterpart. The first observation was made in 1989 by the way of immunostaining against clotting enzyme factor XIII subunit A (FXIIIa), which allowed the identification of highly DCs ("dermal dendrocytes") in the upper dermis of healthy human skin (Cerio et al., 1989). Actually, three phenotypically and functionally distinct subsets of resident cDCs are delineated in the human dermis at the steady state: CD141$^+$ DCs, CD1c$^+$ DCs (both of which being also present in blood, lymph nodes, spleen, and other organs such as lungs, gut, or tonsils), and the more controversial CD14$^+$ DCs (Haniffa et al., 2015; Malissen et al., 2014). *CD141$^+$ DCs* specifically express the C-type lectin-like receptor CLEC9A and the chemokine receptor XCR1. CLEC9A is a receptor for actin filaments exposed at the surface of dead cells, participating to the superior ability of CD141$^+$ DCs to cross-presentation, i.e., the process resulting in presentation of external antigens by the MHC class I molecules to CD8$^+$ T cells. Cross-presentation is an essential mechanism to immunity against tumors and viruses that do not infect antigen-presenting cells. *Dermal CD1c$^+$ DCs* are characterized not only by the expression of CD1c but also CD1a molecules, implicated in the presentation of lipidic and glycolipidic antigens, and are deemed to play a significant role in the defense against mycobacteria, as attested by the occurrence of bacille Calmette-Guérin (BCG) disseminated infection in patients with IRF8 deficiency, who have a marked and selective loss in CD1c$^+$ DCs (Hambleton et al., 2011). Notably, during the early phase of skin infection with *Borrelia burgdorferi*, the agent of Lyme disease that is transmitted by tick bite, rapid upregulation of CD1c has been observed in a way implicating TLR2 signaling, suggesting that CD1c$^+$ DCs play a significant but yet undetermined role in the early phase of this disease (Yakimchuk et al., 2011). Both CD141$^+$ DCs and CD1c$^+$ DCs are continuously migrating to the draining lymph nodes to present collected antigens to the naïve T cells, with different abilities to polarize immune response. In contrast, *dermal CD14$^+$ DCs* found in the steady-state dermis are poor stimulators of allogeneic T cells, secrete IL-10, and induce Treg. They share transcriptomic features of tissue-resident macrophages and are probably derived from monocytes, in the same manner as the prominent population of inflammatory MoDCs observed in lesional skin of patients with psoriasis. CD14 is indeed largely expressed by monocytes and macrophages (O'Keeffe et al., 2015). Other specific inflammatory myeloid DCs have been described in the skin of patients suffering from psoriasis and/or atopic dermatitis: FcεR1$^+$ CD206$^+$ CD207$^-$ HLA-DR$^+$ inflammatory epidermal dendritic cells and tumor necrosis factor alpha-/inducible nitric oxide synthase–producing dermal DCs (TIP-DCs); but these peculiar subtypes will not be further discussed here.

Macrophages

Skin-resident macrophages are among the most abundant inflammatory cells in the dermis, even at the steady state, and are major actors of skin homeostasis. They are believed, at least partially, to share a similar origin with LCs, i.e., a yolk sac-myeloid precursor during the early phase of embryogenesis, even if blood-borne precursors abundantly contribute to the renewal of their population during the whole life (Malissen et al., 2014). Dermal macrophages are large autofluorescent cells with a granular cytoplasm and are classically identified by the following surface cell markers: $CD14^+$ $FXIIIa^+$ $CD1a^-$. A small subset is located around small vessels of the upper dermis, expresses also CD4, and displays a dendritic shape, corresponding to the "dermal dendrocyte" cell population described in the late 80s. On the functional level, macrophages differ notably from DCs with respect to (1) their higher phagocytic activities assorted of powerful microbicide functions, (2) their inability to migrate form the dermis to the draining lymph nodes, and (3) their lower efficiency as antigen-presenting cells, i.e., their inability to prime naïve T cells. In the mice, dermal macrophages have been shown to express molecules that endow them with potent scavenging functions, aimed, for example, at intermediates of self-macromolecules or pathogens. Indeed, dermal macrophages are believed to play an important role in skin protection against infection with *S. aureus*, one of the main human skin surface pathogens. This protection relies crucially on the PRR TLRs, as highlighted by the observation of patients with MyD88 or IRAK-4 deficiencies who suffer from recurrent skin abscesses due to *S. aureus* (Feuerstein et al., 2017). Skin-resident macrophages have been shown to be an important target of the arthropod-borne pathogen *Leishmania major* in the murine model (Von Stebut, 2007). Dermal macrophages are also known to express high levels of IL-10 mRNA, suggesting a role in the regulation of the inflammatory response, as well as in wound healing (Davies et al., 2013).

Beyond this resident population of skin macrophages endowed of self-renewing properties, a large proportion of the macrophages involved in the inflammatory response are blood-borne monocyte-derived macrophages, which have been more studied and are further divided in three distinct polarized populations: classically activated Th1-promoted proinflammatory $CD163^+$ M1 macrophages, Th2-promoted regulatory M2 macrophages, and wound-healing macrophages (Mosser and Edwards, 2008).

Skin-Resident Lymphocytes

The classical model of skin immune surveillance supposes that the main effectors of skin adaptive immunity are lymphocytes recirculating between the skin and the blood. Recent studies have, however, demonstrated that a major population of lymphocytes is in fact permanent residents of the skin, representing a major first-line defense in this tissue. Indeed, it is estimated that in adult human skin, approximately 20 billion T cells are present, i.e., twice the number present in the entire blood volume (Clark et al., 2006). Moreover, beyond this population of "educated" T cells, other previously unknown categories of lymphoid cells with innate immune cell properties have been recently identified in the skin, adding another level of complexity to the skin immune network (Heath and Carbone, 2013). Of note, no B or NK cell counterpart of this resident T cell population is believed to exist, these lymphocyte subsets being only present in the context of inflammatory responses.

Innate Lymphoid Cells

ILCs derive from a common lymphoid progenitor and are currently categorized in three distinct populations on the basis of distinct developmental requirement, transcription factor expression profile, and/or dedicated effector cytokines. Group 1 ILCs (ILC1s) rely on T-bet for their development, produce primarily TNF-α and IFN-γ; group 2 ILCs (ILC2s) are GATA-3-dependent and produce IL-4, IL-5, and IL-13; group 3 ILCs (ILC3s) are ROR-γt-dependent and produce IL-17A and/or IL-22, thus representing an innate counterpart to Th1, Th2, and Th17 T helper cells subsets, respectively (Table 1.2). Their main difference with T cells is an absence of antigen-specific receptors, as they respond to innate cytokine signals and do not rely on cognate interactions with antigen-presenting cells. ILCs can be identified by the absence of common lineage markers (lin⁻) and the expression of CD25, CD90, and CD127. Skin ILCs have been mostly studied in the context of inflammatory diseases (i.e., atopic dermatitis for ILC2s and psoriasis for ILC3s), and thus limited data are available with respect to their specific implication in the defense against pathogens (Kim, 2015; Yang et al., 2017). In noninflamed human skin, only very scarce populations of ILC1s and ILC3s are detected, mainly near the epidermis and in close vicinity to T cells (Bruggen et al., 2016).

Resident Memory T Cells

The discovery of this huge population of skin-resident T cells was made consecutively to elegant xenotransplant experiments showing that normal appearing skin of psoriatic patients grafted on immunodeficient mice gave rise to active psoriasis plaques, implicating that T cells with pathogenic potential where already present in the graft (Boyman et al., 2004). Human resident skin T cells are CD69-expressing long-lived memory CD45RO⁺ T cells (resident memory T cells, T_{REM}) coexpressing the skin-homing adressins CLA and CCR4. T_{RM} in the dermis are predominantly CD4⁺ helper T cells with various profile of cytokine expression (i.e., Th1, Th2, Th17, Th22, Th9, and Treg, see Table 1.2), while CD8⁺ cytotoxic T_{RM} cells are primarily found in the epidermis (Clark and Schlapbach, 2017; Nomura et al., 2014). T_{RM} display effector functions and a diversified T cell receptor repertoire reflecting the variety of antigens encountered in the skin (Clark, 2015; Nomura et al., 2014). Interestingly, it has been showed in the mice model that skin T_{RM} generation following local skin infection with the vaccinia virus was followed not only by seeding of the entire skin with specific T cells (Jiang et al., 2012) but also by the apparition of a lung population of virus-specific T_{RM} that displayed the same potent ability to respond to rechallenge with the virus than the skin T_{RM}, suggesting that skin is a highly pertinent site of vaccination to generate protection in the different epithelial barriers (Liu et al., 2010). In humans, intraepidermal HSV-specific CD8⁺ T_{RM} have been shown to be strongly implicated in the viral control of asymptomatic reactivations in genital herpes (Zhu et al., 2007).

TABLE 1.2 Main Subtypes of CD4+ T_RM in Human Skin

Type	Master Regulator (Inducing Cytokines)	Effector Cytokines	Immune Functions	Host Skin Defenses
Th1	T-bet (IFN-γ, IL-12)	IFN-γ	Activation of M1M	Intracellular bacteria, viruses
Th2	GATA3 (IL-4)	IL-4, IL-5, IL-10, IL-13	Activation of M2M Mac, Bs, Eo	Helminths
Th17	RORγT (IL-6, IL-23)	IL-17A, IL-22, IFN-γ	Recruitment of Ne	Extracellular bacteria, fungi
Th22	AHR (IL-6, TNF-α)	IL-22	Induction of defensins	ND
Th9	PU.1, IRF4 (IL-4, TGF-β)	IL-9	Activation of MC, Eo, Ne	Candida albicans
Treg	FoxP3 (IL-2, TGF-β)	IL-10, TGF-β, IL-35	Tuning of immune response, peripheral tolerance	ND

AHR, aryl hydrocarbon receptor; *Bs*, basophils; *Eo*, eosinophils; *IFN*, interferon; *IL*, interleukin; *M1M*, M1 macrophages; *M2M*, M2 macrophages; *MC*, mast cells; *ND*, non determined; *Ne*, neutrophils; *TGF*, transforming growth factor; *TNF*, tumor necrosis factor.

The implication of skin T_{RM} in skin defense against arthropod-borne pathogens has been very recently demonstrated in the mice model of leishmaniasis, where IFN-γ producing CD4$^+$ T_{RM} are sufficient to protect immune animals by the recruitment of proinflammatory monocytes on the site of infection, without the need of circulating cells (Glennie et al., 2017).

Mast Cells

Mast cells originate from bone marrow multipotent CD34$^+$ hematopoietic precursors that migrate through the blood to populate a large part of the body's tissues, including the skin. Readily recognizable to their granular basophilic cytoplasm, they are frequently found in close vicinity to blood or lymphatic vessels. Mast cells are immune sentinels equipped with a wide array of receptors (TLRs, Fc receptors, complement receptors) dedicated to pathogen detection, as well as effector cells with the ability to secrete a large panel of proinflammatory mediators (histamine, serotonin, proteases such as tryptase and chymase, heparin, chondroitin sulfate A and E) in response to danger signals (Otsuka and Kabashima, 2015). Some of these mediators are major inducers of pruritus, which can be viewed as a potentially efficient strategy to interrupt the blood meal of hematophagous arthropods by the way of reflex itch. Mast cells are also involved in the different phases of wound healing, in part by the proangiogenic activity of some of these mediators (chymase, tryptase, heparin) and by the release of various growth factor (vascular endothelial growth factor, platelet-derived growth factor, nerve growth factor). They participate to the recruitment of effector immune cells including eosinophils and neutrophils and have an antigen-presentation function by the way of constitutively expressed MHC class I molecules, and putatively by the way of MHC class II molecules upregulated under the action of IFN-γ (Moon et al., 2010). Interestingly, mast cells have even direct antimicrobial properties, either by producing ROS or by the release of extracellular traps constituted of DNA, proteases, and LL37 similar to the neutrophil extracellular traps (von Kockritz-Blickwede et al., 2008).

Skin Inflammatory Response to Arthropod Bite

Arthropods are a large family of invertebrate animals that exerts a major burden in global human health across the world, both by the direct consequences of their bite or sting, which can rarely be life-threatening, e.g., in the case of envenomation by *Atrax robustus* spider, and by the ability of some of them to transmit various infectious pathogens during blood meal (see Chapter 2). For instance, the protozoal agent of malaria affects more than 200 million individuals and was responsible for the death of more than 400,000 persons in 2015 (Global Health Organization data). Only the skin inflammatory response to bite from arthropod endowed with vector competences will be discussed here, the interactions of arthropod-borne pathogens with the skin interface being extensively treated in further chapters of this book.

The main arthropod vectors implicated in human disease are mosquitoes, ticks, sand flies, black flies, and reduviid bugs, whose life cycle strongly relies on blood meal from mammalian hosts. This blood meal, which generally requires only short contact with the host, can be schematically subdivided in two important steps. First, the skin barrier needs to be breached, by the mean of either solenophage (mosquitoes) or telmophage (ticks, flies) bite. Solenophage bite means puncture of the skin with a fine needled-shape mouthpart, e.g., *Anopheles* proboscis that directly enters small blood vessels of the dermis to aspirate blood, while telmophage bite involves skin dilacerations with knife- or scissors-like mouthpieces, e.g., *Ixodes* chelicerae and hypostome that result in a blood pool in the dermis, which can be further sucked. Second, the arthropods need to inject saliva in the skin to prevent various host defense mechanisms such as blood clotting and vasoconstriction that would decrease blood flow at the bite site following the blood vessel endothelium damage (anticoagulant and vasodilatator functions); pain that would alert the host (anesthetic functions); and innate immune reactions (immunomodulatory functions), leading to inflammation particularly in the case of arthropods with the longest blood meals (i.e., *Ixodidae*) (Krenn and Aspock, 2012). This step of saliva injection, which is often part of a repeated injection/suction cycle, is not only crucial to the blood meal success but is also the privileged moment where pathogens transmission may occur, as a number of pathogens are preferentially located in salivary glands of arthropods (see Chapters 4 and 5). Remarkably, some of these pathogens have taken great advantage of the immunomodulatory properties of arthropod saliva to enhance their infectivity in the vertebrate host (Bernard et al., 2014; Fontaine et al., 2011).

While antiinflammatory properties of saliva generally allow the hematophagous arthropod to quietly achieve its blood meal, the skin immune system does not stay inert, and various delayed clinical reactions can be observed in the human host in a few hours to days following the arthropod bite. The most frequent is a small pruritic red macule or papule located to the bite site that may last for a few days and is frequently the cause of significant itching, which can result in excoriation and superinfection (Fig. 1.3). Other clinical presentation may be vesicle, bullae, pustule, or nodule. The histopathological presentation of lesions from patients victims of "arthropod assault" (i.e., multiple clustered insect bites) is that of a spongiotic dermatitis with lymphocytic and eosinophilic infiltrate sometimes distributed around the follicles and sweat glands, often accompanied by extravasated erythrocytes as a consequence of external injury to the blood vessels, and sometimes by few neutrophils (Miteva et al., 2009). When a hard tick is removed by punch biopsy in the first hours to days following the bite, an intradermal cavity corresponding to the feeding pit may be seen histologically, accompanied by an inflammatory infiltrate predominantly constituted of macrophages and DCs during the first 24h of attachment, and then also of lymphocytes, plasma cells, and various quantities of eosinophils and neutrophils (Fig. 1.4). Analysis of cytokine and

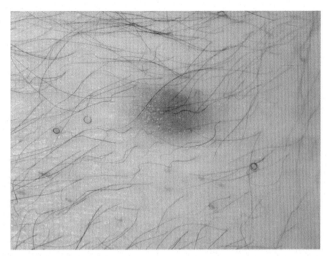

FIGURE 1.3   Persistent skin reaction to tick bite. Slightly elevated pruritic erythematous plaque with few vesicles, still there 14 days following *Ixodes* bite, without any tendency to extend.

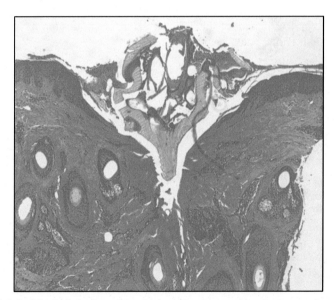

FIGURE 1.4   Histological view of an *Ixodes* biting human skin. *Ixodes* mouthparts are in yellow, penetrating the superficial dermis. Note the scarcity of the inflammatory infiltrate. Magnification × 40, hematoxylin eosin.

chemokine expression during the early phase showed predominant macrophage and neutrophils attractants (CCL2, CCL3, CCM4 and CXCL1, CXCL8, respectively) as well as IL-1β and the Th2 cytokine IL-5, reflecting a strong local innate immune response (Glatz et al., 2017; Patterson et al., 1979).

Exaggerated reactions may be observed in some patients following arthropod bite, such as papular urticaria (also known as "prurigo strophulus" in the French dermatological terminology). Papular urticaria is a common childhood disorder characterized by eruption of symmetrically distributed pruritic papules and papulovesicles following arthropod bite, where individual lesions do not represent local bite but are the consequence of (predominantly type IV) hypersensitivity reaction (Fig. 1.5). It is characterized histologically by a mild epidermal acanthosis and spongiotic dermatitis with a moderate infiltrate of CD45RO⁺ CD4⁺ T cells and CD68⁺ macrophages and some neutrophils and eosinophils, without implication of DCs or B cells (Jordaan and Schneider, 1997). Another example is so-called "mosquito allergy," a heterogeneous clinical and immunological concept used to designate patients with exaggerated reactions to mosquito bites, those reactions ranging from "larger" delayed urticarial papules (without a clear size cutoff to discriminate normal from abnormal reaction…) to authentic anaphylactic IgE-mediated reaction following mosquito bites (Crisp and Johnson, 2013). Life-threatening anaphylactic reactions have also been consistently reported with the saliva of the European pigeon soft tick *Argas reflexus*, and the major protein determinant of this life-threatening complication has even been characterized (Hilger et al., 2005).

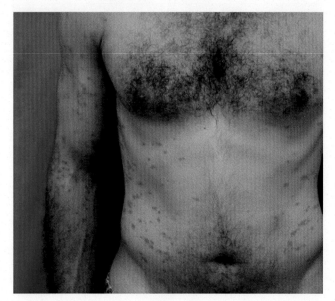

FIGURE 1.5 Papular urticaria. Diffuse eruptive pruritic erythematous papules following insect bite, at distance of the bite site.

Of interest, in the murine model, the intensity of the inflammatory reaction to mosquito bite is correlated with superior infectivity of different transmitted arboviruses. This effect is driven by cutaneous neutrophil-induced extensive inflammatory edema that facilitates virus retention at the inoculation site, thus recruiting local and monocyte-derived macrophages, which become further infected and amplify viral replication (Pingen et al., 2016).

Finally, some arthropods such as hard ticks may leave small chitin part of their mouthparts, either when voluntarily removed or accidentally scratched (Castelli et al., 2008). Granuloma formation of the foreign-body type may then form around and persists for extended duration, reflecting the inability of the immune response to clear the intruder (Fig. 1.6). Less frequently, such granulomas are found in the site of former tick bite, but no remnant of the mouthpart of the tick may be evidenced, questioning the role of long persisting saliva molecules in the genesis of granulomatous inflammation (Hirota et al., 2015). Persistence of tick mouthparts in the skin has also been linked to the rare occurrence of purpuric skin lesions to the bite site, corresponding to a vasculopathic process characterized histologically by the presence of hyaline periodic acid-Schiff–positive intraluminal thrombi indistinguishable from what is observed during skin manifestations of cryoglobulinemia (Galaria et al., 2003).

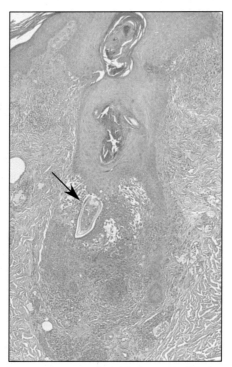

FIGURE 1.6 Foreign-type granuloma around persistent remnants of *Ixodes* mouthpart (*arrow*) in human skin. Magnification×40, hematoxylin eosin.

While a protective role for sensitization of the mammal host to arthropod saliva antigens has been demonstrated in the murine model, i.e., in the case of *Leishmania* transmission to mice by the sand fly (Kamhawi et al., 2000), it remains to be definitely proven in humans. In fact, people who experiment repeated itching episodes associated with tick bite seem protected against Lyme disease in high endemic areas (Burke et al., 2005) and have been shown to harbor antitick IgE antibodies, suggesting that vaccination against arthropod saliva antigens may represent a pertinent strategy to fight against vector-borne diseases (Merino et al., 2013).

Skin Microbiome

Human skin functions as a physical and immune barrier and it accommodates up to 10^6 microorganisms per cm^2 for a total of over 10^{10} bacteria across the $1.8\,m^2$ of skin surface (Belkaid and Segre, 2014; Grice and Segre, 2011). These commensals interact with foreign pathogens directly present in the environment and potentially with those transmitted through arthropod bites. They also play a role on arthropod attractiveness (Verhulst et al., 2011), however, the direct involvement of these commensals in arthropod-borne diseases (ABDs) is still poorly explored (see Chapter 3).

Skin Structure and the Different Microbiological Niches

The skin is made up of very diverse microbiological niches, with associated appendages such as sweat glands, sebaceous glands, and hair follicles giving specific characteristics to different skin sites. Skin microbiota colonizes these various spaces, constituting particular niches or immunoprivileged areas. Consequently, microbial species highly diverse, and population densities can vary by a factor 10,000 depending on temperature, pH, humidity, or the concentration of salts and nutrients. The stratum corneum is well oxygenated and exposed to drying out. Indeed, the corneocytes are dry and poor in energy sources, which favors microbial growth. They are constantly removed by peeling. Microorganisms must therefore relocate to new corneocytes, which will in turn be quickly eliminated. Two types of skin pores are integrated in the corneocyte barrier: the first one, the eccrine sweat gland, delivers not only an aqueous solution providing moisture, minerals (copper, iron, magnesium, zinc, calcium), and amino acids that support microbial proliferation but also AMPs such as dermcidin (Schittek, 2012). The second type is called the pilosebaceous follicle, consisting of the sebaceous glands and the hair follicles, a sebum-rich area, which is low in oxygen, and where the skin microbiota is also present. They produce secretions rich in lipids and AMPs such as cathelicidin, defensin, histone H4, and psoriasin (Gallo and Nakatsuji, 2011). The density of these glands and the qualitative and quantitative characteristics of their secretions affect the skin microbiota. Recent studies have also demonstrated the presence of bacteria in the other layers of the epidermis, as well as in the dermis and subcutaneous tissue (Nakatsuji et al., 2013). Even the adipocytes in the hypoderm are immunologically active by secreting cathelicidin. They participate actively in the defense against *S. aureus* infections (Zhang et al., 2015).

The skin can then be divided in three zones characterized by the prominent type of niches: "dry," "moist," and "oily" (Fig. 1.7). The dry zone is poor in energy sources and humidity, and the microbiota is highly diversified. This type of zone is frequently found on the arms and forearms, thighs, legs, and abdomen. The density of microorganisms lies between 100 and 1000 bacteria/cm^2 and is considered to be low. The moist and hot area, rich in glandular secretions, is prominent in the folds (armpits, perineum, interdigital spaces, etc.). It hosts a diverse and dense microbiota reaching several million microorganisms/cm^2. The oily or sebaceous zone is rich in sebaceous glands and corresponds to the face, scalp, and upper trunk.

The Microbiota

The term microbiota embraces all the microorganisms in a specific environment. Interactions between microbiota and host are complex and often called "symbiotic." In fact, the situation is more complex and knows three levels of interaction: (1) commensalism refers where only the microbiota takes an advantage of the situation, (2) mutualism where both partners mutually benefit from

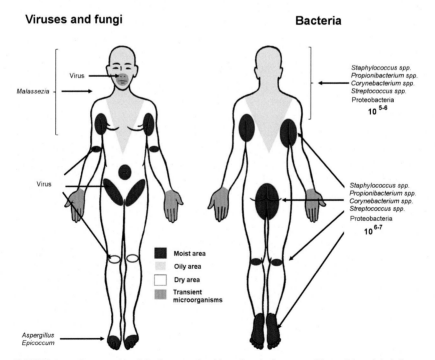

FIGURE 1.7 Cartography of the human microbiota (fungi, viruses, and bacteria) and their location in the different areas of the human body: wet, oily, or dry. *(Adapted from Dermatologie et infections sexuellement transmissibles, 2016. 6ème ed. Elsevier-Masson.)*

the interaction, (3) parasitism at large where the microbiota will cause a deleterious effect. The commensal, on disequilibrium, may become opportunistic infectious organism (Schommer and Gallo, 2013). Essentially, the modification of one or several specific host factors (e.g., age, hormones, drugs, hygiene, cosmetics…) can disrupt the sensitive balance within the skin microbiota: such a break is called dysbiosis.

These cutaneous microorganisms, when in balance, may be considered rather mutualistic than commensal flora, due to the important benefice for the host. The resident flora in the skin includes a variety of microorganisms that have adapted well to the different skin areas: oily, moist, and dry (Table 1.3). They stay there for a long period of time.

Bacteria. Four dominant phyla of bacteria reside on the skin. In one particular review, Actinobacteria (mainly family of Corynebacteriaceae, Propionibacteriaceae, Micrococciaceae) were found to represent 51.8%, Firmicutes (mainly Staphylococcaceae) 24.4%, Proteobacteria 16.5%, and Bacteroidetes 6.3% (Schommer and Gallo, 2013). However, variability between individuals is high, and factors such as geography, lifestyle, and/or ethnicity largely influence the skin microbiota (Bouslimani et al., 2015; Grice and Segre, 2011). The most dominant genders are *Staphylococcus*, *Propionibacterium*, and *Corynebacterium*. Most of the research has so far focused on *Staphylococcus epidermidis* (Cogen, AL et al., 2010; Lai et al., 2010; Otto, 2009). It was found that *S. epidermidis* is able to activate the Toll receptor, TLR2, to minimize inflammation induced during injury involving TLR3 (Lai et al., 2009). *S. epidermidis* produces its own AMPs called bacteriocins (Epidermin, Pep5, and epilancin K7), similar to those produced by the host, which participate in the host defense innate immunity. These peptides, especially phenol-soluble modulin, target not only pathogenic bacteria such as *S. pyogenes* but also *S. aureus*, by enhancing defensin and cathelicidin antimicrobial activity (Cogen et al., 2010). Coryneform bacilli are gram (+), whose main commensal skin genders are *Corynebacterium* spp., *Dermabacter* spp., *Brevibacterium* spp., and *Propionibacterium* spp. The first three genders colonize moist areas, while lipophilic and anaerobic *Propionibacterium* (mainly *P. acnes*) develops in oily areas. *Propionibacterium* is particularly found in sebaceous area and perifollicular spaces. It hydrolyzes triglycerides found in sebum, producing free fatty acids, which results in pH reduction in the skin (SanMiguel and Grice, 2015). *Corynebacterium* produces volatile compounds of the sweat, responsible for odors (Grice, 2014). *Brevibacterium* spp. produces methanethiol, which is responsible for "foot odor" and would be responsible for the attraction of certain insects including mosquitoes (Braks et al., 1999). Interestingly, the diversity of the skin microbiota increases attractiveness to mosquitoes (Verhulst et al., 2011) (See Chapter 3).

Fungi. The lipophilic yeast, *Malassezia*, dominates in the upper part of the body and the arms, while foot sites are colonized by more diverse fungi such as *Aspergillus, Cryptococcus, Epicoccum…* (Findley et al., 2013). *Malassezia* spp.

TABLE 1.3 Main Microorganisms of the Skin Microbiota, Their Abundance, and Their Localization

Gram (+) Cocci		
Staphylococcus coagulase (–)	+++	Dry area
Staphylococcus aureus	++++	Moist area, nostrils
Micrococcus	++	Dry area
Peptococcus		Oily area, anaerobic
Entérocoques		Moist area, perineum
Gram (+) Bacillus		
Corynebacterium	+++	Moist area
Propionibacterium	+++	Oily area, anaerobic
Gram (–) Bacillus		
Acinetobacter	+	Moist area
Proteus, entérobactéries		Perineum
Fungi		
Malassezia	+++	Oily area
Candida	±	Moist area
Trichosporon mucoides	±	Moist area/oily
Aspergillus	+	Moist area (toe web space))
Epicoccum	+	Moist area (plantar skin)
Parasites (Mite)		
Demodex	+	Face
Viruses		
Bactériophages	+	
Polyomavirus		Moist area
Papillomavirus α and γ		

includes several species of lipophilic yeasts. Some species are part of the resident microbiota, while others seem to be pathogenic. For example, *Malassezia globosa* is the yeast responsible for tinea versicolor.

Parasites. They are rare and mainly represented by the mite, *Demodex. D. folliculorum* is the only parasite of the resident microbiota. This is a mite that is found in the pilosebaceous orifices of the face in about 10% of the population.

Viruses. Studies on viruses such as bacteriophages, polyomavirus, and β et γ papillomavirus have only started recently (Duerkop and Hooper, 2013; Foulongne et al., 2012). Viruses are mainly found in moist skin areas (nasal cavity, popliteal fossa, groin, and elbow crease). Identification on the skin is very recent and their precise role remains to be defined. Metagenomics that was used to identify them, only targets DNA viruses, whereas RNA viruses are not identified by this approach. The beneficial nature of this new virobiote remains to be demonstrated since their presence on the skin can be symbiotic with the resident bacteria or be pathogenic, replicating in host cell.

Sterile at birth, human skin microbiota undergoes constant changes throughout life, under the influence of the place of living, lifestyle, medications (antibiotics, immunosuppressive drugs, either used topically or systemically), hormonal changes associated with age, or diseases such as diabetes, kidney failure, HIV infection, or gender (Oh et al., 2014; SanMiguel and Grice, 2015). The skin has developed a variety of mechanisms to control and protect from infectious microorganisms. It is supported by the commensal microbiota that colonizes the skin.

The Microbiota in the Defense Against Infectious Agents

The integrity of the stratum corneum is essential. The layer of keratinocytes bound by a lipid cement forms a barrier, which relegates the commensal microbiota primarily on the surface of the stratum corneum. Since the corneous layer is constantly renewed (peeling, shower, etc.), the microbiota must constantly relocate to new corneocytes, which are themselves eliminated shortly. Subsequently, only microorganisms that can adhere easily to corneocytes and grow fast survive.

Biochemical Defenses

The skin surface is covered with a lipid layer that consists of triglycerides from sebum and keratinocytes. Many microorganisms of the microbiota secrete lipases that cleave triglycerides into free fatty acids (oleic, stearic, or palmitic acids). These free fatty acids are involved in the creation of a "coat" known for its antimicrobial properties, lethal to *S. aureus* and *Streptococcus* (Clarke et al., 2007). These lipids are antibacterial for certain microorganisms or, conversely, they can be an indispensable energy source for *Malassezia* spp., which is a lipophilic yeast.

The pH of the skin, which is in the order of 5.5, is selective for the resident microorganisms. The partially occluded areas (creases) have a neutral or slightly alkaline pH, which promotes a dense and varied microbiota.

Immunological Barriers

The skin as an interface between the body and the environment is in constant contact with microorganisms that may be pathogenic. The commensal microbiota plays an essential role in homeostasis. Resident cells (keratinocytes and fibro-blasts) and immune cells of the skin interact with these commensal microorgan-isms to preserve the skin homeostasis (Gallo and Nakatsuji, 2011) and facilitate wound healing (Clarke et al., 2007). AMPs are essential contributors to cutane-ous immunity in addition to the chemical and physical barrier constituted by the skin. The main sources of AMPs in human skin are keratinocytes, mast cells, neutrophils, and sebocytes as mentioned above. In addition, the resident com-mensal microbiota produces its own AMPs (Gallo and Nakatsuji, 2011).

CONCLUSIONS AND PERSPECTIVES

Research on skin has revealed over the last years that this organ works not only as a physical hurdle but also as an efficient immunological barrier in the protec-tion against environmental menaces and pathogenic bacteria. Discovery of the Toll receptor in 1997 (Medzhitov and Janeway, 1997) has revolutionized the human immunity and pointed out the major role of the resident cells of the epi-dermis, the keratinocytes. Skin cells in general defend very actively the body from external aggressions thanks to their receptors of innate immunity. The recent discovery of the skin microbiome has complicated our understanding of skin immunity (Grice and Segre, 2011). Commensal microbiota substan-tially contributes to the defense against pathogenic bacteria (Belkaid and Segre, 2014). Although a cartography of commensal bacteria has been elaborated by metagenomics (Oh et al., 2014), the role of commensal viruses and fungi in skin immunity is still poorly understood (Schommer and Gallo, 2013). The role of the skin microbiota in cutaneous inflammation and infection started recently to be explored. However, preliminary data show that they might also play a role in ABDs, either by influencing the attractiveness toward the arthropod (Verhulst et al., 2011) or during the direct transmission of pathogens at the skin interface (Naik et al., 2012). A better understanding of these multiple inter-actions between the pathogens inoculated with the arthropod saliva, and host immunity–microbiota complex, should help to develop new strategies in the fight against ABDs (Wikel, 2013; Belkaid and Segre, 2014).

REFERENCES

Albanesi, C., Cavani, A., Girolomoni, G., 1998. Interferon-gamma-stimulated human keratinocytes express the genes necessary for the production of peptide-loaded MHC class II molecules. The Journal of Investigative Dermatology 110, 138–142.

Belkaid, Y., Segre, J., 2014. Dialogue between skin microbiota and immunity. Science 346, 954–959.

Bernard, Q., Jaulhac, B., Boulanger, N., 2014. Smuggling across the border: how arthropod-borne pathogens evade and exploit the host defense system of the skin. Journal of Investigative Dermatology 1–9. https://doi.org/10.1038/jid.2014.36.

Bickers, D.R., Athar, M., 2006. Oxidative stress in the pathogenesis of skin disease. The Journal of Investigative Dermatology 126, 2565–2575.

Bouslimani, A., Porto, C., Rath, C., Wang, M., Guo, Y., Gonzalez, A., Berg-Lyon, D., Ackermann, G., Moeller Christensen, G.J., Nakatsuji, T., Zhang, L., Borkowski, A., Meehan, M., Dorrestein, K., Gallo, R., Bandeira, N., Knight, R., Alexandrov, T., Dorrestein, P., 2015. Molecular cartography of the human skin surface in 3D. Proceedings of the National Academy of Sciences United States of America 112, E2120–E2129.

Boyman, O., Hefti, H.P., Conrad, C., Nickoloff, B.J., Suter, M., Nestle, F.O., 2004. Spontaneous development of psoriasis in a new animal model shows an essential role for resident T cells and tumor necrosis factor-alpha. The Journal of Experimental Medicine 199, 731–736.

Braks, M.A., Anderson, R., Knols, B., 1999. Infochemicals in mosquito host selection: human skin microflora and plasmodium parasites. Parasitology Today 15, 409–413.

Bruggen, M.C., Bauer, W.M., Reininger, B., Clim, E., Captarencu, C., Steiner, G.E., Brunner, P.M., Meier, B., French, L.E., Stingl, G., 2016. In situ mapping of innate lymphoid cells in human skin: evidence for remarkable differences between normal and inflamed skin. The Journal of Investigative Dermatology 136, 2396–2405.

Burke, G., Wikel, S.K., Spielman, A., Telford, S.R., McKay, K., Krause, P.J., Tick-borne Infection Study, G, 2005. Hypersensitivity to ticks and Lyme disease risk. Emerging Infectious Diseases 11, 36–41.

Castelli, E., Caputo, V., Morello, V., Tomasino, R.M., 2008. Local reactions to tick bites. The American Journal of Dermatopathology 30, 241–248.

Cerio, R., Griffiths, C.E., Cooper, K.D., Nickoloff, B.J., Headington, J.T., 1989. Characterization of factor XIIIa positive dermal dendritic cells in normal and inflamed skin. The British Journal of Dermatology 121, 421–431.

Chu, D.H., 2012. Development and structure of skin. In: Goldsmith, L., Katz, S., Gilchrest, B., Paller, A., Leffell, D., Wolff, K. (Eds.), Fitzpatrick's Dermatology in General Medicine, eighth ed. Mc Graw Hill, pp. 58–74.

Clark, R.A., 2015. Resident memory T cells in human health and disease. Science Translational Medicine 7, 269rv261.

Clark, R.A., Chong, B., Mirchandani, N., Brinster, N.K., Yamanaka, K., Dowgiert, R.K., Kupper, T.S., 2006. The vast majority of CLA+ T cells are resident in normal skin. The Journal of Immunology 176, 4431–4439.

Clark, R.A., Schlapbach, C., 2017. TH9 cells in skin disorders. Seminars in Immunopathology 39, 47–54.

Clarke, S., Mohamed, R., Bian, L., Routh, A., Kokai-Kun, J., Mond, J., Tarkowski, A., Foster, S., 2007. The *Staphylococcus aureus* surface protein IsdA mediates resistance to innate defenses of human skin. Cell Host Microbe 1, 199–212.

Clausen, M., Agner, T., 2016. Antimicrobial peptides, infections and the skin barrier. Current Problems in Dermatology 49, 38–46.

Cogen, A.L., Yamasaki, K., Muto, J., Sanche, Z.K., Crotty Alexander, L., Tanios, J., Lai, Y., Kim, J., Nizet, V., Gallo, R., 2010. *Staphylococcus epidermidis* antimicrobial delta-toxin(phenol-soluble modulin-gamma) cooperates with host antimicrobial peptides to kill group A Streptococcus. PLoS One 5, e8557.

Crisp, H.C., Johnson, K.S., 2013. Mosquito allergy. Annals of Allergy, Asthma and Immunology 110, 65–69.

Davies, L.C., Jenkins, S.J., Allen, J.E., Taylor, P.R., 2013. Tissue-resident macrophages. Nature Immunology 14, 986–995.

de Jong, A., Cheng, T.Y., Huang, S., Gras, S., Birkinshaw, R.W., Kasmar, A.G., Van Rhijn, I., Pena-Cruz, V., Ruan, D.T., Altman, J.D., Rossjohn, J., Moody, D.B., 2014. CD1a-autoreactive T cells recognize natural skin oils that function as headless antigens. Nature Immunology 15, 177–185.

Di Meglio, P., Perera, G.K., Nestle, F.O., 2011. The multitasking organ: recent insights into skin immune function. Immunity 35, 857–869.

Dovezenski, N., Billetta, R., Gigli, I., 1992. Expression and localization of proteins of the complement system in human skin. Journal of Clinical Investigation 90, 2000–2012.

Duerkop, B., Hooper, L., 2013. Resident viruses and their interactions with the immune system. Nature Immunology 14, 654–659.

Feuerstein, R., Kolter, J., Henneke, P., 2017. Dynamic interactions between dermal macrophages and *Staphylococcus aureus*. Journal of Leukocyte Biology 101, 99–106.

Findley, K., Oh, J., Yang, J., Conlan, S., Deming, C., Meyer, J., Schoenfeld, D., Nomicos, E., Park, M., Kong, H., Segre, J., 2013. Topographic diversity of fungal and bacterial communities in human skin. Nature 498, 367–370.

Fontaine, A., Diouf, I., Bakkali, N., Missé, D., Pagès, F., Fusai, T., Rogier, C., Almeras, L., 2011. Implication of haematophagous arthropod salivary proteins in host-vector interactions. Parasites Vectors 28, 187.

Foulongne, V., Sauvage, V., Hebert, C., Dereure, O., Cheval, J., Gouilh, M., Pariente, K., Segondy, M., Burguière, A., Manuguerra, J., Caro, V., Eloit, M., 2012. Human skin microbiota: high diversity of DNA viruses identified on the human skin by high throughput sequencing. PLoS One 7, e38499.

Galaria, N.A., Chaudhary, O., Magro, C.M., 2003. Tick mouth parts occlusive vasculopathy: a localized cryoglobulinemic vasculitic response. Journal of Cutaneous Pathology 30, 303–306.

Gallo, R., Hooper, L., 2012. Epithelial antimicrobial defence of the skin and intestine. Nature Reviews Immunology 12, 503–516.

Gallo, R.L., Nakatsuji, T., 2011. Microbial symbiosis with the innate immune defense system of the skin. Journal of Investigative Dermatology 131, 1974–1980. https://doi.org/10.1038/jid.2011.182.

Glatz, M., Means, T., Haas, J., Steere, A.C., Mullegger, R.R., 2017. Characterization of the early local immune response to *Ixodes ricinus* tick bites in human skin. Experimental Dermatology 26, 263–269.

Glennie, N.D., Volk, S.W., Scott, P., 2017. Skin-resident CD4+ T cells protect against Leishmania major by recruiting and activating inflammatory monocytes. PLoS Pathogens 13, e1006349.

Gregorio, J., Meller, S., Conrad, C., Di Nardo, A., Homey, B., Lauerma, A., Arai, N., Gallo, R.L., Digiovanni, J., Gilliet, M., 2010. Plasmacytoid dendritic cells sense skin injury and promote wound healing through type I interferons. The Journal of Experimental Medicine 207, 2921–2930.

Grice, E., 2014. The skin microbiome: potential for novel diagnostic and therapeutic approaches to cutaneous disease. Seminars in Cutaneous Medicine and Surgery 33, 98–103.

Grice, E.A., Segre, J.A., 2011. The skin microbiome. Nature Reviews Microbiology 9, 244–253. https://doi.org/10.1038/nrmicro2537.

Hambleton, S., Salem, S., Bustamante, J., Bigley, V., Boisson-Dupuis, S., Azevedo, J., Fortin, A., Haniffa, M., Ceron-Gutierrez, L., Bacon, C.M., Menon, G., Trouillet, C., McDonald, D., Carey, P., Ginhoux, F., Alsina, L., Zumwalt, T.J., Kong, X.F., Kumararatne, D., Butler, K., Hubeau, M., Feinberg, J., Al-Muhsen, S., Cant, A., Abel, L., Chaussabel, D., Doffinger, R., Talesnik, E., Grumach, A., Duarte, A., Abarca, K., Moraes-Vasconcelos, D., Burk, D., Berghuis, A., Geissmann, F., Collin, M., Casanova, J.L., Gros, P., 2011. IRF8 mutations and human dendritic-cell immunodeficiency. The New England Journal of Medicine 365, 127–138.

Haniffa, M., Gunawan, M., Jardine, L., 2015. Human skin dendritic cells in health and disease. Journal of Dermatological Science 77, 85–92.

Heath, W.R., Carbone, F.R., 2013. The skin-resident and migratory immune system in steady state and memory: innate lymphocytes, dendritic cells and T cells. Nature Immunology 14, 978–985.

Hilger, C., Bessot, J.C., Hutt, N., Grigioni, F., De Blay, F., Pauli, G., Hentges, F., 2005. IgE-mediated anaphylaxis caused by bites of the pigeon tick *Argas reflexus*: cloning and expression of the major allergen Arg r 1. The Journal of Allergy and Clinical Immunology 115, 617–622.

Hirota, K., Kurosawa, Y., Goto, K., Adachi, K., Yoshida, Y., Yamamoto, O., 2015. Tick bite granuloma: recommendations for surgical treatment. Yonago Acta Medica 58, 51–52.

Ito, T., Meyer, K.C., Ito, N., Paus, R., 2008. Immune privilege and the skin. Current Directions in Autoimmunity 10, 27–52.

Jiang, X., Clark, R.A., Liu, L., Wagers, A.J., Fuhlbrigge, R.C., Kupper, T.S., 2012. Skin infection generates non-migratory memory CD8+ T(RM) cells providing global skin immunity. Nature 483, 227–231.

Jordaan, H.F., Schneider, J.W., 1997. Papular urticaria: a histopathologic study of 30 patients. The American Journal of Dermatopathology 19, 119–126.

Kamhawi, S., Belkaid, Y., Modi, G., Rowton, E., Sacks, D., 2000. Protection against cutaneous leishmaniasis resulting from bites of uninfected sand flies. Science 290, 1351–1354.

Kim, B.S., 2015. Innate lymphoid cells in the skin. The Journal of Investigative Dermatology 135, 673–678.

Krenn, H.W., Aspock, H., 2012. Form, function and evolution of the mouthparts of blood-feeding Arthropoda. Arthropod Structure and Development 41, 101–118.

Lai, Y., Cogen, A.L., Radek, K.A., Park, H.J., Daniel, T., Leichtle, A., Ryan, A.F., Di Nardo, A. Gallo, R.L., 2010. Activation of TLR2 by a small molecule produced by Staphylococcus epidermidis increases antimicrobial defense against bacterial skin infections. 130, 2211–2221. https://doi.org/10.1038/jid.2010.123.Activation.

Lai, Y., Di Nardo, A., Nakatsuji, T., Leichtle, A., Yang, Y., Cogen, A.L., Wu, Z.-R., Hooper, L.V., Schmidt, R.R., von Aulock, S., Radek, K.A., Huang, C.-M., Ryan, A.F., Gallo, R.L., 2009. Commensal bacteria regulate Toll-like receptor 3-dependent inflammation after skin injury. Nature Medicine 15, 1377–1382. https://doi.org/10.1038/nm.2062.

Lebre, M.C., van der Aar, A.M., van Baarsen, L., van Capel, T.M., Schuitemaker, J.H., Kapsenberg, M.L., de Jong, E.C., 2007. Human keratinocytes express functional Toll-like receptor 3, 4, 5, and 9. The Journal of Investigative Dermatology 127, 331–341.

Liu, L., Zhong, Q., Tian, T., Dubin, K., Athale, S.K., Kupper, T.S., 2010. Epidermal injury and infection during poxvirus immunization is crucial for the generation of highly protective T cell-mediated immunity. Nature Medicine 16, 224–227.

Malissen, B., Tamoutounour, S., Henri, S., 2014. The origins and functions of dendritic cells and macrophages in the skin. Nature Reviews. Immunology 14, 417–428.

Medzhitov, R., Janeway, C.J., 1997. Innate immunity: impact on the adaptive immune response. Current Opinion in Immunology 9, 4–9.

Meister, M., Tounsi, A., Gaffal, E., Bald, T., Papatriantafyllou, M., Ludwig, J., Pougialis, G., Bestvater, F., Klotz, L., Moldenhauer, G., Tuting, T., Hammerling, G.J., Arnold, B., Oelert, T., 2015. Self-antigen presentation by keratinocytes in the inflamed adult skin modulates T-cell auto-reactivity. The Journal of Investigative Dermatology 135, 1996–2004.

Merino, O., Alberdi, P., Perez de la Lastra, J.M., de la Fuente, J., 2013. Tick vaccines and the control of tick-borne pathogens. Frontiers in Cellular and Infection Microbiology 3, 30. https://doi.org/10.3389/fcimb.2013.00030.

Miteva, M., Elsner, P., Ziemer, M., 2009. A histopathologic study of arthropod bite reactions in 20 patients highlights relevant adnexal involvement. Journal of Cutaneous Pathology 36, 26–33.

Modlin, R.L., Miller, L.S., Bangert, C., Stingl, G., 2012. Innate and adaptative immunity in the skin. In: Medical, M.-H. (Ed.), Fitzpatrick's Dermatology in General Medicine, eighth ed.

Moon, T.C., St Laurent, C.D., Morris, K.E., Marcet, C., Yoshimura, T., Sekar, Y., Befus, A.D., 2010. Advances in mast cell biology: new understanding of heterogeneity and function. Mucosal Immunology 3, 111–128.

Morizane, S., Yamasaki, K., Kabigting, F., Gallo, R., 2010. Kallikrein expression and cathelicidin processing are independently controlled in keratinocytes by calcium, vitamin D(3), and retinoic acid. Journal of Investigative Dermatology 130, 1297–1306.

Mosser, D.M., Edwards, J.P., 2008. Exploring the full spectrum of macrophage activation. Nature Reviews. Immunology 8, 958–969.

Naik, S., Bouladoux, N., Wilhelm, C., Molloy, M., Salcedo, R., Kastenmuller, W., Deming, C., Quinones, M., Koo, L., Conlan, S., Spencer, S., Hall, J., Dzutsev, A., Kong, H., Campbell, D., Trinchieri, G., Segre, J., Belkaid, Y., 2012. Compartmentalized control of skin immunity by resident commensals. Science 337, 1115–1119.

Nakatsuji, T., Chiang, H., Jiang, S., Nagarajan, H., Zengler, K., Gallo, R., 2013. The microbiome extends to subepidermal compartments of normal skin. Nature Communications 4, 1431.

Nakatsuji, T., Gallo, R., 2012. Antimicrobial peptides: old molecules with new ideas. Journal of Investigative Dermatology 123, 887–895.

Nestle, F.O., Di Meglio, P., Qin, J.-Z., Nickoloff, B.J., 2009. Skin immune sentinels in health and disease. Nature Reviews Immunology 9, 679–691. https://doi.org/10.1038/nri2622.

Ng, W.C., Londrigan, S.L., Nasr, N., Cunningham, A.L., Turville, S., Brooks, A.G., Reading, P.C., 2015. The C-type lectin langerin functions as a receptor for attachment and infectious entry of influenza A virus. Journal of Virology 90, 206–221.

Nomura, T., Kabashima, K., Miyachi, Y., 2014. The panoply of alphabetaT cells in the skin. Journal of Dermatological Science 76, 3–9.

Oh, J., Byrd, A., Deming, C., Conlan, S., NISC Comparative Sequencing Program, Kong, H., Segre, J., 2014. Biogeography and individuality shape function in the human skin metagenome. Nature 514, 59–64.

O'Keeffe, M., Mok, W.H., Radford, K.J., 2015. Human dendritic cell subsets and function in health and disease. Cellular and Molecular Life Sciences: CMLS 72, 4309–4325.

Otsuka, A., Kabashima, K., 2015. Mast cells and basophils in cutaneous immune responses. Allergy 70, 131–140.

Otto, M., August 2009. Staphylococcus epidermidis–the "accidental" pathogen. Nature Reviews Immunology 7 (8), 555–567.

Ouchi, T., Kubo, A., Yokouchi, M., Adachi, T., Kobayashi, T., Kitashima, D.Y., Fujii, H., Clausen, B.E., Koyasu, S., Amagai, M., Nagao, K., 2011. Langerhans cell antigen capture through tight junctions confers preemptive immunity in experimental staphylococcal scalded skin syndrome. The Journal of Experimental Medicine 208, 2607–2613.

Pasparakis, M., Haase, I., Nestle, F., 2014. Mechanisms regulating skin immunity and inflammation. Nature Reviews Immunology 14, 289–301.

Patterson, J.W., Fitzwater, J.E., Connell, J., 1979. Localized tick bite reaction. Cutis 24, 168–169, 172.

Peric, M., Koglin, S., Dombrowski, Y., Gross, K., Bradac, E., Büchau, A., Steinmeyer, A., Zügel, U., Ruzicka, T., Schauber, J., 2009. Vitamin D analogs differentially control antimicrobial peptide/"alarmin" expression in psoriasis. PLoS One 4, e6340. https://doi.org/10.1371/journal.pone.0006340.

Pingen, M., Bryden, S.R., Pondeville, E., Schnettler, E., Kohl, A., Merits, A., Fazakerley, J.K., Graham, G.J., McKimmie, C.S., 2016. Host inflammatory response to mosquito bites enhances the severity of arbovirus infection. Immunity 44, 1455–1469.

Pivarcsi, A., Nagy, I., Kemeny, L., 2005. Innate immunity in the skin: how keratinocytes fight against pathogens. Current Immunology Reviews 1, 29–42.

Romani, N., Brunner, P.M., Stingl, G., 2012. Changing views of the role of Langerhans cells. The Journal of Investigative Dermatology 132, 872–881.

Rosenbach, T., Czernielewski, J., Hecker, M., Czarnetzki, B., 1990. Comparison of eicosanoid generation by highly purified human Langerhans cells and keratinocytes. The Journal of Investigative Dermatology 95, 104–107.

SanMiguel, A., Grice, E., 2015. Interactions between host factors and the skin microbiome. Cellular and Molecular Life Sciences 72, 1499–1515.

Schauber, J., Gallo, R., 2009. Antimicrobial peptides and the skin immune defense system. Journal of Clinical Allergy and Immunology 124, R13–R18.

Schittek, B., 2012. The multiple facets of dermcidin in cell survival and host defense. Journal of Innate Immunity 4, 349–360.

Schommer, N.N., Gallo, R.L., 2013. Structure and function of the human skin microbiome. Trends in Microbiology 21, 660–668. https://doi.org/10.1016/j.tim.2013.10.001.

Seneschal, J., Clark, R.A., Gehad, A., Baecher-Allan, C.M., Kupper, T.S., 2012. Human epidermal Langerhans cells maintain immune homeostasis in skin by activating skin resident regulatory T cells. Immunity 36, 873–884.

Streilein, J.W., 1978. Lymphocyte traffic, T-cell malignancies and the skin. The Journal of Investigative Dermatology 71, 167–171.

Verhulst, N., Qiu, Y., Beijleveld, H., Maliepaard, C., Knights, D., Schulz, S., Berg-Lyons, D., Lauber, C., Verduijn, W., Haasnoot, G., Mumm, R., Bouwmeester, H., Claas, F., Dicke, M., van Loon, J., Takken, W., Knight, R., Smallegange, R., 2011. Composition of human skin microbiota affects attractiveness to malaria mosquitoes. PLoS One 6, e28991.

von Kockritz-Blickwede, M., Goldmann, O., Thulin, P., Heinemann, K., Norrby-Teglund, A., Rohde, M., Medina, E., 2008. Phagocytosis-independent antimicrobial activity of mast cells by means of extracellular trap formation. Blood 111, 3070–3080.

Von Stebut, E., 2007. Immunology of cutaneous leishmaniasis: the role of mast cells, phagocytes and dendritic cells for protective immunity. European Journal of Dermatology 17, 115–122.

Wikel, S.K., 2013. Ticks and tick-borne pathogens at the cutaneous interface: host defenses, tick countermeasures, and a suitable environment for pathogen establishment. Frontiers in Microbiology 4, 337. https://doi.org/10.3389/fmicb.2013.00337.

Yakimchuk, K., Roura-Mir, C., Magalhaes, K.G., de Jong, A., Kasmar, A.G., Granter, S.R., Budd, R., Steere, A., Pena-Cruz, V., Kirschning, C., Cheng, T.Y., Moody, D.B., 2011. Borrelia burgdorferi infection regulates CD1 expression in human cells and tissues via IL1-beta. European Journal of Immunology 41, 694–705.

Yang, D., de la Rosa, G., Tewary, P., Oppenheim, J.J., 2009. Alarmins link neutrophils and dendritic cells. Trends in Immunology 30, 531–537. https://doi.org/10.1016/j.it.2009.07.004.

Yang, J., Zhao, L., Xu, M., Xiong, N., 2017. Establishment and function of tissue-resident innate lymphoid cells in the skin. Protein and Cell.

Zhang, L., Guerrero-Juarez, C., Hata, T., Bapat, S., Ramos, R., Plikus, M., Gallo, R., 2015. Innate immunity. Dermal adipocytes protect against invasive *Staphylococcus aureus* skin infection. Science 347, 67–71.

Zhu, J., Koelle, D.M., Cao, J., Vazquez, J., Huang, M.L., Hladik, F., Wald, A., Corey, L., 2007. Virus-specific CD8+ T cells accumulate near sensory nerve endings in genital skin during subclinical HSV-2 reactivation. The Journal of Experimental Medicine 204, 595–603.

Chapter 2

Arthropods: Definition and Medical Importance

Gérard Duvallet[1], Nathalie Boulanger[2], Vincent Robert[3]

[1]Univ. Paul Valéry Montpellier 3, Univ. Montpellier, EPHE, CNRS, IRD, CEFE UMR 5175, Montpellier, France; [2]Université de Strasbourg, Strasbourg, France; [3]MIVEGEC, IRD, CNRS, Université de Montpellier, Montpellier, France

ARTHROPODS OF MEDICAL AND VETERINARY IMPORTANCE

The word "arthropod" comes from the Greek *arthron*, "joint," and *podos*, "foot" or "leg," which together means "jointed leg." Arthropods are invertebrates in the phylum Arthropoda being bilaterally symmetrical, with segmented bodies, external skeletons, and jointed appendages. Arthropods exhibit tagmatosis meaning that segments are grouped into functional units such as head with mouthparts, thorax, and abdomen. The exoskeleton or cuticle mainly consists of chitin, a polymer of glucosamine. Due to the presence of this cuticle, the growth is achieved through a series of molts, even metamorphosis in certain groups such as insects. They possess a complete digestive tract with a mouth with different masticatory organs and salivary glands and a complete digestive tract with an anus. The nervous system is typical annelida-like: cerebroid ganglia, a pair of circumpharyngeal connectives around the anterior gut, and a double ventral nerve cord with paired segmental ganglia and nerves. The circulatory system is open with a body cavity containing the hemolymph and a heart (dorsal vessel in insect). The respiration is accomplished in the aquatic forms by gills and by trachea in terrestrial arthropods. Sexes are separate and the reproduction is typically sexual, although parthenogenesis may exist (Marquardt, 1996).

The number of already known arthropod species is almost two million and account for over 80% of all known living animal species. Estimates of total existing arthropod species vary between 5 and 10 million (Ødegaard, 2000).

Arthropods in general and insects in particular play a capital role in the functioning of ecosystems. Pollination, recycling of dead bodies or feces, natural control of pest populations, etc., are some of the beneficial effects provided by arthropods to the ecosystems (Gillott, 2005).

Skin and Arthropod Vectors. https://doi.org/10.1016/B978-0-12-811436-0.00002-2

Most of arthropods have no relationship with humans and human activities, but the exceptions are of paramount importance for humanity. Some species are pests: pests of cultivated plants and pests of stored products. They can be of public health or medical importance as biological vectors of pathogens (malaria and *Leishmania* parasites, dengue virus, Lyme *Borrelia*...), passive carriers of pathogens (flies on feces), ectoparasites (scabies, lice), nuisance pests (millipedes, bed bugs), envenomation (bees, ants), or intermediate hosts of parasites, finally ingested by humans and animals (Mathison and Pritt, 2014).

The principal arthropod taxa of interest to medical or veterinary entomologists are (Mullen and Durden, 2009): (1) subphylum Chelicerata, which encompasses the class Arachnida: scorpions, spiders, ticks, mites and (2) subphylum Mandibulata with the class Myriapoda: millipedes (Diplopoda), centipedes (Chilopoda); the class Crustacea: crustaceans and the major class Insecta: insects (Fig. 2.1). These arthropods can affect humans and animals, directly (nuisance), indirectly (vectors of pathogens), or can cause injury. The Insecta and the Acari include most of the arthropods of medical and veterinary importance.

Insecta

They are characterized by three distinct parts: the head, the thorax, and the abdomen. They have antennae, masticatory organs adapted to the feeding habits (e.g., long proboscis in mosquito for piercing-sucking), eyes, and three pairs of legs on the thorax (Fig. 2.2A). They can also have wings in adult, but they can be absent such as in fleas and lice. Mouthparts are highly different according to the types of food. Insect development is regulated by metamorphosis controlled by hormones and they generally present after the eggs: larva, pupa, or nymph, and male and female adult stages. The classification relies on three types of development: direct development (the newly hatched insect is a small replica of the adult, there is no pupal stage, e.g., silverfish), incomplete metamorphosis (the pupa differs from the adult only in size, absence of wings, and external genitalia, e.g., louse), and complete metamorphosis (wormlike larva gives after several molts a pupa and then an adult with different feeding habits, e.g., mosquito). Highly specialized insects belong to this group (Neva and Brown, 1994). Within Insecta, orders Hemiptera (kissing bugs), Anoplura (sucking lice), Diptera (mosquitoes, sand flies, black flies, tsetse flies, horse flies, stable flies), and Siphonaptera (fleas) are the most important medically and veterinary (Marquardt, 1996).

Acari

Acari are a taxon of arachnids, which contains ticks and mites. They differ from the insects by an absence of wings and antennae, the presence of four pairs of legs in nymphs and adults, and chelicerae. Head, thorax, and abdomen are fused into a single part called idiosoma (Fig. 2.2B). The mouthparts (one-armed hypostome and two chelicerae) and two pedipalps form the capitulum. The pedipalps

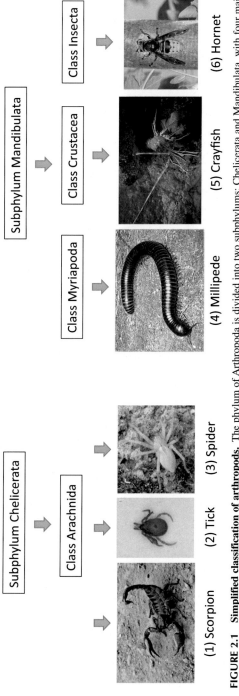

FIGURE 2.1   Simplified classification of arthropods. The phylum of Arthropoda is divided into two subphylums: Chelicerata and Mandibulata, with four main classes. Three illustrations are given for the class Arachnida: Scorpion, tick and spider. One illustration is given for the three other classes Myriapoda, Crustacea, and Insecta. *(1) Scorpion from static.standard.co.uk CC. (2) Tick Ixodes ricinus: female adult. Photo N. Boulanger. (3) Spider Micrommata virescens female, Montpellier. Photo M. T. Goupil. (4) Millipede (Diplopoda) from weblearneng.com CC. (5) Crayfish, by Pcany - CC BY-SA 4.0, https://commons.wikimedia.org/w/index.php?curid=50781549 (6) Insect Hymenoptera Vespa crabo. Photo G. Duvallet.*

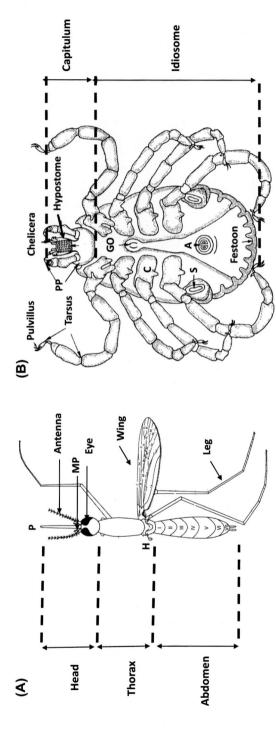

FIGURE 2.2 **Diagrammatic representation of an insect and a hard tick.** (A) Mosquito of culicinae female on a dorsal view. *H*, halter; *MP*, maxillary palp; *P*, proboscis. (*Adapted from "Entomologie Médicale et vétérinaire-Chapitre11: Culicinae-P.248, IRD Edition".*) (B) Hard tick (*Dermacentor* sp.) from a ventral view. *A*, anus; *C*, coxa; *GO*, genital opening; *PP*, pedipalps; *S*, stigma. (*Adapted from Mehlhorn, H., 2001. Encyclopedic Reference of Parasitology, second ed., vol. 1. Springer-Verlag, Berlin-Heidelberg, 678 pages.*)

do not penetrate the tissues and are sensory organs used to detect the host. Ticks are pool feeder. They are divided into two main families: Argasidae or soft ticks and Ixodidae or hard ticks. Argasidae are generally endophilic, living in host nest, stables, or burrow. They have a multihost life cycle. They have two or more nymphal stages. They have rapid blood meals, feeding several times, generally for minutes in nymphs and adults. Feeding and oviposition are cyclical in females. They generally feed at night. During the blood meal, coxal glands located between the first legs secrete a liquid that can contain pathogens. Sexes are very similar; the capitulum is either on the ventral part or protruding from the anterior margin of body in larval stages. The cuticle is smooth and leathery and there is no scutum (Mehlhorn, 2001). Two genera are of public health concern: *Ornithodoros* and *Argas*. Ixodidae are mostly exophilic and are characterized by long blood meals. They undergo one, two, or three host life cycle. During these developments, for each blood meal, ticks can remain on one specific host or change for each stage, the larval, the nymphal, or the adult stages. Each stage sucks only once for several days (3–12 days according to the stage). Females can take 100 to 200 times their size in blood (Mehlhorn, 2011). They have a dorsal hard shield (scutum), covering the whole dorsal region in male and one-third of the dorsal region in other stages. The capitulum is clearly visible at the anterior part. Eyes are absent in *Ixodes* and *Haemaphysalis* and when present are situated dorsally on the anterior lateral margin of the scutum (Mehlhorn, 2001; Neva and Brown, 1994). Ticks are of great medical importance. They can be biological vectors or be toxic by their saliva.

Identification of arthropods is essential to measure the risk of potential human and animal threat. The morphological identification must be referred to specialists (Mathison and Pritt, 2014). Nowadays, the use of molecular biology and mass spectrometry develops rapidly to identify arthropod specimens. The mass spectrometry first used to identify midges (Kaufmann et al., 2011; Steinmann et al., 2013) constitutes a promising tool by the rapidity of the sample identification and its cost. It is more and more extended to arthropod vector identification (Yssouf et al., 2013, 2016).

NUISANCE OF ARTHROPODS

Nuisance can be defined as the unpleasantness or even extreme discomfort caused by the sometimes adventitious presence of one or several arthropods. The density of the arthropods plays its role in the definition. An isolated mosquito bite is not a problem, but hordes of aggressive mosquitoes encountered in northern areas in early summer or in certain tropical islands in the rainy season are a strong nuisance (Fig. 2.3A).

Nuisance also encompasses attacks to health. In the event of huge aggressiveness of hematophagous arthropods, the amount of taken blood (blood spoliation) can lead to severe anemia and even death. Another consequence of the hematophagy is the injection of the arthropod saliva into the vertebrate at the

(A) Direct effect of arthropods: proliferation, nuisance

(C) Arthropods: mechanical transmission and intermediate host

Horse fly

Stomoxys calcitrans

Mosquitoes

(B) Arthropods as carriers of pathogens: phoresis

(D) Arthropods as vector: active transmission

Anopheles

Flies

FIGURE 2.3 Different types of interaction with the vertebrate host. (A) Proliferation: Mosquitoes in Aldabra, the Seychelles. (B) Nuisance of flies on the head and around eyes of horses. (C) A female stable fly *Stomoxys calcitrans* and a female horse fly Philipomyia aprica. (D) Arthropod vector: Female *Anopheles* mosquito, Wikimedia. ((A) *Photo by V. Robert/IRD.* (B) *Photo G. Duvalletand V. Robert/IRD.* (C) *Photo G. Duvallet and F. Baldacchino.*)

time of the blood meal. This saliva is usually inflammatory and immunosuppressive (Kazimírová and Stibrániová, 2013) and can trigger reactions including severe allergies (Commins and Platts-Mills, 2013; Elston, 2010), well documented for ticks.

All arthropods can be allergenic but rarely cause a public health problem. A few ones such as mites in house dust, or their carcasses, can be the cause of serious and frequent respiratory allergies. It is also the case for arthropod dead bodies, cockroach feces, etc.

Some arthropods are pathogenic by themselves as ectoparasites because they are the cause of pathological manifestations. This is the case of the human scabies mites *Sarcoptes scabiei*. They cause itch and the female burrows into the skin. It is also the case of maggots of some flies that develop at the expense of various tissues of the host and are the cause of myiasis (example of the Cayor worm, *Cordylobia anthropophaga*). In veterinary medicine, losses due to ectoparasites of the livestock network in the United States are estimated to more than $2.2 billion per year (Taylor et al., 2012).

Outside the bloodsucking or tissue-invading species, some arthropods induce another form of nuisance related to their sting apparatus. Spiders and centipedes inject saliva with their chelicerae or their forcipules, respectively. Scorpions have a last abdominal segment with a pair of venom glands. This segment bears a stinging structure called the telson. The venom is generally a mixture of neurotoxins effective against other arthropods, the usual preys of scorpions. Stings inflicted to humans constitute a defense mechanism. Dangerous species are present in the Sahara and Mexico. Adult females of Hymenoptera (bees, wasps, hornets, bumble bees) sting with a retractable appendage located at the end of the abdomen. Hymenoptera stings can cause death, primarily in allergic individuals.

Finally, many arthropods use their mouthparts and their saliva to defend when we seized them and inflict painful bites. This is the case of Heteroptera as predatory bugs (genera *Reduvius*, *Belostoma*, and *Rhynocoris*, for example) or sap-sucking bugs (Pentatomidae family). Another group consists of urticating, blistering, and allergenic arthropods. This is the case of urticating caterpillars of moths, which have differentiated setae and spines into tiny harpoons bristles with a central cavity filled with very allergenic venom. Pine processionary caterpillars present these characteristics. In French Guiana, the female of the moth *Hylesia metabus* has abdominal scales filled with allergenic venom. These scales stand permanently and float in the air. Some beetles such as Meloidae (genus *Lytta*) and Staphylinidae (genus *Paederus*) secrete irritating chemicals when the insects are handled or accidentally contact human or animal skin. This is also the case for some millipedes (Myriapoda) of tropical regions.

More marginal, the entomophobia and Ekbom syndrome represent health problems related to the arthropods (Hinkle, 2000). The entomophobia and arachnophobia refer to an unusual and irrational fear for various arthropods. Some people may panic when they encounter these arthropods, such as spiders.

Ekbom syndrome, described in 1938 by a Swedish neurologist (Dr. Ekbom) (Ekbom, 1938), is a psychotic obsession based on the belief of being infested with parasites (delusional parasitosis). Dermatological problems of these patients are attributed to the presence of arthropods. These imaginary attacks by arthropods cannot be handled by entomologists and has to be sent to psychiatrists.

RELATION ARTHROPODS–PATHOGENS

Arthropods can be carrier or vector of infectious pathogens. They play considerable roles in public health (human and animal) and animal production (Kettle, 1995; Mullen and Durden, 2009; Rhodain, 2015). Two distinct modes of transmission exist.

Arthropods as Carriers of Pathogens

Many arthropods can contaminate or spoil food. Insects such as houseflies or cockroaches may regurgitate pathogen-contaminated fluids prior to or during feeding. They may also defecate, contaminating the food with potential pathogens. Subsequent consumption of contaminated food can lead to the transmission of these pathogens to humans or animals. The integument of those insects (particularly mouthparts, legs, and tarsi) carries potential infectious microorganisms that may be transferred to food items or directly to the body of humans and animals (phoresis mechanism). These arthropods may have been in contact with fecal matter, garbage heaps, animal secretions, or other potential sources of pathogens, contributing to health risks such as enteric diseases. This is the case, for example, of flies (Fig. 2.3B) such as housefly (*Musca domestica*), bazaar fly (*Musca sorbens*), bush fly (*Musca vetustissima*), face fly (*Musca autumnalis*), and many other filthy flies, which can transmit bacteria such as *Shigella* sp., *Escherichia coli*, *Chlamydia trachomatis*, or helminths such as the mammalian eyeworm *Thelazia* sp. For example, in cattle, the face fly is known as the main carrier of the bacterium *Moraxella bovis*, etiologic agent of the infectious bovine keratoconjunctivitis. The fly acquires the bacterium from the eyes of cattle and the bacterium can remain viable for several hours on and in the fly.

Arthropods as Vectors

We saw previously that nuisance or pathogenesis are the results of the arthropod itself or its sting; two actors are involved, the arthropod and the vertebrate. However, the concept of vector implies the existence of a third actor, the transmitted infectious agent (Bernard et al., 2015). We have already seen examples of insects as carriers of pathogens. We restrict here the notion of vector to the bloodsucking arthropods, which can transmit a pathogen during a blood meal.

VECTOR DEFINITION

Ticks and mosquitoes are the most important arthropod vectors. If we consider the veterinary medicine, ticks are the most important vectors; if we consider the human medicine, mosquitoes are the most important vectors. The term "mite" includes both ticks and mites, although they are usually classified separately. In some context, ticks are often considered as giant mites. More than 900 species of ticks have been described so far (McCoy and Boulanger, 2016). Concerning mosquitoes, 3500 species grouped into 41 genera have been described. In medical entomology, the term "vector" refers to two distinct definitions according to the authors.

Large Definition of Arthropod Vector

The first definition includes any organism that is involved in the transmission of an infectious agent. It thus designates a vast and heterogeneous set including the "carriers of pathogens" mentioned previously. For example, cockroaches or flies carrying passively on their legs or their dirty mouthparts any kind of bacteria, mosquitoes inoculating viruses or depositing the Bancroftian filariasis nematode on the skin in the vicinity of their bite belong to this definition. Crustacea copepods, intermediate hosts of the guinea worm *Dracunculus medinensis*, such as mollusks, and intermediate hosts of *Schistosoma* spp. are also concerned. This definition also includes leeches as vectors of fish parasites, dogs and bats as vectors of rabies virus transmitted by bites, rats as vectors of *Leptospira* spp., and even migratory birds as vectors of avian influenza virus.

This definition is used by the WHO (2014) for which "vectors are living organisms that can transmit infectious diseases between humans or from animals to humans." We underline that vectors never transmit diseases but infectious agents, which can result or not in diseases.

Stricto sensu Definition of Arthropod Vector

The second definition, which has our preference, designates as vector "any hematophagous arthropod which ensures active transmission (mechanical or biological) of an infectious pathogen from an infected host to a new host." The concept of active transmission requires that the vector, by its behavior, promotes contact between an infectious agent and a vertebrate host. The behavior of the arthropod is here essential, increasing the probability of encounter between the infectious agent and the target vertebrate host.

This definition is more restrictive than the previous one since it concerns only the hematophagous insects and Acari and thus limited to Medical and Veterinary Entomology. This definition excludes all the carrier arthropods seen previously and most intermediate hosts passively releasing into the environment an infectious agent such as mollusks or crustaceans cited in the first definition. Note that there are cases where an insect can meet both definitions. This is the case for example of *Stomoxys calcitrans* (Fig. 2.3C), bloodsucking fly, which can be not only mechanical vector of trypanosomes (second definition) but also intermediate host of helminths of the genus *Habronema* (first definition) (Baldacchino et al., 2013). In the latter case, a horse becomes infected when swallowing a fly infected with this worm.

In the case of a **mechanical vector transmission**, the infectious agents do not multiply or differentiate in contact with the vector. The vector is simply acting as an active flying syringe, in contact at least twice with different vertebrates, on the occasion of interrupted blood meal. The vector becomes infected during the first contact with the host and transmits the pathogen to the second one. The mechanical transmission requires that the delay between the two blood meals from the infected vertebrates to the receptive one is short, to allow the survival of the pathogens. Both mechanisms are possible: either an external coating of the mouthparts of the vector that works as a contaminated needle or a partial regurgitation of the previous blood meal. The vector then works as a contaminated syringe. The stable and horse flies are responsible for numerous mechanical transmissions (Baldacchino et al., 2013, 2014).

In the case of a **biological vector transmission**, three phases are observed: (1) Infection of the vector during a blood meal on an infected vertebrate. The pathogen is ingested with blood or lymph and migrates to the intestinal lumen of the vector. (2) Differentiation and/or multiplication of the pathogen with antigenic variations with very complex development. Generally, viruses, bacteria, and protozoa multiply actively into the vector in the gut, sometimes in the general cavity (hemolymph) or in various organs, and they often end up colonizing the salivary glands. Metazoans (helminths) proceed more by differentiation than by multiplication. (3) Finally the third phase begins when the vector bites and infects the host through a variety of processes. The time necessary for these three phases is designated extrinsic incubation period or prepatent phase, corresponding to the period between the infecting blood meal and the moment when the pathogen can be transmitted again. It usually takes between 6 and 15 days depending on the pathogens and the environmental conditions including temperature. The vector must bring the infectious pathogen (1) at an infectious stage for vertebrate, (2) in a site conducive to its transmission, and (3) in sufficient quantity.

Infectious agents that can be transmitted are either viruses or bacteria including rickettsia or parasites (protozoa and helminths). The infectious agent can be delivered at the bite site with saliva (arboviruses, *Rickettsia* transmitted by ticks or mites, *Plasmodium* transmitted by mosquitoes, *Trypanosoma* transmitted by tsetse flies), by regurgitation of previous meal (*Leishmania* transmitted by sand flies, *Yersinia pestis* transmitted by fleas), by release of the parasite on the skin (filarial nematodes), with the feces (*Trypanosoma cruzi* transmitted by bugs, *Rickettsia* transmitted by lice and fleas), with coxal liquid (liquid secreted by Argasidae ticks), and by arthropod crush (a few *Rickettsia* trapped in the cavity of lice). Most of pathogens (arboviruses, *Plasmodium*, *Leishmania*,

Trypanosoma transmitted by tsetse flies, etc.) are never transmitted alone from the vector to the vertebrate host but are coinoculated with the vector saliva (Fig. 2.3D) (see Chapters 4 and 5).

MAIN INSECTS AND ACARINES OF MEDICAL AND VETERINARY IMPORTANCE AND THE PATHOGENS THEY TRANSMIT

Insects and arachnids of particular interest to medical–veterinary entomologists are presented in Table 2.1. This list includes arthropods that are considered as nuisance, venomous, and vectors. Table 2.2 gives indications about main vectors, various stages, and aspects of hematophagy.

TABLE 2.1 Principal Orders of Insects and Arachnids of Medical-Veterinary Interest

Classes	Orders	Common Names
Insecta	Blattaria	Cockroaches
	Phthiraptera	Lice
	Hemiptera	True bugs, bed bugs, kissing bugs, assassin bugs
	Coleoptera	Beetles
	Siphonaptera	Fleas
	Diptera	Mosquitoes, black flies, horse flies, deer flies, sand flies, tsetse flies, houseflies, stable flies, horn flies, bot flies, blow flies, flesh flies, louse flies, keds, etc.
	Lepidoptera	Moths, butterflies
	Hymenoptera	Wasps, hornets, velvet ants, ants, bees
Arachnida	Scorpionida	Scorpions
	Solpugida	Solpugids, sun spiders, camel spiders, barrel spiders
	Acari	Mites, ticks
	Araneae	Spiders
Chilopoda	Scolopendromorpha	Centipedes
Diplopoda		Millipedes

Adapted from Mullen, G.R., Durden, L.A., 2009. Medical and Veterinary Entomology, second ed. In: Mullen, G.R., Durden L.A. (Eds.). Elsevier, Academic Press, London.

TABLE 2.2 Main Vectors Among Insects and Acari

Classes	Orders	Families	Hematophagous Stages	Biology of Hematophagy	Preimaginal Stages[a]
Insecta	Diptera	Culicidae (mosquitoes)	Adult females	Mainly twilight for *Aedes* Mainly nocturnal for *Anopheles, Culex*, and *Mansonia*	Aquatic (stagnant water)
		Simuliidae (black flies)	Adult females	Diurnal	Aquatic (running water)
		Psychodidae (sand flies)	Adult females	Nocturnal	Soil (humus, litters)
		Tabanidae (horse flies)	Adult females	Diurnal	Semiaquatic
		Ceratopogonidae	Adult females	Mainly twilight but variable according to species	Soil (humus)
		Glossinidae (tsetse flies)	Adult males and females	Diurnal	*In utero* except pupae in soil
	Siphonaptera (fleas)	Several families	Adult males and females	Several blood meals per 24 h	Soil (litters)
	Hemiptera Heteroptera (bugs)	Reduviidae (kissing bugs)	Adult males and females, and immatures	Nocturnal	Soil, burrow bloodsucking
	Anoploura	Pediculicidae (lice)	Adult males and females, and immatures	Several blood meals per 24 h	Hair bloodsucking
Arachnida	Acari	Ixodidae (hard ticks)	Adult females and immatures	Only one blood meal per stage, lasting several days	Vegetation bloodsucking
		Argasidae (soft ticks)	Adult females and immatures	Several blood meals per stage. Mainly nocturnal	Soil, burrows bloodsucking
		Trombiculidae (*Trombicula*)	Larvae	Lymph meal lasting for several days	Soil

[a]*Preimaginal stages: larvae and pupae.*
Adapted From Duvallet, G., de Gentile, L., 2012. Protection Personnelle Antivectorielle. In: Duvallet, G., de Gentile, L. (Eds.). IRD Marseille, France.

The process of transmission can be rapid, generally insect-transmitted pathogens, or delayed like in hard ticks. It is directly correlated to the type of blood meal, rapid for insect and generally long for hard ticks (several days). According to the biting mouthparts, two types of bite can be defined. Telmophagy, likely the most primitive, corresponds to "pool feeding." Arthropods slice the superficial skin and suck the blood that is released from ruptured blood capillaries. Tabanids, black flies, and sand flies possess this type of blood feeding. The second type is solenophagy, which corresponds to a cannulation of blood vessel as observed for mosquitoes (Black and Kondratieff, 2004).

Transmission of Viruses

Transmitted viruses are called arboviruses (arthropod-borne viruses). They develop within the cells of the arthropod tissues. There are more than 500 known arboviruses (Chapters 8 and 10). It is a heterogeneous group, belonging to several large viral families. This group includes, for example, the yellow fever, dengue, and chikungunya viruses, transmitted to humans or other primates by *Aedes* mosquitoes.

Transmission of Bacteria

Vector-transmitted bacteria are also numerous. The plague agent *Y. pestis*, discovered by Alexandre Yersin, is usually transmitted to humans by the bite of the rat flea *Xenopsylla cheopis*. Rickettsiae are obligatory intracellular bacteria, which infect the cytoplasm of eukaryotic cells. They are transmitted by ticks (McCoy and Boulanger, 2016), mites, lice, etc. For example, *Rickettsia prowazekii*, the epidemic typhus agent, has the body louse *Pediculus humanus* for vector. This rickettsia is transmitted by infected louse feces, and inoculation to humans is done when rickettsiae infect the scratched skin in response to louse bites.

Rickettsia conorii, agent of Mediterranean spotted fever, is transmitted by the bite of the tick *Rhipicephalus sanguineus* in southern Europe. *Borreliella* and *Borrelia* can be transmitted by hard ticks (Lyme borreliosis in temperate regions) or soft ticks (relapsing fever in temperate and tropical regions), respectively. The genus *Borrelia* has been recently divided into two genera: *Borreliella* for Lyme borreliosis *(Borrelia burgdorferi* sensu lato complex) and *Borrelia* for relapsing fever (Adeolu and Gupta, 2014).

Transmission of Eukaryotic Parasites

They are many protozoa, unicellular eukaryotes, transmitted by vectors. This is the case of malaria parasites, apicomplexa transmitted by mosquitoes (*Anopheles* for mammal parasites and Culicinae for bird/reptile parasites) and *Leishmania* spp., flagellates responsible of leishmaniosis transmitted by various sand flies. It is also the case of the trypanosomes, agents of human or animal African trypanosomosis, transmitted by tsetse flies.

Some helminths, metazoan parasites, may be transmitted by vectors. *Onchocerca volvulus*, agent of onchocerciosis or river blindness, is transmitted by black flies (*Simulium* spp.). *Wuchereria bancrofti*, responsible of lymphatic filariosis, is transmitted by mosquitoes.

Some pathogens infect only humans, others only animals, and others both humans and animals. In the latter case, the pathogen is defined as zoonotic. A zoonosis is an infection or infestation naturally transmissible from animals to humans, and vice versa (Kahl et al., 2002). Table 2.3 mentions the main pathogens and specifies which ones are zoonoses among major vector-borne diseases.

EPIDEMIOLOGY OF VECTOR-BORNE DISEASES—VECTORIAL SYSTEM

The basic outline of vector ecology may be summarized in an equilateral triangle with the vector, the pathogen, and the host at each tip of the triangle (Bernard et al., 2015). The sides of the triangle are arrows that show the interactions between these elements. This vector system only works if the three elements coexist, are in contact, and tolerate each other. This highlights the phenomenon of coevolution that leads to more or less specificity of interactions between the three actors of arthropod-borne diseases.

The degree of vector competence can be defined between vectors and pathogens (Kahl et al., 2002). A species of *Anopheles* is competent for a species of *Plasmodium*, a tsetse fly is not. The vector competence depends on trophic preferences of the vector, on its behavior and its ability to get infected, and to ensure the development of the pathogen. This competence depends on intrinsic factors under genetic control. Between host and infectious agent, we talk about sensitivity or susceptibility. Humans are likely to be infected by *Plasmodium falciparum* and not by *Trypanosoma congolense*. The contact between the vector and the host will also depend on many environmental (both biotic and abiotic) and socioeconomic factors. We talk about vector capacity as a result of the vector competence and external environmental factors, which make the transmission takes place and be more or less intense (Kahl et al., 2002).

In his book "Parasitism," Combes (2001) highlights filters of encounter and compatibility. The compatibility is the metabolic and immunological system dimension (to accept each other) and thus represents the competence, while the encounter is both the ecological (living in the same place) and the ethological dimension (be in contact) and represents the capacity. All of these notions have been taken recently in a remarkable book of medical entomology (Rhodain, 2015).

INNATE IMMUNITY OF ARTHROPODS

To survive in their hostile environment, such as vertebrates, arthropods have developed a potent immune response. It is triggered by a direct infection with arthropod pathogens (fungi and bacteria) or by the pathogens they harbor and

TABLE 2.3 Main Vector-Borne Diseases

Pathogens	Vector-Borne Diseases	Human or Animal Disease or Zoonosis	Vector	Distribution
Arbovirus				
ASF virus, family Asfarviridae, genus Asfivirus	African swine fever	Animal (Suidae)	Ticks Argasidae *Ornithodoros*	Sub-Saharan Africa, Sardinia
Bluetongue Virus, Reoviridae, Orbivirus	Bluetongue disease	Animal (ungulates)	Ceratopogonidae *Culicoides*	Northern and Sub-Saharan Africa, Europe
CCHF virus, Bunyaviridae, Nairovirus	Crimean–Congo hemorrhagic fever	Zoonosis (mammals, birds)	Ticks	Northern Europe, Sub-Saharan Africa, Asia
Chikungunya virus, Togaviridae, Alphavirus	Chikungunya fever	Humans	Mosquitoes (*Aedes*)	Southern Europe, Sub-Saharan Africa, Asia, South-West Indian Ocean, Oceania
Dengue virus, Flaviviridae, Flavivirus	Dengue fever	Humans	Mosquitoes (*Aedes*)	Southern Europe, Sub-Saharan Africa, South-West Indian Ocean, Asia, Oceania, Northern and Latin America
Japanese encephalitis virus, Flaviviridae, Flavivirus	Japanese encephalitis	Zoonosis	Mosquitoes Culicidae (*Culex tritaeniorhynchus*)	Asia, Oceania
Rift Valley virus, Bunyaviridae, Phlebovirus	Rift Valley fever	Zoonosis (ruminants)	Mosquitoes	Sub-Saharan Africa, South-West Indian Ocean

Continued

Table 2.3 Main Vector-Borne Diseases—cont'd

Pathogens	Vector-Borne Diseases	Human or Animal Disease or Zoonosis	Vector	Distribution
Ross River Virus, Togaviridae, Alphavirus	Ross River fever	Zoonosis (marsupials)	Mosquitoes (*Culex* and *Aedes*)	Oceania
TBEV (tick-borne encephalitis virus) Western subtype, Eastern subtype, Flaviviridae, Flavivirus	European tick-borne encephalitis Asian tick-borne encephalitis	Zoonosis (rodents, ungulates) Zoonosis (ungulates)	Ticks Ixodidae (*Ixodes ricinus*) *Ixodes persulcatus*	Northern Europe East Europe and Asia
West Nile virus, Flaviviridae, Flavivirus	West Nile fever	Zoonosis (birds)	Mosquitoes (*Culex*)	Southern Europe, Northern and Sub-Saharan Africa, Northern and Latin America
Yellow Fever virus, Flaviviridae, Flavivirus	Yellow fever	Zoonosis (primates)	Mosquitoes (*Aedes*)	Sub-Saharan Africa, Latin America
Bacteria				
Anaplasma spp.	Anaplasmosis	Zoonosis (mammals)	Ticks Ixodidae	Europe, Asia, America
Bartonella spp.	Bartonellosis	Zoonosis (mammals, cats, rats)	Depends on *Bartonella* species: fleas, ticks. Sand fly *Lutzomyia*	Northern and Southern Europe, Northern and Sub-Saharan Africa Latin America
Borrelia/Borreliella [a] *burgdorferi sensu lato*	Lyme borreliosis	Zoonosis (mammals, birds, lizards)	Ticks Ixodidae *Ixodes* spp.	Northern and Southern Europe, Northern Africa, Northern America

Borrelia spp.	Tick-borne relapsing fever	Zoonosis (rodents)	Ticks Argasidae *Ornithodoros*	Northern and Sub-Saharan Africa, Southern Europe, America
Borrelia miyamotoi			*Ixodes* spp.	Northern hemisphere
Coxiella burnetii	Q fever	Zoonosis (ungulates, cats, dogs, pigeons)	Ticks Argasidae	Northern and Southern Europe, Northern Africa, Sub-Saharan Africa
Ehrlichia spp.	Ehrlichiosis	Zoonosis	Ticks Ixodidae	Northern America
Ehrlichia ruminantium	Cowdriosis	Animal (ruminants)	Ticks Ixodidae	Northern and Sub-Saharan Africa
Orientia tsutsugamushi	Scrub typhus	Zoonosis	Mites, chiggers	Asia
Rickettsia africae	African tick bite fever	Zoonosis	Tick Ixodidae *Amblyomma*	Sub-Saharan Africa, Guadeloupe
Rickettsia prowazekii	Epidemic typhus	Zoonosis	Lice	Northern and Sub-Saharan Africa, Latin America
Rickettsia rickettsii	Rocky Mountain spotted fever	Zoonosis	Tick Ixodidae *Dermacentor*	Northern America
Rickettsia typhi	Murine typhus	Zoonosis	Rats flea *Xenopsylla cheopis*	Northern and Sub-Saharan Africa, Asia, Latin America
Theileria equi	Equine piroplasmosis	Animal (horses, ruminants, dogs)	Ticks Ixodidae	Southern Europe, Central and Southern America, Asia
Theileria annulata	Tropical theileriosis	Animal (ruminants, dogs)	Ticks Ixodidae	Southern Europe, Northern Africa, Middle East, Asia

Continued

Table 2.3 Main Vector-Borne Diseases—cont'd

Pathogens	Vector-Borne Diseases	Human or Animal Disease or Zoonosis	Vector	Distribution
Theileria parva	Bovine theileriosis	Bovine, buffalos	*R. appendiculatus*	Africa
Yersinia pestis	Plague	Zoonosis	Rats flea	Northern and Sub-Saharan Africa, South-West Indian Ocean, Asia, Northern and Latin America
Protozoan Parasites				
Babesia spp.	Babesiosis	Zoonosis	*Ixodes* sp.	Northern Europe, Northern America
Haemoproteus spp.	Haemoproteosis	Animal (birds, reptiles, amphibians)	Biting midges	All continents except Antarctica
Leishmania spp.	Leishmaniosis	Zoonosis	Phlebotominae	Northern Africa, Asia, Latin America
Leucocytozoon spp.	Leucocytozoonosis	Animal (birds)	Simuliidae	All continents except Antarctica
Plasmodium spp.	Malaria	Humans	Mosquitoes Anophelinae, Culicinae	Sub-Saharan Africa, South-West Indian Ocean, Asia, Oceania, Latin America
		Primates		
		Birds		
		Rodents		

Trypanosoma brucei rhodesiense and *T.b. gambiense*	Human African trypanosomosis (sleeping sickness)	Humans and zoonosis	Tsetse flies	Sub-Saharan Africa
Trypanosoma cruzi	Human American trypanosomosis (Chagas disease)	Zoonosis	Kissing bugs Reduviidae	Latin America
Helminth Parasites				
Loa loa	Loaiosis	Zoonosis	Tabanidae *Chrysops*	Sub-Saharan Africa
Mansonella spp.	Mansonellosis	Humans	Ceratopogonidae *Culicoides*	Sub-Saharan Africa, Latin America
Onchocerca volvulus	Onchocerciosis	Humans	Simuliidae	Sub-Saharan Africa
Wuchereria bancrofti	Lymphatic filariosis	Humans	Mosquitoes	Sub-Saharan Africa, South-West Indian Ocean, Asia, Oceania

[a]A phylogenomic division of the genus Borrelia into two genera has been done: the genus Borrelia contains only the members of the relapsing fever and the genus Borreliella gen. nov. contains the members of the Lyme disease Borrelia (Borrelia burgdorferi sensu lato complex) (Adeolu and Gupta, 2014).
Adapted from Robert V., 2017. Introduction à l'entomologie médicale et vétérinaire. In: Duvallet, G., Fontenille, D., Robert, V. (Eds.), Entomologie médicale et vétérinaire. IRD Editions, Marseille, France.

further transmit as vectors. Endosymbionts do not induce the immune system of arthropods (Fig. 2.4). This immune system is simple and relies solely on innate immunity, characterized by a quick response and an absence of memory. Cellular immunity was first described with phagocytosis relying on hemocytes, the blood cells circulating in hemolymph. Then, humoral immunity was demonstrated with coagulation and melanization processes. Within arthropods, the study of insect immunity has a longer history than arachnid immunity. Pasteur started to work on silkworm diseases in 1870 and Metchnikoff studied the **cellular immunity** of insects. Then, Metalnikov in 1930s at the Pasteur Institute in Paris describes the insect immune response and the presence of bacteriolysins (Brey, 1998). Then the system appeared to be more complicated especially in the most evolved insects. Boman's group studying the hemolymph of *Hyalophora cecropia* in response to bacterial infection was the first to clearly characterize antimicrobial peptides in insect, the cecropin and the attacin (Steiner et al., 1981). Then, *Drosophila melanogaster* becomes a model of choice by its genetics to study insect immunity. The fat body, the equivalent of mammal liver, was then described as well as the role of epithelia in the control of bacterial infection (Lemaitre and Hoffmann, 2007). It allowed the discovery of the Toll pathway (Lemaitre et al., 1996), an analogous of the complement pathway (Lagueux et al., 2000) and a panel of antimicrobial peptides (Uttenweiler-Joseph et al., 1998). The Toll pathway was found to be activated on the interaction of different receptors with pathogens, mainly Gram-positive bacteria and fungi. A second pathway, Imd, was then identified in the control of Gram-negative bacterial infection (Lemaitre and Hoffmann, 2007).

It opened new avenues to define the role of this innate immunity in arthropod vectors. *Anopheles*, vector of malaria parasites, was first studied (Vizioli et al., 2001), then tsetse fly, vector of trypanosomes (Boulanger et al., 2002), sand fly vector of *Leishmania* (Boulanger et al., 2004), and more recently ticks (Hajdusek et al., 2013; Sonenshine et al., 2002). The antimicrobial peptides seem to regulate pathogen population (Boulanger et al., 2006; Hu and Aksoy, 2006). Interestingly, the immune response of the vertebrate host also interferes in the regulation of the innate immunity of the arthropod vector. An elegant study recently described the role of the host interferon (IFN)-γ produced by mice infected by *Borrelia*, responsible of Lyme borreliosis, which triggers an INF-γ-like pathway in *Ixodes* tick leading to the secretion of an antimicrobial peptide, Dae (domesticated amidase effector), which in turn regulates the bacterial infection in tick (Smith et al., 2016).

ARTHROPOD MICROBIOME AND IMPACT ON PATHOGEN TRANSMISSION

Symbiotic/mutualistic obligatory associations are well known for several insects, especially those relying on nutrition, e.g., tsetse and vertebrate blood, termites and wood, etc. Those insects harbor well-described symbionts called

primary symbionts since they have coevolved with their host such as tsetse and *Wigglesworthia* (Aksoy, 1995), termite, and *Trichonympha* (Brune, 2014). **Secondary symbionts** have been also described. They are transient, not present in all the population, and can have diverse origin (vertical or horizontal origin or from the environment) (Weiss and Aksoy, 2011) (Fig. 2.4).

With high genomics sequencing, the human microbiome has been intensively studied these last years (Grice and Segre, 2011). Identification of bacteria, viruses, parasites, and fungi has been done and mapping has been set up (Findley et al., 2013). Recent studies demonstrate the key role of this microbiome at interfaces such as digestive tract, vaginal and urethral mucosa, skin, and especially in different pathologies or inflammatory processes (Cogen et al., 2008; Grice, 2014). It then stimulated researchers to look on the side of the arthropod microbiome and their potential role in the development of pathogens, especially in the gut compartment (Hu and Aksoy, 2006; Wang et al., 2013). The mosquito microbiome has been particularly investigated due to its role in several arthropod-borne diseases (Saraiva et al., 2016), as well as the one of tsetse flies (Wang et al., 2013) and *Ixodes* ticks (Narasimhan et al. 2014, 2017). Interestingly, the microbiome plays a role in vector competence by directly affecting the innate immunity of the arthropod vector (Narasimhan et al., 2017). Gut microbiota and pathogens transmitted by these arthropods activate the innate immune system that then controls the development of parasites or other pathogens (Weiss and Aksoy, 2011).

Wolbachia, the maternally heritable microbe well known to affect reproduction in insect is also involved in the control of certain parasitic (avian malaria, filarial nematodes) and viral (dengue, chikungunya) infections in *Aedes* mosquito. This effect would be due to an upregulation of immune genes (antimicrobial peptides, C-type lectins, complement proteins) by *Wolbachia*. Similarly, *Culex* mosquitoes infected with West Nile virus produced less virus when *Wolbachia* is present (Weiss and Aksoy, 2011).

One interesting approach in the studies on symbionts is their use in **paratransgenesis**. It consists in the modification of the genome of the symbiont, which is first cultured *in vitro*, genetically modified with an antimicrobial molecule, and then reintroduced into the insect vector. The transmitted pathogen is normally targeted by the antimicrobial molecule, and the pathogen transmission is reduced. It has been successfully tested in triatomine bug infected with *T. cruzi*, the agent responsible of Chagas disease (Durvasula et al., 1999). It has also been tested in tsetse, vector of trypanosomes, and its commensal symbiont, *Sodalis* (Hu and Aksoy, 2005; Medlock et al., 2013).

CONCLUSIONS AND PERSPECTIVES

In the control of arthropod-borne diseases, a better understanding of the interaction vector–pathogens is essential to lead to effective control of the diseases (Agarwal et al., 2017; de la Fuente et al., 2017). To facilitate this approach, it

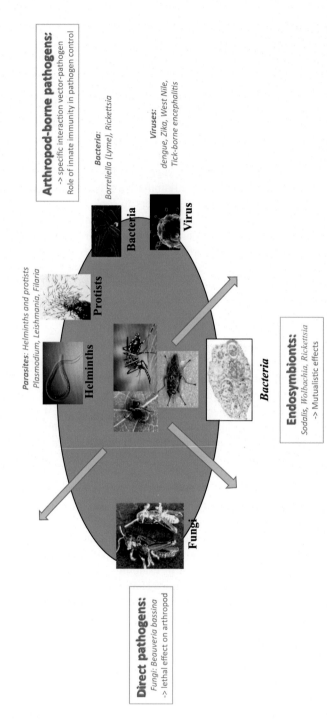

Arthropod-borne pathogens:
-> specific interaction vector-pathogen
Role of innate immunity in pathogen control

Bacteria:
Borreliella (Lyme), Rickettsia

Viruses:
dengue, Zika, West Nile,
Tick-borne encephalitis

Parasites: Helminths and protists
Plasmodium, Leishmania, Filaria

Bacteria

Virus

Protists

Helminths

Bacteria

Fungi

Endosymbionts:
Sodalis, Wolbachia, Rickettsia
-> Mutualistic effects

Direct pathogens:
Fungi: Beauveria bassina
-> lethal effect on arthropod

FIGURE 2.4   Interaction microorganisms—arthropods.   Different types of interaction of arthropods with microorganisms: an arthropod can have symbiotic relationship with certain microorganisms, e.g., *Wolbachia*; some microorganisms can be directly pathogenic to arthropod and kill them like some fungi (*Beauveria bassiana*); some microorganisms develop within the arthropod vector and will be transmitted actively through hematophagous bite (Plasmodium, Leishmania, helminths). All the pictures from Wikimédia.

is necessary to have arthropod colonies in research laboratories to infect them with different pathogens. Some of these vectors can be maintained easily in colonies under laboratory conditions since they have rapid development like mosquitoes, 2 weeks from the eggs to the adults. It is more complicated with some insects such as sand flies and tsetse flies and even more difficult with ticks when the complete development is accomplished in 2 years. To infect them on a regular basis and to control their vector competence are still very tedious.

In parallel, insecticides and acaricides are still widely used but cause serious problems for ecological reasons and for the development of arthropod resistance. New formulations and new products are evaluated to reduce these side effects (Benelli et al., 2017). New strategies of arthropod control have been undertaken targeting the genome (transgenic insects) or the vector competence (modification of arthropod immune system or microbiome) (Medlock et al., 2013; O'Brochta and Handler, 2008; Wilke and Marrelli, 2015). Although significant progresses have been done and are encouraging, efforts on the understanding of arthropod–pathogen interactions need to be pursued. The studies of arthropod microbiome and its role in pathogen transmission have opened new avenues of researches that deserve further investigations (Narasimhan et al., 2017; Weiss and Aksoy, 2011).

REFERENCES

Adeolu, M., Gupta, R., June, 2014. A phylogenomic and molecular marker based proposal for the division of the genus Borrelia into two genera: the emended genus Borrelia containing only the members of the relapsing fever Borrelia, and the genus Borreliella gen. nov. containing the members of the Lyme disease Borrelia (Borrelia burgdorferi sensu lato complex). Antonie Van Leeuwenhoek 105 (6), 1049–1072.

Agarwal, A., Parida, M., Dash, P., August 30, 2017. Impact of transmission cycles and vector competence on global expansion and emergence of arboviruses (in press) Reviews in Medical Virology.

Aksoy, S., 1995. Wigglesworthia gen. nov. and Wigglesworthia glossinidia sp. nov., taxa consisting of the mycetocyte-associated, primary endosymbionts of tsetse flies. International Journal of Systematic Bacteriology 45 (4), 848–851.

Baldacchino, F., Muenworn, V., Desquesnes, M., Desoli, F., Charoenviriyaphap, T., Duvallet, G., 2013. Transmission of pathogens by Stomoxys flies (Diptera, Muscidae): a review. Parasite 26–38.

Baldacchino, F., Desquesnes, M., Mihok, S., Foil, L.D., Duvallet, G., Jittapalapong, S., 2014. Tabanids: neglected subjects of research, but important vectors of disease agents!. Infection, Genetics and Evolution 28, 596–615.

Benelli, G., Maggi, F., Pavela, R., Murugan, K., Govindarajan, M., Vaseeharan, B., Petrelli, R., et al., 2017. Mosquito control with green nanopesticides: towards the one health approach? A review of non-target effects (in press) Environmental Science and Pollution Research International.

Bernard, Q., Jaulhac, B., Boulanger, N., 2015. Skin and arthropods: an effective interaction used by pathogens in vector-borne diseases. European Journal of Dermatology 25 (Suppl. 1), 18–22.

Black, W.C., Kondratieff, B., 2004. Evolution of arthropod disease vectors. In: Marquardt, W. (Ed.), Biol. Dis. Vectors, second ed. Accademic Press-Elsevier, p. 816.

Boulanger, N., Brun, R., Ehret-Sabatier, L., Kunz, C., Bulet, P., 2002. Immunopeptides in the defense reactions of Glossina morsitans to bacterial and Trypanosoma brucei brucei infections. Insect Biochemistry and Molecular Biology 32 (4), 369–375.

Boulanger, N., Lowenberger, C., Volf, P., Ursic, R., Sigutova, L., Sabatier, L., et al., 2004. Characterization of a defensin from the sand fly Phlebotomus duboscqi induced by challenge with bacteria or the protozoan parasite Leishmania major. Infection and Immunity 72 (12), 7140–7146.

Boulanger, N., Bulet, P., Lowenberger, C., 2006. Antimicrobial peptides in the interactions between insects and flagellate parasites. Trends in Parasitology 22 (6), 262–268. Available from: http://www.ncbi.nlm.nih.gov/pubmed/16635587.

Brey, P., 1998. The contributions of the Pasteur school of insect immunity. In: Brey, P.T., Hultmark, D. (Eds.), Mol. Mech. Immune Responses Insects. Chapman & Hall, p. 325.

Brune, A., 2014. Symbiotic digestion of lignocellulose in termite guts. Nature Reviews Microbiology 12, 168–180.

Cogen, A., Nizet, V., Gallo, R., 2008. Skin microbiota: a source of disease or defence? The British Journal of Dermatology 158 (3), 442–455.

Combes, C., 2001. Parasitism: The Ecology and Evolution of Intimate Interactions. University of Chicago Press.

Commins, S.P., Platts-Mills, T.A., 2013. Tick bites and red meat allergy. Current Opinion in Allergy and Clinical Immunology 13 (4), 354–359. Available from: http://www.ncbi.nlm.nih.gov/pubmed/23743512.

de la Fuente, J., Antunes, S., Bonnet, S., Cabezas-Cruz, A., Domingos, A., Estrada-Peña, A., et al., 2017. Tick-pathogen interactions and vector competence: identification of molecular drivers for tick-borne diseases. Frontiers in Cellular Infection Microbiology 7, 114.

Durvasula, R., Gumbs, A., Panackal, A., Kruglov, O., Taneja, J., Kang, A., et al., 1999. Expression of a functional antibody fragment in the gut of *Rhodnius prolixus* via transgenic bacterial symbiont *Rhodococcus rhodnii*. Medical and Veterinary Entomology 13 (2), 115–119.

Duvallet, G., de Gentile, L., 2012. In: Duvallet, G., de Gentile, L. (Eds.), Protection Personnelle Antivectorielle. IRD Marseille, France.

Ekbom, K., 1938. Der Praesenile Dermatozoenwhan. Acta Psychiatrica Scandinavica 13, 227–259.

Elston, D.M., 2010. Tick bites and skin rashes. Current Opinion in Infectious Diseases 23 (2), 132–138. Available from: http://www.ncbi.nlm.nih.gov/pubmed/20071986.

Findley, K., Oh, J., Yang, J., Conlan, S., Deming, C., Meyer, J., et al., 2013. Topographic diversity of fungal and bacterial communities in human skin. Nature 498 (7454), 367–370.

Gillott, C., 2005. Entomology, third ed. Springer Netherlands.

Grice, E., 2014. The skin microbiome: potential for novel diagnostic and therapeutic approaches to cutaneous disease. Seminars in Cutaneous Medicine and Surgery 33 (2), 98–103.

Grice, E.A., Segre, J.A., 2011. The skin microbiome. Nature Reviews Microbiology 9 (4), 244–253.

Hajdusek, O., Sima, R., Ayllon, N., Jalovecka, M., Perner, J., de la Fuente, J., et al., 2013. Interaction of the tick immune system with transmitted pathogens. Frontiers in Cellular Infection Microbiology 3, 26. 2013/07/23. Biological Centre ASCR, Institute of Parasitology Ceske Budejovice, Czech Republic. Available from: http://www.ncbi.nlm.nih.gov/pubmed/23875177.

Hinkle, N.C., 2000. Delusory parasitosis. American Entomologist 46, 17–25.

Hu, Y., Aksoy, S., 2005. An antimicrobial peptide with trypanocidal activity characterized from Glossina morsitans morsitans. Insect Biochemistry and Molecular Biology 35 (2), 105–115.

Hu, C., Aksoy, S., 2006. Innate immune responses regulate trypanosome parasite infection of the tsetse fly Glossina morsitans morsitans. Molecular Microbiology 60 (5), 1194–1204.

Kahl, O., Gern, L., Eisen, L., Lane, R., 2002. Ecology research on Borrelia burgdorferi sensu lato: ecology ans some methodological pitfalls. In: Gray, J., Kahl, O., Lane, R.S., Stanek, G. (Eds.), Lyme Borreliosis Biol. Epidemiol. Control. CABI publishing, p. 335.

Kaufmann, C., Ziegler, D., Schaffner, F., Carpenter, S., Pflüger, V., Mathis, A., 2011. Evaluation of matrix-assisted laser desorption/ionization time of flight mass spectrometry for characterization of Culicoides nubeculosus biting midges. Medical and Veterinary Entomology 25 (1), 32–38.

Kazimírová, M., Stibrániová, I., 2013. Tick salivary compounds: their role in modulation of host defences and pathogen transmission. Frontiers in Cellular Infection Microbiology 3 (43), 1–17.

Kettle, D.S., 1995. Medical and Veterinary Entomology, second ed. CABI, UK.

Lagueux, M., Perrodou, E., Levashina, E.A., Capovilla, M., Hoffmann, J., 2000. Constitutive expression of a complement-like protein in toll and JAK gain-of-function mutants of Drosophila. Proceedings of the National Academy of Sciences USA 97 (21), 11427–11432.

Lemaitre, B., Hoffmann, J., 2007. The host defense of *Drosophila melanogaster*. Annual Review of Immunology 25, 697–743.

Lemaitre, B., Nicolas, E., Michaut, L., Reichhart, J., Hoffmann, J., 1996. The dorsoventral regulatory gene cassette spätzle/Toll/cactus controls the potent antifungal response in Drosophila adults. Cell 86 (6), 973–983.

Marquardt, W., 1996. Introduction to arthropods and vectors. In: Beaty, B., Marquardt, W. (Eds.), Biol. Dis. Vectors, first ed. University Press of Colorado, pp. 1–24.

Mathison, B., Pritt, B., 2014. Laboratory identification of arthropod ectoparasites. Clinical Microbiology Reviews 27 (1), 48–67.

McCoy, K.D., Boulanger, N., 2016. Tiques et maladies à tiques: biologie, écologie évolutive et épidémiologie. IRD Editions, Marseille.

Medlock, J., Atkins, K., Thomas, D., Aksoy, S., Galvani, A., 2013. Evaluating Trypanosomiasis., paratransgenesis as a potential control strategy for African. PLoS Neglected Tropical Diseases 7 (8), e2374.

Mehlhorn, H., 2001. second ed. Encyclopedic Reference of Parasitology, vol. 1. Springer-Verlag, Berlin-Heidelberg. 678 pages.

Mehlhorn, H., 2011. Ixodid ticks: world record holders in starvation and blood engorging. In: Prog. Parasitol. Springer, p. 336.

Mullen, G.R., Durden, L.A., 2009. In: Mullen, G.R., Durden, L.A. (Eds.), Medical and Veterinary Entomology, second ed. Elsevier. Academic Press, London.

Narasimhan, S., Rajeevan, N., Liu, L., Zhao, Y.O., Heisig, J., Pan, J., et al., 2014. Gut microbiota of the tick vector *Ixodes scapularis* modulate colonization of the Lyme disease spirochete. Elsevier Inc. Cell Host and Microbe 15 (1), 58–71. Available from: https://doi.org/10.1016/j.chom.2013.12.001.

Narasimhan, S., Schuijt, T., Abraham, N., Rajeevan, N., Coumou, J., Graham, M., et al., 2017. Modulation of the tick gut milieu by a secreted tick protein favors *Borrelia burgdorferi* colonization. Nature Communications 8 (1), 184.

Neva, F.A., Brown, H.W., 1994. Basic Clinical Parasitology, sixth ed. Appleton and Lange, Norwalk, Connecticut.

Ødegaard, F., 2000. How many species of arthropods? Erwin's estimate revised. Biological Journal of Linnean Society 71, 583–597.

O'Brochta, D., Handler, A., 2008. Perspectives on the state of insect transgenics. Advances in Experimental Medicine and Biology 627, 1–18.

Rhodain, F., 2015. Le parasite, le moustique, l'homme…et les autres. Essai sur l'éco-épidémiologie des maladies à vecteurs. DOCIS, Paris.

Robert, V., 2017. Introduction à l'entomologie médicale et vétérinaire. In: Duvallet, G., Fontenille, D., Robert, V. (Eds.), Entomologie médicale et vétérinaire. IRD Editions, Marseille, France.

Saraiva, R., Kang, S., Simões, M., Angleró-Rodríguez, Y.I., Dimopoulos, G., 2016. Mosquito gut antiparasitic and antiviral immunity. Developmental and Comparative Immunology 64, 53–64.

Smith, A., Navasa, N., Yang, X., Wilder, C., Buyuktanir, O., Marques, A., et al., 2016. Cross-species interferon signaling boosts microbicidal activity within the tick vector. Cell Host and Microbe 20 (1), 91–98.

Sonenshine, D.E., Ceraul, S.M., Hynes, W.E., Macaluso, K.R., Azad, A.F., 2002. Expression of defensin-like peptides in tick hemolymph and midgut in response to challenge with Borrelia burgdorferi, *Escherichia coli* and *Bacillus subtilis*. 2003/10/23. Department of Biological Sciences, Old Dominion University, Norfolk, VA 23529, USA Experimental and Applied Acarology 28, 127–134. Available from: dsonensh@odu.edu. http://www.ncbi.nlm.nih.gov/pubmed/14570122.

Steiner, H., Hultmark, D., Engström, A., Bennich, H., Boman, H., 1981. Sequence and specificity of two antibacterial proteins involved in insect immunity. Nature 292 (5820), 246–248.

Steinmann, I., Pflüger, V., Schaffner, F., Mathis, A., Kaufmann, C., 2013. Evaluation of matrix-assisted laser desorption/ionization time of flight mass spectrometry for the identification of ceratopogonid and culicid larvae. Parasitology 140 (3), 318–327.

Taylor, D.B., Moon, R.D., Mark, D.R., 2012. Economic impact of stable flies (Diptera: Muscidae) on dairy and beef cattle production. Journal of Medical Entomology 49, 198–209.

Uttenweiler-Joseph, S., Moniatte, M., Lagueux, M., Van Dorsselaer, A., Hoffmann, J., Bulet, P., 1998. Differential display of peptides induced during the immune response of Drosophila: a matrix-assisted laser desorption ionization time-of-flight mass spectrometry study. Proceedings of the National Academy of Sciences USA 95 (19), 11342–11347.

Vizioli, J., Bulet, P., Hoffmann, J., Kafatos, F., Müller, H., Dimopoulos, G., 2001. Gambicin: a novel immune responsive antimicrobial peptide from the malaria vector *Anopheles gambiae*. Proceedings of the National Academy of Sciences USA 98 (22), 12630–12635.

Wang, J., Weiss, B.L., Aksoy, S., October 2013. Tsetse fly microbiota: form and function. Frontiers in Cellular Infection Microbiology 3, 69. Available from: http://www.pubmedcentral.nih.gov/articlerender.fcgi?artid=3810596&tool=pmcentrez&rendertype=abstract.

Weiss, B., Aksoy, S., 2011. Microbiome influences on insect host vector competence. Trends in Parasitology 27 (11), 514–522.

Wilke, A., Marrelli, M., 2015. Paratransgenesis: a promising new strategy for mosquito vector control. Parasites and Vectors 8 (342).

Yssouf, A., Flaudrops, C., Drali, R., Kernif, T., Socolovschi, C., Berenger, J.-M., et al., 2013. Matrix-assisted laser desorption ionization-time of flight mass spectrometry for rapid identification of tick vectors. Journal of Clinical Microbiology 51 (2), 522–528. Available from: http://www.pubmedcentral.nih.gov/articlerender.fcgi?artid=3553915&tool=pmcentrez&rendertype=abstract.

Yssouf, A., Almeras, L., Raoult, D., Parola, P., 2016. Emerging tools for identification of arthropod vectors. Future Microbiology 11 (4), 549–566.

Chapter 3

Impact of Skin Microbiome on Attractiveness to Arthropod Vectors and Pathogen Transmission

Niels O. Verhulst[1,2], Nathalie Boulanger[3], Jeroen Spitzen[1]
[1]Wageningen University & Research, Wageningen, The Netherlands; [2]University of Zurich, Zurich, Switzerland; [3]Université de Strasbourg, Strasbourg, France

INTRODUCTION

The detection of vertebrate host for an arthropod vector is vital to provide its hematophagous blood meal. To this end, they either actively search for the host (e.g., mosquito, sand fly, midge) or they live close to their host (e.g., louse, tick). Arthropods have developed sophisticated strategies to locate their target and often use a combination of visual cues, odor attraction (e.g., carbon dioxide, lactic acid, ammonia), and temperature. Once landed on the host skin, they locate a specific site for their bite by chemoreceptors present on their antenna or on mouthparts (Ribeiro, 1996). Then, the feeding stylets penetrate a blood vessel (solenophagy or vessel feeding, e.g., mosquito, tsetse fly) or dilacerate the skin inducing a hemorrhage (telmophagy or pool feeding, e.g., sand fly, black fly). More recently, the mechanisms behind arthropod attraction became increasingly complex with the discovery of the role of the host skin microbiome. This microbiome plays a key role by the metabolome released not only by the microbiota constituted mainly by bacteria but also viruses, fungi, and parasites (Grice and Segre, 2011). New types of interactions are still unraveled. First described in the attractiveness to mosquitoes (Knols and De Jong, 1996), the skin microbiota is more and more studied. This could lead to the discovery of new strategies to fight arthropod-borne diseases such as the development of more efficient repellents. In addition, this skin microbiome was shown to play a major role in skin homeostasis and immunity of the host. It might therefore influence the direct transmission of arthropod-borne diseases at the skin interface as shown for leishmaniasis (Naik et al., 2012). Commensal microbiota

Skin and Arthropod Vectors. https://doi.org/10.1016/B978-0-12-811436-0.00003-4

55

plays a dual role: (1) at distance on arthropod attractiveness (Verhulst et al., 2011b), (2) at the skin interface with cellular and molecular interactions with arthropods during the transmission of pathogens.

Skin Microbiome and Attractiveness to Arthropod Vectors

From the listed blood-feeding arthropods mentioned as potential disease vectors (see Chapter 2), the physical ability to pass on a pathogen does not directly imply high vector competence. The latter is largely influenced by biotic and abiotic factors (e.g., climate change) and shaped by behavioral preferences of the potential vector of interest.

Blood-sucking arthropods use a range of cues to locate and eventually land or climb on their host (Allan, 2010; Cardé, 2015). Exhaled CO_2 is the most general host-derived cue that is best described as a host-seeking activator with observed orientation toward the source ranging up to 50 m distance (Cardé and Gibson, 2010; Gillies, 1980; Marinković et al., 2014). Recently, CO_2 was also associated to induce landing (Webster et al., 2015). The interaction between long range cues such as CO_2 and other body odors becomes apparent in studies where the host cues are studied separately (Brady et al., 1997; Hamilton and Ramsoondar, 1994; Mboera et al., 1997). Orientation toward a host is depending on both wind speed and direction. Visual cues can function as guidance for more efficient and direct movements. Chemical cues play a crucial role in host choice and host finding, and their relative importance differs between generalist or specialist arthropods (Takken and Verhulst, 2013). Where skin volatiles are crucial for host selection in anthropophilic malaria mosquitoes (Takken and Verhulst, 2013), skin volatiles also play a role when vectors are already on the host, such as ticks finding a suitable biting site (Wanzala et al., 2004). In short, the role of skin volatiles largely depends on the scale it is studied and on the mobility of the vector. The emergence of tick-borne diseases has resulted in an increase of studies on their chemical ecology; however, behavioral studies are still scarce. Studies on mosquito vectors provide the majority of gained knowledge on both chemical and behavioral ecology and its relation with disease transmission (Gibson and Torr, 1999; Takken and Knols, 2010).

Zooming in on the host, it has been shown for humans that the skin microbiota plays an important role in the attractiveness of the individual to mosquitoes (Verhulst et al., 2010a, 2011b). Two abiotic factors that contribute directly to the production, composition, and level of attractiveness of skin volatiles are (skin)-temperature and moisture. Skin temperature and moisture have been described as host cues that increase in relative importance when the vector comes close to the host (Cardé and Gibson, 2010; Spitzen et al., 2013; Takken and Knols, 1999).

In this section, we focus on the role of skin volatiles as an important part of the total set of host-derived cues that determine the attractiveness of a host to arthropod vectors. We describe available techniques to collect and identify skin volatiles and how to test their relevance in attractiveness using behavioral assays in both laboratory and field settings. The role of skin microbiomes for the production of volatiles that are exploited by the vectors is explained and discussed based on *in vitro* and *in vivo* experiments. Host preference steered by skin volatiles, intra- and interspecies specific, is discussed and put in perspective in relation to host-vector intervention strategies.

Terminology

Host cues can evoke behavioral responses such as activation, orientation, and landing. We consider "attractiveness" the property of host cues (such as skin volatiles) to initiate orientation of host-seeking arthropods toward the source of the emitter of these cues. It is important to note that attraction does not necessarily follow with a landing response or occupation of the host, it could also lead to lure the vector in the vicinity of the host (cues). The complete set of host-derived cues determines the host preference of the vector. We adopt the definition for host preference as mentioned by Takken and Verhulst (2013) as the trait of blood-seeking arthropods to preferentially select certain host species above others. However, non-preferred hosts are not necessarily unsuitable hosts and can still function as important hosts for the vector, especially in no-choice situations.

Identification of Skin Volatiles and Their Role in Attractiveness

Skin Volatile Collection and Identification

The volatile composition of the human skin has been described in numerous studies (Bernier et al., 1999, 2000; Curran et al., 2005; Gallagher et al., 2008; Logan et al., 2008, 2009; Natsch et al., 2006; Verhulst et al., 2013a). These volatiles can originate from: (1) the skin directly, (2) skin microbiota, (3) skin care products, and perfumes. Although it is known that the skin bacteria play a major role in volatile production, it is difficult to differentiate between volatiles that are released from the skin directly and the volatiles that are products from skin microbiota conversions. Skin care products and perfumes not only have an effect on the volatile profile of the host but also indirectly on the skin microbiota. As a result, the application of these products affects the total volatile profile (Bouslimani et al., 2015; Stoddart, 1990; Wilson, 2008). When collecting volatiles from the skin it is therefore important to consider the protocols to be followed for washing and the use of skin care products and perfumes.

Skin volatiles are normally collected from specific body parts, but some studies have collected whole body volatile profiles by placing volunteers in a foil bag that covers the whole body except the head (Logan, 2008). Before the skin volatiles can be analyzed they need to be collected on a solvent or adsorbent. Solvents can either be applied directly on the skin (Ara et al., 2006; Sastry et al., 1980) or on cotton or gauze pads that have been rubbed over the skin (Natsch et al., 2006; Xiao-Nong et al., 1996). The disadvantage of this method, however, is that it also includes non-volatile compounds that are often not the target of the study. Collection on an adsorbent such as charcoal, Porapak Q, or Tenax is therefore preferred when interested in the headspace volatiles of the skin (Dormont et al., 2013b; Logan et al., 2008, Verhulst et al., 2016). Before gas chromatography (GC) analysis, the volatiles trapped on the adsorbents need to be released by either a solvent or thermal desorption. Solid-phase micro-extraction is an example of a solvent-free method to trap volatiles on small adsorbent fibers that can be released into a GC injector by thermal desorption (Dormont et al., 2013a). Next, the compounds in the volatile profiles can be separated by GC, often followed by mass spectroscopy (MS) for identification.

In total more than 400 compounds have been identified from human skin, the volatility of these compounds varies and as a result not all of them are relevant as host-seeking cues for the arthropod (Dormont et al., 2013a). Although there is a lot of variation in the volatile profiles reported, compounds that are frequently found include lactic acid, sulcatone, nonanal, decanal, and geranylacetone. These compounds are often found to cause a behavioral effect in hematophagous arthropods (Dormont et al., 2013a; Smallegange and Takken, 2010).

Gas Chromatography–Electroantennogram and Single Sensillum Recordings

Arthropods detect odor molecules with receptors that are called sensilla. These sensilla are located on the antennae and maxillary palps of the arthropods. When an odor molecule impacts on a sensilla it is bound to an odor-binding protein that transports the odor molecule to the dendrite where it changes the membrane potential. This electrical signal is then transported to the central nervous system where it may translate into a behavioral output (Zwiebel and Takken, 2004). The electrical signal of the antennae can be recorded as an electroantennogram (EAG) and thereby give information on the ability of an insect to detect a certain volatile.

EAGs have been performed on many different hematophagous arthropods to identify skin volatiles that elicit an antennal response. Mosquitoes have been tested most often, including *Aedes aegypti* (Cook et al., 2011; Logan et al., 2008; McBride et al., 2014), *Culex quinquefasciatus* (Cook et al., 2011; Puri et al., 2006), and *Anopheles gambiae* (Meijerink and van Loon, 1999; Qiu, 2005; Qiu et al., 2004a). Other hematophagous insects include midges (Bhasin and Mordue, 2000; Logan et al., 2009; Mands et al., 2004), tsetse flies (Hall et al., 1984), stable flies (Schofield et al., 1995; Tangtrakulwanich et al., 2011; Warnes and Finlayson, 1986), bedbugs (Singh et al., 2012), sand flies (Sant'ana et al., 2002), and horseflies (Baldacchino et al., 2014). Individual compounds can be tested in an EAG setup, but the technique is more valuable when coupled to GC. Gas chromatography–electroantennogram (GC-EAG) allows the location of EAG-active compounds within complex blends by taking advantage of high-resolution GC with simultaneously recording the response on an antennal preparation of the insect (Logan et al., 2010). Analysis by GC-EAG has led to the identification of behavioral active compounds in several hematophagous insect (Cork et al., 1990; Hall et al., 1984; Logan et al., 2008; Qiu et al., 2004b).

Single sensillum recordings (SSRs) can be used to identify the specific sensillum that responds to a particular chemical (Weeks et al., 2011). Although this technique is more difficult and less suitable for the identification of novel behaviorally active compounds such as GC-EAG, it is often used to unravel the olfactory pathways of insects (Carey et al., 2010; Dougherty et al., 1999; Harraca et al., 2010; Park and Cork, 1999) and can also be coupled to GC for GC-SSR recordings (Ghaninia et al., 2008; McBride et al., 2014).

The Role of the Skin Microbiome in the Production of Attractive Volatiles

Volatile Production by Skin Bacteria

Skin bacteria play an important role in the production of odors released from the skin. Both the intensity of human odor and its volatile composition are correlated with the abundance of certain bacteria on the human skin (Ara et al., 2006; Jackman and Noble, 1983; Leyden et al., 1981; Phelan et al., 1991; Taylor et al., 2003; Xu et al., 2007). In general, the skin is a dry and therefore harsh environment for skin bacteria (see Chapter 1). Therefore, the areas with an abundance of skin glands are the places where most bacteria are found, as these places offer more protection, are humid and contain the metabolic resources required (Stoddart, 1990). Human skin glands can be divided into sebaceous and sweat glands of which the sweat glands comprise eccrine and apocrine glands and mainly produce water. Sebaceous glands produce sebum, which mainly consists of lipids (Stoddart, 1990). These different glands have different bacterial compositions associated with specific volatiles that are released. Corynebacteria are responsible for the typical apocrine odor (Jackman and Noble, 1983; Leyden et al., 1981; Rennie et al., 1991; Taylor et al., 2003), while the feet contain a lot of eccrine glands and their odor is associated with *Bacillus* species. The human scalp has a high density of sebaceous glands (Sastry et al., 1980) and contains numerous propionibacteria that produce volatile fatty acids (Kearney et al., 1984). On the skin, bacteria convert long-chain compounds that are often not volatile into short-chain volatile compounds with their characteristic smell (James et al., 2004a, 2004b). Different species of skin bacteria each have their own specific metabolism and, therefore, generate a characteristic odor profile (James et al., 2004a, 2004b; Verhulst et al., 2010b).

Although it is known that washing the skin reduces odor production on the skin and affects attraction to mosquitoes (Knols and Meijerink, 1997), little is known about the effect of daily hygiene on odor production. Linking mass spectrometry data with sequencing of the microbial communities will elucidate the interactions between the skin microbiota, skincare products, and the volatiles produced (Bouslimani et al., 2015).

The underlying mechanisms that determine the skin bacterial composition and the human odor profile are still largely unknown. Next to the influence of skin products, diet and environment, an individual's genotype may influence both skin bacterial and volatile composition. A study with identical and non-identical twins revealed that there is a genetic component that influences the human odor volatiles that attract mosquitoes (Fernández-Grandon et al., 2015). Finally, the bacteria on the skin also interact with each other and the volatiles produced by one species may inhibit growth of another by quorum sensing (Lemfack et al., 2016).

In Vitro Attractiveness of Skin Bacteria to Arthropods

First evidence that skin bacterial volatiles may be attractive to mosquitoes was provided by Schreck and James (1968). They showed in an *in vitro* experiment that a broth of *Bacillus cereus* derived from a human arm was attractive to female *Ae. aegypti*, and it was suggested that metabolites produced by bacteria on the human skin may contribute to the attractiveness to mosquitoes (James et al., 2004b). *In vitro* experiments with sweat showed that sterilized sweat only becomes attractive to the malaria mosquito *An. gambiae* after incubation with bacteria (Braks et al., 2000). During this incubation period, the human skin microbiota converts fresh sweat into aged sweat (Shelley et al., 1953), which is attractive to *An. gambiae* (Braks et al., 2000).

Instead of sweat, human skin bacteria grown on agar plates can also attract mosquitoes (Verhulst et al., 2009). Tests with the volatiles from individual bacterial species that frequently occur on the human skin revealed that not all odors from skin bacteria are attractive to mosquitoes (Verhulst et al., 2010a). Follow-up experiments in a semi-field setup showed that mosquitoes respond differently to bacterial volatiles and that these volatiles may mediate host preference (Busula et al., 2017).

The identification of volatile compounds produced by skin bacteria can lead to the identification of novel mosquito attractants (Mukabana et al., 2012; Verhulst et al., 2011a) (Fig. 3.1). In addition to the volatiles produced by normal metabolism of the skin bacteria, the volatiles produced by quorum sensing of bacteria can also attract mosquitoes (Zhang et al., 2015). Female *Ae. aegypti* mosquitoes were not attracted anymore to the volatiles from *Staphylococcus epidermidis* when the agr-gene that regulates quorum sensing was removed. This suggests that mosquitoes may "eavesdrop" on the chemical discussions between skin microbes (Zhang et al., 2015).

In Vivo Attractiveness of Skin Bacteria to Arthropods

The *in vitro* experiments described before clearly show that skin bacteria produce volatiles that are attractive to mosquitoes. These studies, however, do

FIGURE 3.1 Skin bacterial headspace sampling. Skin bacteria from feet were grown on blood agar plates and tested for their response to *Anopheles gambiae*. In addition, headspace volatiles were collected on Tenax-TA adsorbent and subsequently analyzed by GC–MS. The black pump provided airflow over the agar plate and through the Tenax-TA tube. *GC*, gas chromatography; *MS*, mass spectroscopy.

not prove that bacteria on the skin, where the substrate is different, produce volatiles that are attractive to mosquitoes. When human feet are washed with a bacterial soap, *An. gambiae* starts to bite other body parts, indicating that bacteria on the skin are responsible for the production of attractive odors (de Jong and Knols, 1995). To study the correlation between skin bacteria and attractiveness in detail, Verhulst et al. (2011b) determined the attractiveness of 48 volunteers to *An. gambiae* by sampling their skin bacterial and volatile profile. They demonstrated that the composition of the skin microbiota affects the degree of attractiveness of human beings to this mosquito species. Bacterial plate counts and 16S rRNA sequencing revealed that individuals that are highly attractive to *An. gambiae* have a significantly higher abundance but lower diversity of bacteria on their skin than individuals that are poorly attractive (Verhulst et al., 2011b). Interestingly, *Staphylococcus* spp. were associated with individuals that were highly attractive and *Pseudomonas* spp. with individuals that were less attractive to mosquitoes (Verhulst et al., 2011b), which nicely matched the results of the *in vitro* experiments with species within this genera (Verhulst et al., 2010a).

Skin Volatiles and Host Preference

For hematophagous arthropods, host location is essential to find their blood host. Although many blood-feeding arthropods express a non-specific, more generalistic host-feeding behavior, many arthropods express a certain degree of host preference, which will favor disease transmission. Host selection is therefore one of the most important determinants of the vectorial capacity of a mosquito (Garrett-Jones and Shidrawi, 1969; Takken and Verhulst, 2013). Understanding the feeding preferences of mosquitoes and other arthropods is therefore essential to determine their potential as a disease vector. For example, *Ae. aegypti* and *An. gambiae s.s.* are two of the most important disease vectors, mainly because they are highly anthropophilic. Also, anthropophilic sand flies are attracted by human odors in addition to CO_2 (Pinto et al., 2001), which increases their potential as a disease vector.

Carbon dioxide is a general host kairomone that is exhaled by all vertebrates and used by almost all hematophagous arthropods to locate their hosts; however, CO_2 alone does not enable an arthropod to distinguish between different hosts. Selecting specific hosts can be beneficial for the arthropod, for example, if this host has less hair, displays less defensive behavior, or their blood is more nutritious (reviewed for mosquitoes by Lyimo and Ferguson, 2009). Both visual and olfactory cues may be used by blood-feeding arthropods, however, to find their specific hosts, for most arthropods, host odors are considered most important especially at increased distance from the emitter (Gibson and Torr, 1999; Takken and Verhulst, 2013).

Differences Between Different Host Species

Skin volatiles have shown to be attractive to many different hematophagous arthropods (Gibson and Torr, 1999) and in addition, the biting preference has been investigated for many of these arthropods. However, the role of skin volatiles in this biting preference has only been studied in detail for mosquitoes. An exception is the study by Mands et al. (2004) on the response of biting midges to hair extracts of different animals. The study showed that especially water buffalo extracts were attractive to the midge *Culicoides impunctatus* in both field and Y-tube assays.

The anthropophilic yellow fever mosquito *Ae. aegypti* is highly attracted to skin rubbings from humans, but not from other animals, which demonstrates the role of skin volatiles for the selection of its host (Steib et al., 2001). Humans have a high amount of lactic acid that is released from their skin compared to other animals, and adding lactic acids to skin rubbings of other animals increases the response of *Ae. aegypti* to a level comparable to the response to human skin rubbings (Steib et al., 2001). *An. gambiae s.s.* is also highly anthropophilic and shows the same attraction to skin volatiles when collected on nylon socks. Human odor was found to be very attractive to *An. gambiae s.s.* in an olfactometer while an aversion to cow odor was observed. Lactic acid is also an important kairomone to *An. gambiae s.s.*, however, it is only attractive when combined with other compounds (Smallegange et al., 2005).

Interestingly, the sister species of *An. gambiae s.s*, *Anopheles quadriannulatus*, and *Anopheles arabiensis* have a wider host range that is more zoophilic (Hunt et al., 1998; Torr et al., 2008; White et al., 1980) or opportunistic (Costantini et al., 1996, 1998). Because these species are less anthropophilic they are not considered as a relevant vector (*An. quadriannulatus*) or only when present in high numbers (*An. arabiensis*). Testing the response of these three mosquito species to the volatiles of three different hosts reveals the importance of skin volatiles in their host-seeking behavior. Skin volatiles of a cow, chicken, and human were collected on nylon socks that were tested in MM-X traps in dual-choice experiments in a climatized room. Thirty *An. gambiae s.s*, *An. quadriannulatus*, and *An. arabiensis* were colored to distinguish them (Verhulst et al., 2013b) and released simultaneously. After 45 min, traps were emptied and mosquitoes counted. *An. gambiae* was significantly more attracted to human odor than to cow or chicken odor, while *An. quadriannulatus* was more attracted to cow odor than to human or chicken odor (Fig. 3.2). *An. arabiensis* did not

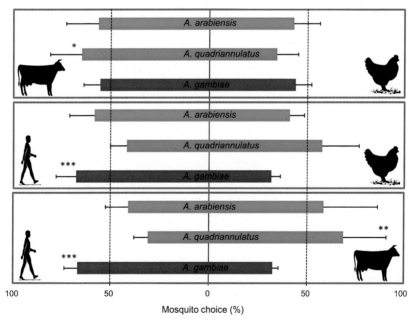

FIGURE 3.2 Mean response of *Anopheles gambiae s.s.*, *Anopheles quadriannulatus*, and *Anopheles arabiensis* to traps baited with odors from human, cow, and chicken. Six times 30 mosquitoes were released per species per treatment in a climatized room with two MM-X traps. *Error bars* represent standard errors of the mean; ***, χ^2-test $P<.001$; **, χ^2-test $P<.01$; *, χ^2-test $P<.05$. Mosquitoes choice=The number of mosquitoes caught in one trap divided by the number of mosquitoes caught in both trapping devices.

show a clear preference. Although *An. gambiae s.s.* has been found to be anthropophilic in different studies with different setups, the preference of *An. quadriannulatus* and *An. arabiensis* varies (Busula et al., 2017; Jaleta et al., 2016; Takken and Verhulst, 2013).

Differences Within the Same Host Species

The interspecific preferences of a hematophagous arthropod will largely determine its efficacy as a disease vector; however, intraspecific differences also occur and will determine which individuals are more at risk. These intraspecific differences may occur be physical (e.g., thicker skin, more hairs), visual (color, body size), differences in odor profile, or behavioral differences. Friesian calves, for example, suffer less from attacks of the stable fly *Stomoxys calcitrans* when they respond more vigorously to these flies (Warnes and Finlayson, 1987).

The differential attractiveness of humans to mosquitoes, as has been studied in detail for several mosquito species and all of them show that mosquitoes prefer to bite some individuals over others (Bernier et al., 2002; Brady et al., 1997; Brouwer, 1960; Knols et al., 1995; Lindsay et al., 1993; Logan et al., 2008; Mayer and James, 1969; Mukabana et al., 2002; Qiu et al., 2006; Schreck et al., 1990; Verhulst et al., 2011b). In addition, there is at least one study that shows that female *C. impunctatus* are differentially attracted to human odors (Logan et al., 2009).

Humans can be ranked for attractiveness to mosquitoes by testing their whole body odor profile (Knols et al., 1995; Lindsay et al., 1993; Mukabana et al., 2002) or by testing volatiles released from a part of the body (Logan et al., 2008; Mayer and James, 1969; Smart and Brown, 1957; Verhulst et al., 2011b). To eliminate the effect of skin temperature or humidity, skin emanations can also be collected on glass beads or cotton pads, which can be tested for attractiveness to mosquitoes (Bernier et al., 1999; Qiu et al., 2006; Schreck et al., 1990; Verhulst et al., 2016). Identification of the volatiles that are responsible for the differential attractiveness of humans has led to the identification of novel attractants and repellents for both mosquitoes and midges (Logan et al., 2008, 2009; Mukabana et al., 2012; Verhulst et al., 2013a).

Although the skin microbiota composition correlates with an individual's attractiveness to mosquitoes, the underlying mechanisms are still largely unknown. It is known, however, that there is a genetic basis for attractiveness to mosquitoes. When 197 monozygotic and 326 dizygotic twin pairs were asked to fill out a questionnaire on mosquito bites, a strong genetic influence on frequency of being bitten by mosquitoes was found (Kirk et al., 2000). In a study by Fernández-Grandon et al. (2015) the response of *Ae. aegypti* to hands of identical and non-identical twins was tested in a Y-tube. They showed that the volatiles from individuals in an identical twin pair showed a significantly higher correlation in attractiveness to mosquitoes than the non-identical twin pairs (Fernández-Grandon et al., 2015). Although it is not known which genes are responsible for this correlation, the human leukocyte antigen genes may be good candidates as they have shown to influence mice and human body odor (Penn and Potts, 1998; Wedekind and Furi, 1997), and a tentative link was found in another study (Verhulst et al., 2013a).

Differences Within an Individual

Within an individual, we can distinguish differences in attractiveness to arthropods over time and differences between body parts of the same individual. Differences in attractiveness to mosquitoes over time seems to be relatively stable as determined by the attractiveness of skin emanations to *An. gambiae s.s.* (Qiu et al., 2006) and in addition, there is a genetic influence on attractiveness to mosquitoes as described before (Fernández-Grandon et al., 2015; Kirk et al., 2000; Verhulst et al., 2013a). The stability of this attractiveness and whether the attractiveness to other hematophagous arthropods is also stable over time remains to be investigated. Although this attractiveness to mosquitoes is relatively stable; age, pregnancy (Ansell et al., 2002; Himeidan et al., 2004; Lindsay et al., 2000), and infection with the malaria parasite *Plasmodium falciparum* (Lacroix et al., 2005; Mukabana et al., 2007) are some of the known physiological factors that do change the attractiveness.

A pathogen infection may thus change an individual's odor profile and therefore attractiveness to hematophagous arthropods. An individual infected with *Plasmodium* parasites becomes more attractive to mosquitoes, thereby enhancing transmission of the parasite. It has been shown that host-seeking *Plasmodium* vectors are able to discriminate between infected and non-infected rodents (De Moraes et al., 2014; Ferguson and Read, 2004), canaries (Cornet et al., 2013a, 2013b), and humans (Batista et al., 2014; Lacroix et al., 2005; Busula et al., 2017). De Moreas et al. (2014) showed that *An. stephensi* mosquitoes can distinguish between the whole body volatiles from healthy- and *Plasmodium*-infected mice, and they identified a set of volatiles that could be involved. However, the mechanisms behind this change and whether this is caused by breath or skin volatiles still needs further investigation. In addition, it would be interesting to investigate whether other infections lead to similar effects in mosquitoes and other arthropods.

Differences in biting sites have been described for many hematophagous arthropods; however, the role of skin emanations in the selection of these sites is probably limited. *Simulium damnosum*, for example, has a preference to bite the underbelly region of cows, which may be related to the tendency of flies to land on shaded parts of the animal. In general, both *Simuliidae* and *Tabanidae* are very visually orientated and attracted to dark regions of the body (Allan et al., 1987). Visual cues also play an important role for *Culicoides* and may differ between species. *Culicoides achrayi*, for example, has a preference for the belly of a cow (Elbers and Meiswinkel, 2014), while in general *Culicoides* species tend to bite the back and flank region of cattle (Ayllón et al., 2014; Bishop et al.,

2008; Elbers and Meiswinkel, 2014). Interestingly, this biting preference may vary between host species. *Culicoides punctatus*, for example, mainly bites on the belly of cows (75.4%) but prefers to bite the back and flanks of ewes (81.7%) (Elbers and Meiswinkel, 2014). Next to visual cues, skin emanations are also used by *Culicoides* to find their host. However, the role of skin emanations in biting-site selection is unknown. Ticks also differ in their biting-site selection and some may favor the head and neck, others the lower extremities, and some do not show any preferred side (Falco and Fish, 1988; Felz and Durden, 1999; Slaff and Newton, 1993). Also for ticks the role of cues such as temperature, moisture, skin texture (thickness, hairs, etc.), and skin emanations remains unknown. Mosquitoes are known to select specific biting sites on their host. *An. gambiae s.s.*, *An. funestus*, and *An. arabiensis* bite most frequently on the feet and ankles (Braack et al., 2015; de Jong and Knols, 1995; Dekker et al., 1998), *Ae. aegypi*, *An. atroparvus*, and *Ae. simpsoni* prefer to bite around the head and shoulder (de Jong and Knols, 1995; Haddow, 1946; Knols, 1996), while *Cx. quinquefasciatus* does not prefer any specific body part (Knols, 1996). The selection of these sites depends on the position of the host and changes when a person lays down (Braack et al., 2015; Dekker et al., 1998). The selection of biting sites has therefore been attributed to either height above ground (Braack et al., 2015) or convection currents (Dekker et al., 1998).

Skin emanations differ between body sites and the distinct odors of, e.g., feet or the armpits can partly be related to their skin bacterial profile (Ara et al., 2006; Leyden et al., 1981; Rennie et al., 1991; Stoddart, 1990; Xu et al., 2007). To tests if skin emanations could play a role in the selection of biting sites, Verhulst et al. (2016) collected skin emanations from the feet, hand, and armpit and tested them for attractiveness to the malaria mosquito *Anopheles coluzzii* in an olfactometer. Skin emanations collected from armpits were less attractive to *An. coluzzii* compared to hands or/and feet, however, the difference may have been caused by deodorant residues. In a subsequent experiment, volunteers were asked to avoid using skin care products for 5 days, and thereafter, no differences in attractiveness of the body parts to mosquitoes were found (Verhulst et al., 2016).

Behavioral Assays in the Laboratory

To relate specific (odor) compounds to eventual host choice by the vector, laboratory experiments can provide a good basis for fundamental research. One of the advantages above fieldwork is the ability to control confounding factors such as climate conditions. Besides, laboratory experiments can be done with a study organism of choice with a fixed number of individuals. Logistically, laboratory experiments are more efficient, especially when compared with (semi-) fieldwork in remote or poorly developed areas. Scientifically sound experiments start with a healthy colony of study organisms and controlled facilities using standardized protocols (Huettel, 1976; Lyimo et al., 1992; Spitzen and Takken, 2005).

FIGURE 3.3 Dual-choice triple olfactometer to examine the response of mosquitoes to skin volatiles. Mosquitoes are released on one side (left) and on the other side they can enter one of the two trapping devices that contain the skin volatiles in addition to, e.g., CO_2.

Dual-choice assays are most commonly used to reveal the relative attractiveness between two different treatments (summarized by Butler, 2006) (Fig. 3.3). Roughly, two approaches can be used here. First, the use of complete blends of odor, for example, testing the response toward skin emanations from an individual (Pates et al., 2001; Verhulst et al., 2011b). Secondly, testing the effect of single compounds in different concentrations (Logan et al., 2009; Smallegange et al., 2005; van Loon et al., 2015). With the latter approach, one can identify crucial compounds that cause host preference. Triple-arm olfactometers have occasionally been used, but are difficult to set up and may require a high number of replicates, depending on the tested treatments (Mukabana et al., 2002; Posey et al., 1998).

Both positive and negative controls are crucial during choice assays given that most treatments will be tested over different days and responses vary, by nature, over days using different individuals. The perception and behavioral response to skin volatiles is affected by the microclimate (on the skin). As such, the odor release point, or end point of the behavioral setup, should be carefully controlled on background temperature and wind speed. One could consider the addition of a heat source to mimic skin temperature and induce landing (Healy et al., 2002; Spitzen et al., 2013).

Dual-choice olfactometers are less suitable for studying the role of repellent compounds. Although few spatial repellents have been described (Achee et al., 2012), the compounds of interest could interfere with the control ports and bias the response data. Also, some repellents may be hard to clean and contaminate the experimental setup, a requisite before conducting the next replicate. An alternative bioassay to test repellents is described by Menger et al. (2014a) who used a synthetic mimic of a human host with or without the repellent and in combination with replaceable cages to prevent contamination.

Another approach to study the effect of individual components on the attractiveness of (skin) volatiles is observing the behavior in wind tunnels (Dekker et al., 2005; Noldus et al., 1988; Takken and Knols, 1990). A good wind tunnel provides a laminar airflow in which detailed flight parameters can be analyzed. Advances in tracking techniques have enabled researchers to reconstruct video recorded flight paths of host-seeking insects in a 3D image plane (Dell et al., 2014; Lacey and Cardé, 2011; Parker et al., 2015; Spitzen et al., 2013). Similar video techniques can also be applied to "landing assays," with the major advantage that the (human) observer, with its accompanied body odors, can leave the experimental room during the course of the experiment.

Behavioral Assays in the (Semi-)Field

Despite the many advantages of laboratory experiments, they are confined with a potential inadequacy. What if the outcomes cannot be implemented or repeated under field conditions? In the semi-field using large screen cages one can release a known number of insects under ambient conditions (Busula et al., 2015; Knols et al., 2002). The use of experimental huts, either inside or outside screen cages, is mostly used for the effects of vector intervention studies (Fig. 3.4). Although we do acknowledge the indirect role of skin volatiles to the decision to "enter or not to enter" houses, other factors such as house design, CO_2 flow, and environmental conditions will be of more importance. For this reason, we will not further discuss the use of experimental houses as a tool for studying skin volatiles.

FIGURE 3.4 The MalariaSphere at the Thomas Odhiambo Campus of the International Centre of Insect Physiology and Ecology, Mbita, Kenya. A simulated mosquito ecosystem in a screen-walled greenhouse with ambient climate conditions as described by Knols et al. (2002). *(Photo courtesy D.J. Menger.)*

Where laboratory tests are often done with a selective number of vector species, testing volatile blends in natural settings includes the competition over these odors by the local vector population. Busula et al. (2015) demonstrated this in a study on natural and synthetic host odors on mosquitoes, where field studies revealed the varying responses between species revealed by field experiments after conducting semi-field studies.

In addition, field studies allow for tests with real life and larger hosts to conduct host preference studies as summarized by Mukabana et al. (2010) or, e.g., for studies on host site selection of ticks in cattle (Wanzala et al., 2004).

Perhaps the most basic method, more often used for mosquito monitoring than for studying the role of individual attractiveness, is the human landing catch method. Although still used, the ethical arguments and observer bias that come along with this method has resulted in the development of alternative methods (Lima et al., 2014; Maliti et al., 2015). The use of synthetic blends and other mimics of host cues have advantages in standardization and repeatability, however, these blends are often species-specific (Busula et al., 2015; Homan et al., 2016).

The Role of Skin Bacterial Volatiles for Other Hematophagous Arthropods

Skin volatiles play an important role in host finding of hematophagous arthropods, especially when they have a strong host preference. An important subset of these volatiles is produced by the skin microbiome, and both *in vivo* and *in vitro* studies have shown that the skin microbiota mediates host finding and host preference (Verhulst et al., 2010a, 2011b). Until now, these studies have only been performed with host-seeking mosquitoes and it would be interesting to know whether other hematophagous arthropods also use these skin bacterial volatiles to find their hosts. This is of interest because the studies with mosquitoes have shown that the identification of these bacterial volatiles can lead to the identification of novel attractants and potentially also repellents (Verhulst et al., 2009).

Host Skin Microbiome and Transmission of Pathogens

The cutaneous interface has a key role in vector-borne diseases as a primary barrier encountered by arthropods (Bernard et al., 2015). Therefore, it is likely that the commensal/mutualistic flora also plays a role in the interaction of arthropod–skin–pathogens yet to be identified (1) by its metabolome, which could have an attractive role on arthropods as shown previously, (2) a direct role in skin inflammation during the arthropod bite, and the inoculation of infectious agents.

Up to now, very few studies have addressed this second aspect on the role of the skin microbiota in the transmission of arthropod-borne diseases. The best explored model is leishmaniasis in an experimental model with murine. Depending on the parasite species, cutaneous or visceral leishmaniasis can develop.

Leishmania is a protozoan parasite transmitted by the bite of a sand fly, which injects promastigote forms into the skin of a vertebrate host. There, the parasites invade macrophages, transform into amastigotes, and actively multiply in the phagolysosomes. A typical lesion appears at the site of inoculation. Acute infection is controlled by a balanced secretion of cytokines secreted by Th1, Th2, and regulatory T cells. Germfree mice have been shown to be more susceptible to *L. major*. Microbiota is essential to control the generation and maintenance of innate cells such as tissue macrophages, dendritic cells, and recruited neutrophils, which participate in the tissue homeostasis (Lopes et al., 2016). An early study performed in 1999 (Oliveira et al., 1999), showed that germfree Swiss mice were more susceptible to *Leishmania* infection, likely due to a lower production of nitric oxide by macrophages and a reduced killing of intracellular parasites. More recently, using another mouse model, C57BL/6 germfree mouse, it was shown that these mice are more permissive to parasite multiplication due to a lower number of CD4$^+$ T cells producing cytokines interferon-γ (IFN-γ) controlling the infection and tumor necrosis factor-alpha (Naik et al., 2012). Interestingly, when they mono-associated *S. epidermidis*, a major skin commensal, IFN-γ secreted by T cells was produced again and it rescued protective immunity. They also showed that IL-1 signaling by keratinocytes is regulated by resident commensals.

Concerning other arthropod-borne diseases, no such study has been accomplished. Due to the role of skin microbiome in skin homeostasis and immunity, its role in the direct transmission of arthropod diseases deserves further investigation.

CONCLUSIONS AND PERSPECTIVES

Recent advances in skin microbial sequencing make it possible to study the effects of internal (physiology, diet, disease, genes) and external (environment, skin care products, perfumes) factors on the skin microbiota. These studies are mainly focused on bacteria; analyses on the virome should also open new perspectives (Schommer and Gallo, 2013). Unraveling these mechanisms and their role in volatile production will help to understand why some individuals are more attractive to arthropod disease vectors and therefore more at risk. In arthropod-borne diseases, an important internal factor is also the infection status of the individual. When an infected individual becomes more attractive to the vector, this will facilitate the transmission of pathogens. Although some work has been done on *Plasmodium*-infected individuals that become more attractive to mosquitoes (Batista et al., 2014; Lacroix et al., 2005), the role of skin volatile composition still needs to be determined.

When studying skin volatile composition and the effects on hematophagous arthropods, external factors are normally excluded as much as possible because they are difficult to control. However, these external factors may be as important as internal factors in the formation of the skin bacterial and skin volatile profile. Compounds from skin care products and perfumes are common on the human skin (Bouslimani et al., 2015) and may influence mosquito behavior (Verhulst et al., 2016). Studying the effect of skin care products on an individual's attractiveness to mosquitoes may lead to specific products that help to reduce a person's attractiveness to mosquitoes and thereby the number of bites received (Verhulst et al., 2016).

Techniques such as GC-EAG and SSR have been valuable for the identification of skin volatiles that can be detected by the olfactory system of a range of hematophagous insects. In addition, they have helped to understand the sensitivity, selectivity, and pathways of the olfactory systems in these arthropods. These studies may also result in novel approaches to disrupt these pathways. In the fruit fly *Drosophila melanogaster*, for example, specific odorants have been identified that act on the CO_2 receptor and thereby disrupt the CO_2 detection (Turner and Ray, 2009). Similar effects have been found for *Culex* mosquitoes, which may lead to the identification of novel volatile compounds that reduce mosquito–human contact (Turner and Ray, 2009).

Identification of skin volatiles that guide anthropophilic mosquitoes to their hosts has led to the identification of novel repellents and attractants. Combining attractants in a trap can be an effective tool to lower mosquito populations and thereby decrease risk (Homan et al., 2016). Next, the addition of a spatial repellent could lead to push–pull systems that are more effective than the odor-baited trap itself (Menger et al., 2014b). Identification of skin volatiles that guide other hematophagous arthropods to their host could lead to the development of similar systems.

Studying the mechanisms that shape the volatile composition released from the skin and the perception of these volatiles by the arthropod will lead to a better understanding of arthropod–host interactions. Identification of the volatiles that play a role in the host-seeking behavior of hematophagous arthropods will lead to novel attractants and repellents that can be applied for personal protection or in novel vector-control tools.

Finally, studies on the direct role of microbiota in the transmission of pathogens in the context of arthropod-borne diseases are only at their beginning. For leishmaniases, they revealed the role of skin microbiota in the priming of the immune system to boost the response to pathogens (Naik et al., 2012). Modulation of skin microbiota could lead to new type of adjuvants to favor efficacy of vaccines against arthropod-borne diseases (Belkaid and Segre, 2014; Grice, 2014; SanMiguel and Grice, 2015).

REFERENCES

Achee, N.L., Bangs, M.J., Farlow, R., Killeen, G.F., Lindsay, S., Logan, J.G., Moore, S.J., Rowland, M., Sweeney, K., Torr, S.J., 2012. Spatial repellents: from discovery and development to evidence-based validation. Malaria Journal 11, 1.

Allan, S.A., Day, J.F., Edman, J.D., 1987. Visual ecology of biting flies. Annual Review of Entomology 32, 297–316.

Allan, S.A., 2010. 15. Chemical Ecology of Tick-Host Interactions. Olfaction in Vector-Host Interactions, vol. 2, p. 327.

Ansell, J., Hamilton, K.A., Walraven, G.E.L., Lindsay, S.W., 2002. Short-range attractiveness of pregnant women to *Anopheles gambiae* mosquitoes. Transactions of the Royal Society of Tropical Medicine and Hygiene 96, 113–116.

Ara, K., Hama, M., Akiba, S., Koike, K., Okisaka, K., Hagura, T., Kamiya, T., Tomita, F., 2006. Foot odor due to microbial metabolism and its control. Canadian Journal of Microbiology 52, 357–364.

Ayllón, T., Nijhof, A.M., Weiher, W., Bauer, B., Allène, X., Clausen, P.-H., 2014. Feeding behaviour of *Culicoides* spp. (Diptera: Ceratopogonidae) on cattle and sheep in northeast Germany. Parasites and Vectors 7, 34.

Baldacchino, F., Manon, S., Puech, L., Buatois, B., Dormont, L., Jay-Robert, P., 2014. Olfactory and behavioural responses of tabanid horseflies to octenol, phenols and aged horse urine. Medical and Veterinary Entomology 28, 201–209.

Batista, E.P., Costa, E.F., Silva, A.A., 2014. *Anopheles darlingi* (Diptera: Culicidae) displays increased attractiveness to infected individuals with *Plasmodium vivax* gametocytes. Parasites and Vectors 7, 251.

Belkaid, Y., Segre, J.A., 2014. Dialogue between skin microbiota and immunity. Science 346, 954–959.

Bernard, Q., Jaulhac, B., Boulanger, N., 2015. Skin and arthropods: an effective interaction used by pathogens in vector-borne diseases. European Journal of Dermatology 25, 18–22.

Bernier, U.R., Booth, M.M., Yost, R.A., 1999. Analysis of human skin emanations by gas chromatography/mass spectrometry. 1.Thermal desorption of attractants for the yellow fever mosquito (*Aedes aegypti*) from handled glass beads. Analytical Chemistry 71, 1–7.

Bernier, U.R., Kline, D.L., Barnard, D.R., Schreck, C.E., Yost, R.A., 2000. Analysis of human skin emanations by gas chromatography/mass spectrometry. 2. Identification of volatile compounds that are candidate attractants for yellow fever mosquito (*Aedes aegypti*). Analytical Chemistry 72, 747–756.

Bernier, U.R., Kline, D.L., Schreck, C.E., Yost, R.A., Barnard, D.R., 2002. Chemical analysis of human skin emanations: comparison of volatiles from humans that differ in attraction of *Aedes aegypti* (Diptera: Culicidae). Journal of the American Mosquito Control Association 18, 186–195.

Bhasin, A., Mordue, W., 2000. Electrophysiological and behavioural identification of host kairomones as olfactory cues for *Culicoides impunctatus* and *C. nubeculosus*. Physiological Entomology 25, 6–16.

Bishop, A.L., McKenzie, H.J., Spohr, L.J., 2008. Attraction of *Culicoides brevitarsis* Kieffer (Diptera: Ceratopogonidae) and *Culex annulirostris* Skuse (Diptera: Culicidae) to simulated visual and chemical stimuli from cattle. Australian Journal of Entomology 47, 121–127.

Bouslimani, A., Porto, C., Rath, C.M., Wang, M., Guo, Y., Gonzalez, A., Berg-Lyon, D., Ackermann, G., Christensen, G.J.M., Nakatsuji, T., 2015. Molecular cartography of the human skin surface in 3D. Proceedings of the National Academy of Sciences 112, E2120–E2129.

Braack, L., Hunt, R., Koekemoer, L.L., Gericke, A., Munhenga, G., Haddow, A.D., Becker, P., Okia, M., Kimera, I., Coetzee, M., 2015. Biting behaviour of African malaria vectors:1. Where do the main vector species bite on the human body? Parasites and Vectors 8, 1–10.

Brady, J., Costantini, C., Sagnon, N., Gibson, G., Coluzzi, M., 1997. The role of body odours in the relative attractiveness of different men to malarial vectors in Burkina Faso. Annals of Tropical Medicine and Parasitology 91, S121–S122.

Braks, M.A.H., Scholte, E.J., Takken, W., Dekker, T., 2000. Microbial growth enhances the attractiveness of human sweat for the malaria mosquito, *Anopheles gambiae* sensu stricto. Chemoecology 10, 129–134.

Brouwer, R., 1960. Variations in human body odour as a cause of individual differences of attraction for malaria mosquitoes. Tropical and Geographical Medicine 12, 186–192.

Busula, A.O., Bousema, T., Mweresa, C.K., Masiga, D., Logan, J.G., Sauerwein, R.W., Verhulst, N.O., Takken, W., de Boer, J.G., 2017. Gametocytemia and attractiveness of *Plasmodium falciparum*–Infected Kenyan children to *Anopheles gambiae* mosquitoes. J. Infect. Dis. 3, 291–295.

Busula, A., Takken, W., Loy, D., Hahn, B., Mukabana, W., Verhulst, N.O., 2015. Mosquito host preferences affect their response to synthetic and natural odour blends. Malaria Journal 14, 133.

Busula, A.O., Takken, W., Boer, J.G.d., Mukabana, W.R., Verhulst, N.O., 2017. Variation in host preference of malaria mosquitoes is mediated by skin bacterial volatiles. Medical and Veterinary Entomology 31, 320–326.

Butler, J.F., 2006. Use of olfactometers for determining attractants and repellents. In: Debboun, M., Frances, S.P., Strickman, D. (Eds.), Insect Repellents: Principles, Methods and Uses. CRC Press, Boca Raton, p. 495.

Cardé, R.T., Gibson, G., 2010. Host finding by female mosquitoes: mechanisms of orientation to host odours and other cues. In: Takken, W., Knols, B.G.J. (Eds.), Olfaction in Vector-Host Interactions. Wageningen Academic Publishers, Wageningen, pp. 115–140.

Cardé, R.T., 2015. Multi-cue integration: how female mosquitoes locate a human host. Current Biology 25, R793–R795.

Carey, A.F., Wang, G., Su, C.-Y., Zwiebel, L.J., Carlson, J.R., 2010. Odorant reception in the malaria mosquito *Anopheles gambiae*. Nature 464, 66–71.

Cook, J., Majeed, S., Ignell, R., Pickett, J., Birkett, M., Logan, J., 2011. Enantiomeric selectivity in behavioural and electrophysiological responses of *Aedes aegypti* and *Culex quinquefasciatus* mosquitoes. Bulletin of Entomological Research 101, 541–550.

Cork, A., Beevor, P., Gough, A., Hall, D., 1990. Gas chromatography linked to electroantennography: a versatile technique for identifying insect semiochemicals. In: Chromatography and Isolation of Insect Hormones and Pheromones. Springer, pp. 271–279.

Cornet, S., Nicot, A., Rivero, A., Gandon, S., 2013a. Both infected and uninfected mosquitoes are attracted toward malaria infected birds. Malaria Journal 12, 179.

Cornet, S., Nicot, A., Rivero, A., Gandon, S., 2013b. Malaria infection increases bird attractiveness to uninfected mosquitoes. Ecology Letters 16, 323–329.

Costantini, C., Gibson, G., Sagnon, N., della Torre, A., Brady, J., Coluzzi, M., 1996. Mosquito responses to carbon dioxide in a West African Sudan savanna village. Medical and Veterinary Entomology 10, 220–227.

Costantini, C., Sagnon, N., della Torre, A., Diallo, M., Brady, J., Gibson, G., Coluzzi, M., 1998. Odor-mediated host preferences of West-African mosquitoes, with particular reference to malaria vectors. American Journal of Tropical Medicine and Hygiene 58, 56–63.

Curran, A.M., Rabin, S.I., Prada, P.A., Furton, K.G., 2005. Comparison of the volatile organic compounds present in human odor using SPME-GC/MS. Journal of Chemical Ecology 31, 0098–0331.

de Jong, R., Knols, B.G.J., 1995. Selection of biting sites on man by two malaria mosquito species. Experientia 51, 80–84.

De Moraes, C.M., Stanczyk, N.M., Betz, H.S., Pulido, H., Sim, D.G., Read, A.F., Mescher, M.C., 2014. Malaria-induced changes in host odors enhance mosquito attraction. Proceedings of the National Academy of Sciences 111, 11079–11084.

Dekker, T., Takken, W., Knols, B.G.J., Bouman, E., Laak, S., Bever, A., Huisman, P.W.T., 1998. Selection of biting sites on a human host by *Anopheles gambiae s.s.*, *An. arabiensis* and *An. quadriannulatus*. Entomologia Experimentalis et Applicata 87, 295–300.

Dekker, T., Geier, M., Carde, R.T., 2005. Carbon dioxide instantly sensitizes female yellow fever mosquitoes to human skin odours. Journal of Experimental Biology 208, 2963–2972.

Dell, A.I., Bender, J.A., Branson, K., Couzin, I.D., de Polavieja, G.G., Noldus, L.P., Pérez-Escudero, A., Perona, P., Straw, A.D., Wikelski, M., 2014. Automated image-based tracking and its application in ecology. Trends in Ecology and Evolution 29, 417–428.

Dormont, L., Bessière, J.-M., Cohuet, A., 2013a. Human skin volatiles: a review. Journal of Chemical Ecology 39, 569–578.

Dormont, L., Bessière, J.-M., McKey, D., Cohuet, A., 2013b. New methods for field collection of human skin volatiles and perspectives for their application in the chemical ecology of human–pathogen–vector interactions. Journal of Experimental Biology 216, 2783–2788.

Dougherty, M.J., Guerin, P.M., Ward, R.D., Hamilton, J.G.C., 1999. Behavioural and electrophysiological responses of the phlebotomine sandfly *Lutzomyia longipalpis* (Diptera: Psychodidae) when exposed to canid host odour kairomones. Physiological Entomology 24, 251–262.

Elbers, A.R.W., Meiswinkel, R., 2014. *Culicoides* (Diptera: Ceratopogonidae) host preferences and biting rates in The Netherlands: comparing cattle, sheep and the black-light suction trap. Veterinary Parasitology 205, 330–337.

Falco, R.C., Fish, D., 1988. Ticks parasitizing humans in a Lyme disease endemic area of southern New York State. American Journal of Epidemiology 128, 1146–1152.

Felz, M.W., Durden, L.A., 1999. Attachment sites of four tick species (Acari: Ixodidae) parasitizing humans in Georgia and South Carolina. Journal of Medical Entomology 36, 361–364.

Ferguson, H.M., Read, A.F., 2004. Mosquito appetite for blood is stimulated by *Plasmodium chabaudi* infections in themselves and their vertebrate hosts. Malaria Journal 3.

Fernández-Grandon, G., Gezan, S., Armour, J., Pickett, J., Logan, J., 2015. Heritability of attractiveness to mosquitoes. PLoS One 10, e0122716.

Gallagher, M., Wysocki, C.J., Leyden, J.J., Spielman, A.I., Sun, X., Preti, G., 2008. Analyses of volatile organic compounds from human skin. British Journal of Dermatology 159, 780–791.

Garrett-Jones, C., Shidrawi, G.R., 1969. Malaria vectorial capacity of a population of *Anopheles gambiae*. Bulletin of The World Health Organization 40, 531–545.

Ghaninia, M., Larsson, M., Hansson, B.S., Ignell, R., 2008. Natural odor ligands for olfactory receptor neurons of the female mosquito *Aedes aegypti*: use of gas chromatography-linked single sensillum recordings. Journal of Experimental Biology 211, 3020–3027.

Gibson, G., Torr, S.J., 1999. Visual and olfactory responses of haematophagous Diptera to host stimuli. Medical and Veterinary Entomology 13, 2–23.

Gillies, M.T., 1980. The role of carbon dioxide in host-finding by mosquitoes (Diptera:Culicidae): a review. Bulletin of Entomological Research 70, 525–532.

Grice, E.A., Segre, J.A., 2011. The skin microbiome. Nature Reviews Microbiology 9, 244–253.

Grice, E.A., 2014. The skin microbiome: potential for novel diagnostic and therapeutic approaches to cutaneous disease. Seminars in Cutaneous Medicine and Surgery 98–103 Frontline Medical Communications.

Haddow, A.J., 1946. The mosquitoes of Bwamba County, Uganda II.- Biting activity with special reference to the influence of microclimate. Bulletin of Entomological Research 36, 33–73.

Hall, D.R., Beevor, P.S., Cork, A., Nesbitt, B.F., Vale, G.A., 1984. 1-Octen-3-ol. A potent olfactory stimulant and attractant for tsetse isolated from cattle odours. International Journal of Tropical Insect Science 5, 335–339.

Hamilton, J.G.C., Ramsoondar, T.M.C., 1994. Attraction of *Lutzomyia longipalpis* to human skin odours. Medical and Veterinary Entomology 8, 375–380.

Harraca, V., Ignell, R., Löfstedt, C., Ryne, C., 2010. Characterization of the antennal olfactory system of the bed bug (*Cimex lectularius*). Chemical Senses 35, 195–204.

Healy, T.P., Copland, M.J.W., Cork, A., 2002. Landing responses of *Anopheles gambiae* elicited by oxocarboxylic acids. Medical and Veterinary Entomology 16, 126–132.

Himeidan, Y.E., Elbashir, M.I., Adam, I., 2004. Attractiveness of pregnant women to the malaria vector, *Anopheles arabiensis*, in Sudan. Annals of Tropical Medicine and Parasitology 98, 631–633.

Homan, T., Hiscox, A., Mweresa, C.K., Masiga, D., Mukabana, W.R., Oria, P., Maire, N., Pasquale, A.D., Silkey, M., Alaii, J., Bousema, T., Leeuwis, C., Smith, T.A., Takken, W., 2016. The effect of mass mosquito trapping on malaria transmission and disease burden (SolarMal): a stepped-wedge cluster-randomised trial. The Lancet 388, 1193–1201.

Huettel, M., 1976. Monitoring the quality of laboratory-reared insects: a biological and behavioral perspective. Environmental Entomology 5, 807–814.

Hunt, R.H., Coetzee, M., Fettene, M., 1998. The *Anopheles gambiae* complex: a new species from Ethiopia. Transactions of the Royal Society of Tropical Medicine and Hygiene 92, 231–235.

Jackman, P.J.H., Noble, W.C., 1983. Normal axillary skin in various populations. Clinical and Experimental Dermatology 8, 259–268.

Jaleta, K.T., Hill, S.R., Birgersson, G., Tekie, H., Ignell, R., 2016. Chicken volatiles repel host-seeking malaria mosquitoes. Malaria Journal 15, 1.

James, A.G., Casey, J., Hyliands, D., Mycock, G., 2004a. Fatty acid metabolism by cutaneous bacteria and its role in axillary malodour. World Journal of Microbiology and Biotechnology 20, 787.

James, A.G., Hyliands, D., Johnston, H., 2004b. Generation of volatile fatty acids by axillary bacteria. International Journal of Cosmetic Science 26, 149–156.

Kearney, J.N., Harnby, D., Gowland, G., Holland, K.T., 1984. The follicular distribution and abundance of resident bacteria on human skin. Journal of General Microbiology 130, 797–801.

Kirk, K.M., Eaves, L.J., Meyer, J.M., Saul, A., Martin, N.G., 2000. Twin study of adolescent genetic susceptibility to mosquito bites using ordinal and comparative rating data. Genetic Epidemiology 19, 178–190.

Knols, B.G.J., De Jong, R., 1996. Limburger cheese as an attractant for the mosquito *Anopheles gambiae s. s.* Parasitology Today 12, 159–161.

Knols, B.G.J., Meijerink, J., 1997. Odors influence mosquito behavior. Science and Medicine 4, 56–63.

Knols, B.G.J., De Jong, R., Takken, W., 1995. Differential attractiveness of isolated humans to mosquitoes in Tanzania. Transactions of the Royal Society of Tropical Medicine and Hygiene 89, 604–606.

Knols, B.G., Njiru, B.N., Mathenge, E.M., Mukabana, W.R., Beier, J.C., Killeen, G.F., 2002. MalariaSphere: a greenhouse-enclosed simulation of a natural *Anopheles gambiae* (Diptera: Culicidae) ecosystem in western Kenya. Malaria Journal 1, 19.

Knols, B.G.J., 1996. Odour-Mediated Host-Seeking Behaviour of the Afro-Tropical Malaria Vector *Anopheles gambiae* Giles. Laboratory for Entomology. Wageningen University, p. 213.

Lacey, E.S., Cardé, R.T., 2011. Activation, orientation and landing of female *Culex quinquefasciatus* in response to carbon dioxide and odour from human feet: 3-D flight analysis in a wind tunnel. Medical and Veterinary Entomology 25, 94–103.

Lacroix, R., Mukabana, W.R., Gouagna, L.C., Koella, J.C., 2005. Malaria infection increases attractiveness of humans to mosquitoes. PLoS Biology 3, 1590–1593.

Lemfack, M.C., Ravella, S.R., Lorenz, N., Kai, M., Jung, K., Schulz, S., Piechulla, B., 2016. Novel volatiles of skin-borne bacteria inhibit the growth of Gram-positive bacteria and affect quorum-sensing controlled phenotypes of Gram-negative bacteria. Systematic and Applied Microbiology 39, 503–515.

Leyden, J.J., McGinley, K.J., Holzle, E., Labows, J.N., Kligman, A.M., 1981. The microbiology of the human axilla and its relationship to axillary odor. Journal of Investigative Dermatology 77, 413–416.

Lima, J.B.P., Rosa-Freitas, M.G., Rodovalho, C.M., Santos, F., Lourenço-de-Oliveira, R., 2014. Is there an efficient trap or collection method for sampling *Anopheles darlingi* and other malaria vectors that can describe the essential parameters affecting transmission dynamics as effectively as human landing catches?-A Review. Memorias do Instituto Oswaldo Cruz 109, 685–705.

Lindsay, S.W., Adiamah, J.H., Miller, J.E., Pleass, R.J., Armstrong, J.R.M., 1993. Variation in attractiveness of human subjects to malaria mosquitoes (Diptera: Culicidae) in The Gambia. Journal of Medical Entomology 30, 308–373.

Lindsay, S., Ansell, J., Selman, C., Cox, V., Hamilton, K., Walraven, G., 2000. Effect of pregnancy on exposure to malaria mosquitoes. The Lancet 355, 1972.

Logan, J.G., Birkett, M.A., Clark, S.J., Powers, S., Seal, N.J., Wadhams, L.J., Mordue, A.J., Pickett, J.A., 2008. Identification of human-derived volatile chemicals that interfere with attraction of *Aedes aegypti* mosquitoes. Journal of Chemical Ecology 34, 308–322.

Logan, J.G., Seal, N.J., Cook, J.I., Stanczyk, N.M., Birkett, M.A., Clark, S.J., Gezan, S.A., Wadhams, L.J., Pickett, J.A., Mordue, A.J., 2009. Identification of human-derived volatile chemicals that interfere with attraction of the scottish biting midge and their potential use as repellents. Journal of Medical Entomology 46, 208–219.

Logan, J.G., Cook, J.I., Mordue, A.J., Kline, D.L., 2010. 10. Understanding and Exploiting Olfaction for the Surveillance and Control of *Culicoides* Biting Midges. Olfaction in Vector-Host Interactions, vol. 2, p. 217.

Logan, J.G., 2008. Why do mosquitoes "choose" to bite some people more than others? Outlooks on Pest Management 19, 280–283.

Lopes, M.M., Carneiro, M., Santos, L., Vieira, L., 2016. Indigenous microbiota and leishmaniasis. Parasite Immunology 38, 37–44.

Lyimo, I.N., Ferguson, H.M., 2009. Ecological and evolutionary determinants of host species choice in mosquito vectors. Trends in Parasitology 25, 189–196.

Lyimo, E., Takken, W., Koella, J., 1992. Effect of rearing temperature and larval density on larval survival, age at pupation and adult size of *Anopheles gambiae*. Entomologia Experimentalis et Applicata 63, 265–271.

Maliti, D.V., Govella, N.J., Killeen, G.F., Mirzai, N., Johnson, P.C., Kreppel, K., Ferguson, H.M., 2015. Development and evaluation of mosquito-electrocuting traps as alternatives to the human landing catch technique for sampling host-seeking malaria vectors. Malaria Journal 14, 1.

Mands, V., Kline, D.L., Blackwell, A., 2004. Culicoides midge trap enhancement with animal odour baits in Scotland. Medical and Veterinary Entomology 18, 336–342.

Marinković, Ž.J., Hackenberger, B.K., Merdić, E., 2014. Maximum radius of carbon dioxide baited trap impact in woodland: implications for host-finding by mosquitoes. Biologia 69, 522–529.

Mayer, M.S., James, J.D., 1969. Attraction of *Aedes aegypti* (L.): responses to human arms, carbon dioxide, and air currents in a new type of olfactometer. Bulletin of Entomological Research 58, 629–643.

Mboera, L.E.G., Knols, B.G.J., Takken, W., dellaTorre, A., 1997. The response of *Anopheles gambiae* sl and *A. funestus* (Diptera: Culicidae) to tents baited with human odour or carbon dioxide in Tanzania. Bulletin of Entomological Research 87, 173–178.

McBride, C.S., Baier, F., Omondi, A.B., Spitzer, S.A., Lutomiah, J., Sang, R., Ignell, R., Vosshall, L.B., 2014. Evolution of mosquito preference for humans linked to an odorant receptor. Nature 515, 222–227.

Meijerink, J., van Loon, J.J.A., 1999. Sensitivity of antennal olfactory neurons of the malaria mosquito, *Anopheles gambiae*, to carboxylic acids. Journal of Insect Fysiology 45, 365–373.

Menger, D., van Loon, J., Takken, W., 2014a. Assessing the efficacy of candidate mosquito repellents against the background of an attractive source that mimics a human host. Medical and Veterinary Entomology 28, 407–413.

Menger, D.J., Otieno, B., De Rijk, M., Mukabana, W.R., Van Loon, J.J., Takken, W., 2014b. A push-pull system to reduce house entry of malaria mosquitoes. Malaria Journal 13, 18.

Mukabana, W.R., Takken, W., Coe, R., Knols, B.G.J., 2002. Host-specific cues cause differential attractiveness of Kenyan men to the African malaria vector *Anopheles gambiae*. Malaria Journal 1, 17.

Mukabana, W.R., Takken, W., Killeen, G.F., Knols, B.G.J., 2007. Clinical malaria reduces human attractiveness to mosquitoes. Proceedings of the Netherlands Entomological Society 18, 125–129.

Mukabana, W.R., Olanga, E.A., Knols, B.G., 2010. Host-Seeking Behaviour of Afrotropical Anophelines: Field and Semi-Field Studies. Olfaction in Vector-Host Interactions, pp. 181–202.

Mukabana, W.R., Mweresa, C.K., Otieno, B., Omusula, P., Smallegange, R.C., van Loon, J.J.A., Takken, W., 2012. A novel synthetic odorant blend for trapping of malaria and other African mosquito species. Journal of Chemical Ecology 38, 235–244.

Naik, S., Bouladoux, N., Wilhelm, C., Molloy, M.J., Salcedo, R., Kastenmuller, W., Deming, C., Quinones, M., Koo, L., Conlan, S., Spencer, S., Hall, J.A., Dzutsev, A., Kong, H., Campbell, D.J., Trinchieri, G., Segre, J.A., Belkaid, Y., 2012. Compartmentalized control of skin immunity by resident commensals. Science 337, 1115–1119.

Natsch, A., Derrer, S., Flachsmann, F., Schmid, J., 2006. A broad diversity of volatile carboxylic acids, released by a bacterial aminoacylase from axilla secretions, as candidate molecules for the determination of human-body odor type. Chemistry and Biodiversity 3, 1–20.

Noldus, L.P.J.J., Lewis, W.J., Tumlinson, J.H., Lenteren van, J.C.v, 1988. In: Voegelé, J., Waage, J.K., van Lenteren, J.C. (Eds.), Olfactometer and Windtunnel Experiments on the Role of Sex Pheromones of Noctuid Moths in the Foraging Behaviour of *Trichogramma* Spp., vol. 43, pp. 223–238.

Oliveira, M.R.d., Tafuri, W.L., Nicoli, J.R., Vieira, E.C., Melo, M.N., Vieira, L.Q., 1999. Influence of microbiota in experimental cutaneous leishmaniasis in Swiss mice. Revista do Instituto de Medicina Tropical de Sao Paulo 41, 87–94.

Park, K.C., Cork, A., 1999. Electrophysiological responses of antennal receptor neurons in female Australian sheep blowflies, *Lucilia cuprina*, to host odours. Journal of Insect Physiology 45, 85–91.

Parker, J.E.A., Angarita-Jaimes, N., Abe, M., Towers, C.E., Towers, D., McCall, P.J., 2015. Infrared video tracking of *Anopheles gambiae* at insecticide-treated bed nets reveals rapid decisive impact after brief localised net contact. Scientific Reports 5, 13392.

Pates, H.V., Takken, W., Stuke, K., Curtis, C.F., 2001. Differential behaviour of *Anopheles gambiae sensu stricto* (Diptera: Culicidae) to human and cow odours in the laboratory. Bulletin of Entomological Research 91, 289–296.

Penn, D., Potts, W.K., 1998. How do major histocompatibility complex genes influence odor and mating preferences? Advances in Immunology 69, 411–436.

Phelan, P.L., Smith, A.W., Needham, G.R., 1991. Mediation of host selection by cuticular hydrocarbons in the honeybee tracheal mite *Acarapis woodi* (Rennie). Journal of Chemical Ecology 17, 463–473.

Pinto, M.C., Campbell-Lendrum, D.H., Lozovei, A.L., Teodoro, U., Davies, C.R., 2001. Phlebotomine sandfly responses to carbon dioxide and human odour in the field. Medical and Veterinary Entomology 15, 132–139.

Posey, K.H., Barbard, D.R., Schreck, C.E., 1998. Triple cage olfactometer for evaluating mosquito (Diptera: Culicidae) attraction responses. Journal of Medical Entomology 35, 330–334.

Puri, S.N., Mendki, M., Sukumaran, D., Ganesan, K., Prakash, S., Sekhar, K., 2006. Electroantennogram and behavioral responses of *Culex quinquefasciatus* (Diptera: Culicidae) females to chemicals found in human skin emanations. Journal of Medical Entomology 43, 207–213.

Qiu, Y.T., Smallegange, R.C., Hoppe, S., van Loon, J.J.A., Bakker, E.J., Takken, W., 2004a. Behavioural and electrophysiological responses of the malaria mosquito *Anopheles gambiae* Giles *sensu stricto* (Diptera: Culicidae) to human skin emanations. Medical and Veterinary Entomology 18, 429–438.

Qiu, Y.T., Smallegange, R.C., Smid, H., Loon, J.J.A.v., Galimard, A.M.S., Posthumus, M.A., Beek, T.A.v., Takken, W., 2004b. GC-EAG analysis of human odours that attract the malaria mosquito *Anopheles gambiae* sensu stricto. Proceedings Experimental and Applied Entomology, N.E.V 59–64.

Qiu, Y.T., Smallegange, R.C., van Loon, J.J.A., Ter Braak, C.J.F., Takken, W., 2006. Interindividual variation in the attractiveness of human odours to the malaria mosquito *Anopheles gambiae s.s.* Medical and Veterinary Entomology 20, 280–287.

Qiu, Y.T., 2005. Sensory and Behavioural Responses of the Malaria Mosquito *Anopheles gambiae* to Human Odours. Laboratory of Entomology. Wageningen University, Wageningen, p. 207.

Rennie, P.J., Gower, D.B., Holland, K.T., 1991. In-vitro and in-vivo studies of human axillary odor and the cutaneous microflora. British Journal of Dermatology 124, 596–602.

Ribeiro, J.M., 1996. Common problems of arthropod vectors of disease. The Biology of Disease Vectors 25–33.

SanMiguel, A., Grice, E.A., 2015. Interactions between host factors and the skin microbiome. Cellular and Molecular Life Sciences 72, 1499–1515.

Sant'ana, A.L., Eiras, A.E., Cavalcante, R.R., 2002. Electroantennographic responses of the *Lutzomyia* (Lutzomyia) *longipalpis* (Lutz & Neiva)(Diptera: Psychodidae) to 1-octen-3-ol. Neotropical Entomology 31, 13–17.

Sastry, S.D., Buck, K.T., Janak, J., Dressler, M., Preti, G., 1980. Volatiles emitted by humans. In: Dermer, O.C., Waller, G.R. (Eds.), Biochemical Applications of Mass Spectrometry. John Wiley & Sons, New York, pp. 1085–1127.

Schofield, S., Cork, A., Brady, J., 1995. Electroantennogram responses of the stable fly, *Stomoxys calcitrans*, to components of host odour. Physiological Entomology 20, 273–280.

Schommer, N.N., Gallo, R.L., 2013. Structure and function of the human skin microbiome. Trends in Microbiology 21, 660–668.

Schreck, C.E., James, J., 1968. Broth cultures of bacteria that attract female mosquitoes. Mosquito News 28, 33.

Schreck, C.E., Kline, D.L., Carlson, D.A., 1990. Mosquito attraction to substances from the skin of different humans. Journal of the American Mosquito Control Association 6, 406–410.

Shelley, W.W.B., Hurley Jr., H.H.J., Nichols, A.A.C., 1953. Axillary odor; experimental study of the role of bacteria, apocrine sweat, and deodorants. A.M.A. Archives of Dermatology and Syphilology 68, 430–446.

Singh, N., Wang, C., Cooper, R., Liu, C., 2012. Interactions among carbon dioxide, heat, and chemical lures in attracting the bed bug, Cimex lectularius L. (Hemiptera: Cimicidae). Psyche: A Journal of Entomology 2012.

Slaff, M., Newton, N.H., 1993. Location of tick (Acari: Ixodidae) attachment sites on humans in North Carolina. Journal of Medical Entomology 30, 485–488.

Smallegange, R.C., Takken, W., 2010. Host-seeking behaviour of mosquitoes - responses to olfactory stimuli in the laboratory. In: Takken, W., Knols, B.G.J. (Eds.), Olfaction in Vector-Host Interactions. Wageningen Academic Publishers, Wageningen, pp. 143–180.

Smallegange, R.C., Qiu, Y.T., van Loon, J.J.A., Takken, W., 2005. Synergism between ammonia, lactic acid and carboxylic acids as kairomones in the host-seeking behaviour of the malaria mosquito Anopheles gambiae sensu stricto (Diptera: Culicidae). Chemical Senses 30, 145–152.

Smart, M.R., Brown, A.W.A., 1957. Studies on the responses of the female Aedes mosquito. Part VII. -The effect of skin temperature, humidity and moisture on the attractiveness of the human hand. Bulletin of Entomological Research 47, 89–101.

Spitzen, J., Takken, W., 2005. Malaria mosquito rearing - maintaining quality and quantity of laboratory-reared insects. Proceedings of the Netherlands Entomological Society Meeting 16, 95–100.

Spitzen, J., Spoor, C.W., Grieco, F., ter Braak, C., Beeuwkes, J., van Brugge, S.P., Kranenbarg, S., Noldus, L.P.J.J., van Leeuwen, J.L., Takken, W., 2013. A 3D analysis of flight behavior of Anopheles gambiae sensu stricto malaria mosquitoes in response to human odor and heat. PLoS One 8, e62995.

Steib, B.M., Geier, M., Boeckh, J., 2001. The effect of lactic acid on odour-related host preference of yellow fever mosquitoes. Chemical Senses 26, 523–528.

Stoddart, D.M., 1990. The Scented Ape: The Biology and Culture of Human Odour. Cambridge University Press, Cambridge, UK.

Takken, W., Knols, B.G.J., 1990. Flight behaviour of Anopheles gambiae Giles (Diptera: Culicidae) in response to host stimuli: a windtunnel study. In: Proceedings Experimental and Applied Entomology, N.E.V., Amsterdam, pp. 121–128.

Takken, W., Knols, B.G.J., 1999. Odor-mediated behavior of afrotropical malaria mosquitoes. Annual Review of Entomology 44, 131–157.

Takken, W., Knols, B.G.J., 2010. Olfaction in Vector-Host Interactions. Wageningen Academic Publishers, Wageningen.

Takken, W., Verhulst, N.O., 2013. Host preferences of blood-feeding mosquitoes. Annual Review of Entomology 58, 433–453.

Tangtrakulwanich, K., Chen, H., Baxendale, F., Brewer, G., Zhu, J., 2011. Characterization of olfactory sensilla of Stomoxys calcitrans and electrophysiological responses to odorant compounds associated with hosts and oviposition media. Medical and Veterinary Entomology 25, 327–336.

Taylor, D., Daulby, A., Grimshaw, S., James, G., Mercer, J., Vaziri, S., 2003. Characterization of the microflora of the human axilla. International Journal of Cosmetic Science 25, 137–145.

Torr, S.J., della Torre, A., Calzetta, M., Costantini, C., Vale, G.A., 2008. Towards a fuller understanding of mosquito behaviour: use of electrocuting grids to compare the odour-orientated responses of Anopheles arabiensis and An. quadriannulatus in the field. Medical and Veterinary Entomology 22, 93–108.

Turner, S.L., Ray, A., 2009. Modification of CO_2 avoidance behaviour in Drosophila by inhibitory odorants. Nature 461, 277–281.

van Loon, J.J.A., Smallegange, R.C., Bukovinszkiné-Kiss, G., Jacobs, F., De Rijk, M., Mukabana, W.R., Verhulst, N.O., Menger, D.J., Takken, W., 2015. Mosquito attraction: crucial role of carbon dioxide in formulation of a five-component blend of human-derived volatiles. Journal of Chemical Ecology 41, 567–573.

Verhulst, N.O., Beijleveld, H., Knols, B.G.J., Takken, W., Schraa, G., Bouwmeester, H.J., Smallegange, R.C., 2009. Cultured skin microbiota attracts malaria mosquitoes. Malaria Journal 8, 302.

Verhulst, N.O., Andriessen, R., Groenhagen, U., Bukovinszkiné Kiss, G., Schulz, S., Takken, W., van Loon, J.J.A., Schraa, G., Smallegange, R.C., 2010a. Differential attraction of malaria mosquitoes to volatile blends produced by human skin bacteria. PLoS One 5, e15829.

Verhulst, N.O., Takken, W., Dicke, M., Schraa, G., Smallegange, R.C., 2010b. Chemical ecology of interactions between human skin microbiota and mosquitoes. FEMS Microbiology Ecology 74, 1–9.

Verhulst, N.O., Mbadi, P., Kiss, G., Mukabana, W., van Loon, J.J.A., Takken, W., Smallegange, R.C., 2011a. Improvement of a synthetic lure for *Anopheles gambiae* using compounds produced by human skin microbiota. Malaria Journal 10, 28.

Verhulst, N.O., Qiu, Y.T., Beijleveld, H., Maliepaard, C., Knights, D., Schulz, S., Berg-Lyons, D., Lauber, C.L., Verduijn, W., Haasnoot, G.W., Mumm, R., Bouwmeester, H.J., Claas, F.H.J., Dicke, M., Loon, J.J.A.v, Takken, W., Knight, R., Smallegange, R.C., 2011b. Composition of human skin microbiota affects attractiveness to malaria mosquitoes. PLoS One 6, e28991.

Verhulst, N.O., Beijleveld, H., Qiu, Y.T., Maliepaard, C., Verduyn, W., Haasnoot, G.W., Claas, F.H.J., Mumm, R., Bouwmeester, H.J., Takken, W., van Loon, J.J.A., Smallegange, R.C., 2013a. Relation between HLA genes, human skin volatiles and attractiveness of humans to malaria mosquitoes. Infection Genetics and Evolution 18, 87–93.

Verhulst, N.O., Loonen, J.A.C.M., Takken, W., 2013b. Advances in methods for colour marking of mosquitoes. Parasites and Vectors 6, 200.

Verhulst, N.O., Weldegergis, B.T., Menger, D., Takken, W., 2016. Attractiveness of volatiles from different body parts to the malaria mosquito *Anopheles coluzzii* is affected by deodorant compounds. Scientific Reports 6, 27141.

Wanzala, W., Sika, N.F.K., Gule, S., Hassanali, A., 2004. Attractive and repellent host odours guide ticks to their respective feeding sites. Chemoecology 14, 229–232.

Warnes, M., Finlayson, L., 1986. Electroantennogram responses of the stable fly, *Stomoxys calcitrans*, to carbon dioxide and other odours. Physiological Entomology 11, 469–473.

Warnes, M.L., Finlayson, L.H., 1987. Effect of host behaviour on host preference in *Stomoxys calcitrans*. Medical and Veterinary Entomology 1, 53–57.

Webster, B., Lacey, E., Cardé, R., 2015. Waiting with bated breath: opportunistic orientation to human odor in the malaria mosquito, *Anopheles gambiae*, is modulated by minute changes in carbon dioxide concentration. Journal of Chemical Ecology 1–8.

Wedekind, C., Furi, S., 1997. Body odour preferences in men and women: do they aim for specific MHC combinations or simply heterozygosity? Proceedings of the Royal Society Biological Sciences Series B 264, 1471–1479.

Weeks, E.N.I., Birkett, M.A., Cameron, M.M., Pickett, J.A., Logan, J.G., 2011. Semiochemicals of the common bed bug, *Cimex lectularius* L. (Hemiptera: Cimicidae), and their potential for use in monitoring and control. Pest Management Science 67, 10–20.

White, G.B., Tessfaye, F., Boreham, P.F.L., Lemma, G., 1980. Malaria vector capacity of *Anopheles arabiensis* and *An. quadriannulatus* in Ethiopia: chromosomal interpretation after 6 years storage of field preparations. Transactions of the Royal Society of Tropical Medicine and Hygiene 74, 683–684.

Wilson, M., 2008. Bacteriology of Humans: An Ecological Perspective. Blackwell Publishing Ltd, Malden, MA.

Xiao-Nong, Z., James, J.L., Andrew, I.S., George, P., 1996. Analysis of characteristic human female axillary odors: qualitative comparison to males. Journal of Chemical Ecology 22, 237–257.

Xu, Y., Dixon, S., Brereton, R., Soini, H., Novotny, M., Trebesius, K., Bergmaier, I., Oberzaucher, E., Grammer, K., Penn, D., 2007. Comparison of human axillary odour profiles obtained by gas chromatography/mass spectrometry and skin microbial profiles obtained by denaturing gradient gel electrophoresis using multivariate pattern recognition. Metabolomics 3, 427–437.

Zhang, X., Crippen, T.L., Coates, C.J., Wood, T.K., Tomberlin, J.K., 2015. Effect of quorum sensing by *Staphylococcus epidermidis* on the attraction response of female adult yellow fever mosquitoes, *Aedes aegypti aegypti* (linnaeus) (diptera: Culicidae), to a blood-feeding source. PLoS One 10.

Zwiebel, L.J., Takken, W., 2004. Olfactory regulation of mosquito-host interactions. Insect Biochemistry and Molecular Biology 34, 645–652.

Chapter 4

Arthropod Saliva and Its Role in Pathogen Transmission: Insect Saliva

INTRODUCTION

The role of arthropod saliva in pathogen transmission was demonstrated for the first time in the context of *Leishmania* transmission (Titus and Ribeiro, 1988). It showed the enhancing effect of sand fly saliva on parasite infectivity and the specificity of this effect since saliva of other arthropod vectors (*Aedes* mosquitoes, triatomine bugs, or *Ixodes* ticks) had no effect on the transmission process. Further studies with sand fly saliva allowed the identification of several molecules targeting host pharmacology and immunology (Belkaid et al., 1998). The model sand fly–*Leishmania* has been extensively studied. Studies on other insect vectors are not so abundant, especially for mosquitoes, despite being major vectors of numerous arthropod-borne diseases such as malaria and dengue. At the end of the nineties, the first description of the effect of mosquito saliva on viruses was published (Limesand et al., 2000; Zeidner et al., 1999) (for a review Titus et al., 2006). More recently, mosquito saliva has been investigated in the context of malaria. In this chapter, we propose to illustrate the role of insect saliva in arthropod transmission by developing data on the role of sand fly saliva in leishmaniasis and of mosquito saliva in malaria.

MOSQUITO SALIVA, SKIN, ALLERGY, AND THE OUTCOME OF MALARIA INFECTION—FROM MICE TO MEN

Claudia Demarta-Gatsi[1,2,3], Salaheddine Mécheri[1,2,3] and Richard E. Paul[4,5]

[1]*Institut Pasteur, Unité de Biologie des Interactions Hôte Parasites, Paris, France;* [2]*Centre National de la Recherche Scientifique ERL9195, Paris, France;* [3]*INSERM U1201, Paris F-75015, France;* [4]*Institut Pasteur, Functional Genetics of Infectious Diseases Unit, Paris, France;* [5]*Centre National de la Recherche Scientifique, Unité de Recherche Associée 3012, Paris, France*

Skin and Arthropod Vectors. https://doi.org/10.1016/B978-0-12-811436-0.00004-6

83

INTRODUCTION

Skin provides the first line of defense against exoantigens including pathogens and allergens and is composed of a complex immune network that subsequently influences the systemic immune response. The evolution of hematophagy has been accompanied by a diversification of salivary components that facilitate blood meal acquisition and overcome triggered skin defenses that consist of hemostasis, pain and itch responses, and immune effector mechanisms (James and Rossignol, 1991; Titus and Ribeiro, 1990). There is compelling evidence that arthropod saliva has profound effects on pathogen transmission (Ribeiro, 1987; Belkaid et al., 1998). An immunomodulatory role of saliva has been reported for arboviruses (Hajnicka et al., 2005; Schneider and Higgs, 2008) and protozoa including *Leishmania* (Belkaid et al., 1998; de Moura et al., 2007), *Trypanosoma*, (Mesquita et al., 2008) and *Plasmodium* (Depinay et al., 2006; Schneider et al., 2011). The association of arthropod-borne pathogens with saliva and the known immune-modulatory capacity of saliva call for elucidation of the role of the skin and saliva in influencing the immediate and long-term immune responses to such pathogens and the subsequent outcome of infection. In this sub-chapter we discuss the importance of this for the specific case of malaria parasites and highlight the value of considering malaria alongside non-infectious skin diseases.

Immunomodulatory Properties of *Anopheles* Mosquito Saliva

Type 1 hypersensitivity is an immediate reaction, mediated by IgE, typical of classical atopy and allergy, including atopic dermatitis (AD), rhinoconjunctivitis, and asthma. Increased production of Th2 cells, which secrete interleukin (IL)-4, IL-5, and IL-13, favors IgE class switching leading to an increased production of IgE in response to antigens or allergens. IgE binds to high-affinity receptors (FcεRI) on mast cells (MCs) and basophils as well as to low-affinity receptors (FcεRII/CD23) notably, on B cells, activated macrophages, and eosinophils, among other cell types. Cross-linking of receptor bound IgE by the antigens triggers the release of pharmacologically active substances from MCs and basophils, including histamine, leukotrienes, and peptides attracting neutrophils and eosinophils. By contrast, delayed-type hypersensitivity (DTH) (Type 4 hypersensitivity), where the reaction takes 2 or 3 days to develop, is antibody-**independent** and mediated by T cells and monocytes/macrophages. CD4+ Th1 cells recognize antigen in a complex with MHC II major histocompatibility complex on the surface of antigen-presenting cells, such as dendritic cells (DCs) and macrophages. These latter secrete IL-12, which stimulates the further proliferation of CD4+ Th1 cells, which secrete IL-2 and IFN-γ, inducing the further release of other Th1 cytokines, thus mediating the immune response. DTH is a major mechanism of defense against various intracellular pathogens, including mycobacteria, fungi, and certain parasites. DTH also occurs in allergic contact dermatitis and in some autoimmune diseases such as multiple sclerosis and coeliac disease.

The induction of specific IgE in response to mosquito saliva has been well documented. Saliva contains pharmacologically active proteins and peptides (Ribeiro, 1987), which provoke a localized allergic reaction in the skin, involving the production of IgE and IgG antibodies (Brummer-Korvenkontio et al., 1994; Chen et al., 1998) and dermal hypersensitivity reactions (French and West, 1971; Reunala et al., 1994). Intradermal injections with *Aedes albopictus* salivary gland extracts (SGEs) in humans elicit both an immediate and a delayed response (Oka et al., 1989). Mosquito bites thus result in both immediate and delayed local cutaneous reactions (Peng et al., 1996; Reunala et al., 1991). While these broad categories of hypersensitivity immune responses are indeed pertinent to the immune response to saliva, recent work suggests that this response is more complex and that MCs play an important central role in both these two responses.

It is known that the ability of DCs to direct the development of naïve T cells into Th1, Th2, or regulatory T cells is largely dependent on the signals that they receive in the peripheral tissues at the time of antigen capture. Thymic stromal lymphopoietin (TSLP) is a master regulator of allergic inflammation in the skin (Esnault et al., 2008; Soumelis et al., 2002), is produced by epithelial cells (such as keratinocytes and MCs), and has an important role in conditioning DC maturation. The marked increase in TSLP expression in response to inflammation leads to macrophage activation, DC maturation, induction of inflammatory Th2 cells, and eventual chemoattraction of a suite of innate immune cells, such as eosinophils, neutrophils, and MCs, with resulting pathological effects. TSLP-activated DCs induce inflammatory Th2 cells that produce the classical Th2 cytokines IL-4, IL-5, and IL-13 and large amounts of TNF (Fig. 4.1). During the sensitization phase of the DTH response, DCs capture the antigen, migrate to draining lymph nodes, and undergo a maturation process required for the activation of naive T cells. In addition to TSLP, MCs also produce histamine and other inflammatory mediators that affect the maturation of adjacent DCs, which then fail to ultimately elicit fully activated effector T cells. Notably, the MCs induce the production of IL-10 via histamine (Elenkov et al., 1998). IL-10 is an important regulator of the DTH response (Ferguson et al., 1994), limiting the associated inflammation and tissue damage (Grimbaldeston et al., 2007). Absence of IL-10 leads to extended DTH and, conversely, high levels suppress the DTH reaction.

In addition to the well-recognized IgE-dependent activation of MCs, an IgE-independent activation of MCs was demonstrated; MC-dependent inflammatory responses at skin sites exposed to *Anopheles stephensi* mosquito bites were observed in naive mice, indicating that the saliva has the capacity to directly trigger MC activation in the absence of saliva-induced specific antibodies. *Anopheles* mosquito bites thus also result in a local **IgE-independent** degranulation of skin MCs (Demeure et al., 2005). The mosquito bite induced cellular infiltrate and hyperplasia of the draining lymph nodes with an increase in T lymphocytes, B lymphocytes, DCs, neutrophils, and monocytes/macrophages,

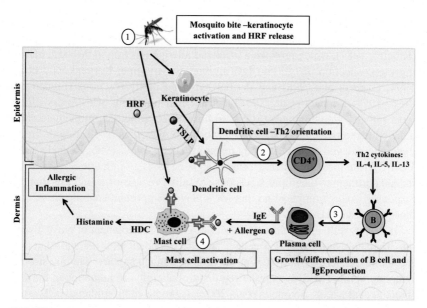

FIGURE 4.1 Key immunological pathways and effectors initiated in the immunological response to mosquito bites and subsequent allergic inflammation. *HDC*, histidine decarboxylase; *HRF*, histamine releasing factor; *TSLP*, thymic stromal lymphopoietin. (*Adapted from Paul, R.E., Sakuntabhai, A., 2016. Atopic dermatitis – need for a sub-Saharan perspective. European Medical Journal Allergy Immunology 1, 58–64.*)

in short, a classic DTH response. MCs are a source of TNF-α and macrophage inflammatory protein 2 (MIP-2), which are both mediators of neutrophil recruitment and T cell-mediated DTH (Biedermann et al., 2000; Malaviya et al., 1996). In a model of contact hypersensitivity, MIP-2 was found to be abundant but only in the presence of MCs and shown to be involved in DC migration (Biedermann et al., 2000; Enk et al., 1993). The early induction of MIP-2, following mosquito bites, was followed by production of IL-10 in the draining lymph nodes and subsequent downregulation of T cell-mediated immune responses mediated by IL-10 (Depinay et al., 2006). Among saliva components that could trigger skin MC activation, histamine releasing factor (HRF), a highly conserved protein encoded by all eukaryotic cells, including all *Plasmodium* parasite species with both intracellular and extracellular functions, was found to be present (personal observation).

The upregulation of IL-10 expression after exposure to mosquito saliva has been observed across a range of mosquito species and is thus a key generalized immune response (Depinay et al., 2006; Schneider and Higgs, 2008). IL-10 inhibits the synthesis of IFN-γ, IL-2, and TNF-β (Brady et al., 2003), antagonizes IL-12, downregulates MHC class II expression by monocytes, and inhibits antigen presentation by several antigen-presenting cells (Enk et al., 1993; Macatonia et al., 1993). Increased IL-10 production can thus induce lasting

T cell inactivation with clear consequences for the development of an effective immune response against any co-inoculated pathogens (Ejrnaes et al., 2006). IL-10 and other inflammatory mediators released by MCs following response to mosquito saliva likely lead to a dysfunctional DTH response and subsequent poorly effective antigen-specific T cell responses. This would have a significant impact on any antigens co-present at the time of saliva inoculation.

In conclusion, the immunological response to saliva leads to both Type 1 and Type 4 hypersensitivity reactions that are associated with an antibody-dependent IgE-mediated reaction and an antibody-independent T cell reaction that is abnegated by induction of anti-inflammatory cytokine IL-10. MCs are key to linking these two hypersensitivity responses. The consequences of this deviation toward a Th2 response on the one hand and a dysfunctional Th1 response on the other for the outcome of malaria parasite are now discussed.

Modulation of Malaria Parasite Infection Outcome by Saliva

It is recognized that the type of immune balance driven by the parasite operates at a very early-stage postparasite delivery. Early divergence in the immune response against malaria parasites may elevate or diminish the likelihood of progressing to the more severe forms of malaria (Mitchell et al., 2005). The response of sentinel cells, such as DCs, determines the evolution of the immune response and can lead to protection, tolerance, or immunopathology (Banchereau et al., 2000). Orientation of the immune response toward a Th1 profile is crucial for immunity to intracellular pathogens (Mosmann and Coffman, 1989), whereas orientation toward a Th2 profile drives immunity to extracellular pathogens and antigens resulting in class switching giving rise to IgE-producing B cells (Zhu and Paul, 2008). DCs that are oriented to a Th2 phenotype by an antigen are more susceptible to orient the immune response toward a Th2 profile when confronted by a second antigen (de Jong et al., 2002).

Mouse Models

In experimental models, across a wide range of insect vectors and their associated pathogens, insect saliva was found to enhance pathology, infection level, and the production of Th2 cytokines (IL-4 and IL-10) (Ockenfels et al., 2014). Conversely, prior exposure to insect bites decreases subsequent infection severity; repeated exposure to bites from uninfected insects eventually leads to the development of a Th1 response to salivary antigens and by consequence to the pathogen (Donovan et al., 2007; Kamhawi et al., 2000). Creation of a Th1-biased environment rather than a Th2-biased one is seemingly crucial in dictating the outcome of a subsequent infection (Rogers et al., 2006). Thus, saliva could orient the response mounted against the malaria parasite.

Inflammatory cells, such as DCs and monocytes, play an important role in anti-parasite activity (Pierrot et al., 2007; Waters et al., 1987) and any reduction in their recruitment and/or maturation undermines any effective anti-parasite response. The influx of key cells of the innate immune response at the site of sporozoite inoculation is significant in the early response to *Plasmodium* both for reducing the number of viable parasites that migrate to the liver and in initiating the adaptive immune response. Mosquito feeding in mouse models was shown to lead to elevated parasitemia and increased disease severity (cerebral malaria). These effects occur following dysregulation of immune signaling and a reduction in the recruitment of key inflammatory cells into the inoculation site (Schneider et al., 2011). The immediate consequences of such dysregulation can be inferred from the crucial anti-parasite role played by DCs in the cutaneous lymph nodes; it is here that the first cohort of T cells is primed by DCs after an infectious mosquito bite, and the protective anti-sporozoite CD8$^+$ T cell response originates predominantly from these cutaneous draining lymph nodes (Chakravarty et al., 2007).

With many pathogens, early cytokine responses involving IL-4 and IL-10 increase host susceptibility, whereas responses involving IL-12 and IFN-γ are important for resistance. The immunosuppressive role of IL-10 upregulated by saliva was shown to exacerbate the infection and disease; early IL-10 expression was associated with increased T regulatory cell proliferation, suppression of Th1 cytokines, as well as amplified parasitemia and mortality (Wu et al., 2007). Mosquito saliva also stimulates IL-4 expression with resulting increased severity of infection (Fonseca et al., 2007; Schneider and Higgs, 2008; Schneider et al., 2011). Thus both the Type 1 hypersensitivity response, as suggested by IL-4 expression plus the defective Type 4 hypersensitivity response abrogated by IL-10 contribute to increased infection severity and poor development of an effective immune response.

Mouse models of malaria have therefore generated increasing evidence that the immunomodulation of the Type 1 and Type 4 hypersensitivity responses by saliva creates an immunological environment that leads to exacerbation of disease and dysfunctional development of immunity. Mouse model studies have revealed much about the immunomodulatory role of saliva and its impact on the outcome of malaria parasite infection. Interestingly, despite using different species of mosquito and different parasite species, there are consistent effects suggesting that there exist generalizable phenomena potentially pertinent to human malaria. Extending from mouse models to natural infections in humans living in malaria endemic settings is a necessary but challenging step.

Human Malaria

The acquisition of immunity to *Plasmodium falciparum*, the etiological agent of human lethal tertian malaria, develops very slowly and is not sterilizing. Even in zones where the transmission intensity is high, children "only" develop a clinical immunity (Doolan et al., 2009), whereby they tolerate elevated parasite densities without showing clinical symptoms. Although clinical immunity is accompanied by a reduction in parasite density, effective anti-parasite immunity develops much more slowly (Marsh and Snow, 1997) with individuals achieving a state of premunition, whereby they maintain low-grade parasite densities in an asymptomatic state (Perignon and Druilhe, 1994). Although the immune effectors of clinical immunity are still poorly defined, there is strong evidence that acquired anti-parasite immunity is IgG-dependent (Cohen et al., 1961; Wilson and McGregor, 1973) and cytophilic immunoglobulins (IgG1 and IgG3), which are capable of eliminating the parasites by opsonization and/or by antibody-dependent cellular immunity, play an important role in premunition (Perignon and Druilhe, 1994).

In contrast to IgG, the immunological consequences of total and specific titers of IgE, characteristic of allergy, and a Th2 response have been subject to relatively little interest in the context of natural infections in endemic settings and have only recently considered the potential role of mosquito saliva. Individuals living in malaria endemic regions have elevated total and *P. falciparum* parasite-specific IgE levels, and these latter correlate very strongly with anti-SGE IgE titers (Lawaly et al., 2012). Several findings suggest that IgE could play a detrimental role during malaria disease development. Elevated levels are observed in severe acute clinical episodes and cerebral malaria as compared to uncomplicated malaria, suggesting a pathogenic role of IgE (Desowitz, 1989; Perlmann et al., 1997). Immunohistological studies on brain sections revealed the presence of IgE deposits in brain microvessels and on infected erythrocytes from cerebral malaria patients as well as in placentas infected with *P. falciparum* (Maeno et al., 2000). By contrast, although IgE levels were much reduced among patients with uncomplicated malaria in comparison to those suffering from severe malaria (Perlmann et al., 1999), high levels in asymptomatic infections are seemingly protective against subsequent clinical episodes (Bereczky et al., 2004). Interestingly, plasma anti-parasite IgE levels correlated with IL-10 titers in uncomplicated malaria patients (Duarte et al., 2007). Plasma IgE titers, however, do not necessarily equate to functional activity. Indeed, anti-parasite IgE titers per se were not found to correlate with disease severity, but functional activity, as revealed through ex vivo MC degranulation, did show a correlation: functional activity was higher in asymptomatic and uncomplicated malaria patients than in severe malaria groups (Duarte et al., 2007; Guiyedi et al., 2015).

While there does seem to be some evidence that a Th2 phenotype influences infection severity, the immunological significance of mosquito saliva may go beyond disease severity. In malaria endemic settings, there is considerable variation in the transmission intensity and the majority of infections occur in individuals not naïve to infection, who can carry parasites without showing symptoms for a considerable length of time and who are exposed continually to mosquito bites. If parasites have evolved an adaptive response to benefit from the immune-modulatory effects of saliva, more subtle effects on the outcome of infection might be expected.

Gametocyte Production, Recirculation, and Seasonality in Mosquito Biting

Gametocytes are specialized sexual parasite stages required for transmission from man to the mosquito. Gametocyte production is associated with non-specific immune responses occurring during febrile episodes of symptomatic infections and can be induced following the addition of lymphocytes from naturally infected Gambian children (Oesterholt et al., 2009). Parasites increase their conversion rate to gametocytes in individuals with acquired immunity (Drakeley et al., 2006), suggesting key roles for immunoglobulin titers. Hemoglobinopathies have been found to be associated with increased gametocyte production and infectiousness to mosquitoes (Gouagna et al., 2010; Grange et al., 2015), again suggesting that factors impeding the asexual proliferation of the parasite lead to conversion to sexual transmission stages.

In malaria endemic areas of seasonal transmission, individuals can carry *P. falciparum* parasites at very low often sub-patent densities without symptoms for the duration of the non-transmission season. Such infections are the reservoir of infection from which mosquitoes become infected following their re-appearance once the seasonal rains arrive. Seasonal cues triggering the production of transmission stages have been observed in other apicomplexa in birds (Atkinson and van Riper, 1991; Weatherhead and Bennett, 1991) as well as in the classical vernal relapses of *Plasmodium vivax* malaria (Swellengrebel and De Buck, 1938). In malaria endemic regions of seasonal mosquito activity, chronic infection parasites may respond to the effects of anopheline bites by producing gametocytes to transmit rapidly after the expansion of the anopheline population (Paul et al., 2004).

Following the seasonal expansion of the mosquito population with the rains, mosquito bites were found to be strongly positively associated with an increase in parasite density in chronic pre-existing asymptomatic infections (Lawaly et al., 2012). This seasonal trend has been noted before in very different settings in Liberia, Nigeria, and Thailand (Barber and Olinger, 1931; Miller, 1958; Rosenberg et al., 1990). However, there was a strong negative impact of mosquito bites on the production of gametocytes, suggesting parasites were investing preferentially in asexual rather than gametocyte stages. Individual anti-mosquito SGE IgE titer was also found to be strongly positively correlated with anti-parasite IgE titer. This is consistent with the hypothesis that mosquito bites pre-dispose individuals to develop an IgE anti-parasite response, potentially by orientation of the immune response to a Th2 profile (Zeidner et al., 1999). Such an orientation of the immune response would be expected to lead to a reduced Th1 type environment resulting in a lower acquisition of asexual parasite-targeting effectors and thus a more fertile ground for asexual parasite survival. Although IgG3 anti-parasite titers did not impact on gametocyte prevalence or parasite density, they increased with age, which itself had a significant negative impact on parasite density. Interestingly, IgG3 anti-parasite titers were negatively correlated with the seasonal decrease in IgE anti-parasite titers. Such a seasonal decrease might be indicative of exhaustion of circulating IgE, potentially being bound to effector cells. High levels of antiparasite IgG3 might interfere with this and thus abnegate the role of anti-parasite IgE. This clearly suggests the possibility of competitive interference of antiparasite IgE by antiparasite IgG3, with potential consequences on the parasite infection outcome.

Parasites need to produce gametocytes to transmit to mosquitoes and they are generated from the circulating asexual parasite population. Consequently, parasites are faced with a trade-off between, on the one hand, producing sufficient asexual parasites to maintain an effective population size to withstand immunological destruction, and, on the other, generating sufficient gametocytes to be able to transmit. Chronic infection parasites persist at very low densities, often only detectable by polymerase chain reaction. The parasite must therefore generate gametocytes from a very low-density asexual population. Accelerated parasite replication following anopheline mosquito bites would provide parasites with a sufficient biomass to generate gametocytes at high enough densities to ensure transmission. Investment in asexual stages would thus be at a cost to gametocyte production, hence the observed negative impact of parasite density on gametocyte prevalence. Interestingly, bird malaria model experiments revealed the same increase in asexual parasite density in chronic infections following uninfected mosquito bites and although gametocyte density was unchanged there was an increase in subsequent transmission rates to mosquitoes (Cornet et al., 2014). Despite the very differing immune systems of birds and humans and the different parasite life cycles, the similarity in the detectable effects of mosquito bites on parasite replication does suggest that parasites do respond to the immunological consequences of mosquito bites and that transmission and not just gametocyte density offers a finer measure of this. Gametocytes are known to be sequestered in the skin (Chardome and Janssen, 1952), often at higher densities than in the circulating peripheral blood and thus may offer a source of gametocytes not measurable in the blood and which will be localized in the skin where the immunological effects of saliva are immediate.

Clinical and Parasitological Outcome of Infection

The continual underlying thesis is that saliva generates an imbalance in the Th1/Th2 response, inducing an IgE response and a dysfunctional Th1 response. This is reminiscent of allergy and the relationship between allergy and malaria has been the subject of few studies and generally in the context of the hygiene hypothesis (from the perspective of the impact of malaria on atopy) (Haileamlak et al., 2005; Lell et al., 2001). In this context, in contrast to attempting to assess directly the immunological impact of saliva on infection severity in the complex conditions of endemic settings, the impact of an imbalanced Th1/Th2 terrain was explored. Such a Th1/Th2 imbalance is characteristic of atopy and thus atopic individuals might be expected to respond differentially to mosquito bites, parasitic infection, and the immune-modulatory role of saliva. Orientation of the immune response toward a Th2 profile by allergic diseases such as asthma or AD would result in a poor Th1 response and thus amplify the effects of saliva and hence the immunological response to infection.

In a birth cohort of children living in malaria endemic settings, there was an association of asthma and AD with susceptibility to clinical *P. falciparum* episodes (Herrant et al., 2013). This allergy-associated risk of malaria occurred after the age at which premunition occurred, suggesting that allergy delayed the development of clinical immunity to malaria. In particular, children with clinically defined asthma and especially AD had increased risk of presenting with *P. falciparum* malaria episodes beyond the average age of premunition and also had higher parasite density during clinical episodes, suggesting a reduced ability to control parasite replication. The higher parasite density during symptomatic episodes observed in the asthma group suggests impaired development of acquired immunity. Impaired acquisition of immunity to malaria in children with asthma or AD may stem from their imbalanced Th1/Th2 response. Interestingly, only mosquito saliva, a known major local allergen, was found to be a significant risk factor of AD, inducing a specific IgE response at significantly higher titers in individuals with AD. In light of the strong positive correlation between saliva and parasite IgE titers, this result strongly suggests that a Th2 IgE environment is indeed impairing control of the parasite and undermining the development of anti-parasite immunity.

In conclusion, the early response of sentinel cells, such as DCs and MCs, determines the evolution of the immune response. Saliva provokes a localized allergic reaction in the skin and induces the production of IgE and IgG antibodies. DCs that are primed by saliva to elicit a Th2 phenotype are more susceptible to orienting the immune response toward a Th2 profile when confronted by a second antigen. The orientation of the immune response toward a Th1 profile is crucial for immunity to intracellular pathogens, whereas orientation toward a Th2 profile drives immunity to extracellular pathogens and antigens, resulting in class switching, giving rise to IgE-producing B cells. Anti-saliva and parasite IgE titers were positively correlated with asexual parasite densities but negatively correlated with gametocyte (transmission stage) parasites. Anti-saliva IgE titers were found to be strongly associated with the occurrence of AD, which was found to reduce the rate of development of clinical immunity in a birth cohort study. Thus, an atopic Th2 terrain, exacerbated by mosquito bites, influences the course of a single parasite infection and the long-term ability to develop immunity against the parasite.

SHARED PATHWAYS AND ASSOCIATIONS OF ALLERGY WITH MALARIA PATHOGENESIS (FIG. 4.2)

Several studies conducted in humans suggest a strong relationship between clinical susceptibility to malaria and severe allergic-type responses (Demeure et al., 2005; Griffiths et al., 2005; Sakuntabhai et al., 2008). Indeed, the contribution of host genetic factors to the risk of severe outcome was highlighted in different studies during *P. falciparum* infection in African populations. Genetic evidence on the overlap of these pathologies with respect to a Th1/Th2 imbalance comes

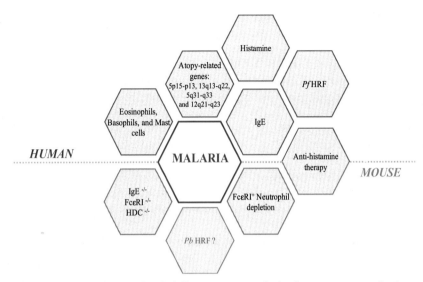

FIGURE 4.2   Malaria and allergic inflammatory cascade do share common mechanisms.
Several pieces of evidence, obtained in human and malaria experimental models studies, support
the concept that clinical susceptibility to malaria may be related to allergic-associated inflammatory
response.

from genetic studies (Linkage and genome-wide association studies [GWAS]).
GWAS studies have increasingly found that the same loci are often found to be
associated with several diseases (Hindorff et al., 2009), suggesting that genes
with a pleiotropic effect occur frequently and play a key role in basic physiopathological mechanisms underlying a number of diseases. This has been extensively found in genetic studies of malaria susceptibility in mouse models and
humans. A mouse model for human atopic disease (NC/Jic) was found to be susceptible to murine malaria and a major quantitative trait locus (derm1) for atopic
disease mapped close to the region controlling parasitemia (char1 or pymr) on
mouse chromosome 9 (Kohara et al., 2001; Ohno et al., 2001). Likewise, several
linkage and association studies of malaria in human populations have repeatedly
highlighted the association of a Th2 phenotype with disease severity. In particular, genome-wide linkage analyses have identified several linkage regions
(5p15-p13, 13q13-q22, 5q31-q33, and 12q21-q23) all of which overlap with
those that have been previously identified to be involved in asthma/atopy and
especially IgE levels (Friberg et al., 2006; Jang et al., 2005). Regions 13q13-q22
and 12q21-q23 contain genes known to increase total serum IgE levels, namely
PHF11 and *STAT6* genes (Chen and Khurana Hershey, 2007; Jang et al., 2005).
The Stat6 protein plays a central role in exerting IL-4-mediated responses. The
IL4 gene is located on 5q31 together with genes responsible for expression of
the cytokines IL-5 and IL-13 that are major cytokines promoting the differentiation of Th2 cells involved in the allergic response. This region was found to be

linked to parasite density (Garcia et al., 1998; Rihet et al., 1998; Sakuntabhai et al., 2008) and consistently reported to be linked to asthma-related phenotypes (Bouzigon et al., 2010). Genetic variants in *IL4* and *IL13* genes and *STAT6* have been consistently found to be associated with total IgE levels (Holloway et al., 2010; Weidinger et al., 2008). Of particular interest is the 5p13 linkage region; this includes several genes involved in innate immunity and notably, the interleukin 7 receptor (IL7R). IL7R plays a crucial role in signal transduction of TSLP, increased expression of which has been strongly associated with AD as well as other allergic diseases, including asthma and allergic rhinitis (Esnault et al., 2008; Soumelis et al., 2002) (Fig. 4.1). Additionally, skin immunobiology and specifically the Type 1 hypersensitivity response has been extensively studied with respect to AD and which may throw some light on the allergenic nature of mosquito saliva and its immune-modulatory capacity in an IgE-dependent context.

In humans, in addition to genetic analyses, several other lines of evidence support the concept that susceptibility to malaria and atopy may be related to the same immunological defect and/or influence one another. Elevated plasma levels of IgE and anti-plasmodial IgE Abs that bind to basophils and MCs have been associated with severity to *P. falciparum* infections in adults and children living in malaria endemic areas in different African and Asian countries (Perlmann et al., 1994, 1999; Troye-Blomberg et al., 1999). Indeed, IgE levels were reduced among patients with uncomplicated malaria in comparison to those suffering from severe malaria (Seka-Seka et al., 2004) and high IgE levels were found during cerebral *P. falciparum* malaria related to the deepness of the coma compared to uncomplicated malaria (Luty et al., 1994; Maeno et al., 2000). During the allergic inflammatory response, the IgE Abs bind to the high-affinity receptor FcεRI, and the formation of the antigen–IgE–FcεRI complex activates the cells and stimulates the release of inflammatory mediators such as histamine. Activation of MCs (Furuta et al., 2006) may exacerbate malaria pathogenesis through histamine secretion. Increased levels of histamine in plasma and tissue, derived from basophils and MCs, were correlated with systemic *P. falciparum* disease complications within infected children and in animal models (Bhattacharya et al., 1988; Enwonwu et al., 2000; Srichaikul et al., 1976). Chlorpheniramine, a H1R antagonist reversed resistance to chloroquine and amodiaquine both *in vivo* and *in vitro* (Sowunmi et al., 2007). Moreover, astemizole, another H1R antagonist, was identified as an anti-malarial agent in a clinical drug library screen (Chong et al., 2006).

In experimental models, a major role of the allergic cascade including IgE, FcεRI, and histamine release in the development of the severe forms of malaria disease has also been demonstrated. Mice deficient in histamine due to disruption of the histidine decarboxylase gene (HDC$^{-/-}$) or treated with antihistamines or deficient for the H1 and H2 receptors of histamine are more resistant to experimental cerebral malaria, suggesting a harmful role of histamine pathway signaling through these receptors (Beghdadi et al., 2008). The involvement of histamine in the severity of the disease was confirmed by the absence of neuro-pathology observed in histamine-deficient mice. This resistance is characterized by the preservation of the integrity of the blood–brain barrier with a decrease in the expression of ICAM-1 by endothelial cells and the absence of sequestration of infected red blood cells or adhesion of leukocytes to the cerebral blood vessels (Beghdadi et al., 2008). In addition to the impact of IgE through binding to the high-affinity receptors of MCs and basophils, recent studies have implicated an active role of a neutrophil sub-population expressing induced α chain of FcεRI. FcεRIα is essentially absent in neutrophils in naive mice but becomes induced during *Plasmodium* infection. Targeted disruption of IgE or the α chain of FcεRI led to resistance to the development of cerebral malaria (Porcherie et al., 2011). This effect occurred in MC deficient and basophil-depleted mice, and a neutrophil sub-population was identified as the effector cell (Porcherie et al., 2011): Transfer of FcεRI^{+2}, but not FcεRI^{-2}, neutrophils conferred susceptibility to cerebral malaria in otherwise resistant FcεRI-α-KO mice (Porcherie et al., 2011). This sub-population induced on malaria infection is reminiscent of the specific expression of FcεRI, as a trimeric FcεRIαγ2 receptor, in monocytes, eosinophils, and neutrophils of atopic asthmatics, with associated asthma-related pathology (Saffar et al., 2007). Altogether, these findings demonstrate that components of the allergic cascade including IgE, FcεRI, histamine, and the newly identified FcεRI-positive neutrophil population, are intimately associated with severe forms of malaria disease. This may have practical consequences on the management of malaria disease by introducing novel anti-allergic therapeutic approaches used individually or in combination with classical antimalarial chemotherapy.

Several studies have documented the importance of an alternative mechanism, independent of the classical pathway implicating the IgE–FcεRI complex during histamine release, where cells are directly activated by a *Plasmodium*-derived protein called HRF shown to be implicated in the release of pro-inflammatory histamine during late-phase allergy and malaria infection (MacDonald et al., 2001). Additionally, high levels of *Pf*HRF have been detected in the serum of mildly and severely *P. falciparum*-infected Malawian children, suggesting that *Plasmodium* HRF may affect host immune responses and contribute to the pathogenesis (Janse et al., 2006; MacDonald et al., 2001). Recently, in experimental models using two HRF-deficient (*hrfΔ*) murine parasites (*Pb*ANKA and *Pb*NK65), the importance of HRF in enhancing the virulence of the parasite was demonstrated. Indeed, infection with *Pb*ANKA-*hrfΔ* sporozoites led to a decrease in the frequency of cerebral malaria due to the impairment of the development of the mutant parasites in liver stages as a consequence of the upregulation of IL-6. Infection with *Pb*NK65-*hrfΔ* resulted in parasite clearance leading to a long-lasting protection and immune memory as reflected by an upregulation of IL-6 and the enhancement of antibody-mediated phagocytosis (Demarta-Gatsi et al., 2016; Mathieu et al., 2015).

CONCLUDING REMARKS

An increasing wealth of data from a broad spectrum of sources is highlighting the importance of an allergic-type immune response for the outcome of a malaria infection and that the immune-modulatory capacity of saliva, not least at the site of the skin bite, plays a major role. The extent to which the malaria parasite is itself contributing to this allergic immunological environment is unclear, but it is clearly benefiting from it. Further exploration into the extent to which parasites contribute directly to such an allergic-type response would be of significant interest and there is clearly a case for considering malaria in the context of allergic diseases.

ROLE OF SAND FLY SALIVA ON *LEISHMANIA* INFECTION AND THE POTENTIAL OF VECTOR SALIVARY PROTEINS AS VACCINES

Shaden Kamhawi[1], Jesus G. Valenzuela[1], Iliano V. Coutinho-Abreu[1] and Cláudia Ida Brodskyn[2]
[1]*National Institute of Allergy and Infectious Diseases, National Institutes of Health;*
[2]*Instituto Gonçalo Moniz, FIOCRUZ-Bahia, Salvador, Brazil*

INTRODUCTION

Leishmaniasis is a neglected tropical disease, and around 350 million people are at risk of infection with 2 million becoming infected per year, mainly in the poorest communities (WHO/Leishmaniasis, 2014). Clinical manifestations of infection range from cutaneous ulcers to a visceral form, considered one of the most severe and fatal if not treated (Desjeux, 2004).

All forms of leishmaniasis are transmitted by the bite of infected phlebotomine sand flies, Phlebotomus for Old World and Lutzomyia for the New World. As infected females feed on mammalian hosts, they inject saliva, which contains different proteins shown to inhibit hemostatic system and facilitate their blood feeding. Saliva of sand flies was shown to contain a vasodilator (Ribeiro et al., 1989, 1999; Ribeiro and Modi, 2001) that prevents vasoconstriction, an apyrase that works as an inhibitor of platelet aggregation by destroying the platelet agonist adenosine diphosphate (Charlab et al., 1999; Hamasaki et al., 2009; Valenzuela et al., 2001b), an inhibitor of the blood coagulation cascade (Charlab et al., 1999) and an inhibitor of the classical pathway of the complement cascade (Cavalcante et al., 2003). These biological activities inhibit the efficiency and redundancy of the hemostatic system consequently facilitating the acquisition of a blood meal by sand flies.

Considering that about 30 phlebotomine species transmit leishmaniasis (Desjeux, 2004), having only partial information on salivary proteins present in *Phlebotomus papatasi* and *Lutzomyia longipalpis* (Charlab et al., 1999;

Valenzuela et al., 2001b) till the beginning of the century reflected the difficulty of purifying and characterizing the different proteins present in saliva. At this point, transcriptomics, a new technology impacted the vector biology field, enabling salivary glands transcriptomes from different blood-feeding arthropods including the salivary gland transcriptome of *Lu. longipalpis* to be completed (Ribeiro and Francischetti, 2003; Valenzuela et al., 2004). This was followed by the completion of salivary gland transcriptomes from several relevant sand fly vectors including *Phlebotomus ariasi* (Oliveira et al., 2006), *Phlebotomus perniciosus* and *Phlebotomus argentipes* (Anderson et al., 2006), *Phlebotomus duboscqi* (Kato et al., 2013), *Phlebotomus arabicus* (Hostomska et al., 2009), *Phlebotomus tobbi* and *Phlebotomus sergenti* (Rohousova et al., 2012), *P. papatasi* (Abdeladhim et al., 2012), *Lutzomyia ayacuchensis* (Kato et al., 2013), *Lutzomyia intermedia* (de Moura et al., 2013), and *Phlebotomus orientalis* (Vlkova et al., 2014).

Most salivary proteins are soluble and only a few nanograms or femtograms of each molecule are injected into the skin of the host. *Leishmania*-infected sand flies regurgitate parasites together with the salivary proteins into the bite wound and release different proteins with immunomodulatory properties, which facilitate the establishment of the infection by *Leishmania*. Conversely, distinct salivary proteins elicit a robust adaptive immune response that adversely affects *Leishmania* establishment. Herein, we will review the immunomodulatory properties of saliva or salivary proteins and consider their contribution to the establishment of *Leishmania* infection. We will also discuss the prospect of using them as potential candidates for vaccines against *leishmania*sis.

Immunomodulatory Properties of Salivary Proteins

The exact role of most of these immunomodulatory components in blood feeding is not well understood and needs to be further elucidated. Nevertheless, their potential use in biomedicine makes them an attractive target as therapeutic molecules for inflammatory diseases. For leishmaniasis, these immunomodulatory activities may facilitate parasite transmission and its establishment.

In 1988, Titus and Ribeiro described for the first time that the injection of *Leishmania major* in the presence of homogenized of salivary glands (salivary gland sonicate—SGS) from *Leishmania longipalpis* resulted in lesions that were routinely 5–10 times as large and contained as much as 5000 times as many parasites as controls. With inocula consisting of low numbers of *L. major*, parasites were detected at the site of injection only when the inoculum also contained *L. longipalpis* salivary gland material (Titus and Ribeiro, 1988). Using a different species of *Leishmania* (*L. braziliensis*), Samuelson et al. (1991) observed the same effect after injecting parasites with saliva from *L. longipalpis* into the footpad of BALB/c mice.

It was also shown that *L. longipalpis* saliva induced lipid body formation in murine macrophages *in vitro* and *in vivo* and these lipid bodies were linked with the production of prostaglandin E2 (Araujo-Santos et al., 2010), a molecule that could act on parasite dissemination. Human dendritic cells (DC), neutrophils, and monocytes were also affected by incubation with *L. longipalpis* saliva. *L. longipalpis* saliva induced apoptosis of neutrophils leading to an increase in parasite burden (Prates et al., 2011), altered the expression of co-stimulatory molecules in DC, macrophages, and monocytes (Costa et al., 2004) and down-regulated the production of TNF and IL-12p40 in lipopolysaccharide-stimulated monocytes (Costa et al., 2004). Further, the incubation of *P. papatasi, P. sergenti*, or *L. longipalpis* saliva reduced proliferation of mitogen-activated murine splenocytes and inhibited the production of the Th1 cytokine IFN-γ (Rohousova et al., 2005). Maxadilan and *P. papatasi* SGS decreased IFN-γ and IL-12p40 production by human peripheral blood mononuclear cells (PBMCs) and increased IL-6 secretion *in vitro* (Rogers and Titus, 2003).

Identification of Different Sand Fly Saliva

Various molecules have been identified from salivary glands of different sand fly species.

Maxadilan

It is a potent vasodilator, was the first molecule characterized from sand fly saliva (Lerner et al., 1991; Ribeiro et al., 1989). It inhibits proliferation of mouse lymphocytes *in vitro* (Qureshi et al., 1996) and decreases TNF production *in vivo* (Bozza et al., 1998) and in lipopolysaccharide-treated mouse macrophages (Soares et al., 1998). Importantly, maxadilan alone exacerbated *L. major* infection (Morris et al., 2001) due to its capacity to upregulate the production of IL-10 and TGF-β and to suppress IL-12p40, TNF, and NO production (Brodie et al., 2007). SGS from *P. duboscqi*, a vector for *L. major*, was chemoattractive for mouse monocytes (Anjili et al., 1995), whereas SGS from *P. papatasi*, another vector of *L. major*, inhibited macrophage activation by IFN-γ (Hall and Titus, 1995) and downregulated inducible nitric oxide synthase (iNOS) expression, thereby reducing NO production in murine macrophages (Waitumbi and Warburg, 1998).

Lundep: LJL138

More recently, other biological activities of sand fly saliva have been reported and the proteins responsible for them have been identified (Abdeladhim et al., 2014). LJL138 is a salivary protein present in the transcriptome of *Lu. longipalpis* and encodes an endonuclease. This protein, now called Lundep, was recently established as a potent endonuclease (Chagas et al., 2014). The DNase activity of Lundep may contribute to the anti-thrombotic and anti-inflammatory functions of *Lu. longipalpis* saliva by hydrolyzing the DNA of neutrophil extracellular traps (NETs) at the bite site. As an endonuclease, this activity had a direct impact on the development of *Leishmania* parasites. *Leishmania* parasites evade killing by neutrophils either by blocking the oxidative burst and entering a non-lytic compartment unable to fuse with lysosomes or by resisting the leishmanicidal activity of parasite-induced NETs (Gabriel et al., 2010). Chagas et al. (2014) demonstrated that Lundep can effectively facilitate the survival of *L. major* parasites. Survival of *Leishmania* in neutrophils has been reported as a mechanism for silent uptake by macrophages favoring establishment of infections (Peters et al., 2008).

Nucleosides

Besides proteins, sand fly saliva also contains nucleosides. *P. papatasi* saliva includes large amounts of adenosine and 5′-AMP. These components are known as anti-inflammatory mediators, downregulating TNF and IL-12p40 and increasing IL-10 production by DC in an experimental model of collagen-induced arthritis (Carregaro et al., 2011). They also block the antigen presentation by DC and interfere with Th17 activation, leading to a suppression of the inflammatory immune response (Carregaro et al., 2011).

Yellow Proteins

For many years the function of yellow proteins in sand flies, namely LJM11, LJM17, and LJM111, was unknown. Yellow proteins are the most abundant families of salivary proteins and they are encountered in all sand fly transcriptomes. Recently, it was shown that all three members of the yellow family of proteins from saliva of the sand fly *Lu. longipalpis* bind biogenic amines, including serotonin, catecholamines, and histamine (Xu et al., 2011). Biogenic amines are a group of pro-hemostatic and pro-inflammatory mediators that potentially obstruct feeding. In addition to binding biogenic amines (Xu et al., 2011), LJM111 was shown to have another function. LJM111 was characterized as a potent anti-inflammatory molecule. Recombinant LJM111 was shown to inhibit IL-17, TNF, and IFN-γ production by leukocytes obtained from lymph nodes after *in vitro* stimulation with methylated bovine serum albumin (mBSA). LJM111 also inhibited neutrophil migration following mBSA-challenge of Ovalbumin (OVA)-immunized mice used as a model of OVA-induced neutrophil migration (Grespan et al., 2012). Furthermore, LJM111 reduced the hypernociception (pain) in a model of arthritis and inhibited the production of pro-inflammatory molecules consequently reducing *in vivo* neutrophil recruitment (Grespan et al., 2012). DCs that immigrate to inflamed joints produce proinflammatory mediators that lead to expansion and differentiation of Th1 and/or Th17 cells, which play a pathologic role in arthritis (Carregaro et al., 2011). LJM111 also affected the maturation of DCs leading to increased IL-10 production and reduced synthesis of TNF (Grespan et al., 2012).

Lufaxin: LJL143

Only recently, after expression and testing of a number of recombinant salivary proteins from the *Lu. longipalpis* salivary gland transcriptome was LJL143 identified as the protein responsible for the anticoagulant activity in saliva. This 32 kDa protein was consequently renamed Lufaxin (*Lutzomyia* Factor Xa (FXa) inhibitor) (Collin et al., 2012). Importantly, homologues of the transcript coding for this protein are present in all sand flies studied to date, suggesting that these Lufaxin-like proteins are the putative anticoagulants of all New world and Old world sand flies (Abdeladhim et al., 2012; Anderson et al., 2006; de Moura et al., 2013; Hostomska et al., 2009; Kato et al., 2006, 2013; Oliveira et al., 2006; Valenzuela et al., 2004; Vlkova et al., 2014). Importantly, Lufaxin was also shown to interfere with inflammation. FXa activates receptors PAR1 or PAR2 (protease-activated receptors) in different cell types enabling and promoting inflammation and immune modulation (Collin et al., 2012). Lufaxin acts on PAR receptors inhibiting the activity of FXa and attenuating inflammation leading to the prevention of arterial thrombosis in a mouse model (Collin et al., 2012).

SP15

The SP15-like odorant binding family of proteins are contact activation inhibitors and are also present in salivary glands of some sand flies, mainly in *Phlebotomus* species (Alvarenga et al., 2013). Recently, PdSP15 from saliva of the sand fly *P. duboscqi*, a member of this family of proteins, was shown to inhibit the activation of FXII and FXI and the cleavage of FXI by FXIIa or thrombin (Alvarenga et al., 2013). It is also responsible for inhibiting other anionic surface-mediated reactions (Alvarenga et al., 2013). This family of proteins may prevent downstream effects of mast cell activation caused by the bite of the sand fly including the formation of bradykinin, an inducer of pain.

Salivary Anticomplement of *Lutzomyia longipalpis*

This protein described recently in *Lu. longipaplpis* saliva was identified as a classical pathway complement inhibitor. Salivary anticomplement of *Lu. longipalpis* (SALO), an 11 kDa protein, has no homology to proteins of any other organism apart from New World sand flies. Both rSALO and SGS inhibited C4b deposition and cleavage of C4 (Ferreira et al., 2016). rSALO, however, neither inhibit the protease activity of C1s nor the enzymatic activity of FXa, uPA, thrombin, kallikrein, trypsin, and plasmin. Importantly, rSALO did not inhibit the alternative or the lectin pathway of complement (Ferreira et al., 2016).

Hyaluronidase

This enzyme activity was originally described in saliva of the sand fly *Lu. longipalpis* (Charlab et al., 1999) and was later shown to be present in the saliva of Old World sand flies (Cerna et al., 2002; Rohousova et al., 2012; Vlkova et al., 2014; Volfova et al., 2008). The salivary hyaluronidase plays an important role in blood meal acquisition by degrading hyaluronan (HA), abundant in host skin, and probably increases tissue permeability for other salivary components

(Volfova et al., 2008). Using commercially available hyaluronidase, Volfova et al. (2008) demonstrated that the activity of this enzyme facilitated transmission and establishment of *L. major* parasites. Importantly, fragments of HA were shown to have immunomodulatory properties; they affect maturation and migration of DC, induction of iNOS, chemokine secretion by macrophages, and proliferation of activated T cells (Volfova et al., 2008).

IN VIVO MODEL STUDIES AND IMPACT OF SAND FLY SALIVA

In *in vivo* studies done by Belkaid et al. (1998), a model of infection was developed in which co-inoculation of mice with *L. major* parasites plus *P. papatasi* SGS transformed C57BL/6 mice—naturally resistant to *L. major* infection—into a non-healing phenotype. This fact was associated with an increase in Th2 cytokines such as IL-4 and IL-5. cytometric bead array mice co-inoculated with *L. major* parasites and *P. papatasi* SGS also increased the expression of IL-4 and diminished the production of IFN-γ, IL-12, and iNOS favoring parasite proliferation (Mbow et al., 1998). Following these observations, a series of studies explored mechanisms involving enhanced *Leishmania* infection in the presence of sand fly SGS: on co-inoculation of *L. amazonensis* plus *L. longipalpis* SGS, larger lesions developed and were associated with elevated IL-10 production by draining lymph node cells stimulated *in vitro* with SGS (Norsworthy et al., 2004). IL-10 suppresses effector functions of monocytes and macrophages and NO and H_2O_2 production (Moore et al., 2001). Moreover, *L. longipalpis* SGS recruited macrophages *in vitro*, promoting parasite survival and proliferation (Zer et al., 2001), dependent on production of CCL2 (Teixeira et al., 2005). This observation was later confirmed *in vivo* following exposure of mice to *L. longipalpis* bites (Silva et al., 2005) or stimulation of the peritoneal cavity with *L. major* plus *L. longipalpis* SGS (Monteiro et al., 2007), which also resulted in IL-10 production.

All the above described properties raised the idea that salivary sand fly proteins could be used as vaccines since the immune response directed against them would neutralize their effects. Alternately, mounting a Th1 response against salivary proteins would create an inhospitable environment for *Leishmania*, not allowing the establishment of infection.

Salivary Proteins as Potential Candidates for Vaccines Against *Leishmania* Infections

There are several reports in the literature demonstrating that exposure to saliva induces an immune response in different species of animals that confers protection against leishmaniasis. The first line of evidence was obtained when animals (mice) immunized with *P. papatasi* saliva (Belkaid et al., 1998) or previously exposed to bites of uninfected *P. papatasi* (Kamhawi et al., 2000) were protected against *L. major* infection initiated by needle challenge or infected sand flies, respectively. Protection from leishmaniasis conferred by previous exposure to sand fly saliva or to bites was later reported for other sand fly and parasite species (Teixeira et al., 2014). The protective effect of sand fly saliva has been attributed to either the neutralization of the exacerbate effect of saliva (Belkaid et al., 1998; Morris et al., 2001) or to the induction of a Th1 cellular immune response that is detrimental for the establishment of the *Leishmania* parasite (Kamhawi et al., 2000).

Although the majority of studies reproduced the protective effect of previous exposure to saliva, for the New World sand fly *Lu. intermedia* the opposite effect was observed where exposure to its saliva resulted in exacerbation of the disease (de Moura et al., 2013; de Moura et al., 2010; Weinkopff et al., 2014). Mice pre-exposed to *Lu. intermedia* salivary gland extract (SGE) and challenged with *L. braziliensis* plus SGE showed a significant decrease in CXCL10 expression paralleled by an increase in IL-10 expression (de Moura et al., 2010). More recently, Carvalho et al. (2015) showed that individuals living in an endemic area for *L. braziliensis* in Brazil, who presented antibodies against *Lu. intermedia*, also displayed higher levels of IL-10 when their PBMCs were stimulated with vector saliva, pointing out the anti-inflammatory effect of previous exposure to saliva of this specific vector. Interestingly, exposure of mice to saliva from *Lutzomyia whitmani*, another vector responsible for transmitting *L. braziliensis*, induced protection against an infection caused by *L. braziliensis* plus *Lu. whitmani* saliva (Gomes et al., 2016).

Trials of Vaccinations in Animal Models

Different salivary proteins from different species of sand flies have been tested as potential candidates for vaccines against cutaneous and visceral *leishmaniasis* caused by various *Leishmania* species (Gomes and Oliveira, 2012).

PpSP15

It was the first salivary protein to be identified as a potential *Leishmania* vaccine (Oliveira et al., 2008; Valenzuela et al., 2001b). PpSP15 was identified by SDS–PAGE (sodium dodecyl sulfate–polyacrylamide gel electrophoresis) separation of salivary gland proteins from *P. papatasi* (Valenzuela et al., 2001a). The proteins were grouped into three fractions and only the one containing PpSP15 protected mice against *L. major* infection. Immunization of mice with PpSP15 plasmid induced a delayed-type hypersensitivity response (DTH) that was correlated to protection (Valenzuela et al., 2001a). The authors also demonstrated that PpSP15-specific protection was cell-mediated and antibody-independent. This important study established that DTH-induction in immunized animals is the hallmark biomarker of a protective response induced by salivary proteins against *Leishmania* parasites. Oliveira et al. (2008) validated the protective effect of PpSP15 against *L. major* in mice. Furthermore, in addition to the induction of a DTH, PpSP15-immunized mice expressed IFN-γ and IL-12 2 h after bites and showed an accelerated development of a *Leishmania*-specific immunity (Oliveira et al., 2008). Importantly, the DTH-inducing salivary protein PpSP44 produced a Th2 response in immunized mice that exacerbated *L. major* infections (Oliveira et al., 2008). Therefore, only Th1-DTH-inducing salivary proteins were considered as vaccine candidates for *leishmania*sis (Fig. 4.3).

Maxadilan

Morris et al. (2001) demonstrated that mice vaccinated with maxadilan, the vasodilator in *Lu. longipalpis* saliva, developed both cellular immunity and antibodies against this salivary protein that protected animals against *L. major* infection. The authors proposed that maxadilan had an exacerbatory effect on *L. major* infection that was neutralized by protective anti-maxadilan antibodies (Morris et al., 2001). Maxadilan, a protein of 6.8 kDa, is absent from *Phlebotomus* species and has only been described from some *Lutzomyia* vectors to date (de Moura et al., 2013; Valenzuela et al., 2004).

Based on information from the salivary transcriptomes, plasmids encoding the most abundant secreted salivary proteins were selected for screening. Classical vaccinology was used to identify salivary vaccine candidates in rodents where vaccination of groups of mice or hamsters with different vaccine candidates is feasible. Using this experimental approach, it was demonstrated that LJM19, a 10.7 kDa protein from saliva of *Lu. longipalpis* induces a strong DTH in immunized hamsters challenged with *L. longipalpis* saliva (Gomes et al., 2008), supporting the data from *P. papatasi* saliva. In addition, LJM19-immunized hamsters challenged with *L. infantum* and saliva of *Lu. longipalpis* were protected from death due to visceral *leishmania*sis, and this immune response was characterized by an increase in IFN-γ production and inhibition of IL-10 and TGF-β (Gomes et al., 2008). Immunization of hamsters with LJM19 plasmid followed by challenge with *L. braziliensis* in the presence of *Lu. intermedia* saliva, the vector responsible for the transmission of this parasite presented smaller cutaneous lesions compared to the non-immunized controls (Tavares et al., 2011).

LJM11

This 43 kDa salivary protein from *Lu. longipalpis* that belongs to the yellow family of proteins (Valenzuela et al., 2004) induced a strong DTH in immunized mice and protected them from an infection caused by *L. major* transmitted by infected sand flies (Gomes et al., 2012; Xu et al., 2011). CD4[+] T cells from LJM11-immunized mice produced IFN-γ, or both IFN-γ and TNF, at the bite site 48 h after challenge with infected sand flies (Gomes et al., 2012). Although previous exposure to *Lu. intermedia* saliva does not protect mice against an infection by *Leishmania* plus saliva, it was shown that Linb 11, a small molecule of 4,5 Kda found in its saliva, did protect mice against an infection by *L. braziliensis* and saliva from *Lu. intermedia* (de Moura et al., 2013). This underlines

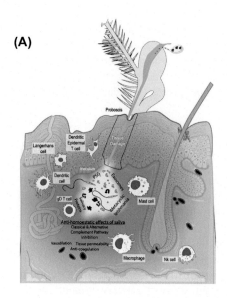

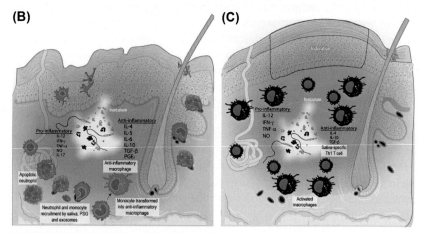

FIGURE 4.3 **The physiological and immunomodulatory effect of sand fly saliva at the site of a *Leishmania*-infected bite.** (A) Skin at steady state is disrupted by an infected sand fly bite that injects a complex infectious inoculum composed of metacyclic promastigotes, exosomes, promastigote secretory gel (PSG), and saliva into host skin. Saliva counteracts host homeostasis. (B) Skin of a saliva-naïve individual, neutrophils, and monocytes are recruited to the site of bite in an anti-inflammatory environment that favors parasite survival. (C) Skin of a saliva-exposed individual, a delayed-type hypersensitivity adaptive immune response recruits saliva-specific Th1 T cells that activate macrophages, a hostile environment for nearby parasites.

the fact that initiating an appropriate Th1-biased immune response to a distinct salivary protein can override the absence of protection to total saliva.

Although all the above-mentioned candidates protected rodents against an infection by *Leishmania*, it is necessary to test them in dogs, natural reservoirs of *L. infantum*, and in non-human primates (NHP) to validate their ability to protect humans. In this case, classical vaccinology is not feasible, due to the high cost and difficult logistics. To overcome this obstacle, Collin et al. (2009) developed a methodology named RAS (reverse antigenic screen). In this study, the authors exposed a group of dogs to repeated bites of uninfected *Lu. longipalpis* sand flies then challenged them with different plasmids encoding distinct proteins from the vector's saliva. They observed that two proteins LJM17 and LJL143 induced a strong DTH in these animals. Immunizing dogs with a prime–boost strategy of plasmids and recombinant proteins induced a Th1 immune response, with a large production of IFN-γ and IL-12 by PBMC restimulated *in vitro* with the recombinant proteins or total saliva. Additionally, saliva-stimulated lymphocytes from immunized dogs also reduced the percentage of autologous-infected macrophages. Besides that, biopsies taken from the DTH site of bite 48 h after exposing immunized dogs to uninfected *Lu. longipalpis* sand flies showed a cellular infiltrate dominated by CD3$^+$ T cells and macrophages. These infiltrating cells expressed high levels of IFN-γ and IL-12 corroborating the result obtained from PBMC stimulation (Collin et al., 2009).

From Rodents to Nonhuman Primates

Using the RAS approach, Oliveira et al. (2015) observed that PdSP15 from *P. duboscqi* saliva induced a DTH response in NHP with a higher production of IFN-γ. The authors also demonstrated that NHP exposed to uninfected *P. duboscqi* sand fly bites or immunized with the salivary protein PdSP15 were protected against cutaneous leishmaniasis initiated by infected bites. Uninfected sand fly exposed 7 of 10 PdSP15-immunized rhesus macaques displayed a significant reduction in disease and parasite burden compared to controls. Protection correlated to an early appearance of *Leishmania*-specific CD4$^+$IFN-γ$^+$ lymphocytes, suggesting that immunity to saliva or PdSP15 augments the host immune response to the parasites while maintaining minimal pathology. Sera and PBMCs from individuals naturally exposed to *P. duboscqi* bites recognized PdSP15, demonstrating its immunogenicity in humans.

Collectively, these results demonstrate that salivary proteins from different sand flies seem promising as vaccines against *Leishmania* infections and the idea to combine parasite antigens and salivary proteins could produce an effective vaccine. Some groups have tested combinations of sand fly saliva or salivary proteins with *Leishmania* antigens or attenuated *Leishmania* parasites. In the study by Aguiar-Soares et al. (2014), a vaccine composed of *L. braziliensis* antigens adjuvanted with saponin and *Lu. longipalpis* SGE (LBSapSal vaccine) was tested in dogs. This combination vaccine elicited both anti-*Leishmania* and anti-saliva humoral and cellular immune responses and resulted in a reduction of the splenic parasite load in immunized dogs as compared to control groups (Aguiar-Soares et al., 2014). In another study mice immunized with a vaccine combination of a single sand fly salivary protein (PpSP15) with live nonpathogenic *L. tarentolae* expressing the cysteine proteases A and B (Ltar CPA/CPB) displayed better immunity and protection against cutaneous leishmaniasis compared to animals vaccinated with PpSP15 alone or with nonpathogenic *L. tarentolae* expressing the Ltar CPA/CPB without the salivary protein (Zahedifard et al., 2014). It is important to note that this increase in protection was only observed when animals were first primed with PpSP15 DNA and then boosted with PpSP15 DNA and live Ltar CPA/CPB (Zahedifard et al., 2014). Apparently priming with the salivary protein in a combination vaccine is important since in a separate study animals vaccinated simultaneously with KMP11 and the salivary protein LJM19 showed no improvement in the protective efficacy over the KMP11 or LJM19 vaccines alone (da Silva et al., 2011). These studies suggest that priming with a sand fly salivary protein may be required for a vaccine that envisions a combination of a sand fly salivary protein and a *Leishmania* antigen.

Another important point is the need for a stringent challenge to test vaccines against *Leishmania* infection. Although vector saliva is important for the establishment of *Leishmania*, other products from the vector contribute equally to the success of the infection. Peters et al. (2009), showed that vaccines that protected mice against a needle challenge failed when the animals were exposed to infected sand flies, pointing out the importance of natural transmission in testing vaccines for *Leishmania* and potentially other vector-borne diseases. Similarly, transmission of *L. mexicana* by infected *Lu. longipalpis* also enhanced the infection in mice due to the presence of the promastigote secretory gel released with the parasites (Rogers et al., 2004). Recently, Atayde et al. (2015) showed that *Leishmania* constitutively secretes exosomes within the lumen of the sand fly midgut. Through egestion experiments, the authors demonstrated that *Leishmania* exosomes are part of the sand fly inoculum and are co-egested with the parasite during the vector's bite. The co-inoculation of mice footpads with *L. major* plus midgut-isolated or *in vitro*-isolated *L. major* exosomes resulted in a significant increase in footpad swelling. These co-injections produced exacerbated lesions through over induction of IL-17a.

The Immune Response to Sand Fly Salivary Proteins in Humans

To consider salivary proteins as potential candidate vaccines to be used in humans, it is essential to know the immune response elicited by these proteins in individuals. Several studies have evaluated the systemic or local immune response induced by the bites of different species of sand flies. Vinhas et al. (2007), observed that volunteers experimentally exposed to uninfected *Lu. longipalpis* bites still responded to sand fly salivary proteins even after 1 year after the last exposure. *In vitro* restimulation of PBMC from these volunteers with sand fly saliva resulted in an increase in the frequency of CD4 and CD8 T cells expressing CD25, as well as an increase in IFN-γ and IL10 production. In Tunisia, restimulation of PBMC from individuals living in an endemic area for cutaneous

Leishmania with saliva from *P. papatasi* increased production of IL-10 by CD8[+] T cells and IFN-γ by CD4[+]T cells (Abdeladhim et al., 2011). In the study developed by Oliveira et al. (2013), the authors observed the duration and nature of the DTH to sand fly saliva in humans from an endemic area of Mali through volunteers exposed to bites of colony bred *P. duboscqi* sand flies. A DTH response to bites was observed in 75% of individuals aged 1–15 years, decreasing gradually to 48% by age 45, and dropping to 21% thereafter. Dermal biopsies obtained from DTH+ individuals were dominated by T lymphocytes and macrophages and displayed an abundant expression of IFN-γ in the absence of Th2 cytokines establishing the Th1 nature of this DTH response. However, PBMCs from 98% of individuals responded to sand fly saliva (Oliveira et al., 2013), and of these only 23% were polarized to a Th1 and 25% showed a Th2 response. This study demonstrated the durability and Th1 nature of a DTH to sand fly bites in humans living in a cutaneous leishmaniasis-endemic area. Importantly, a systemic Th2 response may explain why some individuals remain susceptible to disease. This suggests that not all individuals respond in the same way to sand fly salivary proteins and these differences may account for the different outcomes of *leishmania*sis in a population that is constantly bitten by sand flies.

CONCLUDING REMARKS

Salivary proteins from sand flies are important for the establishment of infection by *Leishmania*. Although most of the mechanisms induced by salivary proteins lead to the development of anti-inflammatory responses and a better establishment of the parasite in the host, this mostly applies to their effect on the innate immune system. Certainly, there is a strong body of evidence demonstrating that induction of an appropriate adaptive immune response to saliva or distinct salivary proteins confers a strong protection against *Leishmania* infection by creating an inhospitable environment that impairs parasite proliferation. As such, we need to be prudent in distinguishing between the exacerbatory role of salivary molecules in *Leishmania* infections and their properties as immunogens of value as antigens for leishmaniasis vaccines. Considering that there are no available vaccines against any form of leishmaniasis for humans, we need to explore all the resources within our reach to develop an effective product.

REFERENCES

Abdeladhim, M., Ben Ahmed, M., Marzouki, S., Belhadj Hmida, N., Boussoffara, T., Belhaj Hamida, N., Ben Salah, A., Louzir, H., 2011. Human cellular immune response to the saliva of *Phlebotomus papatasi* is mediated by IL-10-producing CD8[+] T cells and Th1-polarized CD4[+] lymphocytes. PLoS Neglected Tropical Diseases 5, e1345.

Abdeladhim, M., Jochim, R.C., Ben Ahmed, M., Zhioua, E., Chelbi, I., Cherni, S., Louzir, H., Ribeiro, J.M., Valenzuela, J.G., 2012. Updating the salivary gland transcriptome of *Phlebotomus papatasi* (Tunisian strain): the search for sand fly-secreted immunogenic proteins for humans. PLoS One 7, e47347.

Abdeladhim, M., Kamhawi, S., Valenzuela, J.G., 2014. What's behind a sand fly bite? The profound effect of sand fly saliva on host hemostasis, inflammation and immunity. Infection, Genetics and Evolution 28, 691–703.

Aguiar-Soares, R.D., Roatt, B.M., Ker, H.G., Moreira, N., Mathias, F.A., Cardoso, J.M., Gontijo, N.F., Bruna-Romero, O., Teixeira-Carvalho, A., Martins-Filho, O.A., Correa-Oliveira, R., Giunchetti, R.C., Reis, A.B., 2014. LBSapSal-vaccinated dogs exhibit increased circulating T-lymphocyte subsets (CD4(+) and CD8(+)) as well as a reduction of parasitism after challenge with Leishmania infantum plus salivary gland of *Lutzomyia longipalpis*. Parasites and Vectors 7, 61.

Alvarenga, P.H., Xu, X., Oliveira, F., Chagas, A.C., Nascimento, C.R., Francischetti, I.M., Juliano, M.A., Juliano, L., Scharfstein, J., Valenzuela, J.G., Ribeiro, J.M., Andersen, J.F., 2013. Novel family of insect salivary inhibitors blocks contact pathway activation by binding to polyphosphate, heparin, and dextran sulfate. Arteriosclerosis, Thrombosis, and Vascular Biology 33, 2759–2770.

Anderson, J.M., Oliveira, F., Kamhawi, S., Mans, B.J., Reynoso, D., Seitz, A.E., Lawyer, P., Garfield, M., Pham, M., Valenzuela, J.G., 2006. Comparative salivary gland transcriptomics of sandfly vectors of visceral leishmaniasis. BMC Genomics 7, 52.

Anjili, C.O., Mbati, P.A., Mwangi, R.W., Githure, J.I., Olobo, J.O., Robert, L.L., Koech, D.K., 1995. The chemotactic effect of *Phlebotomus duboscqi* (Diptera: psychodidae) salivary gland lysates to murine monocytes. Acta Tropica 60, 97–100.

Araujo-Santos, T., Prates, D.B., Andrade, B.B., Nascimento, D.O., Clarencio, J., Entringer, P.F., Carneiro, A.B., Silva-Neto, M.A., Miranda, J.C., Brodskyn, C.I., Barral, A., Bozza, P.T., Borges, V.M., 2010. *Lutzomyia longipalpis* saliva triggers lipid body formation and prostaglandin E(2) production in murine macrophages. PLoS Neglected Tropical Diseases 4, e873.

Atayde, V.D., Aslan, H., Townsend, S., Hassani, K., Kamhawi, S., Olivier, M., 2015. Exosome secretion by the parasitic protozoan leishmania within the sand fly midgut. Cell Reports 13, 957–967.

Atkinson, C., van Riper, C., 1991. Pathogenicity and epizootiology of avian haematozoa: plasmodium, leucocytozoon, and haemoproteus. In: Loye, J., Zuk, M. (Eds.), In Bird-Parasite Interactions. Oxford University Press, Oxford, pp. 19–48.

Bancherereau, J., Briere, F., Caux, C., Davoust, J., Lebecque, S., Liu, Y.J., Pulendran, B., Palucka, K., 2000. Immunobiology of dendritic cells. Annual Review of Immunology 18, 767–811.

Barber, M., Olinger, M., 1931. Studies on malaria in southern Nigeria. Annals of Tropical Medicine and Parasitology 25, 461–501.

Beghdadi, W., Porcherie, A., Schneider, B.S., Dubayle, D., Peronet, R., Huerre, M., Watanabe, T., Ohtsu, H., Louis, J., Mecheri, S., 2008. Inhibition of histamine-mediated signaling confers significant protection against severe malaria in mouse models of disease. The Journal of Experimental Medicine 205, 395–408.

Belkaid, Y., Kamhawi, S., Modi, G., Valenzuela, J., Noben-Trauth, N., Rowton, E., Ribeiro, J., Sacks, D.L., 1998. Development of a natural model of cutaneous leishmaniasis: powerful effects of vector saliva and saliva preexposure on the long-term outcome of Leishmania major infection in the mouse ear dermis. The Journal of Experimental Medicine 188, 1941–1953.

Bereczky, S., Montgomery, S.M., Troye-Blomberg, M., Rooth, I., Shaw, M.A., Farnert, A., 2004. Elevated anti-malarial IgE in asymptomatic individuals is associated with reduced risk for subsequent clinical malaria. International Journal for Parasitology 34, 935–942.

Bhattacharya, U., Roy, S., Kar, P.K., Sarangi, B., Lahiri, S.C., 1988. Histamine & kinin system in experimental malaria. The Indian Journal of Medical Research 88, 558–563.

Biedermann, T., Kneilling, M., Mailhammer, R., Maier, K., Sander, C.A., Kollias, G., Kunkel, S.L., Hultner, L., Rocken, M., 2000. Mast cells control neutrophil recruitment during T cell-mediated delayed-type hypersensitivity reactions through tumor necrosis factor and macrophage inflammatory protein 2. The Journal of Experimental Medicine 192, 1441–1452.

Bouzigon, E., Forabosco, P., Koppelman, G.H., Cookson, W.O., Dizier, M.H., Duffy, D.L., Evans, D.M., Ferreira, M.A., Kere, J., Laitinen, T., Malerba, G., Meyers, D.A., Moffatt, M., Martin, N.G., Ng, M.Y., Pignatti, P.F., Wjst, M., Kauffmann, F., Demenais, F., Lewis, C.M., 2010. Meta-analysis of 20 genome-wide linkage studies evidenced new regions linked to asthma and atopy. European Journal of Human Genetics 18, 700–706.

Bozza, M., Soares, M.B., Bozza, P.T., Satoskar, A.R., Diacovo, T.G., Brombacher, F., Titus, R.G., Shoemaker, C.B., David, J.R., 1998. The PACAP-type I receptor agonist maxadilan from sand fly saliva protects mice against lethal endotoxemia by a mechanism partially dependent on IL-10. European Journal of Immunology 28, 3120–3127.

Brady, M.T., MacDonald, A.J., Rowan, A.G., Mills, K.H., 2003. Hepatitis C virus non-structural protein 4 suppresses Th1 responses by stimulating IL-10 production from monocytes. European Journal of Immunology 33, 3448–3457.

Brodie, T.M., Smith, M.C., Morris, R.V., Titus, R.G., 2007. Immunomodulatory effects of the *Lutzomyia longipalpis* salivary gland protein maxadilan on mouse macrophages. Infection and Immunity 75, 2359–2365.

Brummer-Korvenkontio, H., Lappalainen, P., Reunala, T., Palosuo, T., 1994. Detection of mosquito saliva-specific IgE and IgG4 antibodies by immunoblotting. The Journal of Allergy and Clinical Immunology 93, 551–555.

Carregaro, V., Sa-Nunes, A., Cunha, T.M., Grespan, R., Oliveira, C.J., Lima-Junior, D.S., Costa, D.L., Verri Jr., W.A., Milanezi, C.M., Pham, V.M., Brand, D.D., Valenzuela, J.G., Silva, J.S., Ribeiro, J.M., Cunha, F.Q., 2011. Nucleosides from *Phlebotomus papatasi* salivary gland ameliorate murine collagen-induced arthritis by impairing dendritic cell functions. Journal of Immunology 187, 4347–4359.

Carvalho, A.M., Cristal, J.R., Muniz, A.C., Carvalho, L.P., Gomes, R., Miranda, J.C., Barral, A., Carvalho, E.M., de Oliveira, C.I., 2015. Interleukin 10-dominant immune response and increased risk of cutaneous leishmaniasis after natural exposure to *Lutzomyia* intermedia sand flies. The Journal of Infectious Diseases 212, 157–165.

Cavalcante, R.R., Pereira, M.H., Gontijo, N.F., 2003. Anti-complement activity in the saliva of phlebotomine sand flies and other haematophagous insects. Parasitology 127, 87–93.

Cerna, P., Mikes, L., Volf, P., 2002. Salivary gland hyaluronidase in various species of phlebotomine sand flies (Diptera: psychodidae). Insect Biochemistry and Molecular Biology 32, 1691–1697.

Chagas, A.C., Oliveira, F., Debrabant, A., Valenzuela, J.G., Ribeiro, J.M., Calvo, E., 2014. Lundep, a sand fly salivary endonuclease increases Leishmania parasite survival in neutrophils and inhibits XIIa contact activation in human plasma. PLoS Pathogens 10, e1003923.

Chakravarty, S., Cockburn, I.A., Kuk, S., Overstreet, M.G., Sacci, J.B., Zavala, F., 2007. CD8+ T lymphocytes protective against malaria liver stages are primed in skin-draining lymph nodes. Nature Medicine 13, 1035–1041.

Chardome, M., Janssen, P.J., 1952. Inquiry on malarial incidence by the dermal method in the region of Lubilash, Belgian Congo. Annales de la Societe belge de medecine tropicale (1920) 32, 209–211.

Charlab, R., Valenzuela, J.G., Rowton, E.D., Ribeiro, J.M., 1999. Toward an understanding of the biochemical and pharmacological complexity of the saliva of a hematophagous sand fly *Lutzomyia longipalpis*. Proceedings of the National Academy of Sciences of the United States of America 96, 15155–15160.

Chen, W., Khurana Hershey, G.K., 2007. Signal transducer and activator of transcription signals in allergic disease. The Journal of Allergy and Clinical Immunology 119, 529–541 quiz 542–523.

Chen, Y.L., Simons, F.E., Peng, Z., 1998. A mouse model of mosquito allergy for study of antigen-specific IgE and IgG subclass responses, lymphocyte proliferation, and IL-4 and IFN-gamma production. International Archives of Allergy and Immunology 116, 269–277.

Chong, C.R., Chen, X., Shi, L., Liu, J.O., Sullivan Jr., D.J., 2006. A clinical drug library screen identifies astemizole as an antimalarial agent. Nature Chemical Biology 2, 415–416.

Cohen, S., Mc, G.I., Carrington, S., 1961. Gamma-globulin and acquired immunity to human malaria. Nature 192, 733–737.

Collin, N., Gomes, R., Teixeira, C., Cheng, L., Laughinghouse, A., Ward, J.M., Elnaiem, D.E., Fischer, L., Valenzuela, J.G., Kamhawi, S., 2009. Sand fly salivary proteins induce strong cellular immunity in a natural reservoir of visceral leishmaniasis with adverse consequences for Leishmania. PLoS Pathogens 5, e1000441.

Collin, N., Assumpcao, T.C., Mizurini, D.M., Gilmore, D.C., Dutra-Oliveira, A., Kotsyfakis, M., Sa-Nunes, A., Teixeira, C., Ribeiro, J.M., Monteiro, R.Q., Valenzuela, J.G., Francischetti, I.M., 2012. Lufaxin, a novel factor Xa inhibitor from the salivary gland of the sand fly *Lutzomyia longipalpis* blocks protease-activated receptor 2 activation and inhibits inflammation and thrombosis in vivo. Arteriosclerosis, Thrombosis, and Vascular Biology 32, 2185–2198.

Cornet, S., Nicot, A., Rivero, A., Gandon, S., 2014. Evolution of plastic transmission strategies in avian malaria. PLoS Pathogens 10, e1004308.

Costa, D.J., Favali, C., Clarencio, J., Afonso, L., Conceicao, V., Miranda, J.C., Titus, R.G., Valenzuela, J., Barral-Netto, M., Barral, A., Brodskyn, C.I., 2004. *Lutzomyia longipalpis* salivary gland homogenate impairs cytokine production and costimulatory molecule expression on human monocytes and dendritic cells. Infection and Immunity 72, 1298–1305.

da Silva, R.A., Tavares, N.M., Costa, D., Pitombo, M., Barbosa, L., Fukutani, K., Miranda, J.C., de Oliveira, C.I., Valenzuela, J.G., Barral, A., Soto, M., Barral-Netto, M., Brodskyn, C., 2011. DNA vaccination with KMP11 and *Lutzomyia longipalpis* salivary protein protects hamsters against visceral leishmaniasis. Acta Tropica 120, 185–190.

de Jong, E.C., Vieira, P.L., Kalinski, P., Schuitemaker, J.H., Tanaka, Y., Wierenga, E.A., Yazdanbakhsh, M., Kapsenberg, M.L., 2002. Microbial compounds selectively induce Th1 cell-promoting or Th2 cell-promoting dendritic cells in vitro with diverse th cell-polarizing signals. Journal of Immunology 168, 1704–1709 (Baltimore, Md.:1950).

de Moura, T.R., Oliveira, F., Novais, F.O., Miranda, J.C., Clarencio, J., Follador, I., Carvalho, E.M., Valenzuela, J.G., Barral-Netto, M., Barral, A., Brodskyn, C., de Oliveira, C.I., 2007. Enhanced *Leishmania braziliensis* infection following pre-exposure to sandfly saliva. PLoS Neglected Tropical Diseases 1, e84.

de Moura, T.R., Oliveira, F., Rodrigues, G.C., Carneiro, M.W., Fukutani, K.F., Novais, F.O., Miranda, J.C., Barral-Netto, M., Brodskyn, C., Barral, A., de Oliveira, C.I., 2010. Immunity to *Lutzomyia* intermedia saliva modulates the inflammatory environment induced by *Leishmania braziliensis*. PLoS Neglected Tropical Diseases 4, e712.

de Moura, T.R., Oliveira, F., Carneiro, M.W., Miranda, J.C., Clarencio, J., Barral-Netto, M., Brodskyn, C., Barral, A., Ribeiro, J.M., Valenzuela, J.G., de Oliveira, C.I., 2013. Functional transcriptomics of wild-caught *Lutzomyia* intermedia salivary glands: identification of a protective salivary protein against *Leishmania braziliensis* infection. PLoS Neglected Tropical Diseases 7, e2242.

Demarta-Gatsi, C., Smith, L., Thiberge, S., 2016. Protection against malaria in mice is induced by blood stage-arresting histamine-releasing factor (HRF)-deficient parasites. The Journal of Experimental Medicine 213, 1419–1428.

Demeure, C.E., Brahimi, K., Hacini, F., Marchand, F., Peronet, R., Huerre, M., St-Mezard, P., Nicolas, J.F., Brey, P., Delespesse, G., Mecheri, S., 2005. Anopheles mosquito bites activate cutaneous mast cells leading to a local inflammatory response and lymph node hyperplasia. Journal of Immunology 174, 3932–3940 (Baltimore, Md.:1950).

Depinay, N., Hacini, F., Beghdadi, W., Peronet, R., Mecheri, S., 2006. Mast cell-dependent down-regulation of antigen-specific immune responses by mosquito bites. Journal of Immunology 176, 4141–4146 (Baltimore, Md.:1950).

Desjeux, P., 2004. Leishmaniasis: current situation and new perspectives. Comparative Immunology, Microbiology and Infectious Diseases 27, 305–318.

Desowitz, R.S., 1989. Plasmodium-specific immunoglobulin E in sera from an area of holoendemic malaria. Transactions of the Royal Society of Tropical Medicine and Hygiene 83, 478–479.

Donovan, M.J., Messmore, A.S., Scrafford, D.A., Sacks, D.L., Kamhawi, S., McDowell, M.A., 2007. Uninfected mosquito bites confer protection against infection with malaria parasites. Infection and Immunity 75, 2523–2530.

Doolan, D.L., Dobano, C., Baird, J.K., 2009. Acquired immunity to malaria. Clinical Microbiology Reviews 22, 13–36 (Table of Contents).

Drakeley, C., Sutherland, C., Bousema, J.T., Sauerwein, R.W., Targett, G.A., 2006. The epidemiology of Plasmodium falciparum gametocytes: weapons of mass dispersion. Trends in Parasitology 22, 424–430.

Duarte, J., Deshpande, P., Guiyedi, V., Mecheri, S., Fesel, C., Cazenave, P.A., Mishra, G.C., Kombila, M., Pied, S., 2007. Total and functional parasite specific IgE responses in *Plasmodium falciparum*-infected patients exhibiting different clinical status. Malaria Journal 6, 1.

Ejrnaes, M., Filippi, C.M., Martinic, M.M., Ling, E.M., Togher, L.M., Crotty, S., von Herrath, M.G., 2006. Resolution of a chronic viral infection after interleukin-10 receptor blockade. The Journal of Experimental Medicine 203, 2461–2472.

Elenkov, I.J., Webster, E., Papanicolaou, D.A., Fleisher, T.A., Chrousos, G.P., Wilder, R.L., 1998. Histamine potently suppresses human IL-12 and stimulates IL-10 production via H2 receptors. Journal of Immunology 161, 2586–2593 (Baltimore, Md.:1950).

Enk, A.H., Angeloni, V.L., Udey, M.C., Katz, S.I., 1993. Inhibition of Langerhans cell antigen-presenting function by IL-10. A role for IL-10 in induction of tolerance. Journal of Immunology 151, 2390–2398 (Baltimore, Md.:1950).

Enwonwu, C.O., Afolabi, B.M., Salako, L.O., Idigbe, E.O., Bashirelah, N., 2000. Increased plasma levels of histidine and histamine in falciparum malaria: relevance to severity of infection. Journal of Neural Transmission 107, 1273–1287 (Vienna, Austria:1996).

Esnault, S., Rosenthal, L.A., Wang, D.S., Malter, J.S., 2008. Thymic stromal lymphopoietin (TSLP) as a bridge between infection and atopy. International Journal of Clinical and Experimental Pathology 1, 325–330.

Ferguson, T.A., Dube, P., Griffith, T.S., 1994. Regulation of contact hypersensitivity by interleukin 10. The Journal of Experimental Medicine 179, 1597–1604.

Ferreira, V.P., Fazito Vale, V., Pangburn, M.K., Abdeladhim, M., Mendes-Sousa, A.F., Coutinho-Abreu, I.V., Rasouli, M., Brandt, E.A., Meneses, C., Lima, K.F., Nascimento Araujo, R., Pereira, M.H., Kotsyfakis, M., Oliveira, F., Kamhawi, S., Ribeiro, J.M., Gontijo, N.F., Collin, N., Valenzuela, J.G., 2016. SALO, a novel classical pathway complement inhibitor from saliva of the sand fly *Lutzomyia longipalpis*. Scientific Reports 6, 19300.

Fonseca, L., Seixas, E., Butcher, G., Langhorne, J., 2007. Cytokine responses of CD4+ T cells during a *Plasmodium chabaudi* chabaudi (ER) blood-stage infection in mice initiated by the natural route of infection. Malaria Journal 6, 77.

French, F.E., West, A.S., 1971. Skin reaction specificity of guinea pig immediate hypersensitivity to bites of four mosquito species. The Journal of Parasitology 57, 396–400.

Friberg, C., Bjorck, K., Nilsson, S., Inerot, A., Wahlstrom, J., Samuelsson, L., 2006. Analysis of chromosome 5q31-32 and psoriasis: confirmation of a susceptibility locus but no association with SNPs within SLC22A4 and SLC22A5. The Journal of Investigative Dermatology 126, 998–1002.

Furuta, T., Kikuchi, T., Iwakura, Y., Watanabe, N., 2006. Protective roles of mast cells and mast cell-derived TNF in murine malaria. Journal of Immunology 177, 3294–3302 (Baltimore, Md.:1950).

Gabriel, C., McMaster, W.R., Girard, D., Descoteaux, A., 2010. *Leishmania donovani* promastigotes evade the antimicrobial activity of neutrophil extracellular traps. Journal of Immunology 185, 4319–4327.

Garcia, A., Marquet, S., Bucheton, B., Hillaire, D., Cot, M., Fievet, N., Dessein, A.J., Abel, L., 1998. Linkage analysis of blood *Plasmodium falciparum* levels: interest of the 5q31-q33 chromosome region. The American Journal of Tropical Medicine and Hygiene 58, 705–709.

Gomes, R., Oliveira, F., 2012. The immune response to sand fly salivary proteins and its influence on leishmania immunity. Frontiers in Immunology 3, 110.

Gomes, R., Teixeira, C., Teixeira, M.J., Oliveira, F., Menezes, M.J., Silva, C., de Oliveira, C.I., Miranda, J.C., Elnaiem, D.E., Kamhawi, S., Valenzuela, J.G., Brodskyn, C.I., 2008. Immunity to a salivary protein of a sand fly vector protects against the fatal outcome of visceral leishmaniasis in a hamster model. Proceedings of the National Academy of Sciences of the United States of America 105, 7845–7850.

Gomes, R., Oliveira, F., Teixeira, C., Meneses, C., Gilmore, D.C., Elnaiem, D.E., Kamhawi, S., Valenzuela, J.G., 2012. Immunity to sand fly salivary protein LJM11 modulates host response to vector-transmitted leishmania conferring ulcer-free protection. The Journal of Investigative Dermatology 132, 2735–2743.

Gomes, R., Cavalcanti, K., Teixeira, C., Carvalho, A.M., Mattos, P.S., Cristal, J.R., Muniz, A.C., Miranda, J.C., de Oliveira, C.I., Barral, A., 2016. Immunity to *Lutzomyia whitmani* saliva protects against experimental *Leishmania braziliensis* infection. PLoS Neglected Tropical Diseases 10, e0005078.

Gouagna, L.C., Bancone, G., Yao, F., Yameogo, B., Dabire, K.R., Costantini, C., Simpore, J., Ouedraogo, J.B., Modiano, D., 2010. Genetic variation in human HBB is associated with *Plasmodium falciparum* transmission. Nature Genetics 42, 328–331.

Grange, L., Loucoubar, C., Telle, O., Tall, A., Faye, J., Sokhna, C., Trape, J.F., Sakuntabhai, A., Bureau, J.F., Paul, R., 2015. Risk factors for *Plasmodium falciparum* gametocyte positivity in a longitudinal cohort. PLoS One 10, e0123102.

Grespan, R., Lemos, H.P., Carregaro, V., Verri Jr., W.A., Souto, F.O., de Oliveira, C.J., Teixeira, C., Ribeiro, J.M., Valenzuela, J.G., Cunha, F.Q., 2012. The protein LJM 111 from *Lutzomyia longipalpis* salivary gland extract (SGE) accounts for the SGE-inhibitory effects upon inflammatory parameters in experimental arthritis model. International Immunopharmacology 12, 603–610.

Griffiths, M.J., Shafi, M.J., Popper, S.J., Hemingway, C.A., Kortok, M.M., Wathen, A., Rockett, K.A., Mott, R., Levin, M., Newton, C.R., Marsh, K., Relman, D.A., Kwiatkowski, D.P., 2005. Genomewide analysis of the host response to malaria in Kenyan children. The Journal of Infectious Diseases 191, 1599–1611.

Grimbaldeston, M.A., Nakae, S., Kalesnikoff, J., Tsai, M., Galli, S.J., 2007. Mast cell-derived interleukin 10 limits skin pathology in contact dermatitis and chronic irradiation with ultraviolet B. Nature Immunology 8, 1095–1104.

Guiyedi, V., Becavin, C., Herbert, F., Gray, J., Cazenave, P.A., Kombila, M., Crisanti, A., Fesel, C., Pied, S., 2015. Asymptomatic *Plasmodium falciparum* infection in children is associated with increased auto-antibody production, high IL-10 plasma levels and antibodies to merozoite surface protein 3. Malaria Journal 14, 162.

Haileamlak, A., Dagoye, D., Williams, H., Venn, A.J., Hubbard, R., Britton, J., Lewis, S.A., 2005. Early life risk factors for atopic dermatitis in Ethiopian children. The Journal of Allergy and Clinical Immunology 115, 370–376.

Hajnicka, V., Vancova, I., Kocakova, P., Slovak, M., Gasperik, J., Slavikova, M., Hails, R.S., Labuda, M., Nuttall, P.A., 2005. Manipulation of host cytokine network by ticks: a potential gateway for pathogen transmission. Parasitology 130, 333–342.

Hall, L.R., Titus, R.G., 1995. Sand fly vector saliva selectively modulates macrophage functions that inhibit killing of Leishmania major and nitric oxide production. Journal of Immunology 155, 3501–3506.

Hamasaki, R., Kato, H., Terayama, Y., Iwata, H., Valenzuela, J.G., 2009. Functional characterization of a salivary apyrase from the sand fly, *Phlebotomus duboscqi*, a vector of Leishmania major. Journal of Insect Physiology 55, 1044–1049.

Herrant, M., Loucoubar, C., Bassene, H., Goncalves, B., Boufkhed, S., Diene Sarr, F., Fontanet, A., Tall, A., Baril, L., Mercereau-Puijalon, O., Mecheri, S., Sakuntabhai, A., Paul, R., 2013. Asthma and atopic dermatitis are associated with increased risk of clinical *Plasmodium falciparum* malaria. BMJ Open 3.

Hindorff, L.A., Sethupathy, P., Junkins, H.A., Ramos, E.M., Mehta, J.P., Collins, F.S., Manolio, T.A., 2009. Potential etiologic and functional implications of genome-wide association loci for human diseases and traits. Proceedings of the National Academy of Sciences of the United States of America 106, 9362–9367.

Holloway, J.W., Yang, I.A., Holgate, S.T., 2010. Genetics of allergic disease. The Journal of Allergy and Clinical Immunology 125, S81–S94.

Hostomska, J., Volfova, V., Mu, J., Garfield, M., Rohousova, I., Volf, P., Valenzuela, J.G., Jochim, R.C., 2009. Analysis of salivary transcripts and antigens of the sand fly *Phlebotomus arabicus*. BMC Genomics 10, 282.

James, A.A., Rossignol, P.A., 1991. Mosquito salivary glands: parasitological and molecular aspects. Parasitology Today 7, 267–271 (Personal ed.).

Jang, N., Stewart, G., Jones, G., 2005. Polymorphisms within the PHF11 gene at chromosome 13q14 are associated with childhood atopic dermatitis. Genes and Immunity 6, 262–264.

Janse, C.J., Ramesar, J., Waters, A.P., 2006. High-efficiency transfection and drug selection of genetically transformed blood stages of the rodent malaria parasite *Plasmodium berghei*. Nature Protocols 1, 346–356.

Kamhawi, S., Belkaid, Y., Modi, G., Rowton, E., Sacks, D., 2000. Protection against cutaneous leishmaniasis resulting from bites of uninfected sand flies. Science 290, 1351–1354 (New York, N.Y.).

Kato, H., Anderson, J.M., Kamhawi, S., Oliveira, F., Lawyer, P.G., Pham, V.M., Sangare, C.S., Samake, S., Sissoko, I., Garfield, M., Sigutova, L., Volf, P., Doumbia, S., Valenzuela, J.G., 2006. High degree of conservancy among secreted salivary gland proteins from two geographically distant *Phlebotomus duboscqi* sandflies populations (Mali and Kenya). BMC Genomics 7, 226.

Kato, H., Jochim, R.C., Gomez, E.A., Uezato, H., Mimori, T., Korenaga, M., Sakurai, T., Katakura, K., Valenzuela, J.G., Hashiguchi, Y., 2013. Analysis of salivary gland transcripts of the sand fly *Lutzomyia ayacuchensis*, a vector of Andean-type cutaneous leishmaniasis. Infection, Genetics and Evolution 13, 56–66.

Kohara, Y., Tanabe, K., Matsuoka, K., Kanda, N., Matsuda, H., Karasuyama, H., Yonekawa, H., 2001. A major determinant quantitative-trait locus responsible for atopic dermatitis-like skin lesions in NC/Nga mice is located on Chromosome 9. Immunogenetics 53, 15–21.

Lawaly, R., Konate, L., Marrama, L., Dia, I., Diallo, D., Diene Sarr, F., Schneider, B.S., Casademont, I., Diallo, M., Brey, P.T., Sakuntabhai, A., Mecheri, S., Paul, R., 2012. Impact of mosquito bites on asexual parasite density and gametocyte prevalence in asymptomatic chronic *Plasmodium falciparum* infections and correlation with IgE and IgG titers. Infection and Immunity 80, 2240–2246.

Lell, B., Borrmann, S., Yazdanbakhsh, M., Kremsner, P.G., 2001. Atopy and malaria. Wiener Klinische Wochenschrift 113, 927–929.

Lerner, E.A., Ribeiro, J.M., Nelson, R.J., Lerner, M.R., 1991. Isolation of maxadilan, a potent vasodilatory peptide from the salivary glands of the sand fly *Lutzomyia longipalpis*. The Journal of Biological Chemistry 266, 11234–11236.

Limesand, K., Higgs, S., Pearson, L., Beaty, B., 2000. Potentiation of vesicular stomatitis New Jersey virus infection in mice by mosquito saliva. Parasite Immunology 22, 461–467.

Luty, A.J., Mayombo, J., Lekoulou, F., Mshana, R., 1994. Immunologic responses to soluble exoantigens of *Plasmodium falciparum* in Gabonese children exposed to continuous intense infection. The American Journal of Tropical Medicine and Hygiene 51, 720–729.

Macatonia, S.E., Doherty, T.M., Knight, S.C., O'Garra, A., 1993. Differential effect of IL-10 on dendritic cell-induced T cell proliferation and IFN-gamma production. Journal of Immunology 150, 3755–3765 (Baltimore, Md.:1950).

MacDonald, S.M., Bhisutthibhan, J., Shapiro, T.A., Rogerson, S.J., Taylor, T.E., Tembo, M., Langdon, J.M., Meshnick, S.R., 2001. Immune mimicry in malaria: *Plasmodium falciparum* secretes a functional histamine-releasing factor homolog in vitro and in vivo. Proceedings of the National Academy of Sciences of the United States of America 98, 10829–10832.

Maeno, Y., Perlmann, P., PerlmannH, Kusuhara, Y., Taniguchi, K., Nakabayashi, T., Win, K., Looareesuwan, S., Aikawa, M., 2000. IgE deposition in brain microvessels and on parasitized erythrocytes from cerebral malaria patients. The American Journal of Tropical Medicine and Hygiene 63, 128–132.

Malaviya, R., Ikeda, T., Ross, E., Abraham, S.N., 1996. Mast cell modulation of neutrophil influx and bacterial clearance at sites of infection through TNF-alpha. Nature 381, 77–80.

Marsh, K., Snow, R.W., 1997. Host-parasite interaction and morbidity in malaria endemic areas. Philosophical Transactions of the Royal Society of London. Series B, Biological Sciences 352, 1385–1394.

Mathieu, C., Demarta-Gatsi, C., Porcherie, A., Brega, S., Thiberge, S., Ronce, K., Smith, L., Peronet, R., Amino, R., Menard, R., Mecheri, S., 2015. *Plasmodium berghei* histamine-releasing factor favours liver-stage development via inhibition of IL-6 production and associates with a severe outcome of disease. Cellular Microbiology 17, 542–558.

Mbow, M.L., Bleyenberg, J.A., Hall, L.R., Titus, R.G., 1998. *Phlebotomus papatasi* sand fly salivary gland lysate down-regulates a Th1, but up-regulates a Th2, response in mice infected with Leishmania major. Journal of Immunology 161, 5571–5577.

Mesquita, R.D., Carneiro, A.B., Bafica, A., Gazos-Lopes, F., Takiya, C.M., Souto-Padron, T., Vieira, D.P., Ferreira-Pereira, A., Almeida, I.C., Figueiredo, R.T., Porto, B.N., Bozza, M.T., Graca-Souza, A.V., Lopes, A.H., Atella, G.C., Silva-Neto, M.A., 2008. *Trypanosoma cruzi* infection is enhanced by vector saliva through immunosuppressant mechanisms mediated by lysophosphatidylcholine. Infection and Immunity 76, 5543–5552.

Miller, M.J., 1958. Observations on the natural history of malaria in the semi-resistant West African. Transactions of the Royal Society of Tropical Medicine and Hygiene 52, 152–168.

Mitchell, A.J., Hansen, A.M., Hee, L., Ball, H.J., Potter, S.M., Walker, J.C., Hunt, N.H., 2005. Early cytokine production is associated with protection from murine cerebral malaria. Infection and Immunity 73, 5645–5653.

Monteiro, M.C., Lima, H.C., Souza, A.A., Titus, R.G., Romao, P.R., Cunha, F.Q., 2007. Effect of *Lutzomyia longipalpis* salivary gland extracts on leukocyte migration induced by Leishmania major. The American Journal of Tropical Medicine and Hygiene 76, 88–94.

Moore, K.W., de Waal Malefyt, R., Coffman, R.L., O'Garra, A., 2001. Interleukin-10 and the interleukin-10 receptor. Annual Review of Immunology 19, 683–765.

Morris, R.V., Shoemaker, C.B., David, J.R., Lanzaro, G.C., Titus, R.G., 2001. Sandfly maxadilan exacerbates infection with Leishmania major and vaccinating against it protects against L. major infection. Journal of Immunology 167, 5226–5230.

Mosmann, T.R., Coffman, R.L., 1989. TH1 and TH2 cells: different patterns of lymphokine secretion lead to different functional properties. Annual Review of Immunology 7, 145–173.

Norsworthy, N.B., Sun, J., Elnaiem, D., Lanzaro, G., Soong, L., 2004. Sand fly saliva enhances *Leishmania amazonensis* infection by modulating interleukin-10 production. Infection and Immunity 72, 1240–1247.

Ockenfels, B., Michael, E., McDowell, M.A., 2014. Meta-analysis of the effects of insect vector saliva on host immune responses and infection of vector-transmitted pathogens: a focus on leishmaniasis. PLoS Neglected Tropical Diseases 8, e3197.

Oesterholt, M.J., Alifrangis, M., Sutherland, C.J., Omar, S.A., Sawa, P., Howitt, C., Gouagna, L.C., Sauerwein, R.W., Bousema, T., 2009. Submicroscopic gametocytes and the transmission of antifolate-resistant *Plasmodium falciparum* in Western Kenya. PLoS One 4, e4364.

Ohno, T., Ishih, A., Kohara, Y., Yonekawa, H., Terada, M., Nishimura, M., 2001. Chromosomal mapping of the host resistance locus to rodent malaria (*Plasmodium yoelii*) infection in mice. Immunogenetics 53, 736–740.

Oka, K., Ohtaki, N., Yasuhara, T., Nakajima, T., 1989. A study of mosquito salivary gland components and their effects on man. The Journal of Dermatology 16, 469–474.

Oliveira, F., Kamhawi, S., Seitz, A.E., Pham, V.M., Guigal, P.M., Fischer, L., Ward, J., Valenzuela, J.G., 2006. From transcriptome to immunome: identification of DTH inducing proteins from a *Phlebotomus ariasi* salivary gland cDNA library. Vaccine 24, 374–390.

Oliveira, F., Lawyer, P.G., Kamhawi, S., Valenzuela, J.G., 2008. Immunity to distinct sand fly salivary proteins primes the anti-Leishmania immune response towards protection or exacerbation of disease. PLoS Neglected Tropical Diseases 2, e226.

Oliveira, F., Traore, B., Gomes, R., Faye, O., Gilmore, D.C., Keita, S., Traore, P., Teixeira, C., Coulibaly, C.A., Samake, S., Meneses, C., Sissoko, I., Fairhurst, R.M., Fay, M.P., Anderson, J.M., Doumbia, S., Kamhawi, S., Valenzuela, J.G., 2013. Delayed-type hypersensitivity to sand fly saliva in humans from a leishmaniasis-endemic area of Mali is Th1-mediated and persists to midlife. The Journal of Investigative Dermatology 133, 452–459.

Oliveira, F., Rowton, E., Aslan, H., Gomes, R., Castrovinci, P.A., Alvarenga, P.H., Abdeladhim, M., Teixeira, C., Meneses, C., Kleeman, L.T., Guimaraes-Costa, A.B., Rowland, T.E., Gilmore, D., Doumbia, S., Reed, S.G., Lawyer, P.G., Andersen, J.F., Kamhawi, S., Valenzuela, J.G., 2015. A sand fly salivary protein vaccine shows efficacy against vector-transmitted cutaneous leishmaniasis in nonhuman primates. Science Translational Medicine 7, 290ra290.

Paul, R.E., Sakuntabhai, A., 2016. Atopic dermatitis – need for a sub-Saharan perspective. European Medical Journal Allergy and Immunology 1, 58–64.

Paul, R.E., Diallo, M., Brey, P.T., 2004. Mosquitoes and transmission of malaria parasites - not just vectors. Malaria Journal 3, 39.

Peng, Z., Yang, M., Simons, F.E., 1996. Immunologic mechanisms in mosquito allergy: correlation of skin reactions with specific IgE and IgG antibodies and lymphocyte proliferation response to mosquito antigens. Annals of Allergy, Asthma and Immunology 77, 238–244.

Perignon, J.L., Druilhe, P., 1994. Immune mechanisms underlying the premunition against *Plasmodium falciparum* malaria. Memorias do Instituto Oswaldo Cruz 89 (Suppl. 2), 51–53.

Perlmann, H., Helmby, H., Hagstedt, M., Carlson, J., Larsson, P.H., Troye-Blomberg, M., Perlmann, P., 1994. IgE elevation and IgE anti-malarial antibodies in *Plasmodium falciparum* malaria: association of high IgE levels with cerebral malaria. Clinical and Experimental Immunology 97, 284–292.

Perlmann, P., Perlmann, H., Flyg, B.W., Hagstedt, M., Elghazali, G., Worku, S., Fernandez, V., Rutta, A.S., Troye-Blomberg, M., 1997. Immunoglobulin E, a pathogenic factor in *Plasmodium falciparum* malaria. Infection and Immunity 65, 116–121.

Perlmann, P., Perlmann, H., ElGhazali, G., Blomberg, M.T., 1999. IgE and tumor necrosis factor in malaria infection. Immunology Letters 65, 29–33.

Peters, N.C., Egen, J.G., Secundino, N., Debrabant, A., Kimblin, N., Kamhawi, S., Lawyer, P., Fay, M.P., Germain, R.N., Sacks, D., 2008. In vivo imaging reveals an essential role for neutrophils in leishmaniasis transmitted by sand flies. Science 321, 970–974.

Peters, N.C., Kimblin, N., Secundino, N., Kamhawi, S., Lawyer, P., Sacks, D.L., 2009. Vector transmission of leishmania abrogates vaccine-induced protective immunity. PLoS Pathogens 5, e1000484.

Pierrot, C., Adam, E., Hot, D., Lafitte, S., Capron, M., George, J.D., Khalife, J., 2007. Contribution of T cells and neutrophils in protection of young susceptible rats from fatal experimental malaria. Journal of Immunology 178, 1713–1722 (Baltimore, Md.:1950).

Porcherie, A., Mathieu, C., Peronet, R., Schneider, E., Claver, J., Commere, P.H., Kiefer-Biasizzo, H., Karasuyama, H., Milon, G., Dy, M., Kinet, J.P., Louis, J., Blank, U., Mecheri, S., 2011. Critical role of the neutrophil-associated high-affinity receptor for IgE in the pathogenesis of experimental cerebral malaria. The Journal of Experimental Medicine 208, 2225–2236.

Prates, D.B., Araujo-Santos, T., Luz, N.F., Andrade, B.B., Franca-Costa, J., Afonso, L., Clarencio, J., Miranda, J.C., Bozza, P.T., Dosreis, G.A., Brodskyn, C., Barral-Netto, M., Borges, V.M., Barral, A., 2011. *Lutzomyia longipalpis* saliva drives apoptosis and enhances parasite burden in neutrophils. Journal of Leukocyte Biology 90, 575–582.

Qureshi, A.A., Asahina, A., Ohnuma, M., Tajima, M., Granstein, R.D., Lerner, E.A., 1996. Immunomodulatory properties of maxadilan, the vasodilator peptide from sand fly salivary gland extracts. The American Journal of Tropical Medicine and Hygiene 54, 665–671.

Reunala, T., Lappalainen, P., Brummer-Korvenkontio, H., Coulie, P., Palosuo, T., 1991. Cutaneous reactivity to mosquito bites: effect of cetirizine and development of anti-mosquito antibodies. Clinical and Experimental Allergy 21, 617–622.

Reunala, T., Brummer-Korvenkontio, H., Rasanen, L., Francois, G., Palosuo, T., 1994. Passive transfer of cutaneous mosquito-bite hypersensitivity by IgE anti-saliva antibodies. The Journal of Allergy and Clinical Immunology 94, 902–906.

Ribeiro, J.M., Francischetti, I.M., 2003. Role of arthropod saliva in blood feeding: sialome and post-sialome perspectives. Annual Review of Entomology 48, 73–88.

Ribeiro, J.M., Modi, G., 2001. The salivary adenosine/AMP content of *Phlebotomus argentipes* Annandale and Brunetti, the main vector of human kala-azar. The Journal of Parasitology 87, 915–917.

Ribeiro, J.M., Vachereau, A., Modi, G.B., Tesh, R.B., 1989. A novel vasodilatory peptide from the salivary glands of the sand fly *Lutzomyia longipalpis*. Science 243, 212–214.

Ribeiro, J.M., Katz, O., Pannell, L.K., Waitumbi, J., Warburg, A., 1999. Salivary glands of the sand fly *Phlebotomus papatasi* contain pharmacologically active amounts of adenosine and 5′-AMP. The Journal of Experimental Biology 202, 1551–1559.

Ribeiro, J.M., 1987. Role of saliva in blood-feeding by arthropods. Annual Review of Entomology 32, 463–478.

Rihet, P., Traore, Y., Abel, L., Aucan, C., Traore-Leroux, T., Fumoux, F., 1998. Malaria in humans: *Plasmodium falciparum* blood infection levels are linked to chromosome 5q31-q33. American Journal of Human Genetics 63, 498–505.

Rogers, K.A., Titus, R.G., 2003. Immunomodulatory effects of maxadilan and *Phlebotomus papatasi* sand fly salivary gland lysates on human primary in vitro immune responses. Parasite Immunology 25, 127–134.

Rogers, M.E., Ilg, T., Nikolaev, A.V., Ferguson, M.A., Bates, P.A., 2004. Transmission of cutaneous leishmaniasis by sand flies is enhanced by regurgitation of fPPG. Nature 430, 463–467.

Rogers, M.E., Sizova, O.V., Ferguson, M.A., Nikolaev, A.V., Bates, P.A., 2006. Synthetic glyco-vaccine protects against the bite of leishmania-infected sand flies. The Journal of Infectious Diseases 194, 512–518.

Rohousova, I., Volf, P., Lipoldova, M., 2005. Modulation of murine cellular immune response and cytokine production by salivary gland lysate of three sand fly species. Parasite Immunology 27, 469–473.

Rohousova, I., Subrahmanyam, S., Volfova, V., Mu, J., Volf, P., Valenzuela, J.G., Jochim, R.C., 2012. Salivary gland transcriptomes and proteomes of *Phlebotomus tobbi* and *Phlebotomus sergenti*, vectors of leishmaniasis. PLoS Neglected Tropical Diseases 6, e1660.

Rosenberg, R., Andre, R.G., Ketrangsee, S., 1990. Seasonal fluctuation of *Plasmodium falciparum* gametocytaemia. Transactions of the Royal Society of Tropical Medicine and Hygiene 84, 29–33.

Saffar, A.S., Alphonse, M.P., Shan, L., Hayglass, K.T., Simons, F.E., Gounni, A.S., 2007. IgE modulates neutrophil survival in asthma: role of mitochondrial pathway. Journal of Immunology 178, 2535–2541 (Baltimore, Md.:1950).

Sakuntabhai, A., Ndiaye, R., Casademont, I., Peerapittayamongkol, C., Rogier, C., Tortevoye, P., Tall, A., Paul, R., Turbpaiboon, C., Phimpraphi, W., Trape, J.F., Spiegel, A., Heath, S., Mercereau-Puijalon, O., Dieye, A., Julier, C., 2008. Genetic determination and linkage mapping of *Plasmodium falciparum* malaria related traits in Senegal. PLoS One 3, e2000.

Samuelson, J., Lerner, E., Tesh, R., Titus, R., 1991. A mouse model of *Leishmania braziliensis braziliensis* infection produced by coinjection with sand fly saliva. The Journal of Experimental Medicine 173, 49–54.

Schneider, B.S., Higgs, S., 2008. The enhancement of arbovirus transmission and disease by mosquito saliva is associated with modulation of the host immune response. Transactions of the Royal Society of Tropical Medicine and Hygiene 102, 400–408.

Schneider, B.S., Mathieu, C., Peronet, R., Mecheri, S., 2011. *Anopheles stephensi* saliva enhances progression of cerebral malaria in a murine model. Vector Borne and Zoonotic Diseases 11, 423–432 (Larchmont, N.Y.).

Seka-Seka, J., Brouh, Y., Yapo-Crezoit, A.C., Atseye, N.H., 2004. The role of serum immunoglobulin E in the pathogenesis of *Plasmodium falciparum* malaria in Ivorian children. Scandinavian Journal of Immunology 59, 228–230.

Silva, F., Gomes, R., Prates, D., Miranda, J.C., Andrade, B., Barral-Netto, M., Barral, A., 2005. Inflammatory cell infiltration and high antibody production in BALB/c mice caused by natural exposure to *Lutzomyia longipalpis* bites. The American Journal of Tropical Medicine and Hygiene 72, 94–98.

Soares, M.B., Titus, R.G., Shoemaker, C.B., David, J.R., Bozza, M., 1998. The vasoactive peptide maxadilan from sand fly saliva inhibits TNF-alpha and induces IL-6 by mouse macrophages through interaction with the pituitary adenylate cyclase-activating polypeptide (PACAP) receptor. Journal of Immunology 160, 1811–1816.

Soumelis, V., Reche, P.A., Kanzler, H., Yuan, W., Edward, G., Homey, B., Gilliet, M., Ho, S., Antonenko, S., Lauerma, A., Smith, K., Gorman, D., Zurawski, S., Abrams, J., Menon, S., McClanahan, T., de Waal-Malefyt Rd, R., Bazan, F., Kastelein, R.A., Liu, Y.J., 2002. Human epithelial cells trigger dendritic cell mediated allergic inflammation by producing TSLP. Nature Immunology 3, 673–680.

Sowunmi, A., Gbotosho, G.O., Happi, C.T., Adedeji, A.A., Bolaji, O.M., Fehintola, F.A., Fateye, B.A., Oduola, A.M., 2007. Enhancement of the antimalarial efficacy of amodiaquine by chlorpheniramine in vivo. Memorias do Instituto Oswaldo Cruz 102, 417–419.

Srichaikul, T., Archararit, N., Siriasawakul, T., Viriyapanich, T., 1976. Histamine changes in *Plasmodium falciparum* malaria. Transactions of the Royal Society of Tropical Medicine and Hygiene 70, 36–38.

Swellengrebel, N., De Buck, A., 1938. Malaria in The Netherlands Amsterdam. Scheltema & Holkema Ltd.

Tavares, N.M., Silva, R.A., Costa, D.J., Pitombo, M.A., Fukutani, K.F., Miranda, J.C., Valenzuela, J.G., Barral, A., de Oliveira, C.I., Barral-Netto, M., Brodskyn, C., 2011. *Lutzomyia longipalpis* saliva or salivary protein LJM19 protects against *Leishmania braziliensis* and the saliva of its vector, *Lutzomyia* intermedia. PLoS Neglected Tropical Diseases 5, e1169.

Teixeira, C.R., Teixeira, M.J., Gomes, R.B., Santos, C.S., Andrade, B.B., Raffaele-Netto, I., Silva, J.S., Guglielmotti, A., Miranda, J.C., Barral, A., Brodskyn, C., Barral-Netto, M., 2005. Saliva from *Lutzomyia longipalpis* induces CC chemokine ligand 2/monocyte chemoattractant protein-1 expression and macrophage recruitment. Journal of Immunology 175, 8346–8353.

Teixeira, C., Gomes, R., Oliveira, F., Meneses, C., Gilmore, D.C., Elnaiem, D.E., Valenzuela, J.G., Kamhawi, S., 2014. Characterization of the early inflammatory infiltrate at the feeding site of infected sand flies in mice protected from vector-transmitted Leishmania major by exposure to uninfected bites. PLoS Neglected Tropical Diseases 8, e2781.

Titus, R.G., Ribeiro, J.M., 1988. Salivary gland lysates from the sand fly *Lutzomyia longipalpis* enhance Leishmania infectivity. Science 239, 1306–1308.

Titus, R.G., Ribeiro, J.M., 1990. The role of vector saliva in transmission of arthropod-borne disease. Parasitology Today 6, 157–160 (Personal ed.).

Titus, R.G., Bishop, J.V., Mejia, J.S., 2006. The immunomodulatory factors of arthropod saliva and the potential for these factors to serve as vaccine targets to prevent pathogen transmission. Parasite Immunology 28, 131–141.

Troye-Blomberg, M., Perlmann, P., Mincheva Nilsson, L., Perlmann, H., 1999. Immune regulation of protection and pathogenesis in *Plasmodium falciparum* malaria. Parassitologia 41, 131–138.

Valenzuela, J.G., Belkaid, Y., Garfield, M.K., Mendez, S., Kamhawi, S., Rowton, E.D., Sacks, D.L., Ribeiro, J.M., 2001a. Toward a defined anti-Leishmania vaccine targeting vector antigens: characterization of a protective salivary protein. The Journal of Experimental Medicine 194, 331–342.

Valenzuela, J.G., Belkaid, Y., Rowton, E., Ribeiro, J.M., 2001b. The salivary apyrase of the blood-sucking sand fly *Phlebotomus papatasi* belongs to the novel Cimex family of apyrases. The Journal of Experimental Biology 204, 229–237.

Valenzuela, J.G., Garfield, M., Rowton, E.D., Pham, V.M., 2004. Identification of the most abundant secreted proteins from the salivary glands of the sand fly *Lutzomyia longipalpis*, vector of *Leishmania chagasi*. The Journal of Experimental Biology 207, 3717–3729.

Vinhas, V., Andrade, B.B., Paes, F., Bomura, A., Clarencio, J., Miranda, J.C., Bafica, A., Barral, A., Barral-Netto, M., 2007. Human anti-saliva immune response following experimental exposure to the visceral leishmaniasis vector, *Lutzomyia longipalpis*. European Journal of Immunology 37, 3111–3121.

Vlkova, M., Sima, M., Rohousova, I., Kostalova, T., Sumova, P., Volfova, V., Jaske, E.L., Barbian, K.D., Gebre-Michael, T., Hailu, A., Warburg, A., Ribeiro, J.M., Valenzuela, J.G., Jochim, R.C., Volf, P., 2014. Comparative analysis of salivary gland transcriptomes of *Phlebotomus orientalis* sand flies from endemic and non-endemic foci of visceral leishmaniasis. PLoS Neglected Tropical Diseases 8, e2709.

Volfova, V., Hostomska, J., Cerny, M., Votypka, J., Volf, P., 2008. Hyaluronidase of bloodsucking insects and its enhancing effect on leishmania infection in mice. PLoS Neglected Tropical Diseases 2, e294.

Waitumbi, J., Warburg, A., 1998. *Phlebotomus papatasi* saliva inhibits protein phosphatase activity and nitric oxide production by murine macrophages. Infection and Immunity 66, 1534–1537.

Waters, L.S., Taverne, J., Tai, P.C., Spry, C.J., Targett, G.A., Playfair, J.H., 1987. Killing of *Plasmodium falciparum* by eosinophil secretory products. Infection and Immunity 55, 877–881.

Weatherhead, P., Bennett, G., 1991. Ecology of red-winged Blackbird parasitism by haematozoa. Canadian Journal of Zoology 69, 2352–2359.

Weidinger, S., Gieger, C., Rodriguez, E., Baurecht, H., Mempel, M., Klopp, N., Gohlke, H., Wagenpfeil, S., Ollert, M., Ring, J., Behrendt, H., Heinrich, J., Novak, N., Bieber, T., Kramer, U., Berdel, D., von Berg, A., Bauer, C.P., Herbarth, O., Koletzko, S., Prokisch, H., Mehta, D., Meitinger, T., Depner, M., von Mutius, E., Liang, L., Moffatt, M., Cookson, W., Kabesch, M., Wichmann, H.E., Illig, T., 2008. Genome-wide scan on total serum IgE levels identifies FCER1A as novel susceptibility locus. PLoS Genetics 4, e1000166.

Weinkopff, T., de Oliveira, C.I., de Carvalho, A.M., Hauyon-La Torre, Y., Muniz, A.C., Miranda, J.C., Barral, A., Tacchini-Cottier, F., 2014. Repeated exposure to *Lutzomyia* intermedia sand fly saliva induces local expression of interferon-inducible genes both at the site of injection in mice and in human blood. PLoS Neglected Tropical Diseases 8, e2627.

WHO/Leishmaniasis, 2014. In: Leishmaniasis: Magnitude of the Problem. http://www.who.int/leishmaniasis/burden/magnitude/burden_magnitude/en/index.html.

Wilson, R.J., McGregor, I.A., 1973. Immunoglobulin characteristics of antibodies to malarial S-antigens in man. Immunology 25, 385–398.

Wu, Y., Wang, Q.H., Zheng, L., Feng, H., Liu, J., Ma, S.H., Cao, Y.M., 2007. *Plasmodium yoelii*: distinct CD4(+)CD25(+) regulatory T cell responses during the early stages of infection in susceptible and resistant mice. Experimental Parasitology 115, 301–304.

Xu, X., Oliveira, F., Chang, B.W., Collin, N., Gomes, R., Teixeira, C., Reynoso, D., My Pham, V., Elnaiem, D.E., Kamhawi, S., Ribeiro, J.M., Valenzuela, J.G., Andersen, J.F., 2011. Structure and function of a "yellow" protein from saliva of the sand fly *Lutzomyia longipalpis* that confers protective immunity against Leishmania major infection. The Journal of Biological Chemistry 286, 32383–32393.

Zahedifard, F., Gholami, E., Taheri, T., Taslimi, Y., Doustdari, F., Seyed, N., Torkashvand, F., Meneses, C., Papadopoulou, B., Kamhawi, S., Valenzuela, J.G., Rafati, S., 2014. Enhanced protective efficacy of nonpathogenic recombinant *Leishmania tarentolae* expressing cysteine proteinases combined with a sand fly salivary antigen. PLoS Neglected Tropical Diseases 8, e2751.

Zeidner, N.S., Higgs, S., Happ, C.M., Beaty, B.J., Miller, B.R., 1999. Mosquito feeding modulates Th1 and Th2 cytokines in flavivirus susceptible mice: an effect mimicked by injection of sialokinins, but not demonstrated in flavivirus resistant mice. Parasite Immunology 21, 35–44.

Zer, R., Yaroslavski, I., Rosen, L., Warburg, A., 2001. Effect of sand fly saliva on Leishmania uptake by murine macrophages. International Journal for Parasitology 31, 810–814.

Zhu, J., Paul, W.E., 2008. CD4 T cells: fates, functions, and faults. Blood 112, 1557–1569.

Chapter 5

Tick Saliva and Its Role in Pathogen Transmission

Sarah Bonnet[1], Mária Kazimírová[2], Jennifer Richardson[3], Ladislav Šimo[1]

[1]UMR BIPAR 956 INRA-ANSES-ENVA, Maisons-Alfort, France; [2]Institute of Zoology, Slovak Academy of Sciences, Bratislava, Slovakia; [3]UMR 1161 Virologie INRA-ANSES-ENVA, Maisons-Alfort, France

INTRODUCTION

Ticks are obligate hematophagous arthropods that belong to two main families: hard and soft ticks. These two major tick families have evolved different feeding strategies. Feeding in hard—or *Ixodidae*—ticks is a slow and complex process, taking several days for repletion and detachment alone, whereas soft—or *Argasidae*—ticks usually complete a blood meal in less than 1 h, except larval stages of certain species, which feed for longer periods or not at all (Mans, 2011).

The long feeding period of ticks necessitates extended control over the host's hemostasis and immunity, both systemically and locally in the skin. During the feeding process, ticks inject saliva and absorb blood in an alternating pattern into and from the wound. They are pool feeders, sucking all the fluids that are exuded into the wound generated by the bite. They begin attachment by cutting and tearing the host skin with their chelicerae, allowing the penetration of their hypostome and laceration of dermal tissues and capillaries, creating a hemorrhagic pool from which they collect the nutritive fluids. Both the hypostome and cement secreted by the salivary glands (SGs) contribute to anchoring the tick in its host. Owing to de novo synthesis of expandable cuticle, hard ticks are able to imbibe enough blood to increase in size as much as 100 times for certain females (Sonenshine et al., 2014). Ticks have larval, nymphal, and adult stages, all of which generally require a blood meal or even several for soft ticks to complete their development and reproduction. A two- or three-host life cycle, each of which including host-seeking, feeding, and off-host moulting (or egg-laying for the female), is the most common developmental pattern for the majority of hard ticks of medical and veterinary importance, with the exception of *Rhipicephalus (Boophilus) microplus*, which is a one-host tick species.

Skin and Arthropod Vectors. https://doi.org/10.1016/B978-0-12-811436-0.00005-8

121

Indeed, ticks likely possess the most complex feeding biology of all hematophagous arthropods, and their capacity to imbibe a very large quantity of blood over a relatively long period largely explains their remarkable success as disease vectors (Bonnet and Liu, 2012). This success is bolstered by their longevity and high reproductive potential, as well as by the very large host spectrum for several species. They indeed represent the most important vectors of pathogens that affect animals worldwide and are second only to mosquitoes where humans are concerned (Dantas-Torres et al., 2012). In addition, they surpass all other arthropods in the variety of pathogenic microorganisms—which include helminths, viruses, bacteria, and protozoa—they are able to transmit (Rizzoli et al., 2014). For most tick-borne pathogens (TBP), including RNA and DNA viruses, bacteria such as *Anaplasma* spp., *Borrelia* spp., *Rickettsia* spp., and the protozoa *Babesia* spp. and *Theileria* spp., transmission to the vertebrate host occurs via the saliva injected during the tick bite, underscoring the importance of both SGs and saliva in the transmission process. Usually, invasion of SGs and/or pathogen multiplication occurs within the SGs following the stimulus of blood feeding, which might reduce the stress that the presence and multiplication of a pathogen would otherwise induce in questing ticks. Several studies have reported that tick SGs differentially express transcripts and proteins in response to pathogen infection, and tick SG factors are likely implicated in pathogen transmission (Liu and Bonnet, 2014). Tick saliva also contains a large number of various non-proteinaceous substances and secreted proteins that are differentially expressed in tick developmental stages, sexes, and during the course of blood feeding. To assure their feeding, ticks have evolved a complex and sophisticated pharmacological armament against the potentially harmful processes represented by host hemostasis, inflammation, and adaptive immunity, and many proteins present in tick saliva dampen these host defenses and facilitate the flow of blood to assure adequate feeding (Francischetti et al., 2009; Kazimirova and Stibraniova, 2013; Kotal et al., 2015; Ribeiro, 1995). Lastly, it has been demonstrated that these molecules create a favorable environment for transmission, survival, and propagation of TBP within the vertebrate host (Kazimírová and Štibrániová, 2013; Wikel, 1999).

Tick Salivary Glands

Tick SGs are an extremely versatile organ playing multiple essential functions during both on- and off-host periods to meet the tick's varying physiological requirements. Their crucial importance in development and transmission of TBP makes them without a doubt the most studied tissue among the various internal organs of ticks.

Structure of Salivary Glands

The paired SGs form a grapelike cluster of acini (or otherwise called alveoli) located anterolaterally on both sides of the tick body cavity and represent the largest glands in the tick's body (Fig. 5.1). A single SG in *Rhipicephalus*

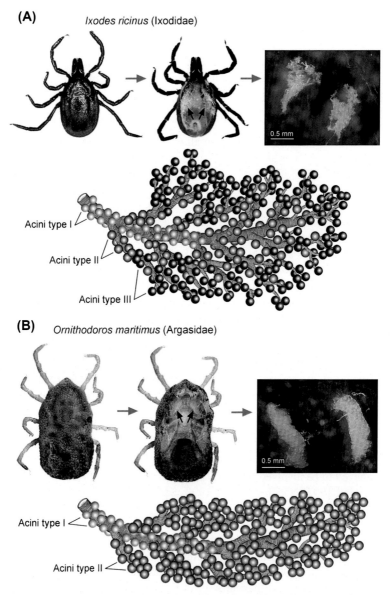

FIGURE 5.1   Female salivary glands of the two tick families Ixodidae (hard ticks) and Argasidae (soft ticks). (A) *Ixodes ricinus* (Ixodidae). (B) *Ornithodoros maritimus* (Argasidae). A schematic drawing of the salivary gland is shown below the photographs. Three types (I, II, and III) of salivary gland acini are present in Ixodidae, while only two types (I and II) are found in Argasidae. Of note, the majority of internal organs have been removed to show the position of the salivary glands (*black arrows* in A and B) in the tick body. Also of note, the size of the ticks is not to scale. The *O. maritimus* ticks were a generous gift from the laboratory of Dr. Karen D. McCoy, French National Centre for Scientific Research (CNRS, Montpellier, France). *(Photo author: Dr. Ladislav Šimo, French National Institute for Agricultural Research (INRA, UMR-BIPAR, Maisons-Alfort, France).)*

appendiculatus consists of approximately 1400 acini of type I–III and 1350 acini of type I–IV in females and males, respectively (Walker et al., 1985). The main excretory ducts exit each of the SGs and continue anteriorly into the mouthparts (part of the capitulum), where they merge into a single salivarium. The secondary salivary ducts, which are smaller in diameter, connect directly to the main SG duct and branch more distally into thinner lobular (tertiary) ducts (Sonenshine, 1991). Two major types of SG acini, agranular and granular, are commonly recognized in both argasid and ixodid ticks. While only one type of agranular acinus (type I) is present in all tick species, a single type (type II) of granular acinus is typically found in argasid ticks and multiple types (types II, III, and the male-specific type IV) are commonly found in ixodid ticks (Fig. 5.1) (Sonenshine, 1991). Agranular type I acini join exclusively the anterior part of the main salivary duct and occasionally proximal regions of the associated secondary ducts. The granular types II and III acini occupy more distally located secondary and tertiary ducts, respectively (Fig. 5.1) (Binnington, 1978; Coons and Roshdy, 1973; Fawcett et al., 1981a; Sonenshine, 1991; Walker et al., 1985).

The agranular type I acini are the least abundant acini in SGs (Needham and Teel, 1986) and no significant change in their mean diameter has been observed during tick feeding (Fawcett et al., 1986). The cellular organization of agranular acini is similar in both argasid and ixodid ticks (Needham et al., 1990; Sonenshine, 1991). Four distinct cell types that lack large secretory granules have been found in these structures: a single central lamellate cell, multiple peripheral lamellate cells, peritubular cells, and one circumlumenal cell. Both the central lamellate cell, which is the only cell that lies in direct contact with the acinar lumen, and the peripheral lamellate cells comprise multiple convoluted outer membrane infoldings that greatly increase the surface exposed to hemolymph (Fawcett et al., 1986). The basal peritubular and circumlumenal cells line the acinar duct, which opens directly into the main SG duct (Coons and Roshdy, 1973; Fawcett et al., 1986; Needham et al., 1990). The type I acini are believed to be involved in the uptake of water vapor in fasting ticks (Gaede and Knülle, 1997; Needham et al., 1990; Sauer et al., 1995) (see Function of Salivary Glands).

The type II (in both argasid and ixodid ticks) and type III (in ixodid ticks) granular acini are characterized by the presence of diverse glandular cells, most of them containing a dense accumulation of secretory granules (Fawcett et al., 1986; Sonenshine, 1991). In general, the structure of granular acini is much more complex in ixodid than in argasid ticks (Sonenshine, 1991). In ixodid ticks, both type II and III acini contain a valve separating the short acinar duct from the associated secondary lobular ducts, while this structure is absent in argasid ticks (Sonenshine, 1991). From a histological point of view, a number of different (7–9) granular cell types have been commonly recognized in granular acini, although the number may vary among the different tick species (Binnington, 1978; Sonenshine, 1991). In general, granular cell types *a, b,* and *c* in type II acini and *d, e,* and *f* in type III acini have been described in ixodid ticks (Binnington, 1978). Moreover, several non-granular cell types, such as epithelial, a single adlumenal, and multiple ablumenal interstitial cells, have been detected between the individual granular cells (Fawcett et al., 1986). In addition, multiple neck cells surrounding the basal part of the acinar duct have been found in both type II and III acini (Binnington, 1978; Coons and Kaufman, 1988; Fawcett et al., 1986). The single adlumenal cell, also called the cap or myoepithelial cell (Krolak et al., 1982; Meredith and Kaufman, 1973), lines the luminal surface of the acinus in weblike fashion. This cell has been shown to be highly innervated and suggested to possess myoepithelial properties that contribute to expulsion of the acinar contents into the connecting ducts (Coons et al., 1994; Krolak et al., 1982; Šimo et al., 2014), as will be discussed in a later section (see Physiology of Salivary Gland Secretion).

In ixodid females, both type II and III acini undergo remarkable morphological transformation over the course of feeding, without any change in cell number, resulting in overall increase in the mass of the SGs (Binnington, 1978; Fawcett et al., 1986). In *Ixodes scapularis* females, the mean diameter of type II and III acini is about 45 μm when unfed but more than 150 μm in the final feeding stages (Šimo et al., 2013). These dramatic changes in proportion are due to hypertrophy of the majority of the acinar cells, as a result of enlargement of their cytoplasm and nuclei and often accompanied by mitochondrial proliferation (Binnington, 1978; Coons and Kaufman, 1988; Sauer et al., 1995; Walker et al., 1985). The interstitial ablumenal cells form a labyrinth in extracellular spaces, which is more obvious in type III acini (Coons and Kaufman, 1988; Fawcett et al., 1981a, 1981b). In particular, in type III acini, ablumenal cells interdigitate with adjacent apical *f* cells, together forming a complex epithelium that transports fluid and electrolytes into the acinus from the hemolymph (Fawcett et al., 1981a; Meredith and Kaufman, 1973). Due to fluid uptake, the lumen of these acini greatly expands and the acinar cells become flatter (Kim et al., 2014). By contrast, in type II acini, which are thought to be involved in secretion of proteins and lipids rather than fluid, the cell bodies enlarge and the lumen remains proportionally smaller.

The structure of the male-specific type IV acinus is similar to that of types II and III but with an additional granular cell type *g* that fills with secretory granules over the feeding period (Binnington, 1978; Walker et al., 1985). The role of type IV acini in male SGs has been associated with reproduction since their secretions contribute to the efficient transfer of spermatophore to the female genital aperture during mating (Feldman-Muhsam et al., 1970).

Function of Salivary Glands

The multifunctional tick SGs play an important physiological role essential for tick survival during both on- and off-host stages. Their functional activities include (1) absorption of moisture from unsaturated atmosphere during the fasting period, (2) control of osmoregulation during tick feeding (ixodid ticks), (3) production of a cement-like substance to anchor the mouthparts within the host's skin (in most ixodid tick species) and secretion of toxins (in some tick species), and (4) secretion of biologically active molecules that facilitate acquisition of the blood meal and modulate the host immune responses. In addition, the tick SGs are an important site for (5) TBP development, and SG secretions represent the principal route for their transmission. Here we briefly discuss the first three functional activities, while the last two are elaborated later in this chapter.

1. Most tick species spend the vast majority of their life in the fasting period off the host. During this life stage, the conservation of water is critical for them to avoid death due to desiccation. It has been well documented that in unfed ixodid ticks the agranular type I acini of SGs are actively involved in uptake of water vapor from the surrounding environment by secretion of highly hygroscopic saliva onto the surface of the hypostome, which lies in proximity to the preoral canal (Gaede and Knülle, 1997; Knulle and Rudolph, 1982; Needham et al., 1990; Needham and Teel, 1986; Sonenshine, 1991). Here, the atmospheric moisture is absorbed by the highly salty crystalized deposit and subsequently swallowed back (Gaede and Knülle, 1997; Yoder et al., 2006). The absorptive function of SGs was elegantly confirmed by a recent study in which a fluorescent dye (rhodamine 123) imbibed by desiccated ticks was shown to accumulate exclusively in type I acini (Kim et al., 2016a).

2. A female ixodid tick is capable of increasing its weight up to 100-fold when attached to a host (Balashov, 1972; Sauer et al., 1995; Sonenshine, 1991). Homeostasis during tick feeding requires extensive elimination of excess water and ions. In contrast to insects in which the Malpighian tubules carry out these processes, in ixodid ticks this function is mainly accomplished by their SGs. In particular, water and ions from the digested blood meal are transferred via the gut wall into the hemolymph where they are taken up by SGs and secreted through the salivarium back to the host. During this process the nutrient portion of the blood meal concentrates and the level of solutes and water in hemolymph remains constant (Sauer et al., 1995). In *Dermacentor andersoni* females 74% of the water and 96% of the sodium are expelled back to the host via this route (Kaufman and Phillips, 1973). The transport and secretion of fluids and small molecules by SGs is supported by the marked enlargement of the acinar lumen (in the most abundant type III acini), in proportion to surrounding cells, which periodically fills with fluid over the feeding period (Kim et al., 2014; Meredith and Kaufman, 1973). The epithelium of these acini has been suggested to be the site for transport of water and electrolytes from hemolymph into the SGs (Fawcett et al., 1981b; Meredith and Kaufman, 1973). In addition, a role for the sodium potassium pump (Na/K-ATPase) in formation of the sodium-rich primary saliva in type III acini has been recently proposed (Kim et al., 2016a). In argasid ticks, the mechanism by which excess water is eliminated is different and is accomplished by the coxal glands, which are unique to this tick family (Binnington, 1975).

3. When attached to a host, tick SGs secrete a cement-like substance allowing them to anchor their hypostome firmly in the host skin (Kemp et al., 1982) and simultaneously protect the mouthparts from the host immune system (Alekseev et al., 1995). Of note, this feature has been observed exclusively in the ixodid lineage, although in the Prostriata (only the *Ixodes* genus) not all species produce cement (Kemp et al., 1982; Mans, 2014). In all members of the Metastriata (all Ixodid genera except the *Ixodes* genus) the size of the cement cone, fashioned from polymerized and hardened glycine-rich proteins, lipids, and certain carbohydrates, is related to the size of their mouthparts (Kemp et al., 1982; Moorhouse and Tatchell, 1966). It has been well documented that in ticks with short mouthparts (*Dermacentor* spp., *Rhipicephalus* spp., *Haemaphysalis* spp.) an external cement cone anchors the external portion of the hypostome and the palps, whereas in ticks with long mouthparts (*Amblyomma* spp., *Aponomma* spp., and *Hyalomma* spp.) the cone encases the entire length of the hypostome—including its external and skin-inserted portions—and palps (Kemp et al., 1982). In the SGs, synthesis of cement is believed to be associated with *a* cells in type II acini and *e* and *d* cells in type III acini (Chinery, 1973; Jaworski et al., 1992). Recent proteomic analyses of the cement cone from *Amblyomma americanum* blood-fed on artificial membrane have revealed multiple classes of proteins including glycine-rich proteins, serine protease inhibitors, metalloproteases, and unidentified secreted proteins (Bullard et al., 2016). Some of these proteins have been shown to have antigenic properties and are considered to be promising anti-tick vaccine candidates (Bishop et al., 2002; Mulenga et al., 1999; Shapiro et al., 1987).

About 81 tick species around the world have been shown to be associated with toxicoses causing paralysis (Gothe, 1999). This phenomenon has been most frequently attributed to ixodid ticks, though toxicity of SG components has been described for some argasid ticks as well (Gothe, 1999; Hall-Mendelin et al., 2011; Mans et al., 2003). Tick-derived toxins originate from distinct cells in the SGs. As regards the best-known example, the Australian paralysis tick *Ixodes holocyclus*, *e* and *b* cells in type II and III acini, respectively, have been suggested to be a potential source of these substances (Binnington and Stone, 1981; Stone et al., 1989).

The additional functions of SGs deployed during feeding, such as secretion of a variety of pharmacologically active substances facilitating the blood meal uptake or modulating host immune responses, are discussed below in this chapter.

Physiology of Salivary Gland Secretion

The tick SGs are a highly innervated tissue (Fawcett et al., 1986; Kaufman and Harris, 1983; Šimo et al., 2014a, b; Šimo et al., 2008, 2009, 2012, 2014) and to date there is no data regarding hormonal control of this organ (Sauer et al., 2000). While earlier studies of SG physiology were largely based on pharmacological approaches, more recent studies in the post-genomic era have led to breakthroughs in the molecular understanding of the processes underlying regulation of this organ. Tick SG secretion is a complex dynamic process involving multiple biological pathways, and over the last four decades a number of components that regulate tick SG secretion have been discovered (Bowman and Sauer, 2004; Kaufman, 1978; Kim et al., 2014; Sauer et al., 1995, 2000). An extensive effort has been made to identify the agents devoted to the control of SG secretion in ixodid ticks. Among the multiple pharmacological/biological components that have been considered, catecholamines have been shown to be the most potent activators of SG fluid secretion. In particular, dopamine-mediated tick salivary secretion has been well documented in both *in vivo* and *in vitro* assays in various tick species (Kaufman, 1976; Lindsay and Kaufman, 1986; McSwain et al., 1992; Šimo et al., 2012). It has been suggested that dopamine activates two independent signaling pathways in SGs: the cAMP (cyclic adenosine monophosphate)-dependent protein phosphorylation cascade that induces fluid secretion and the calcium-dependent pathway that activates cytosolic phospholipase A2 and thereby the release of arachidonic acid and ensuing secretion of prostaglandin E_2 (PGE_2). Subsequently, PGE_2 may be secreted into the saliva or may act locally within the SGs themselves to induce the exocytosis of anticoagulants and possibly other proteins via Ca^{2+} mobilization (Qian et al., 1998; Sauer et al., 2000).

For over three decades, researcher believed that in the SGs dopamine was released from the synaptic boutons arising from dopaminergic neurons located in the tick central nervous system, the synganglion (Bowman and Sauer, 2004; Sauer et al., 2000; Sonenshine, 1991). Although there is some evidence that dopamine is present in nerves reaching the tick SGs (Binnington and Stone, 1977; Kaufman and Harris, 1983), several studies have suggested that SG cells themselves may represent the major pool of dopamine acting as an autocrine/paracrine signal in this tissue (Kaufman et al., 1999; Koči et al., 2014; Šimo et al., 2011). Recently, two different dopamine receptors have been found to be expressed in the SGs of *I. scapularis*: the D1 dopamine receptor and the invertebrate-specific D1-like dopamine receptor (InvD1L) (Šimo et al., 2011, 2014b). Immunohistochemistry has revealed the presence of these transmembrane proteins in both type II and III acini. In particular, the D1 receptor has been strongly detected in scattered patches in the cell junctions on the luminal surface of type II and III acini, while InvD1L has been evidenced in the axon terminals (with unknown origin) lining the acinar valve and lumen in the same type of acini. Pharmacological profiling of these receptors in a heterologous expression system has revealed that activation of the D1 receptor preferentially triggers the cAMP-dependent downstream pathway, while activation of the InvD1L resulted exclusively in mobilization of intracellular calcium (Šimo et al., 2011, 2014b). Based on the use of specific agonists/antagonists selectively activating or blocking either of these receptors in *in vitro* assays, a distinct physiological role for each of these receptors has been proposed; namely, the epithelial D1 receptor may regulate the inward transport of fluid into the acini while the axonal InvD1L receptor may control the expulsion of the acinar context into the connecting ducts, presumably via the myoepithelial cell (Kim et al., 2014). Gene knockdown of an aquaporin water channel by RNA inference (RNAi) completely inhibited dopamine-stimulated secretion in isolated *Ixodes ricinus* SGs (Campbell et al., 2010).

Besides dopamine, several others compounds have been shown to exert either a direct or indirect effect on salivary secretion; however, their precise mode of action via their cognate receptors has not as yet been elucidated. Briefly, octopamine has also been shown to induce saliva in *A. americanum*, albeit at a threefold lower magnitude than dopamine (Lindsay and Kaufman, 1986; Needham and Pannabecker, 1983; Pannabecker and Needham, 1985). While treatment with γ-aminobutyric acid does not trigger salivation in isolated SGs, it does potentiate the effect of dopamine in mediating fluid secretion (Lindsay and Kaufman, 1986). Ergot alkaloids, which are exogenous molecules in ticks, also stimulate fluid secretion via an unknown receptor (Kaufman and Wong, 1983). A cholinomimetic alkaloid, pilocarpine, has been frequently used as a highly potent activator of tick oral secretion *in vivo* but has failed to stimulate fluid secretion in isolated SGs (Hsu and Sauer, 1975; Kaufman, 1978; McSwain et al., 1992).

Recent discovery of several neuropeptides in the axonal projections reaching different parts of the tick SGs has shed new light on the physiology of this tissue (Roller et al., 2015; Šimo et al., 2008, 2009, 2012, 2014a,b). More particularly, two neuropeptides, myoinhibitory peptide (MIP) and SIFamide, have been colocalized in the pair of giant neuronal cells in the synganglion that innervates all of the basal regions of the type II and III acini of the SGs. The SIFamide receptor was found to be expressed close to the acinar valve, suggesting its role in control of this structure (Šimo et al., 2009, 2013). Two other putative neuropeptides, orcokinin and pigment dispersing factor, were evidenced by immunostaining in axon terminals that line exclusively the lumen of type II acini (Roller et al., 2015; Šimo et al., 2009, 2012). Thus, depending on the physiological requirement, it appears that ticks are capable of selectively controlling particular types of acini (and likely individual cells within the acini) via their neuropeptidergic network.

Although a number of the various mechanisms underlying tick SG secretion have been studied in detail, the majority of the pathways involved in the regulation of this tissue remain enigmatic. Complete elucidation of the complex physiology of the tick SGs has a strong potential to bring to light new strategies to impede tick feeding and transmission of TBP.

Tick Saliva

Tick Saliva Composition

The prolonged period of tick attachment to their host has sparked great interest in studying tick SG secretions during feeding. Molecular definition of the composition of tick saliva is a prerequisite for a full understanding of the mechanisms implicated in not only tick feeding but also in pathogen transmission. This may lead to the discovery of new protective antigens and markers of exposure to ticks that may be of utility in epidemiological studies.

Advances in global profiling have had a profound impact on the number of tick salivary components that may be studied at one time. Indeed, transcript and protein profiling in tick saliva have been applied to several species of hard and soft ticks (Table 5.1). These studies have led to the discovery of several factors that contribute to successful feeding and evasion of the host immune and hemostatic defenses. The first proteomic studies that addressed tick saliva date from the first decade of the 21st century (Madden et al., 2004). Since then, the salivary transcriptomes and proteomes of diverse hard tick species have been established, whereas comparatively few studies have concerned the soft ticks (Table 5.1). In addition, comparative analyses of tick SGs have reported that molecular expression varied according to tick life stage, sex, or behavior (Anatriello et al., 2010; Diaz-Martin et al., 2013b; Ribeiro et al., 2006), as well as in relation to pathogen infection (Liu and Bonnet, 2014). It should be noted, however, that a minority of the salivary proteins have been functionally annotated, and the assigned function has been verified for fewer than 5% of the annotated proteins (Francischetti et al., 2009). On sequencing the transcriptome of the *I. ricinus* SGs, Liu et al. have shown that most of the identified transcripts could be assigned to "oxidation reduction" biological processes (12.8%), with the most abundant term for the cellular components category being "integral to membrane" (11.4%) and the most abundant term for the molecular function category being "binding proteins" (63.2%) (Liu et al., 2014b). In the same tick species, Schwarz et al. reported that 13% of the sequenced contigs from SGs were classified as encoding secreted proteins and, among these, protease inhibitors were the largest group (2.6% of all contigs) (Schwarz et al., 2013). In addition to these global analyses, several studies have focused on specific components of tick saliva in relation to their role in tick feeding and transmission of TBP, which are reviewed below.

Role of Saliva During Tick Feeding

Vertebrates react to skin injury by formation of a hemostatic plug, vasoconstriction, inflammation, and tissue remodeling related to wound healing. If unchecked, these processes would disrupt tick feeding, cause rejection of the ticks, and arrest their further development. To ensure completion of their blood meal, however, ticks have evolved countermeasures to host responses in the form of a large array of bioactive salivary molecules that inhibit hemostasis, block pain and itch, and modulate innate and adaptive immune responses, angiogenesis, and wound healing of their hosts (Chmelař et al., 2016b; Francischetti et al., 2009; Kazimírová and Štibrániová, 2013; Kotál et al., 2015; Mans, 2011; Štibrániová et al., 2013; Valdes, 2014; Wikel, 2013). The composition of tick saliva is complex and its components display pleiotropy and redundancy in their

TABLE 5.1 Overview of Studies on the Tick Sialome: Comparison of Developmental Stages, Sexes, Feeding Stages, and Species

Tick Species	Developmental Stage, Sex, Feeding Stage	Technique Used	References
Transcriptomic Studies			
Argasidae			
Ornithodoros coriaceus	Adults	LCS	Francischetti et al. (2008a)
Ornithodorus parkeri	Last instar nymphs, adults	LCS	Francischetti et al. (2008b)
Argas monolakensis	Fed and unfed adult females	LCS	Mans et al. (2008a)
Ixodidae			
Ixodes scapularis	Females, partially fed	LCS	Valenzuela et al. (2002)
	Females, unfed, 6–12 h, 18–24 h, and 3–4 days after attachment	LCS	Ribeiro et al. (2006)
	Nymphs, unfed, 3-days fed	LCS	McNally et al. (2012)
	Females, 24 h fed, immunogenic proteins	LCS, NGS	Lewis et al. (2015)
Ixodes pacificus	Females, questing	LCS	Francischetti et al. (2005b)
Ixodes ricinus	Females, unfed, 24 h, 4 and 7 days after attachment	LCS	Chmelař et al. (2008)
	Nymphs, females, early and later after attachment, cofeeding, fed on different laboratory hosts	NGS	Schwarz et al. (2013)
	Nymphs, females, 3–36 h fed	NGS	Schwarz et al. (2014)
	Females, partially fed	NGS	Liu et al. (2014b)
	Nymphs, adults, partially fed	NGS	Kotsyfakis et al. (2015)

Continued

TABLE 5.1 Overview of Studies on the Tick Sialome: Comparison of Developmental Stages, Sexes, Feeding Stages, and Species—cont'd

Tick Species	Developmental Stage, Sex, Feeding Stage	Technique Used	References
Amblyomma variegatum	Females, partially fed	LCS	Nene et al. (2002)
	Females	LCS	Lambson et al. (2005)
	Females, partially fed, collected from free-roaming zebu cattle	LCS	Ribeiro et al. (2011)
Amblyomma cajennense	Females, partially fed	LCS	Batista et al. (2008)
	Females feeding on natural hosts, 3–4 days postattachment	LCS	Maruyama et al. (2010)
	Nymphs, females, partially fed on different hosts	LCS	Garcia et al. (2014)
Amblyomma americanum	Females, partially fed	LCS	Aljamali et al. (2009)
	Males, females, 24 and 96h fed, immunogenic proteins	LCS, NGS	Radulović et al. (2014)
	Females, partially fed	NGS	Karim and Ribeiro (2015)
Amblyomma maculatum	Females, partially fed	LCS	Karim et al. (2011)
Amblyomma triste, Amblyomma parvum	Nymphs, females, partially fed on different hosts	LCS	Garcia et al. (2014)
Haemaphysalis longicornis	Females, partially fed	LCS	Nakajima et al. (2005)
Haemaphysalis flava	Adults, semi- and fully engorged	NGS	Xu et al. (2015)
Dermacentor variabilis	Females, 4-day fed	DD-PCR	Macaluso et al. (2003)

TABLE 5.1 Overview of Studies on the Tick Sialome: Comparison of Developmental Stages, Sexes, Feeding Stages, and Species—cont'd

Tick Species	Developmental Stage, Sex, Feeding Stage	Technique Used	References
Dermacentor andersoni	Females, 18–24 h after attachment	LCS	Alarcon-Chaidez et al. (2007)
	Females, partially fed	LCS	Mudenda et al. (2014)
Rhipicephalus appendiculatus	Females, 4-day fed	LCS	Nene et al. (2004)
	Females	LCS	Lambson et al. (2005)
	Males, females, partially fed	NGS	de Castro et al. (2016)
Rhipicephalus (Boophilus) microplus	Females, males, rapidly engorging	LCS	Santos et al. (2004)
	Males, partially fed	SSH	Zivkovic et al. (2010)
	Larvae, adults, partially fed	SSH	Lew-Tabor et al. (2010)
	Nymphs, females, males, partially fed on different bovine strains, expression profiles of transcripts for putative anti-hemostatic proteins	LCS	Carvalho et al. (2010)
	Females feeding on natural hosts, 3–4 days postattachment	LCS	Maruyama et al. (2010)
Rhipicephalus sanguineus	Females, 3–5 day fed	LCS	Anatriello et al. (2010)
	Females feeding on natural hosts, 3–4 days postattachment	LCS	Maruyama et al. (2010)
Rhipicephalus haemaphysaloides	Males, females, partially fed	SSH	Xiang et al. (2012)
	Females, partially fed, cysteine proteases	NGS	Yu et al. (2015)

Continued

TABLE 5.1 Overview of Studies on the Tick Sialome: Comparison of Developmental Stages, Sexes, Feeding Stages, and Species—cont'd

Tick Species	Developmental Stage, Sex, Feeding Stage	Technique Used	References
Rhipicephalus pulchellus	Females, males, partially fed	NGS	Tan et al. (2015)
Hyalomma marginatum rufipes	Females, partially fed	LCS	Francischetti et al. (2011)
Proteomic Studies			
Argasidae			
Ornithodoros moubata	Adults	DGE–HPLC MS/MS	Oleaga et al. (2007)
	Females, males	LC–MS/MS	Diaz-Martin et al. (2013b)
Ornithodoros erraticus	Adults	DGE–HPLC MS/MS	Oleaga et al. (2007)
O. parkeri	Last instar nymphs, adults	DGE–HPLC MS/MS	Francischetti et al. (2008b)
O. coriaceus	Adults	DGE–HPLC MS/MS	Francischetti et al. (2008a)
Ixodidae			
Ixodes scapularis	Nymphs, 24 and 66h fed	DIGE–HPLC	Narasimhan et al. (2007)
	Females, 24, 48, 72, 96, and 120h fed, engorged but not detached	LC–MS/MS	Kim et al. (2016b)
Amblyomma americanum	Females	RPLC	Ribeiro et al. (1992)
	Females, partially fed	DGE-MALDI-TOF MS	Madden et al. (2004)
	Nymphs, adults	DGE	Villar et al. (2014)
Amblyomma maculatum	Females, partially fed	DGE-MALDI-TOF MS	Madden et al. (2004)

TABLE 5.1 Overview of Studies on the Tick Sialome: Comparison of Developmental Stages, Sexes, Feeding Stages, and Species—cont'd

Tick Species	Developmental Stage, Sex, Feeding Stage	Technique Used	References
Amblyomma variegatum	Females, fed, collected from free-roaming zebu cattle	nanoRPLC-MS/MS	Ribeiro et al. (2011)
	Adults	DGE	Villar et al. (2014)
Amblyomma cajennense	Adults	DGE	Villar et al. (2014)
Haemaphysalis longicornis	Saliva, fully engorged nymphs and females	LC–MS/MS	Tirloni et al. (2015)
Dermacentor andersoni	Females, partially fed	2D-LC–MS/MS	Mudenda et al. (2014)
Rhipicephalus sanguineus	Females, partially fed	DG-HPLC MS/MS	Oliveira et al. (2013)
Rhipicephalus (boophilus) microplus	Females, different feeding stages	LC–MS/MS	Tirloni et al. (2014)

DD-PCR, differential display polymerase chain reaction; *DGE*, dimensional gel electrophoresis; *DIGE*, differential in-gel electrophoresis; *HPLC*, high-performance liquid chromatography; *LCS*, sequencing of cDNA library clones; *MALDI-TOF*, matrix-assisted laser desorption/ionization time-of-flight; *MS*, mass spectrometry; *NGS*, next-generation sequencing Techniques; *RPLC*, reverse phase liquid chromatography; *SSH*, suppression subtractive hybridization.

effect on host responses (see examples in Table 5.2). Nevertheless, in spite of increasing knowledge of the compounds secreted in tick saliva, only a small number of salivary molecules have been functionally characterized (Table 5.2), the function of the majority remaining completely unknown.

Effects of Tick Feeding on Host Hemostasis

Hemostasis is a series of physiological events that control blood loss after vascular injury by termination of bleeding from damaged blood vessels (vasoconstriction), formation of a platelet plug, blood coagulation, and fibrinolysis (Hoffman et al., 2009). Ticks have developed effective mechanisms to block different arms of the hemostatic system of their hosts; saliva of a single species of tick typically contains multiple molecules that inhibit hemostasis, of which some target individual host factors while others display multiple functions (Chmelar et al., 2012; Francischetti et al., 2009; Koh and Kini, 2009; Mans and Neitz, 2004; Steen et al., 2006; Valenzuela, 2004) (Table 5.2).

TABLE 5.2 Examples of Biological Activities of Tick Saliva and Salivary Molecules in Modulation of Host Defense Reactions

Tick Species	Molecule	Function/Target And/Or Activity	References
Argasidae			
Ornithodoros spp.	Apyrase	Platelet aggregation inhibition/ATP, ADP	Mans et al. (1998)
Ornithodoros moubata	Moubatin	Platelet aggregation inhibition/collagen receptor	Waxman and Connolly (1993)
	Disaggregin	Platelet aggregation inhibition/integrin antagonist	Karczewski et al. (1994)
	Ornithodorin	Anticoagulation/thrombin	van de Locht et al. (1996)
	TAP	Anticoagulation/FXa	Waxman et al. (1990)
	OmCI	Complement inhibition/C5, prevention of interaction of C5 with C5 convertase	Nunn et al. (2005)
	Enolase	Stimulation of fibrinolysis/plasminogen receptor	Diaz-Martin et al. (2013a)
Ornithodoros savignyi	Savignygrin	Platelet aggregation inhibition/integrin antagonist	Mans et al. (2002b)
	Savignin	Anticoagulation/thrombin	Nienaber et al. (1999)
	TAP-like protein	Anticoagulation/FXa	Joubert et al. (1998)
	TSGP2, TSGP3	Complement inhibition/C5, prevention of interaction of C5 with C5 convertase, binding to the neutrophil chemoattractant leukotriene B4	Mans and Ribeiro (2008a)

TABLE 5.2 Examples of Biological Activities of Tick Saliva and Salivary Molecules in Modulation of Host Defense Reactions—cont'd

Tick Species	Molecule	Function/Target And/ Or Activity	References
Argas monolakensis	Monobin	Anticoagulation/ thrombin	Mans and Ribeiro (2008)
	Monogrin	Platelet aggregation inhibition/integrin antagonist	
Ixodidae	Cystatins	Immunomodulation	Schwarz et al. (2012)
Ixodes scapularis	Prostacyclin	Vasodilation	Ribeiro and Mather (1998)
	tHRF	Vasodilation	Dai et al. (2010)
	Apyrase	Platelet aggregation inhibition/ATP, ADP	Ribeiro et al. (1985)
Ixodes scapularis, Ixodes pacificus	Ixodegrin	Platelet aggregation inhibition/integrin antagonist	Francischetti et al. (2005b)
Ixodes scapularis	Ixolaris	Anticoagulation/TF pathway inhibitor	Francischetti et al. (2002)
	Penthalaris	Anticoagulation/TF pathway inhibitor	Francischetti et al. (2004)
	Ixolaris	Tumor growth inhibitor/ inhibition of direct TFFVIIaPAR2 signaling and anticoagulation	Carneiro-Lobo et al. (2012)
	Salp14	Anticoagulation/intrinsic pathway	Narasimhan et al. (2002)
	TIX-5	Anticoagulation/ inhibitor of FXa-mediated FV activation	Schuijt et al. (2013)
	Isac	Complement inhibition/ alternative complement pathway, interacts with C3 convertase	Valenzuela et al. (2000)

Continued

TABLE 5.2 Examples of Biological Activities of Tick Saliva and Salivary Molecules in Modulation of Host Defense Reactions—cont'd

Tick Species	Molecule	Function/Target And/Or Activity	References
	Salp20	Complement inhibition/alternative complement pathway, C3 convertase, inhibition of alternative complement pathway -dependent pathogenesis in the mouse	Hourcade et al. (2016) and Tyson et al. (2007)
	Salp15	Immunomodulation/impairs IL-2 production and T cell proliferation; binds *B. burgdorferi* OspC, protects the spirochete from antibody-mediated killing	Anguita et al. (2002) and Ramamoorthi et al. (2005)
	IL-2 binding protein	Immunomodulation/inhibits proliferation of human T cells and CTLL-2 cells	Gillespie et al. (2001)
	ISL 929 and ISL 1373	Immunomodulation/impairs adherence of polymorphonuclear leukocytes	Guo et al. (2009)
	Sialostatin L, L2	Immunomodulation/Inhibits cathepsin L activity	Kotsyfakis et al. (2006)
		Inhibition of IL-9 production from Th9 cells and mast cells	Horka et al. (2012)
	Metalloprotease	Fibrinolysis	Francischetti et al. (2003)
		Wound healing, angiogenesis/inhibits angiogenesis	Francischetti et al. (2005a)
	IxscS-1E1	Anticoagulation, platelet aggregation inhibition/ADP, thrombin, trypsin	Ibelli et al. (2014)

TABLE 5.2 Examples of Biological Activities of Tick Saliva and Salivary Molecules in Modulation of Host Defense Reactions—cont'd

Tick Species	Molecule	Function/Target And/Or Activity	References
Ixodes ricinus	Ir-CPI	Anticoagulation/intrinsic pathway, fibrinolysis	Decrem et al. (2009)
	IRS-2	Immunomodulation/ inhibits host inflammation/α-chymotrypsin, chymase, cathepsin G, trypsin, mMCP-4	Chmelar et al. (2011)
		Inhibition of the IL-6/STAT-3 signaling pathway in DCs/impairs Th17 differentiation	Páleníková et al. (2015)
		Platelet aggregation inhibition/cathepsin G, thrombin	Chmelar et al. (2011)
	Irac I, II, Isac paralogues	Complement inhibition/ alternative complement pathway, interacts with C3 convertase	Daix et al. (2007)
	Iris	Immunomodulation/ modulates T lymphocyte and macrophage responsiveness, induces Th2 type responses	Leboulle et al. (2002)
		Anticoagulation/ thrombin, FXa	Prevot et al. (2006)
	BIP	Immunomodulation/ inhibitor of B cell proliferation	Hannier et al. (2004)
	Ir-LBP	Immunomodulation/ impairs neutrophil functions	Beaufays et al. (2008)
	Metalloproteases	Wound healing, angiogenesis/ involvement in tissue remodeling or disruption through digestion of structural components	Decrem et al. (2008a)

Continued

TABLE 5.2 Examples of Biological Activities of Tick Saliva and Salivary Molecules in Modulation of Host Defense Reactions—cont'd

Tick Species	Molecule	Function/Target And/ Or Activity	References
	saliva	Immunomodulation/ upregulation of MCP-1, TCA-3 and MIP-2 in mouse splenocytes	Langhansová et al. (2015)
	Salp15-like protein	Immunomodulation/ inhibition of IL-10 production, impairment of B cell activity.	Liu et al. (2014a)
Ixodes persulcatus	Salp16 Iper1, Salp16 Iper2	Immunomodulation/ inhibition of bovine neutrophil migration induced by IL-8	Hidano et al. (2014) and Liu et al. (2014b)
Amblyomma americanum	Prostaglandins	Vasodilation	Bowman et al. (1996)
	Americanin	Anticoagulation/ thrombin	Zhu et al. (1997)
	AamS6	Anti-hemostatic/inhibitor of serine- and papain-like cysteine proteases	Mulenga et al. (2013a)
	AamAV422	Anticoagulation, complement inhibition	Mulenga et al. (2013b)
	Serpin19	Anticoagulation/factors Xa and XIa, trypsin, plasmin	Kim et al. (2015b)
	Calreticulin	Anticoagulation/binding of calcium ions	Jaworski et al. (1995)
	MIF homolog	Inhibition of macrophage migration	Jaworski et al. (2001)
Amblyomma variegatum	SGE, fraction AV 16/3	Platelet aggregation inhibition/thrombin	Kazimírová et al. (2002)
	Variegin	Anticoagulation/ Thrombin	Koh et al. (2007)
	Amregulin	Immunomodulation/ inhibition of the secretion of inflammatory factors TNF-alpha, IL-1, IL-8, and IFN-gamma	Tian et al. (2016)

TABLE 5.2 Examples of Biological Activities of Tick Saliva and Salivary Molecules in Modulation of Host Defense Reactions—cont'd

Tick Species	Molecule	Function/Target And/ Or Activity	References
Amblyomma cajennense	Amblyomin-X	Anticoagulation/FXa	Batista et al. (2010)
		Antitumor/cytotoxicity effects in several tumor cell lines, proteasome inhibitor, induces cell cycle arrest	Chudzinski-Tavassi et al. (2016)
	saliva	Immunomodulation/host immune cells, including DCs	Carvalho-Costa et al. (2015)
	saliva	Complement inhibition/classical pathway	Franco et al. (2016)
Amblyomma maculatum	metalloproteases	Degradation of pro-inflammatory peptides/bradykinin	Jelinski (2016)
Haemaphysalis longicornis	Longicornin	Platelet aggregation inhibition/collagen	Cheng et al. (1999)
	Madanin-1; Madanin-2	Anticoagulation/thrombin	Iwanaga et al. (2003)
	Chimadanin	Anticoagulation/thrombin	Nakajima et al. (2006)
	Haemaphysalin	Anticoagulation/FXII/XIIa	Kato et al. (2005)
	Longistatin	Fibrinolysis	Anisuzzaman et al. (2011a)
	Haemangin	Wound healing, angiogenesis/inhibits of angiogenesis	Islam et al. (2009)
	HLTnI	Wound healing, angiogenesis/inhibits angiogenesis	Fukumoto et al. (2006)

Continued

TABLE 5.2 Examples of Biological Activities of Tick Saliva and Salivary Molecules in Modulation of Host Defense Reactions—cont'd

Tick Species	Molecule	Function/Target And/Or Activity	References
Dermacentor variabilis	Variabilin	Platelet aggregation inhibition/integrin antagonist	Wang et al. (1996)
	saliva	Immunomodulation/increases both basal and PDGF-stimulated migration of macrophage IC-21 cells	Kramer et al. (2011)
	PGE$_2$	Wound healing/subvert the ability of macrophages to secrete pro-inflammatory mediators and recruit fibroblasts to the feeding lesion	Poole et al. (2013a)
	saliva	Wound healing/effects on the migratory and invasive activities of metastatic cancer cells	Poole et al. (2013b)
Dermacentor reticulatus	SHBP	Histamine and serotonin binding	Sangamnatdej et al. (2002)
	SGE	Vasodilation/vasoconstriction	Pekárikováa et al. (2015)
Dermacentor andersoni	SGE	Anticoagulation/FV, FVII	Gordon and Allen (1991)
	P36	Immunomodulation/T cell inhibitor	Bergman et al. (2000)
Rhipicephalus appendiculatus	65 kDa protein	Anticoagulation/prothrombinase complex	Limo et al. (1991)
	RaHBP(M), RaHBP(F)	Histamine-binding proteins	Paesen et al. (1999)
	TdPI	Tryptase inhibitor	Paesen et al. (2007)
	Japanin	Immunomodulation/reprograms DC responses	Preston et al. (2013)
	SGE	Vasodilation/vasoconstriction	Pekáriková et al. (2015)

TABLE 5.2 Examples of Biological Activities of Tick Saliva and Salivary Molecules in Modulation of Host Defense Reactions—cont'd

Tick Species	Molecule	Function/Target And/Or Activity	References
Rhipicephalus sanguineus	Evasin-1 Evasin-3 Evasin-4	Chemokine binding/chemokines CCL3, CCL4, CCL18 Chemokine binding/chemokines CXCL8 and CXCL1 Chemokine binding/chemokines CCL5 and CCL11	Deruaz et al. (2008) and Frauenschuh et al. (2007)
	Ado, PGE_2	Immunomodulation/modulation of host inflammatory responses	Oliveira et al. (2011)
Rhipicephalus hemaphysaloides	Rhipilin-1 Rhipilin-2	Kunitz-type serine protease inhibitors/coagulation common pathway Kunitz-type serine protease inhibitors/trypsin and elastase	Cao et al. (2013) and Gao et al. (2011)
Rhipicephalus (Boophilus) microplus	BmAP	Anticoagulation/thrombin	Horn et al. (2000)
	Microphilin	Anticoagulation/thrombin	Ciprandi et al. (2006)
	RmS-15	Anticoagulation/thrombin	Xu et al. (2016)
	SGE	Immunomodulation/bovine macrophages	Brake and de Leon (2012)
	SGE	Complement inhibition/classical and alternative pathway	Silva et al. (2016)
Boophilus calcaratus	Calcaratin	Anticoagulation/thrombin	Motoyashiki et al. (2003)

Continued

TABLE 5.2 Examples of Biological Activities of Tick Saliva and Salivary Molecules in Modulation of Host Defense Reactions—cont'd

Tick Species	Molecule	Function/Target And/Or Activity	References
Hyalomma asiaticum	BIF	Immunomodulation/inhibits lipopolysaccharide-induced proliferation of B cells	Yu et al. (2006)
	Hyalomin A, B	Immunomodulation suppresses host inflammatory responses (modulation of cytokine secretion, detoxification of free radicals)	Wu et al. (2010)
Hyalomma marginatum rufipes	Hyalomin-1	Anticoagulation/thrombin	Jablonka et al. (2015)
Hyalomma excavatum	SGE	Wound healing/PDGF-binding	Slovák et al. (2014)

AamAV422, Amblyomma americanum AV422 protein; *AamS6, Amblyomma americanum* tick serine protease inhibitor 6; *Ado*, adenosine; *BIF*, B-cell inhibitory factor; *BIP*, B-cell inhibitory protein; *BmAP, Rhipicephalus (Boophilus) microplus* anticoagulant protein; *DC*, dendritic cell; *HLTnI*, troponin I-like molecule; *IFN*, interferon; *IL*, interleukin; *Iper, I. persulcatus; Irac, I. ricinus* anticomplement; *Ir-CPI*, coagulation contact phase inhibitor from *I. ricinus; Iris, I. ricinus* immunosuppressor; *Ir-LBP, Ixodes ricinus* salivary LTB4-binding lipocalin; *IRS, I. ricinus* serpin; *Isac, I. scapularis* salivary anticomplement; *ISL 929 and ISL 1373, I. scapularis* salivary proteins 929 and 1373; *IxscS-1E1*, blood meal-induced *I. scapularis* (Ixsc) tick saliva serine protease inhibitor (serpin [S]); *MCP-1*, monocyte chemoattractant protein-1; *MIF*, macrophage migration inhibitory factor; *MIP-2*, macrophage inflammatory protein 2; *OmCI, O. moubata* complement inhibitor; *P36*, 36-kDa immunosuppressant protein; *PDGF*, platelet-derived growth factor; PGE_2, prostaglandin E_2; *RaHBP(M), RaHBP(F)*, female (F) and male (M) *Rhipicephalus appendiculatus* histamine-binding protein; *RmS-15, Rhipicephalus (Boophilus) microplus* serpin 15; *SGE*, salivary gland extract; *SHBP*, serotonin- and histamine-binding protein; *TAP*, tick anticoagulant peptide; *TCA-3*, thymus-derived chemotactic agent 3; *TdPI*, tick-derived peptidase inhibitor; *TF*, tissue factor, Salp, salivary protein; *tHRF*, tick histamine release factor; *TIX-5*, tick inhibitor of factor Xa toward factor V; *TNF*, tumor necrosis factor; *TSGP2, TSGP3*, soft tick lipocalins.

Investigation of the anti-hemostatic strategies of ticks has suggested that they evolved independently in hard and soft ticks (Mans, 2011; Mans et al., 2008b).

Impact on vasodilation/vasoconstriction. Following injury of blood vessels by tick mouthparts, arachidonic acid is released by activated platelets and is converted into thromboxane A_2, a platelet-aggregating, platelet-degranulating, and vasoconstricting substance. Activated platelets release serotonin, which, together with thromboxane A_2, is responsible for early vasoconstriction in local inflammation caused by tissue injury. Ticks secrete vasodilators into the feeding pool to counteract vasoconstrictors produced by the host at the site of tissue injury. To date, only non-proteinaceous, lipid-derived vasodilators, such as prostacyclin and prostaglandins, have been identified in tick saliva (Bowman et al., 1996; Ribeiro et al., 1988). It has, however, been suggested that a tick histamine release factor (tHRF) from *I. scapularis* saliva (Dai et al., 2010) and a serine proteinase inhibitor (serpin) IRS-2 from *I. ricinus* saliva, the latter of which inhibits cathepsin G and chymase (Chmelar et al., 2011), may also modulate vascular permeability (Chmelar et al., 2012). Salivary gland extract (SGE) from female *Dermacentor reticulatus* and *R. appendiculatus ticks* has recently been found to induce constriction of rat femoral artery, whereas SGE from males of these tick species induced vasoconstriction or vasodilation, depending on the duration of feeding (Pekáriková et al., 2015). The identity of the active compound(s) involved in vasoconstriction has not been determined, but data suggest that they are not prostaglandins, and the biological function of tick vasoconstrictors requires further investigation.

Inhibition of platelet aggregation. Platelet aggregation represents the initial stage of hemostasis. Following vascular injury, platelets adhere to the subendothelial tissue and become activated by agonists (collagen, thrombin, adenosine diphosphate (ADP), thromboxane A_2). Agonists bind to specific receptors on the surface of platelets and initiate a highly complex chain of intracellular chemical reactions leading to platelet aggregation and formation of a hemostatic plug. Platelet aggregation is targeted by ticks at several stages (Francischetti, 2010). Soft ticks (Mans et al., 1998, 2002a; Ribeiro et al., 1991) and a few hard tick species (Liyou et al., 1999; Ribeiro et al., 1985) were found to target ADP via salivary apyrase (an adenosine triphosphate (ATP) diphosphohydrolase enzyme) by hydrolysis of the phosphodiester bonds of ATP and ADP. In contrast, saliva of *A. americanum* does not contain apyrase activity (Ribeiro et al., 1992), but salivary prostaglandins of this species were found to inhibit platelet aggregation by preventing ADP secretion during platelet activation (Bowman et al., 1995). The *Ornithodoros moubata*-derived lipocalin called moubatin was found to prevent activation of platelets by collagen, and the tick adhesion inhibitor inhibited the adhesion of platelets to matrix collagen (Karczewski et al., 1995; Waxman and Connolly, 1993). Longicornin, another inhibitor of collagen-mediated platelet aggregation whose mechanism is similar to that of moubatin, was identified in *Haemaphysalis longicornis* (Cheng et al., 1999). Thrombin is the key enzyme in thrombosis and hemostasis and contains three functional sites—the active site, the anion-binding exosite I that mediates binding of thrombin to fibrinogen, the platelet receptor and thrombomodulin, and the anion-binding exosite II (heparin-binding site). Tick salivary antithrombins, which are involved in the inhibition of the fibrin clot formation, also inhibit thrombin-induced platelet aggregation (Hoffmann et al., 1991; Kazimírová et al., 2002; Nienaber et al., 1999). The IRS-2 serpin from *I. ricinus* was found to inhibit both cathepsin G- and thrombin-induced platelet aggregation (Chmelar et al., 2011), and the serpin IxscS-1E1 from *I. scapularis* inhibits ADP and thrombin-induced platelet aggregation (Ibelli et al., 2014). Post-activation inhibitors of platelet aggregation derived from tick saliva target the platelet fibrinogen receptor. The tick-derived disintegrin-like peptides savignygrin from *Ornithodoros savignyi* (Mans et al., 2002b) and variabilin from *Dermacentor variabilis* (Wang et al., 1996) contain the Arg-Gly-Asp (RGD) integrin recognition motif and can inhibit platelet aggregation by preventing the binding of other ligands to the platelet receptor. By contrast, disaggregin, a fibrinogen receptor antagonist from *O. moubata*, lacks the RGD motif and inhibits platelet aggregation by preventing ligand binding by a mechanism distinct from that of disintegrin-like peptides (Karczewski et al., 1994). Ixodegrins from *Ixodes pacificus* and *I. scapularis* display sequence similarity to variabilin and are integrin antagonists (Francischetti, 2010; Francischetti et al., 2005b). Should inhibition of platelet aggregation fail, ticks have also adopted strategies to disaggregate platelet aggregates, either by displacement of fibrinogen from its receptor by competitive binding (e.g., savignygrin Mans et al., 2002c) or fibrinolysis (Anisuzzaman et al., 2011b; Decrem et al., 2009; Diaz-Martin et al., 2013a).

Inhibition of blood coagulation. The blood coagulation cascade involves a series of enzymatic reactions during which an inactive proenzyme (coagulation factor) is converted to an active form, which then activates the next proenzyme. Thrombin is involved in the final common pathway of the coagulation cascade and converts fibrinogen into fibrin but also regulates the activity of other blood coagulation factors and stimulates platelet aggregation. Ticks have evolved a wide range of compounds that target different coagulation factors, with Kunitz-type proteinase inhibitors being the most abundant class (Chmelar et al., 2012; Koh and Kini, 2009).

Thrombin inhibitors derived from tick saliva belong to at least seven structural classes that target the enzyme at different sites and through various mechanisms (Koh and Kini, 2009). Ornithodorin (van de Locht et al., 1996), savignin (Nienaber et al., 1999), and monobin (Mans et al., 2008a) from soft ticks are Kunitz-type antithrombins, and Iris (immunosuppressive protein) and IRS-2 from *I. ricinus* (Chmelar et al., 2011; Leboulle et al., 2002) and RmS-15 from *R. (B.) microplus* (Xu et al., 2016) are serpins. Antithrombins belonging to the hirudin-like/madanin/variegin superfamily include madanin-1 and 2 from *H. longicornis* (Iwanaga et al., 2003), variegin from *Amblyomma variegatum* (Koh et al., 2007), and hyalomin-1 from *Hyalomma marginatum rufipes* (Jablonka et al., 2015). The antithrombins microphilin (Ciprandi et al., 2006) and BmAP (Horn et al., 2000) from *R. (B.) microplus* and calcaratin from *Boophilus calcaratus* (Motoyashiki et al., 2003) cannot be classified in any of the previously mentioned groups.

The Kunitz-type tick anticoagulant peptide (TAP) from *O. moubata* (Waxman et al., 1990), the TAP-like protein from *O. savignyi* (Joubert et al., 1998), amblyomin-X from *Amblyomma cajennense* (Batista et al., 2010), and Salp14 (a basic tail-secreted protein) from *I. scapularis* (Narasimhan et al., 2002) are all inhibitors of factor Xa.

Ixolaris, a two-domain Kunitz-type inhibitor of ETC, and penthalaris, containing five Kunitz domains, both of which display homology to the tissue factor pathway inhibitor, have been detected in *I. scapularis* (Francischetti et al., 2002, 2004). SGE from *D. andersoni* has been described to inhibit factor V and factor VII (Gordon and Allen, 1991). TIX-5 (tick inhibitor of factor Xa toward factor V) has been identified in SGs of nymphal *I. scapularis* and was found to prolong activation of the coagulation system by specifically inhibiting the factor Xa-mediated activation of factor V (Schuijt et al., 2013). Moreover, immunization with TIX-5 impaired tick feeding, suggesting that neutralization of TIX-5 diminished the anticoagulatory activity of ticks needed for optimal feeding.

A plasma kallikrein–kinin system inhibitor, haemaphysalin, was identified in *H. longicornis* (Kato et al., 2005), and a contact phase inhibitor (Ir-CPI) that impairs the intrinsic coagulation pathway and fibrinolysis was found in *I. ricinus* (Decrem et al., 2009). Fibrinolytic activity mediated by a metalloprotease was detected in saliva of *I. scapularis* (Francischetti et al., 2003). Longistatin, a plasminogen activator from *H. longicornis*, was found to hydrolyze fibrinogen and delay fibrin clot formation (Anisuzzaman et al., 2011a). Tick-derived serine protease inhibitors with similarity to proteins of the serpin family may also interact with host hemostasis. Iris, an immunomodulatory serpin from SGs of *I. ricinus* was the first tick serpin that was reported to interfere with host hemostasis and immune response and impair the contact phase-activated pathway of coagulation and fibrinolysis (Prevot et al., 2006). Tick-derived calcium-binding proteins with sequence homology to calreticulin (CRT) may also play a role in modulating host hemostasis, by sequestration of calcium ions, which are required as coagulation enzyme cofactors (Jaworski et al., 1995).

Effects of Tick Feeding on Host Immune Responses

Cellular innate immune responses and the complement system are the first lines of defense of vertebrate hosts against invading pathogens. The complement proteins in serum can be activated by different pathways that lead to the generation of molecules with various activities in inflammation and opsonization and lysis of invading pathogens. The adaptive immune response is triggered when activated antigen-presenting cells migrate to lymphoid tissues where they present antigens to T cells, which play a central role in cellular immune responses at the site of infection or assist in the activation of B cells and the generation of an antigen-specific humoral response (Janeway et al., 1999).

It is well known that ticks have developed various strategies to evade the immune responses of their hosts and ensure completion of feeding and further development (Brossard and Wikel, 2008; Chmelař et al., 2016b; Gillespie et al., 2001; Kotál et al., 2015; Valenzuela, 2004). However, and despite a relatively broad knowledge of tick-induced host immunomodulation, the number of functionally characterized immunomodulatory molecules from tick SGs is still limited (Table 5.2).

Innate immune responses and complement. Host innate immune responses are the first line of defense against local injury and comprise complement, acute phase proteins, neutrophils, macrophages, mast cells, basophils, eosinophils, dendritic cells (DCs), and natural killer (NK) cells. Complement components, prostaglandins, leukotrienes, antimicrobial peptides, chemokines, and cytokines contribute to the recruitment of inflammatory cells to the site of injury (Andrade et al., 2005). As countermeasures against inflammation, ticks secrete molecules that modulate the pro-inflammatory functions of most cell types infiltrating the tick attachment site, e.g., neutrophils (Guo et al., 2009; Ribeiro et al., 1990), NK cells (Kubeš et al., 1994), macrophages (Kopecký and Kuthejlová, 1998; Kramer et al., 2011), T cells (Bergman et al., 2000; Ramachandra and Wikel, 1992), and DCs (Carvalho-Costa et al., 2015; Cavassani et al., 2005; Páleníková et al., 2015; Skallová et al., 2008). These molecules suppress the production of pro-inflammatory Th1 cytokines and up-regulate anti-inflammatory Th2 cytokines, leading to Th2 polarization of the host immune response (Ferreira and Silva, 1999; Heinze et al., 2014; Mejri et al., 2001). The tick-induced Th2 cytokine profile is likely to suppress local inflammation and thereby ensure completion of tick feeding but may also enhance the transmission of TBP (Schoeler and Wikel, 2001; Wikel and Alarcon-Chaidez, 2001). Ticks are also able to evade host immune responses by means of proteins that mimic host proteins. A macrophage migration inhibitory factor identified in SGs of *A. americanum* (Jaworski et al., 2001) probably protects feeding ticks from macrophage attack.

The skin represents the main organ at the tick–host–pathogen interface. Soon after attachment of *I. scapularis* nymphs, a neutrophil-dominated immune response is elicited, as evidenced by analysis of both gene expression and histopathology (Heinze et al., 2012a). Similarly, during infestation of mice with *D. andersoni* nymphs, inflammation was inhibited at the tick bite site in the early phase. Infiltration of inflammatory cells and replication of epithelial cells were, however, observed to increase during the later phase of tick feeding (Heinze et al., 2014). Ir-LBP, a lipocalin from *I. ricinus*, was found to inhibit neutrophil chemotaxis and impair neutrophil function in inflammation at the tick bite site (Beaufays et al., 2008). Moreover, due to an antialarmin effect on human primary keratinocytes, saliva of *I. ricinus* inhibited cutaneous innate immunity and migration of immune cells to the tick bite site (Marchal et al., 2011).

Chemokines are key mediators of the host inflammatory response against pathogens and trigger the recruitment of specific leukocyte populations. Ticks have developed strategies to manipulate the host cytokine network. Anti-CXCL8 (IL-8) activity has been reported in the saliva of several hard tick species (Hajnická et al., 2001) and for salivary proteins Salp16, Iper1, and Iper2 from *Ixodes persulcatus* (Hidano et al., 2014). Furthermore, tick salivary compounds can inhibit functions of the pro-inflammatory cytokine IL-2 and the chemokines CCL2 (monocyte chemotactic protein-1 [MCP-1]), CCL3 (macrophage inflammatory protein 1-alpha [MIP-1α]), CCL5 (RANTES), and CCL11 (eotaxin) (Hajnická et al., 2005) (Vancova et al., 2010b). Evasins, a family of chemokine-binding proteins, have been identified in SGs of *Rhipicephalus sanguineus* (Frauenschuh et al., 2007) and show selectivity for different chemokines: evasin-1 binds to CCL3, CCL4 (macrophage inflammatory protein-1 beta [MIP-1β]), and CCL18 (chemokine [C-C motif] ligand 18); evasin-3 binds to CXCL8 and CXCL1 (chemokine [C-X-C motif] ligand 1); and evasin-4 binds to CCL5 and CCL11 (Deruaz et al., 2008; Frauenschuh et al., 2007). Evasin-3-like activities have also been demonstrated for SGE of *A. variegatum*, *R. appendiculatus*, and *D. reticulatus* (Vančová et al., 2010a). The immunoregulatory peptides hyalomin-A and -B from SGs of *Hyalomma asiaticum asiaticum* were found to suppress host inflammatory responses either by inhibiting secretion of tumor necrosis factor (TNF)-alpha, MCP-1, and interferon (IFN)-gamma or by increasing secretion of the immunosuppressive cytokine IL-10 (Wu et al., 2010). Anti-inflammatory effects due to suppression of the production of TNF-alpha, IL-1, IL-8, and IFN-gamma *in vitro* have been reported for the salivary protein amregulin from *A. variegatum* (Tian et al., 2016). Lastly, non-proteinaceous substances such as purine nucleoside adenosine and prostaglandin PGE$_2$ present in saliva of *R. sanguineus* have been found to inhibit the production of pro-inflammatory IL-12p40 and TNF-alpha and stimulate the production of anti-inflammatory IL-10 by murine DCs (Oliveira et al., 2011).

Modulation of wound healing and angiogenesis are other strategies adopted by ticks to suppress host inflammatory responses (Francischetti, 2010; Hajnická et al., 2011). It has been demonstrated that hard tick salivary molecules bind to the transforming growth factor-β1, the platelet-derived growth factor (PDGF), the fibroblast growth factor-2, and the hepatocyte growth factor in a species-specific manner with the presence of PDGF activity appearing to be associated with species possessing long mouthparts (Slovák et al., 2014). *D. variabilis* saliva was found to suppress basal and PDGF-stimulated fibroblast migration and reduce extracellular signal-regulated kinase activity stimulated by PDGF (Kramer et al., 2008). Tick salivary molecules with similarities to disintegrin metalloproteases and thrombospondin are able to impair cell–matrix interactions and angiogenesis (Francischetti et al., 2005a; Fukumoto et al., 2006; Valenzuela et al., 2002). The *I. scapularis* salivary proteins ISL 929 and ISL 1373, which bear homology to the cysteine-rich domain of disintegrin metalloproteinases, reduce the expression of β2 integrins and impair the adherence of polymorphonuclear leukocytes (PMN) (Guo et al., 2009). A troponin I-like molecule (Fukumoto et al., 2006) and a Kunitz-type protein haemangin (Islam et al., 2009) from SGs of *H. longicornis* also impair angiogenesis and wound healing.

Bradykinin and histamine are important mediators of itch and pain and can stimulate host grooming to the detriment of feeding ticks. Ticks have developed effective mechanisms to prevent host reactions to itch and pain, e.g., through salivary kininases (Ribeiro and Mather, 1998) or metalloproteases (Jelinski, 2016), which hydrolyze bradykinin (Ribeiro and Mather, 1998). The amine-binding proteins of the lipocalin family, such as the histamine-binding proteins

RaHBP(M) and RaHBP(F)-1,2 from *R. appendiculatus* (Paesen et al., 1999) and the serotonin- and histamine-binding protein from *D. reticulatus* (Sangamnatdej et al., 2002), have been found to impair the functions of histamine and serotonin, respectively. The tick-derived protease inhibitor from *R. appendiculatus* (Paesen et al., 2007) suppresses the activity of human β-tryptases, i.e., mast cell-specific serine proteases with roles in inflammation and tissue remodeling.

The complement system links the innate and adaptive immune responses of the host and is activated through three pathways (alternative, classical, and lectin). The alternative pathway is the major line of defense against pathogens and is also involved in resistance to ticks (Wikel, 1979). The molecules Isac, Salp20, and Isac-1 from *I. scapularis* (Tyson et al., 2007; Valenzuela et al., 2000) and the Isac orthologs IRAC I and II from *I. ricinus* (Couvreur et al., 2008; Daix et al., 2007) specifically inhibit formation of the C3 convertase of the alternative pathway by blocking binding of complement factor B to complement C3b. The lipocalin *O. moubata* complement inhibitor specifically targets the C5 activation step in the complement cascade (Nunn et al., 2005). Inhibition of the classical pathway of the complement system by saliva and SGE of *A. cajennense* has also been reported by Franco et al. (2016).

Acquired Immune Responses

Some vertebrate species develop resistance to repeated tick feeding, while others remain susceptible to tick infestation. Immunoglobulin and T cell-mediated immune responses are induced in the host during initial exposure to ticks. Salivary immunogens are taken up and processed by professional antigen-presenting cells in the skin and following migration to draining lymph nodes are presented to lymphocytes for induction of antibody- and cell-mediated responses (Andrade et al., 2005; Schoeler and Wikel, 2001). The activation of antigen-specific T cells is evidenced by the delayed-type hypersensitivity response—with influx of lymphocytes, macrophages, basophils, and eosinophils—which may develop at the tick feeding site. Plasma cells secrete antibodies with specificity for tick antigens; those of the IgE class are bound by basophils and mast cells and, when cross-linked by cognate antigen, trigger the release of pharmacologic mediators of intermediate-type sensitivity. Memory B and T lymphocytes are generated.

In tick-resistant hosts (e.g., guinea pigs, rabbits), the presence of antibodies and effector T lymphocytes with specificity for tick antigens assures a rapid secondary response to infestation that impairs tick feeding. Susceptible hosts, such as mice, generally do not develop acquired resistance to repeated tick feeding (Schoeler et al., 1999), possibly in relation to a cytokine response of mixed Th1/Th2 profile, and enhanced activity of regulatory T cells during secondary tick infestation (Heinze et al., 2012b).

Tick-induced immunosuppression in the host is characterized by decreased primary antibody responses to T cell-dependent antigens and polarization of the host immune response to a Th2-type profile. This polarization is generally characterized by down-regulation of pro-inflammatory Th1 cytokines (IL-2, IFN-gamma) and up-regulation of Th2 cytokines (IL-4, IL-5, IL-6, IL-10, IL-13) (Brossard and Wikel, 2008; Gillespie et al., 2000; Schoeler and Wikel, 2001; Wikel and Alarcon-Chaidez, 2001). Several T cell inhibitory molecules have been identified in tick saliva or SGs: P36 in *D. andersoni* (Bergman et al., 2000), Iris in *I. ricinus* (Leboulle et al., 2002), and Salp15 in *I. scapularis* (Anguita et al., 2002). The expression of Iris and Salp15 was found to be induced by feeding. Iris suppresses T cell proliferation, induces a Th2-type immune response, and inhibits the production of pro-inflammatory cytokines IL-6 and TNF-alpha. Salp15 binds to CD4 molecules on the surface of CD4$^+$ T (helper) cells, resulting in inhibition of T cell receptor-mediated signaling and leading to reduced IL-2 production and impaired T cell proliferation (Anguita et al., 2002; Garg et al., 2006). Furthermore, Salp15 impairs DC function by inhibiting Toll-like receptor (TLR)- and *Borrelia burgdorferi*-induced production of pro-inflammatory cytokines by DCs and DC-induced T cell activation (Hovius et al., 2008a). Transcripts encoding proteins of the Salp family with similarity to Salp15 have also been identified in SGs of *I. pacificus*, *I. ricinus*, and *I. persulcatus* (Hojgaard et al., 2009; Hovius et al., 2007; Mori et al., 2010).

Other salivary immunomodulatory proteins that facilitate tick feeding and pathogen transmission have also been detected in *I. scapularis*. These include a secreted IL-2 binding protein, which suppresses T cell proliferation and in all likelihood many other IL-2 dependent activities of diverse immune effector cells (Gillespie et al., 2001), and the salivary cysteine protease inhibitors sialostatin L and sialostatin L2, which possess inhibitory activity against cathepsin L (Kotsyfakis et al., 2006). Sialostatin L inhibits proliferation of cytotoxic T lymphocytes and lipopolysaccharide-induced maturation of DCs (Kotsyfakis et al., 2006; Sa-Nunes et al., 2009), while sialostatin L2 stimulates the growth of *B. burgdorferi* in murine skin by a yet unknown mechanism (Kotsyfakis et al., 2010).

Ticks can also impair the production of specific anti-tick antibodies by suppression of B cell responses. B-cell inhibitory proteins BIP and BIF have been identified in *I. ricinus* and *H. asiaticum asiaticum*, respectively (Hannier et al., 2004; Yu et al., 2006).

Moreover, tick saliva contains immunoglobulin (IgG)-binding proteins that bind ingested host IgG, which succeed in penetrating the midgut and facilitate their excretion by salivation during feeding. These proteins are believed to protect ticks from ingested host IgG (Wang and Nuttall, 1999).

A novel mechanism described for tick-mediated immunomodulation involves Japanin, an SG protein from *R. appendiculatus* belonging to a new clade of lipocalins (Preston et al., 2013). Japanin modifies the expression of co-stimulatory and co-inhibitory molecules and secretion of pro-inflammatory, anti-inflammatory. and T cell polarizing cytokines and inhibits the differentiation of DCs from monocytes.

Tick Saliva and Pathogen Transmission

Beyond their harmfulness as blood-feeding ectoparasites, ticks are known as vectors of a large number of microorganisms (viruses, bacteria, protozoa) that can cause disease in humans and animals. After ingestion with infected host blood, the established itinerary for a tick-borne microorganism is penetration of the gut and invasion of the hemocoel and inner organs, including SGs. Multiplication of microorganisms takes place in certain inner organs, of which SGs play a key role in transmission of pathogenic microorganisms to vertebrate hosts. Transmission, however, cannot be reduced to a simple process of syringe inoculation. During their long co-evolution with ticks and vertebrate hosts, microorganisms have developed various strategies to exploit tick salivary molecules to ensure transmission, local infection, and systemic dissemination (Brossard and Wikel, 2004; Nuttall and Labuda, 2004; Ramamoorthi et al., 2005; Wikel, 2013). Moreover, the molecular interactions between ticks, pathogens, and hosts reflect not only evolutionary conflict but also cooperation, with certain interactions providing benefit for two or even all three players (de la Fuente et al., 2016).

The phenomenon by which pathogen transmission is promoted by tick saliva (saliva-assisted transmission, or SAT) has been observed for a number of tick–pathogen associations. The underlying molecular mechanisms, however, remain largely unknown, and only a limited number of tick molecules associated with pathogen transmission have been identified (e.g., (Kazimírová and Štibrániová, 2013; Liu and Bonnet, 2014; Nuttall and Labuda, 2004; Ramamoorthi et al., 2005; Wikel, 2013) see also Table 5.3). Mediators of SAT may vary according to tick species—and indeed may define vector competence for certain pathogen/tick associations—and to the transmitted pathogen (Nuttall and Labuda, 2004).

Not only do salivary components influence transmission of pathogens but also the presence of pathogens in SGs can modify tick salivary composition. Modern high-throughput techniques have been applied to the systematic characterization of tick salivary components, and notably, as regards tick–host–pathogen interactions, have revealed a large set of differentially expressed genes in SGs, some of which encoding mediators of SAT, depending on the presence or absence of microorganisms (Chmelař et al., 2016a).

Impact of Salivary Components on Pathogen Transmission

Saliva-assisted transmission (SAT). SAT was first demonstrated for Thogotovirus (THOV). Indeed, when *R. appendiculatus* nymphs were fed on animals that had been inoculated with a mixture of THOV and SGE derived from female ticks, acquisition of virus was higher than when ticks were fed on animals inoculated with the virus alone (Jones et al., 1989). Enhanced infection of ticks when fed on animals experimentally inoculated with pathogens along with tick saliva (or SGE)—which constitutes direct evidence of SAT (Nuttall and Labuda, 2004)—has subsequently been reported for several pathogens,

TABLE 5.3 Reported Cases of Saliva-Assisted Transmission of Tick-Borne Microorganisms

Microorganism	Tick Species	SAT Factor, Effect	References
Viruses	**Argasidae**		
African swine fever virus	*Ornithodoros porcinus*	Salivary gland extract (SGE), modulation of the immune response in pigs, increased macrophage recruitment in the dermis, probably promotes viral infection	Bernard et al. (2016)
	Ixodidae		
Thogotovirus	*Rhipicephalus appendiculatus*	Non-viremic transmission (NVT)	Jones et al. (1987)
		SGE, enhanced transmission and infectivity	Jones et al. (1989)
Tick-borne encephalitis virus (TBEV)	*Ixodes persulcatus*	Co-feeding transmission	Alekseev and Chunikhin (1990)
	Ixodes ricinus	NVT	Labuda et al. (1993a)
		SGE, enhanced transmission and infectivity	Labuda et al. (1993b)
		Saliva, *in vitro* modulation of infection rate of dendritic cell (DC) and production of cytokines	Fialová et al. (2010)
	Ixodes scapularis	Sialostatin L2, interference with interferon (IFN) action, enhanced replication of TBEV in cells (DCs)	Lieskovská et al. (2015a)
Powassan virus	*I. scapularis*	SGE, virus dose-dependent saliva-activated transmission	Hermance and Thangamani (2015)
Palma, Bhanja virus	*R. appendiculatus*	NVT	Labuda et al. (1997a)
	Dermacentor marginatus		

Continued

TABLE 5.3 Reported Cases of Saliva-Assisted Transmission of Tick-Borne Microorganisms—cont'd

Microorganism	Tick Species	SAT Factor, Effect	References
Bacteria/Extracellular			
Borrelia burgdorferi sensu stricto	*I. ricinus*	Co-feeding transmission	Gern and Rais (1996)
		SGE, accelerating effect on spirochete proliferation in the host	Macháčková et al. (2006)
		Saliva, increased spirochete load in host skin, increased transmission to ticks	Horká et al. (2009)
		SGE and Salp15, inhibition of *in vitro* keratinocyte inflammation induced by *Borrelia*	Marchal et al. (2011)
	I. scapularis	Co-feeding transmission	Piesman and Happ (2001)
		Salivary gland (SG) lysate, increase of spirochete load in target organs	Zeidner et al. (2002)
Borrelia afzelii	*I. ricinus*	SGE, inhibition of killing of spirochetes by murine macrophages, reduction of the production of two major defense molecules of phagocytes, superoxide and NO	Kuthejlová et al. (2001)
		Co-feeding transmission	Richter et al. (2002)
		SGE, accelerating effect on spirochete proliferation in the host, suppression of proinflammatory cytokines	Pechová et al. (2002)
		SGE, anti-inflammatory effect, enhancement of *Borrelia* transmission	Severinová et al. (2005)

TABLE 5.3 Reported Cases of Saliva-Assisted Transmission of Tick-Borne Microorganisms—cont'd

Microorganism	Tick Species	SAT Factor, Effect	References
		Saliva, impairment of DC functions	Slámová et al. (2011)
		SGE, attenuation of the initial signal transduction pathway of type I IFN	Lieskovska and Kopecky (2012b)
		Saliva, interference with signaling pathways activated by Toll-like receptor (TLR)-2 ligand and spirochetes in DCs	Lieskovská and Kopecký (2012a)
		Borrelia strain-specific variation in co-feeding transmission	Tonetti et al. (2015)
Borrelia lusitaniae	I. ricinus	SG lysate, increase of spirochete load in target organs	Zeidner et al. (2002)
B. burgdorferi sensu lato	I. ricinus	Saliva, modulation of innate immunity in murine skin	Kern et al. (2011)
		BIP, inhibition of B lymphocyte proliferation induced by the Borrelia lipoproteins OspA and OspC	Hannier et al. (2003)
		Salp15 Iric-1, Salp15 homolog, binding to Borrelia OspC	Hovius et al. (2008b)
	I. ricinus/I. scapularis	Salp15 protects serum-sensitive isolates of Borrelia against complement-mediated killing	Schuijt et al. (2008)
	I. scapularis	Salp15, immunosuppressive functions, binds to Borrelia OspC, protects the spirochetes from antibody-mediated killing, facilitates transmission and replication of the spirochetes	Ramamoorthi et al. (2005)

Continued

TABLE 5.3 Reported Cases of Saliva-Assisted Transmission of Tick-Borne Microorganisms—cont'd

Microorganism	Tick Species	SAT Factor, Effect	References
		Salp25D, antioxidant, facilitates the acquisition of spirochetes by the tick from an infected mammalian host	Narasimhan et al. (2007)
		Salp20, inhibits complement, facilitates *Borrelia* survival	Tyson et al. (2007)
		Salp15, binds to DC-SIGN, triggers a novel Raf-1/MEK-dependent signaling pathway, modulates TLR-induced DC activation	Hovius et al. (2008a)
		Tick histamine release factor, probably modulates vascular permeability and increases blood flow to the tick bite site, facilitating tick engorgement	Dai et al. (2010)
		Tick salivary lectin pathway inhibitor, facilitates pathogen transmission	Schuijt et al. (2011)
		Sialostatins L and L2, influence on DCs exposed to *B. burgdorferi* and TLR ligands	Lieskovská et al. (2015)
Bacteria/Intracellular			
Bartonella henselae	*I. ricinus*	*I. ricinus* serine protease inhibitor, facilitates both bacteria multiplication in SGs and tick engorgement	Liu et al. (2014b)
Francisella tularensis	*I. ricinus*	SGE, accelerates proliferation of the bacteria in the host	Kročová et al. (2003)

TABLE 5.3 Reported Cases of Saliva-Assisted Transmission of Tick-Borne Microorganisms—cont'd

Microorganism	Tick Species	SAT Factor, Effect	References
Anaplasma phagocytophilum	I. scapularis	Salp16, facilitates migration of the pathogen to SGs	Sukumaran et al. (2006)
		Saliva, inhibits inflammatory cytokine secretion during rickettsial transmission	Chen et al. (2012)
		Sialostatin L2, targets caspase-1 activity during host stimulation with A. phagocytophilum, activates macrophages, inhibits inflammasome formation during infection	Chen et al. (2014)
Ehrlichia muris-like agent	I. scapularis	Co-feeding transmission	Karpathy et al. (2016)
Rickettsia conorii	Rhipicephalus sanguineus	Co-feeding transmission	Zemtsova et al. (2010)
		Saliva, inhibition of inflammation, facilitates transmission	Milhano et al. (2015)
	Amblyomma americanum/A. maculatum	Co-feeding transmission	Wright et al. (2015)
Rickettsia parkeri	A. maculatum	Tick feeding, enhancement of pathogenicity of the rickettsiae	Banajee et al. (2015)

including tick-borne encephalitis virus (TBEV) (Alekseev and Chunikhin, 1990; Labuda et al., 1993a), African swine fever virus (Bernard et al., 2016), Powassan virus (Hermance and Thangamani, 2015), *Borrelia burgdorferi* sensu lato (Macháčková et al., 2006; Pechová et al., 2002; Zeidner et al., 2002), *Francisella tularensis* (Kročová et al., 2003), and *Rickettsia conorii* (Milhano et al., 2015) (see also Table 5.3). Furthermore, injection of *B. burgdorferi* s.l. spirochetes into mice together with *I. ricinus* or *I. scapularis* SGE increased the level of bacteremia in mice, enhanced the infection of feeding ticks, and suppressed the production of pro-inflammatory cytokines in draining lymph nodes (Pechová et al., 2002; Zeidner et al., 2002). The presence of SGE has also been reported to reduce the production of two major anti-microbial molecules—superoxide and nitric oxide—involved in phagocytosis and to inhibit killing of *Borrelia afzelii* by macrophages *in vitro* (Kuthejlová et al., 2001). Moreover, SGE reduced adhesion of PMN via downregulation of beta2-integrins and diminished their capacity to take up and kill *Borrelia*, thus facilitating *Borrelia* transmission (Montgomery et al., 2004).

Non-viraemic transmission (NVT). Jones et al. observed that transmission of THOV to uninfected *R. appendiculatus* ticks was more efficient while co-feeding with infected ticks on non-viremic guinea pigs than while feeding on viremic hamsters, thus identifying a novel mode of arbovirus transmission independent of systemic infection of the host (Jones et al., 1987). NVT of pathogens from infected to uninfected ticks while co-feeding on the same host has been considered to be indirect evidence of SAT (Nuttall and Labuda, 2004). Co-feeding transmission occurs when the pool of saliva produced by ticks feeding in close proximity in the host skin creates a favorable environment for pathogen exchange (Randolph, 2011; Voordouw, 2015). Co-feeding transmission seems to be more efficient for tick-borne viruses than bacteria and has been found to occur for TBEV even in the presence of virus-specific neutralizing antibodies (Labuda et al., 1997b), probably representing an immune evasion strategy for the virus. Immunomodulation of the tick attachment site by tick salivary compounds has been suggested to play a crucial role in NVT. During co-feeding on mice, transmission of TBEV from infected to uninfected *I. ricinus* ticks has been shown to be correlated with local skin infection at the site of tick feeding. Cellular infiltration was evidenced at these sites, and the presence of viral antigen in migratory Langerhans cells and neutrophils and infectious virus in migratory monocyte/macrophages suggested that migratory cells may serve as vehicles for transmission of virus from infected to uninfected co-feeding ticks (Labuda et al., 1996). These findings were supported by *in vitro* studies in which treatment of DCs with tick saliva increased the proportion of virus-infected cells and decreased virus-induced production of TNF-alpha and IL-6 and virus-induced apoptosis (Fialová et al., 2010). Sialostatin L2 derived from *I. scapularis* saliva was found to suppress the IFN response and enhance TBEV replication in DCs (Lieskovská et al., 2015a).

Co-feeding transmission has also been demonstrated for both extracellular (*B. burgdorferi* s.l.) and intracellular bacteria (*R. conorii*, *Rickettsia parkeri*, *Ehrlichia muris*-like agent; see Table 5.3). In laboratory models, it has been established that duration of infection in the host, density, and distance between ticks are important factors affecting efficiency of transmission of *B. burgdorferi* s.l. by co-feeding (Gern and Rais, 1996; Piesman and Happ, 2001; Richter et al., 2002). Furthermore, strain-specific variation in co-feeding transmission of *B. afzelii* by *I. ricinus* has been observed (Jacquet et al., 2016; Tonetti et al., 2015).

The host immune responses to *B. burgdorferi*, the dissemination of *Borrelia* in the host, and its infectivity for ticks have all been found to be affected by the presence of the vector. For example, mice infected with *B. burgdorferi* by *I. ricinus* were more infective for subsequently attached ticks than those experimentally inoculated with *Borrelia* (Gern et al., 1993). BALB/c mice developed a Th2 immune response against *B. burgdorferi* after tick inoculation and a mixed Th1/Th2 response after syringe inoculation. In addition, IL-4 produced in draining lymph nodes of the murine host following tick bites inhibited the production of anti-borrelial IgG2a antibodies (Christe et al., 2000). Injection of *I. ricinus* saliva together with *Borrelia* reduced the numbers of leukocytes and T lymphocytes in the infected murine epidermis at early time points after infection and decreased the total cell count in draining lymph nodes (Severínová et al., 2005). *I. ricinus* saliva was found to inhibit maturation of murine DCs *in vitro* (Skallová et al., 2008). In another study, treatment of murine DCs with tick saliva diminished phagocytosis of *B. afzelii*, reduced the production of Th1 and Th2 cytokines, and impaired their capacity to induce proliferation and IL-2 production in *Borrelia*-specific CD4+ T cells (Slámová et al., 2011). In addition, *I. ricinus* saliva modulated IFN-gamma signaling pathways in DCs (Lieskovská and Kopecký, 2012b) and pathways activated by a TLR2 ligand in *Borrelia*-stimulated DCs (Lieskovská and Kopecký, 2012a).

Innate immunity in host skin, particularly antimicrobial peptides of the cathelicidin and defensin families, is also modulated in the course of *Borrelia* infection by tick saliva. During transmission of *Borrelia* by *I. ricinus*, expression of inflammatory genes was suppressed, suggesting that tick saliva may facilitate establishment of *Borrelia* in host skin (Kern et al., 2011). Tick saliva also diminished cytokine production by keratinocytes in response to TLR2/TLR3 ligands during *Borrelia* transmission (Bernard et al., 2016).

Repeated infestation of laboratory animals with pathogen-free *I. scapularis* nymphs afforded protection against infection with tick-transmitted *B. burgdorferi*, suggesting that immunity against tick antigens can interfere with transmission of *Borrelia* (Nazario et al., 1998; Wikel et al., 1997). Furthermore, immunization of guinea pigs with *I. scapularis* SG proteins produced within the first day of tick attachment impaired *B. burgdorferi* transmission from ticks to hosts, probably by eliciting acquired immunity against tick antigens secreted during the first 24 h of feeding (Narasimhan et al., 2007).

Moreover, certain tick salivary compounds inhibit secretion of inflammatory cytokines during transmission of bacteria belonging to the order of Rickettsiales since saliva of *I. scapularis* was observed to impair the pro-inflammatory cytokine response by murine macrophages to infection with *Anaplasma phagocytophilum* (Chen et al., 2012).

Tick molecules involved in pathogen transmission. A few tick molecules implicated in pathogen transmission have been identified (Table 5.3); some facilitate the traffic and replication of the pathogen in the vector, while others enhance the transmission of pathogens to the host. *B. burgdorferi* s.l. displays distinct phenotypic plasticity (Radolf et al., 2012). Within the infected tick, *Borrelia* spirochetes express OspA and bind to the midgut wall by means of a tick expressed protein (Pal et al., 2004). Following tick attachment, spirochetes begin to express outer surface protein C (OspC) and move from the midgut through the hemolymph to the SGs. *Borrelia* spirochetes are transmitted to the host in tick saliva along with various salivary molecules that modulate T cells (Salp15), complement (Isac, Salp20), macrophages, neutrophils, and B cell activities (BIP), and other components of the host immune system, and help *Borrelia* to infect and disseminate in the mammalian host.

BIP, a B-cell inhibitory protein from *I. ricinus* SGs, suppressed B lymphocyte proliferation induced by the *B. burgdorferi* OspC, suggesting that BIP may play an important role in enhancing *Borrelia* transmission by the tick (Hannier et al., 2003).

The secreted salivary protein Salp15, considered to be the first tick mediator of SAT to be discovered, was first identified in the SGs of *I. scapularis*. Salp15 has been shown to bind to mammalian CD4 (Garg et al., 2006) and inhibit activation of CD4+ T lymphocytes (Anguita et al., 2002). Despite its immunosuppressive functions, Salp15 is an immunoprotective antigen since RNAi silencing drastically reduces the capacity of tick-borne spirochetes to infect mice (Ramamoorthi et al., 2005), and passive immunization of mice afforded significant protection from *I. scapularis*-transmitted *B. burgdorferi* (Dai et al., 2009). In the tick SGs, spirochetes bind, via OspC, to Salp15, which protects them from antibody- and complement-mediated killing (Schuijt et al., 2008) and facilitates their transmission and replication in the host skin (Ramamoorthi et al., 2005). Salp15 Iric-1, a Salp15 homolog, has been identified in *I. ricinus*. The protein appears to protect *Borrelia burgdorferi* sensu stricto, *Borrelia garinii*, and *B. afzelii* in a differential manner from antibody-mediated killing in the host (Hovius et al., 2008b).

Salp25D, an immunodominant salivary protein from *I. scapularis*, is important during tick acquisition of *B. burgdorferi* and acts as an antioxidant that facilitates pathogen survival (Das et al., 2001; Narasimhan et al., 2007).

Salp20, a salivary protein from *I. scapularis*, inhibits the alternative complement pathway by binding properdin and causing dissociation of the C3 convertase (Tyson et al., 2007; Hourcade et al., 2016). Salp20 partially protects *B. burgdorferi* from lysis by normal human serum, suggesting that, together with the plasma activation factor H, Salp20 may protect *Borrelia* from components of the host's complement cascade during transmission.

A tick histamine release factor (tHRF) is secreted into tick saliva and has been shown capable of binding mammalian basophils and triggering release of histamine (Dai et al., 2010; Mulenga et al., 2003b). It is up-regulated in *I. scapularis* SGs during the rapid feeding phase and probably facilitates tick engorgement and *B. burgdorferi* infection by increasing vascular permeability and hence blood flow to the tick bite site (Dai et al., 2010). Immunization of mice with the recombinant protein and silencing tHRF interfered with tick feeding and decreased *Borrelia* burden in mice at 7 days after infection in skin and at 3 weeks in heart and joints (Dai et al., 2010). Mulenga et al. (2003b) proposed that tHRF was required for vasodilation during the rapid feeding phase when large volumes of blood are required, thereby supporting enhanced replication and dissemination of *Borrelia* from the midgut. While reduced burden in mice may simply have been secondary to reduced burden in ticks, the authors suggested that the vasodilatory activity of histamine might also enhance systemic dissemination of *Borrelia* from the bite site (Dai et al., 2010).

The tick salivary lectin pathway inhibitor (TSLPI) from SGs of *I. scapularis* was found to interfere with the human lectin complement cascade and impair neutrophil phagocytosis and chemotaxis and to protect *Borrelia* from killing by the lectin complement pathway (Schuijt et al., 2011).

I. ricinus serine protease inhibitor (IrSPI), a protein from *I. ricinus* SGs belonging to the BPTI (bovine pancreatic trypsin inhibitor)/Kunitz family of serine protease inhibitors, probably impairs host hemostasis and facilitates tick feeding and *Bartonella henselae* transmission. RNAi silencing impaired tick feeding and reduced *B. henselae* load in the tick SGs (Liu et al., 2014b).

A. phagocytophilum was found to induce expression of the *I. scapularis* Salp16 gene in tick SGs during feeding. RNA interference-mediated silencing of Salp16 gene expression diminished migration of the bacteria ingested via host blood meal to tick SGs, which demonstrated the specific requirement of the pathogen for a tick salivary protein for persistence within the vector (Sukumaran et al., 2006). It was recently demonstrated that the *I. scapularis* salivary protein sialostatin L2 inhibited inflammasome formation during *A. phagocytophilum* infection by impairing caspase-1 activity (Chen et al., 2014).

Impact of Pathogens on Gene Expression in Salivary Glands

An extensive literature attests to the influence that acquisition of diverse pathogenic microorganisms has on gene expression in ticks (Table 5.4). Among the myriad genes whose level of expression is altered, some encode secreted proteins that are expressed in SGs. In these cases, colonization of the SGs may be expected to have an impact on salivary composition and thus on processes that occur at the tick/host interface, including transmission of pathogens from the tick to the vertebrate host or to co-feeding ticks. As the transcriptional response to infection has been shown to be tissue-specific (Ayllón et al., 2015; Mercado-Curiel et al., 2011), only studies that explicitly addressed the impact of infection on the SG transcriptome or the salivary proteome will be considered here.

The impact of infection on the SG transcriptome has been addressed for several species of ticks as regards diverse pathogens belonging to different classes of organisms, including viruses, bacteria, and protozoa (Table 5.4). Published reports have concerned *D. variabilis* and *Rickettsia montanensis* (Macaluso et al., 2003), *R. appendiculatus* and *Theileria parva* (Nene et al., 2004), *R. microplus* and *Anaplasma marginale* (Mercado-Curiel et al., 2011; Zivkovic et al., 2010), *I. scapularis* and *A. phagocytophilum* (Ayllón et al., 2015) or

TABLE 5.4 Functional Transcriptomic/Proteomic Studies of Interactions Between the Tick Salivary Glands and Tick-Borne Pathogens

Tick Species	Pathogen	Technique Used	Reference
Transcriptomic Studies			
Dermacentor variabilis, females	*Rickettsia montanensis*	DD-PCR	Macaluso et al. (2003)
Ixodes scapularis, nymphs	*Borrelia burgdorferi*	LCS	Ribeiro et al. (2006)
Rhipicephalus appendiculatus, females	*Theileria parva*	LCS	Nene et al. (2004)
Rhipicephalus microplus, males	*Anaplasma marginale*	SSH	Zivkovic et al. (2010)
R. microplus	*A. marginale*	SSH, microarray	Mercado-Curiel et al. (2011)
I. scapularis, nymphs	Langat virus		McNally et al. (2012)
R. microplus, males, 5 selected genes (GST, COXIII, DYN, SYN, PHOS)	*A. marginale*	RT-qPCR	Bifano et al. (2014)
Ixodes ricinus	*Bartonella henselae*	NGS	Liu et al. (2014b)
I. scapularis, females	*Anaplasma phagocytophilum*	NGS	Ayllón et al. (2015)
I. ricinus, lipocalins	*Borrelia afzelii*	RT-qPCR	Valdes et al. (2016)
I. scapularis, histones and histone-modifying enzymes	*A. phagocytophilum*	RT-qPCR	Cabezas-Cruz et al. (2016)
Proteomic Studies			
I. scapularis, females	*A. phagocytophilum*	LC–MS/MS	Ayllón et al. (2015)
I. scapularis, females Host proteins	*A. phagocytophilum*	LC–MS/MS	Villar et al. (2016)
I. scapularis, nymphs	*B. burgdorferi*	DIGE–MALDI-TOF MS/MS	Dai et al. (2010)
I. ricinus, females	*B. burgdorferi*	LC–MS/MS	Cotté et al. (2014)

DD-PCR, differential display polymerase chain reaction; *DIGE,* differential in-gel electrophoresis; *HPLC,* high-performance liquid chromatography; *LCS,* cDNA library clones sequencing; *MALDI-TOF,* matrix-assisted laser desorption/ionization time-of-flight; *MS,* mass spectrometry; *RT-qPCR,* real-time quantitative PCR; *SSH,* suppression subtractive hybridization.

Langat virus (McNally et al., 2012), and *I. ricinus* and *B. henselae* (Liu et al., 2014b). Diverse methods have been employed, from groundbreaking studies based on differential display polymerase chain reaction (PCR) (Macaluso et al., 2003) and sequencing of expressed sequence tags (Nene et al., 2004) to later studies based on suppression subtractive hybridization (Zivkovic et al., 2010) and microarrays (McNally et al., 2012; Mercado-Curiel et al., 2011) and to recent studies that exploit next-generation sequencing technologies (Ayllon et al., 2015; Liu et al., 2014b). Moreover, the impact of infection on the SG proteome or indeed on the proteome of saliva itself has been investigated for infection of *I. scapularis* with *A. phagocytophilum* (Ayllón et al., 2015) or *B. burgdorferi* (Dai et al., 2010) and for infection of *I. ricinus* with *B. burgdorferi* (Cotté et al., 2014).

Taken together, these studies have shown that acquisition of pathogenic microorganisms can exert a profound influence on gene expression in ticks. Nevertheless, and as regards the SG transcriptome, the number of regulated genes and the magnitude of up- or down-regulation have been reported to be relatively modest or extensive, depending on the study. While it is likely that different characteristics of the experimental strategy employed—such as the sensitivity with which transcripts are detected—affect the appreciation of the degree to which the transcriptome is affected, it is probable that the impact on gene expression reflects the type of relationship established by the tick/pathogen pair in question. It has, for example, been suggested by Mercado-Curiel et al. (2011) that pathogens that are highly adapted to their tick, such as *A. marginale* for *R. microplus*, and whose presence imposes a minimal fitness cost, have little effect on the SG transcriptome (Mercado-Curiel et al., 2011; Zivkovic et al., 2010), whereas pathogens that have a dramatic effect on tick fitness—such as *B. bovis* for *Rhipicephalus annulatus* (Ouhelli et al., 1987)—are likely to have a much greater impact.

In studies of the SG transcriptome, and as described earlier, multiple families of genes have typically been shown to be regulated. While regulation of many of these gene families occurs in multiple species of ticks in response to infection, the complexity of the observations and the lack of functional annotation render interpretation difficult. Indeed, the families tend to have numerous members, whose precise role is generally unknown, and of which some are up-regulated and others down-regulated for each species of tick in response to a given pathogen. Furthermore, while many of the genes that are regulated in different species of ticks may well represent functional orthologues, such association requires further functional annotation. It is likely, moreover, that the identity of regulated genes is affected by the class—virus, bacteria, or protozoa—or lifestyle—intra-versus extracellular—of the pathogen and potentially by the degree of adaptation existing between pathogen and tick. Indeed, on infection of *R. microplus* by *A. marginale*, the majority of regulated genes—of which there were few—had no known counterparts, with stress-related genes and defensins being conspicuously absent. The authors speculated that the small set of regulated genes might encode proteins with novel activity that could facilitate transmission of pathogen without affecting tick fitness (Mercado-Curiel et al., 2011).

Similar to vertebrates, ticks are protected against invading microorganisms by an innate immune system that is capable of mounting both humoral- and cell-mediated immune responses (Hynes, 2014). During co-evolution, however, tick-borne microorganisms have developed various strategies to evade the immune responses of their vectors to survive, persist, and be transmitted. While the immune responses elicited in mammals on infection with TBP have been extensively documented, the interactions established between the tick immune system and invading microorganisms are largely unknown. Modern high-throughput techniques have provided insight into some of the mechanisms by which TBP may manipulate their vectors (Smith and Pal, 2014). For example, down-regulation of immune-related genes has been observed in *I. scapularis* on infection with Langat virus (McNally et al., 2012), in *I. ricinus* on infection with *B. henselae* (Liu et al., 2014b), and in BME26 cells derived from *R. (B.) microplus* embryos when infected with *A. marginale* (Rosa et al., 2016). These observations suggest that TBP that have co-evolved with the vector may be able to manipulate the expression of immune-related genes and suppress tick immune responses. Manipulation of epigenetic modifications (Cabezas-Cruz et al., 2016) and regulation of the level of vertebrate host proteins in tick midgut and SGs, the latter effect probably resulting from modification of proteolytic pathways (Villar et al., 2016), as described recently for infection of *I. scapularis* with *A. phagocytophilum*, have been proposed to be additional mechanisms by which TBP may manipulate their vectors to facilitate infection while ensuring tick survival.

In most transcriptional studies, the presumed regulation has been validated—by quantitative reverse transcription-PCR—for only a few selected genes, and functional characterization is often limited to the role of the cognate proteins in the colonization of the tick by the pathogen. Their potential role in transmission of the pathogen from the tick to the vertebrate host has been less frequently addressed, and as far as the authors can ascertain, never as regards co-feeding ticks. Nevertheless, some of the SGs genes whose level of expression is affected by infection encode proteins that are predicted to be secreted, and in some instances their presence in saliva has been directly ascertained or deduced from the presence of specific antibodies in animals exposed to tick infestation. Evidence that transmission of pathogen from ticks to vertebrate hosts may be influenced by changes in salivary composition induced by infection exists for a small number of tick proteins, including IrSPI, CRT, Salp15, tHRF, and 5.3-kDa antimicrobial peptides.

CRT is a calcium-binding protein found in diverse organisms and to which multiple intra- and extracellular functions have been attributed (for review see Gold et al., 2006). Tick-encoded CRT has been identified in the saliva of *A. americanum* and *D. variabilis* females (Jaworski et al., 1995) and in *R. microplus* (Ferreira et al., 2002) and *I. ricinus* ticks (Cotté et al., 2014). The gene encoding

CRT has been shown to be up-regulated on infection of *R. annulatus* with *Babesia bigemina* (Antunes et al., 2012) and CRT itself shown to be up-regulated in the salivary proteome of *I. ricinus* on infection with *B. burgdorferi* (Cotté et al., 2014). When rabbits were immunized with a fusion protein comprising the CRT of *A. americanum*, subsequent infestation with this tick gave rise to necrotic feeding lesions (Jaworski et al., 1995). While it has been suggested that salivary CRT might facilitate tick feeding and pathogen transmission through inhibition of thrombosis and complement, the precise role of CRT remains elusive. Indeed, while the CRT of *A. americanum* was shown to bind to C1q, the first component of the classical complement pathway, it did not inhibit activation of the complement cascade (Kim et al., 2015a).

Histamine release factor is an evolutionarily conserved protein that in mammals induces the release of histamine from mast cells and basophils. tHRF is up-regulated in the SGs of *I. scapularis* on infection with *B. burgdorferi* (Dai et al., 2010) and in the ovaries of *D. variabilis* on infection with *R. montanensis* (Mulenga et al., 2003a).

Members of the 5.3-kDa family of peptides possess anti-microbial properties (Liu et al., 2012; Pichu et al., 2009). Their expression is induced by the tick Janus kinase/signal transducers and activators of transcription signal transduction pathway (Liu et al., 2012), and they are secreted into saliva (Pichu et al., 2009). Members of the 5.3-kDa family were upregulated in SGs of *I. scapularis* during infection with *A. phagocytophilum* (Liu et al., 2012) and Langat virus (McNally et al., 2012). The transcriptome of *I. scapularis* nymphs was also enriched in such transcripts on infection with *B. burgdorferi* (Ribeiro et al., 2006). RNAi knockdown of one member, encoded by gene-15, increased *A. phagocytophilum* burden in SGs of *I. scapularis* and in blood of mice on which gene-15-deficient ticks fed (Liu et al., 2012). While reduced burden in mice may have been secondary to reduced burden in ticks, the authors of the study suggested that the salivary AMP secreted into the bite site may have impaired *A. phagocytophilum* survival at the bite site (Liu et al., 2012).

The genes encoding proteins from the Salp15 family are preferentially induced during engorgement in the SGs of *B. burgdorferi*-infected *I. scapularis* (Ramamoorthi et al., 2005) and on infection of *I. ricinus* by *B. henselae* (Liu et al., 2014b).

Tick Saliva Antigens for Epidemiology and Control

Biologic Markers of Tick Exposure

The rapid evolution of tick distribution and density has created an urgent need for more effective methods for surveillance and risk assessment for tick-borne diseases (TBD) (Heyman et al., 2010; Leger et al., 2013; Medlock et al., 2013). Risk assessment for human and animal populations will determine public and veterinary public health priorities and instruct the implementation of appropriate countermeasures or complementary studies.

To date, most surveillance strategies of TBD have focused on pathogen detection in vertebrate hosts or in ticks or on modeling/forecasting approaches (Hai et al., 2014). Although host antibody responses to pathogens are commonly used in epidemiologic research to evaluate exposure and disease risk, few studies have addressed exposure to tick bites. Nonetheless, anti-tick immunity was first described almost a century ago (Trager, 1939), and many tick salivary proteins have been shown to be immunogenic in vertebrate hosts (Wikel, 1996). Indeed, it has been well recognized since the early 1990s that during the bite ticks inject salivary proteins that elicit antibody responses, and that these antibody responses could be used as biomarkers of host exposure to tick bites (Schwartz et al., 1990). The greatest challenge, however, lies in identification of antigenic markers within tick saliva that may be used to reveal host/vector contact and, moreover, that allow discrimination among the different tick species to which the host has been exposed. The specificity of the saliva-based immunological test is indeed a critical prerequisite, especially in areas with high biodiversity as regards hematophagous arthropod species (Fontaine et al., 2011). It is thus necessary to pursue efforts to obtain a comprehensive inventory of salivary components of the various species of ticks.

In 1990, Schwartz et al. reported that outdoor workers who were seropositive for Lyme disease agent had significantly higher anti-*Ixodes dammini* antibody levels than seronegative controls. Based on this observation, the authors proposed, for the first time, to use antibodies against sonicated tick SGs as markers of tick exposure in humans. Unfortunately, however, the lack of specificity of the assay as regards exposure to other arthropods limited its epidemiologic utility. Later, a recombinant form of CRT was used to evaluate exposure to *I. scapularis* bites, and results showed that this test had higher specificity but lower sensitivity than the test based on whole SGs (Sanders et al., 1999). Furthermore, antibodies against this *I. scapularis* protein have been sought in a longitudinal study to assess the impact of educational interventions on tick exposure (Malouin et al., 2003). In humans, such antibodies were found to be present in sera following tick exposure and to persist for as long as a year and a half (Alarcon-Chaidez et al., 2006). Cross-reactivity, however, has been reported between recombinant CRT from *I. scapularis* and *A. americanum* (Sanders et al., 1998b), illustrating the challenge of developing cognate ELISA tests of the desired specificity. Indeed, initiatives are still being undertaken to identify discriminant antigenic proteins in tick saliva that could be used to evaluate exposure to different tick species (Sanders et al., 1998a; Vu Hai et al., 2013a, 2013b). Antibodies against the protein Rs24p have also been proposed as biomarkers of *R. sanguineus* exposure in dogs, but anti-Rs24p antibodies were detected only for a short period after tick infestation (Liu and Bonnet, 2014; Medlock et al., 2013).

Lastly, and in addition to the assessment of host/vector contact and efficacy of anti-vector measures, these candidate biomarkers of exposure to ticks could be helpful in the diagnosis of TBD. In fact, it is well known that self-reported tick exposure is a poor correlate of true tick exposure as, for example, subjects with Lyme disease may deny a preceding tick bite. The development of reliable markers would enable documentation of an antecedent tick bite in patients for whom TBD is suspected.

Vaccine Candidates

Animals repeatedly exposed to tick bites develop immunity with, notably, production of antigen-specific antibodies directed against multiple tick salivary proteins (Vu Hai et al., 2013b), which impair tick attachment and curtail the blood meal, with a negative impact on the reproduction rate (Brossard and Wikel, 1997; Wikel and Alarcon-Chaidez, 2001). In addition, it has been reported that repeated exposure to ticks likely affords a measure of protection against animals and humans from TBD (Bell et al., 1979; Burke et al., 2005; Krause et al., 2009; Wikel et al., 1997). Such studies naturally spawned the hypothesis that tick salivary components could represent effective vaccine candidates against TBD. Moreover, and in light of limited understanding of immunity to TBP, TBP strain diversity and, more generally, the transmission of multiple TBP by the same tick species, vaccine strategies that target conserved processes in vector infestation and vector capacity, are increasingly being sought (Nuttall et al., 2006; Willadsen, 2004). This class of vaccine holds the promise of affording broad protection against multiple TBD transmitted by the same vector. Salivary proteins that play a key role in tick feeding represent good vaccine candidates, as immunity against such proteins may block the feeding process prior to pathogen transmission, which typically occurs many hours or even days after tick attachment. Moreover, the utilization of so-called "exposed" antigens present in saliva, rather than "concealed" tick antigens to which the host is never naturally exposed, may allow natural boosting of the host response on exposure to ticks (Nuttall et al., 2006). Lastly, the anti-tick vaccine approach is compatible with the inclusion of multiple antigens, including those from pathogens, as demonstrated in the *Borrelia* sp./*I. scapularis* infection model for OsPA and Salp15 (Steere et al., 1998) so as to reinforce protection against defined pathogens or extend protection to pathogens transmitted by different tick species.

Currently, vaccines derived from Bm86, a midgut protein of *R. microplus*, are the only ectoparasite vaccines that are commercially available (in Australia and Cuba) but provide a powerful proof of principle for the feasibility of creating and deploying anti-tick vaccines (Willadsen et al., 1995). Vaccine-elicited antibodies are believed to lyse the tick's gut wall, thus interfering with feeding and subsequent egg production. In this case, reduction in tick burden is probably feasible because the targeted tick species feeds principally on the host species for which the vaccine is intended, i.e., *R. microplus* and cattle. Indeed, for several species of ticks responsible for important TBD, such as *Ixodes* sp., which feed on multiple hosts and notably on wildlife, vaccines that exert a direct effect on tick blood meal acquisition or vector competence must be sought. With this aim in view, salivary antigens represent natural vaccine candidates.

Vaccines that provide immunity against salivary proteins are likely to reduce pathogen transmission, as has been reported for Salp15 (Dai et al., 2009), TSLPI (Schuijt et al., 2011) and tHRF (Dai et al., 2010) in the transmission of *Borrelia* sp. to mice by *I. scapularis*, for Salp25D in the acquisition of *Borrelia* sp. by ticks (Wagemakers et al., 2016), or for 64TRP (a cement protein) in the transmission of TBE to mice by *I. ricinus* (Labuda et al., 2006). Immunity against tick salivary proteins may also diminish host/vector contact by limiting blood intake or diminishing the duration of the blood meal, as reported for *H. longicornis* (Anisuzzaman et al., 2012; Imamura et al., 2005; Mulenga et al., 1999; Zhang et al., 2011), *O. moubata* and *Ornithodoros erraticus* (Astigarraga et al., 1995; Garcia-Varas et al., 2010), *A. americanum* (de la Fuente et al., 2010), and *R. microplus* (Ali et al., 2015; Andreotti et al., 2002; Merino et al., 2013). For *I. ricinus*, vaccination against metalloproteases and a serpin (iris) derived from SGs also interfered with completion of the blood meal and subsequently affected their reproductive fitness (Decrem et al., 2008b; Prevot et al., 2007). Finally, Imamura et al. also reported an increased mortality of *H. longicornis* and *R. appendiculatus* during their blood meal on cattle immunized with salivary serpins (Imamura et al., 2005, 2006).

CONCLUSION

To facilitate the uptake of their voluminous and drawn-out blood meals, ticks have developed multiple ingenious strategies, perhaps the most extraordinary of which being the subversion of the host response mediated by saliva introduced into the host during feeding. Indeed, tick saliva comprises myriad salivary components that play essential roles in counteracting host hemostatic and immunological defenses. In addition, immunomodulatory components, which dampen host defenses to assure adequate feeding, create a favorable context for survival and propagation of TBP. These salivary components may be exploited both as immunological markers for meaningful assessment of exposure to tick bites and as vaccine candidates to protect against TBD. The deployment of anti-tick vaccines designed to reduce transmission of TBP and reduce reliance on acaricides and repellents would represent a major improvement over current control measures as regards environmental conservation and occupational exposure to TBP. First, the use of an anti-tick vaccine would cause neither contamination of the environment and foodstuffs nor harm to off-target species. Second, anti-tick vaccines are expected to provide broad protection against diverse TBP. They may be combined with vaccines against individual TBP, when these are characterized and cognate vaccines are available, to afford greater protection, but are also—and most significantly—expected to protect against uncharacterized TBP, whether these are currently in circulation or as yet to emerge. Lastly, a better understanding of the complex physiology of the essential organ represented by tick SGs may also lead to the conception of hitherto unimagined strategies for controlling ticks and TBD.

ACKNOWLEDGMENTS

SB, JR, and LS were supported by the Institut National de la Recherche Agronomique. MK was supported by the Slovak Research and Development Agency (contract no. APVV-0737-12).

REFERENCES

Alarcon-Chaidez, F., Ryan, R., Wikel, S., Dardick, K., Lawler, C., Foppa, I.M., Tomas, P., Cushman, A., Hsieh, A., Spielman, A., et al., 2006. Confirmation of tick bite by detection of antibody to ixodes calreticulin salivary protein. Clinical and Vaccine Immunology 13, 1217–1222.

Alarcon-Chaidez, F.J., Sun, J., Wikel, S.K., 2007. Transcriptome analysis of the salivary glands of *Dermacentor andersoni* Stiles (Acari: Ixodidae). Insect Biochemistry and Molecular Biology 37, 48–71.

Alekseev, A.N., Chunikhin, S.P., 1990. The exchange of the tick-borne encephalitis virus between ixodid ticks feeding jointly on animals with a subthreshold level of viremia. Meditsinskaia Parazitologiia I Parazitarnye Bolezni 48–50.

Alekseev, A.N., Burenkova, L.A., Podboronov, V.M., Chunikhin, S.P., 1995. Bacteriocidal qualities of ixodid tick (Acarina: Ixodidae) salivary cement plugs and their changes under the influence of a viral tick-borne pathogen. Journal of Medical Entomology 32, 578–582.

Ali, A., Parizi, L.F., Guizzo, M.G., Tirloni, L., Seixas, A., Vaz Ida Jr., S., Termignoni, C., 2015. Immunoprotective potential of a *Rhipicephalus* (*Boophilus*) microplus metalloprotease. Veterinary Parasitology 207, 107–114.

Aljamali, M.N., Hern, L., Kupfer, D., Downard, S., So, S., Roe, B.A., Sauer, J.R., Essenberg, R.C., 2009. Transcriptome analysis of the salivary glands of the female tick *Amblyomma americanum* (Acari: Ixodidae). Insect Molecular Biology 18, 129–154.

Anatriello, E., Ribeiro, J.M., de Miranda-Santos, I.K., Brandao, L.G., Anderson, J.M., Valenzuela, J.G., Maruyama, S.R., Silva, J.S., Ferreira, B.R., 2010. An insight into the sialotranscriptome of the brown dog tick, *Rhipicephalus sanguineus*. BMC Genomics 11, 450.

Andrade, B.B., Teixeira, C.R., Barral, A., Barral-Netto, M., 2005. Haematophagous arthropod saliva and host defense system: a tale of tear and blood. Anais da Academia Brasileira de Ciencias 77, 665–693.

Andreotti, R., Gomes, A., Malavazi-Piza, K.C., Sasaki, S.D., Sampaio, C.A., Tanaka, A.S., 2002. BmTI antigens induce a bovine protective immune response against *Boophilus microplus* tick. International Immunopharmacology 2, 557–563.

Anguita, J., Ramamoorthi, N., Hovius, J.W.R., Das, S., Thomas, V., Persinski, R., Conze, D., Askenase, P.W., Rincón, M., Kantor, F.S., et al., 2002. Salp15, an *Ixodes scapularis* salivary protein, inhibits CD4(+) T cell activation. Immunity 16, 849–859.

Anisuzzaman, I.M.K., Alim, M.A., Miyoshi, T., Hatta, T., Yamaji, K., Matsumoto, Y., Fujisaki, K., Tsuji, N., 2011a. Longistatin, a plasminogen activator, is key to the availability of blood-meals for ixodid ticks. PLoS Pathogens 7, e1001312.

Anisuzzaman, K.I.M., Abdul Alim, M., Miyoshi, T., Hatta, T., Yamaji, K., Matsumoto, Y., Fujisaki, K., Tsuji, N., 2011b. Longistatin, a novel plasminogen activator from vector ticks, is resistant to plasminogen activator inhibitor-1. Biochemical and Biophysical Research Communications 413, 599–604.

Anisuzzaman, I.M.K., Alim, M.A., Miyoshi, T., Hatta, T., Yamaji, K., Matsumoto, Y., Fujisaki, K., Tsuji, N., 2012. Longistatin is an unconventional serine protease and induces protective immunity against tick infestation. Molecular and Biochemical Parasitology 182, 45–53.

Antunes, S., Galindo, R.C., Almazan, C., Rudenko, N., Golovchenko, M., Grubhoffer, L., Shkap, V., do Rosario, V., de la Fuente, J., Domingos, A., 2012. Functional genomics studies of *Rhipicephalus* (*Boophilus*) *annulatus* ticks in response to infection with the cattle protozoan parasite, *Babesia bigemina*. International Journal for Parasitology 42, 187–195.

Astigarraga, A., Oleaga-Perez, A., Perez-Sanchez, R., Encinas-Grandes, A., 1995. A study of the vaccinal value of various extracts of concealed antigens and salivary gland extracts against *Ornithodoros erraticus* and *Ornithodoros moubata*. Veterinary Parasitology 60, 133–147.

Ayllón, N., Villar, M., Galindo, R.C., Kocan, K.M., Šíma, R., Lopez, J.A., Vazquez, J., Alberdi, P., Cabezas-Cruz, A., Kopáček, P., de la Fuente, J., et al., 2015. Systems biology of tissue-specific response to *Anaplasma phagocytophilum* reveals differentiated apoptosis in the tick vector *Ixodes scapularis*. PLoS Genetics 11, e1005120.

Balashov, Y.S., 1972. Bloodsucking Ticks (Ixodoidea) Vectors of Disease of Man and Animals. Entomol Soc Amer Misc Publ.

Banajee, K.H., Embers, M.E., Langohr, I.M., Doyle, L.A., Hasenkampf, N.R., Macaluso, K.R., 2015. *Amblyomma maculatum* feeding augments *Rickettsia parkeri* infection in a rhesus macaque model: a pilot study. PLoS One 10.

Batista, I.F.C., Chudzinski-Tavassi, A.M., Faria, F., Simons, S.M., Barros-Batestti, D.M., Labruna, M.B., Leao, L.I., Ho, P.L., Junqueira-de-Azevedo, I.L.M., 2008. Expressed sequence tags (ESTs) from the salivary glands of the tick *Amblyomma cajennense* (Acari: Ixodidae). Toxicon 51, 823–834.

Batista, I.F., Ramos, O.H., Ventura, J.S., Junqueira-de-Azevedo, I.L., Ho, P.L., Chudzinski-Tavassi, A.M., 2010. A new factor Xa inhibitor from *Amblyomma cajennense* with a unique domain composition. Archives of Biochemistry and Biophysics 493, 151–156.

Beaufays, J., Adam, B., Menten-Dedoyart, C., Fievez, L., Grosjean, A., Decrem, Y., Prevot, P.P., Santini, S., Brasseur, R., Brossard, M., et al., 2008. Ir-LBP, an *Ixodes ricinus* tick salivary LTB4-binding lipocalin, interferes with host neutrophil function. PLoS One 3, e3987.

Bell, J.F., Stewart, S.J., Wikel, S.K., 1979. Resistance to tick-borne *Francisella tularensis* by tick-sensitized rabbits: allergic klendusity. The American Journal of Tropical Medicine and Hygiene 28, 876–880.

Bergman, D.K., Palmer, M.J., Caimano, M.J., Radolf, J.D., Wikel, S.K., 2000. Isolation and molecular cloning of a secreted immunosuppressant protein from *Dermacentor andersoni* salivary gland. The Journal of Parasitology 86, 516–525.

Bernard, J., Hutet, E., Paboeuf, F., Randriamparany, T., Holzmuller, P., Lancelot, R., Rodrigues, V., Vial, L., Le Potier, M.F., 2016. Effect of *O. Porcinus* tick salivary gland extract on the African swine fever virus infection in domestic pig. PLoS One 11.

Bifano, T.D., Ueti, M.W., Esteves, E., Reif, K.E., Braz, G.R.C., Scoles, G.A., Bastos, R.G., White, S.N., Daffre, S., 2014. Knockdown of the *Rhipicephalus microplus* cytochrome c oxidase subunit III gene is associated with a failure of *Anaplasma marginale* transmission. PLoS One 9.

Binnington, K.C., Stone, B.F., 1977. Distribution of catecholamines in the cattle tick *Boophilus microplus*. Comparative Biochemistry and Physiology C Comparative Pharmacology 58, 21–28.

Binnington, K.C., Stone, B.F., 1981. Developmental changes in morphology and toxin content of the salivary gland of the australian paralysis tick *Ixodes holocyclus*. International Journal for Parasitology 11, 343–351.

Binnington, K.C., 1975. Secretory coxal gland, active during apolysis in ixodid and argasid ticks (Acarina). International Journal of Insect Morphology and Embryology 4, 183–191.

Binnington, K.C., 1978. Sequential changes in salivary gland structure during attachment and feeding of the cattle tick, *Boophilus microplus*. International Journal for Parasitology 8, 97–115.

Bishop, R., Lambson, B., Wells, C., Pandit, P., Osaso, J., Nkonge, C., Morzaria, S., Musoke, A., Nene, V., 2002. A cement protein of the tick *Rhipicephalus appendiculatus*, located in the secretory e cell granules of the type III salivary gland acini, induces strong antibody responses in cattle. International Journal for Parasitology 32, 833–842.

Bonnet, S.I., Liu, X.Y., 2012. Laboratory artificial infection of hard ticks: a tool for the analysis of tick-borne pathogen transmission. Acariologia 52 (4), 453–464.

Bowman, A.S., Sauer, J.R., 2004. Tick salivary glands: function, physiology and future. Parasitology (129 Suppl.), S67–S81.

Bowman, A.S., Sauer, J.R., Zhu, K., Dillwith, J.W., 1995. Biosynthesis of salivary prostaglandins in the lone star tick, *Amblyomma americanum*. Insect Biochemistry and Molecular Biology 25, 735–741.

Bowman, A.S., Dillwith, J.W., Sauer, J.R., 1996. Tick salivary prostaglandins: presence, origin and significance. Parasitology Today 12, 388–396.

Brake, D.K., de Leon, A.A.P., 2012. Immunoregulation of bovine macrophages by factors in the salivary glands of *Rhipicephalus microplus*. Parasites and Vectors 5.

Branco, V.G., Iqbal, A., Alvarez-Flores, M.P., Sciani, J.M., de Andrade, S.A., Iwai, L.K., Serrano, S.M.T., Chudzinski-Tavassi, A.M., 2016. Amblyomin-X having a Kunitz-type homologous domain, is a noncompetitive inhibitor of FXa and induces anticoagulation in vitro and in vivo. Biochimica Et Biophysica Acta-proteins and Proteomics 1864, 1428–1435.

Brossard, M., Wikel, S.K., 1997. Immunology of interactions between ticks and hosts. Medical and Veterinary Entomology 11, 270–276.

Brossard, M., Wikel, S.K., 2004. Tick immunobiology. Parasitology (129 Suppl.), S161–S176.

Brossard, M., Wikel, S.K., 2008. Tick immunobiology. In: Bowman, A.S., Nuttall, P.A. (Eds.), Ticks: Biology, Disease and Control. Cambridge University Press, pp. 186–204.

Bullard, R., Allen, P., Chao, C.-C., Douglas, J., Das, P., Morgan, S.E., Ching, W.-M., Karim, S., 2016. Structural characterization of tick cement cones collected from in vivo and artificial membrane blood-fed lone star ticks (*Amblyomma americanum*). Ticks and Tick-Borne Diseases 7, 880–892.

Burke, G., Wikel, S.K., Spielman, A., Telford, S.R., McKay, K., Krause, P.J., Tick-borne Infection Study G, 2005. Hypersensitivity to ticks and Lyme disease risk. Emerging Infectious Diseases 11, 36–41.

Cabezas-Cruz, A., Alberdi, P., Ayllón, N., Valdés, J.J., Pierce, R., Villar, M., de la Fuente, J., 2016. *Anaplasma phagocytophilum* increases the levels of histone modifying enzymes to inhibit cell apoptosis and facilitate pathogen infection in the tick vector *Ixodes scapularis*. Epigenetics 11, 303–319.

Campbell, E.M., Burdin, M., Hoppler, S., Bowman, A.S., 2010. Role of an aquaporin in the sheep tick *Ixodes ricinus*: assessment as a potential control target. International Journal for Parasitology 40, 15–23.

Cao, J., Shi, L., Zhou, Y.Z., Gao, X., Zhang, H.S., Gong, H.Y., Zhou, J.L., 2013. Characterization of a new kunitz-type serine protease inhibitor from the hard tick *Rhipicephalus hemaphysaloides*. Archives of Insect Biochemistry and Physiology 84, 104–113.

Carneiro-Lobo, T.C., Schaffner, F., Disse, J., Ostergaard, H., Francischetti, I.M.B., Monteiro, R.Q., Ruf, W., 2012. The tick-derived inhibitor Ixolaris prevents tissue factor signaling on tumor cells. Journal of Thrombosis and Haemostasis 10, 1849–1858.

Carvalho, W.A., Maruyama, S.R., Franzin, A.M., Abatepaulo, A.R.R., Anderson, J.M., Ferreira, B.R., Ribeiro, J.M.C., More, D.D., Maia, A.A.M., Valenzuela, J.G., et al., 2010. *Rhipicephalus (Boophilus) microplus*: clotting time in tick-infested skin varies according to local inflammation and gene expression patterns in tick salivary glands. Experimental Parasitology 124, 428–435.

Carvalho-Costa, T.M., Mendes, M.T., da Silva, M.V., da Costa, T.A., Tiburcio, M.G., Anhe, A.C., Rodrigues Jr., V., Oliveira, C.J., 2015. Immunosuppressive effects of *Amblyomma cajennense* tick saliva on murine bone marrow-derived dendritic cells. Parasites and Vectors 8, 22.

Cavassani, K.A., Aliberti, J.C., Dias, A.R., Silva, J.S., Ferreira, B.R., 2005. Tick saliva inhibits differentiation, maturation and function of murine bone-marrow-derived dendritic cells. Immunology 114, 235–245.

Chen, G., Severo, M.S., Sohail, M., Sakhon, O.S., Wikel, S.K., Kotsyfakis, M., Pedra, J.H., 2012. *Ixodes scapularis* saliva mitigates inflammatory cytokine secretion during *Anaplasma phagocytophilum* stimulation of immune cells. Parasites and Vectors 5, 229.

Chen, G., Wang, X.W., Severo, M.S., Sakhon, O.S., Sohail, M., Brown, L.J., Sircar, M., Snyder, G.A., Sundberg, E.J., Ulland, T.K., et al., 2014. The tick salivary protein sialostatin L2 inhibits caspase-1-mediated inflammation during *Anaplasma phagocytophilum* infection. Infection and Immunity 82, 2553–2564.

Cheng, Y., Wu, H., Li, D., 1999. An inhibitor selective for collagen-stimulated platelet aggregation from the salivary glands of hard tick *Haemaphysalis longicornis* and its mechanism of action. Science in China. Series C, Life Sciences 42, 457–464.

Chinery, W.A., 1973. The nature and origin of the "cement" substance at the site of attachment and feeding of adult *Haemaphysalis spinigera* (ixodidae). Journal of Medical Entomology 10, 355–362.

Chmelař, J., Anderson, J.M., Mu, J., Jochim, R.C., Valenzuela, J.G., Kopecký, J., 2008. Insight into the sialome of the castor bean tick, *Ixodes ricinus*. BMC Genomics 9, 233.

Chmelař, J., Oliveira, C.J., Rezacova, P., Francischetti, I.M., Kovarova, Z., Pejler, G., Kopacek, P., Ribeiro, J.M., Mares, M., Kopecky, J., et al., 2011. A tick salivary protein targets cathepsin G and chymase and inhibits host inflammation and platelet aggregation. Blood 117, 736–744.

Chmelař, J., Calvo, E., Pedra, J.H., Francischetti, I.M., Kotsyfakis, M., 2012. Tick salivary secretion as a source of antihemostatics. Journal of Proteomics 75, 3842–3854.

Chmelař, J., Kotál, J., Karim, S., Kopacek, P., Francischetti, I.M., Pedra, J.H., Kotsyfakis, M., 2016a. Sialomes and mialomes: a systems-biology view of tick tissues and tick-host interactions. Trends in Parasitology 32, 242–254.

Chmelař, J., Kotál, J., Kopecký, J., Pedra, J.H., Kotsyfakis, M., 2016b. All for one and one for all on the tick-host battlefield. Trends in Parasitology 32, 368–377.

Christe, M., Rutti, B., Brossard, M., 2000. Cytokines (IL-4 and IFN-gamma) and antibodies (IgE and IgG2a) produced in mice infected with *Borrelia burgdorferi* sensu stricto via nymphs of *Ixodes ricinus* ticks or syringe inoculations. Parasitology Research 86, 491–496.

Chudzinski-Tavassi, A.M., Morais, K.L.P., Fernandes Pacheco, M.T., Mesquita Pasqualoto, K.F., de Souza, J.G., 2016. Tick salivary gland as potential natural source for the discovery of promising antitumor drug candidates. Biomedicine and Pharmacotherapy 77, 14–19.

Ciprandi, A., de Oliveira, S.K., Masuda, A., Horn, F., Termignoni, C., 2006. *Boophilus microplus*: its saliva contains microphilin, a small thrombin inhibitor. Experimental Parasitology 114, 40–46.

Coons, L.B., Kaufman, W.R., 1988. Evidence that developmental changes in type III acini in the tick *Amblyomma hebraeum* (Acari: Ixodidae) are initiated by a hemolymph-borne factor. Experimental and Applied Acarology 4, 117–139.

Coons, L.B., Roshdy, M.A., 1973. Fine structure of the salivary glands of unfed male *Dermacentor variabilis* (Say) (Ixodoidea: Ixodidae). The Journal of Parasitology 59, 900–912.

Coons, L.B., Lessman, C.A., Ward, M.W., Berg, R.H., Lamoreaux, W.J., 1994. Evidence of a myoepithelial cell in tick salivary glands. International Journal for Parasitology 24, 551–562.

Cotté, V., Sabatier, L., Schnell, G., Carmi-Leroy, A., Rousselle, J.-C., Arsène-Ploetze, F., Malandrin, L., Sertour, N., Namane, A., Ferquel, E., et al., 2014. Differential expression of *Ixodes ricinus* salivary gland proteins in the presence of the *Borrelia burgdorferi* sensu lato complex. Journal of Proteomics 96, 29–43.

Couvreur, B., Beaufays, J., Charon, C., Lahaye, K., Gensale, F., Denis, V., Charloteaux, B., Decrem, Y., Prevot, P.P., Brossard, M., et al., 2008. Variability and action mechanism of a family of anti-complement proteins in *Ixodes ricinus*. PLoS One 3, e1400.

Dai, J., Wang, P., Adusumilli, S., Booth, C.J., Narasimhan, S., Anguita, J., Fikrig, E., 2009. Antibodies against a tick protein, Salp15, protect mice from the Lyme disease agent. Cell Host and Microbe 6, 482–492.

Dai, J., Narasimhan, S., Zhang, L., Liu, L., Wang, P., Fikrig, E., 2010. Tick histamine release factor is critical for *Ixodes scapularis* engorgement and transmission of the Lyme disease agent. PLoS Pathogens 6, e1001205.

Daix, V., Schroeder, H., Praet, N., Georgin, J.P., Chiappino, I., Gillet, L., de Fays, K., Decrem, Y., Leboulle, G., Godfroid, E., et al., 2007. Ixodes ticks belonging to the *Ixodes ricinus* complex encode a family of anticomplement proteins. Insect Molecular Biology 16, 155–166.

Dantas-Torres, F., Chomel, B.B., Otranto, D., 2012. Ticks and tick-borne diseases: a one health perspective. Trends in Parasitology 28, 437–446.

Das, S., Banerjee, G., DePonte, K., Marcantonio, N., Kantor, F.S., Fikrig, E., 2001. Salp25D, an *Ixodes scapularis* antioxidant, is 1 of 14 immunodominant antigens in engorged tick salivary glands. The Journal of Infectious Diseases 184, 1056–1064.

de Castro, M.H., de Klerk, D., Pienaar, R., Latif, A.A., Rees, D.J.G., Mans, B.J., 2016. De novo assembly and annotation of the salivary gland transcriptome of *Rhipicephalus appendiculatus* male and female ticks during blood feeding. Ticks and Tick-Borne Diseases 7, 536–548.

de la Fuente, J., Manzano-Roman, R., Naranjo, V., Kocan, K.M., Zivkovic, Z., Blouin, E.F., Canales, M., Almazan, C., Galindo, R.C., Step, D.L., et al., 2010. Identification of protective antigens by RNA interference for control of the lone star tick, *Amblyomma americanum*. Vaccine 28, 1786–1795.

de la Fuente, J., Villar, M., Cabezas-Cruz, A., Estrada-Peña, A., Ayllón, N., Alberdi, P., 2016. Tick-host-pathogen interactions: conflict and cooperation. PLoS Pathogens 12, e1005488.

Decrem, Y., Beaufays, J., Blasioli, V., Lahaye, K., Brossard, M., Vanhamme, L., Godfroid, E., 2008a. A family of putative metalloproteases in the salivary glands of the tick *Ixodes ricinus*. The FEBS Journal 275, 1485–1499.

Decrem, Y., Mariller, M., Lahaye, K., Blasioli, V., Beaufays, J., Zouaoui Boudjeltia, K., Vanhaeverbeek, M., Cerutti, M., Brossard, M., Vanhamme, L., et al., 2008b. The impact of gene knock-down and vaccination against salivary metalloproteases on blood feeding and egg laying by *Ixodes ricinus*. International Journal for Parasitology 38, 549–560.

Decrem, Y., Rath, G., Blasioli, V., Cauchie, P., Robert, S., Beaufays, J., Frere, J.M., Feron, O., Dogne, J.M., Dessy, C., et al., 2009. Ir-CPI, a coagulation contact phase inhibitor from the tick *Ixodes ricinus*, inhibits thrombus formation without impairing hemostasis. The Journal of Experimental Medicine 206, 2381–2395.

Deruaz, M., Frauenschuh, A., Alessandri, A.L., Dias, J.M., Coelho, F.M., Russo, R.C., Ferreira, B.R., Graham, G.J., Shaw, J.P., Wells, T.N., et al., 2008. Ticks produce highly selective chemokine binding proteins with antiinflammatory activity. The Journal of Experimental Medicine 205, 2019–2031.

Diaz-Martin, V., Manzano-Roman, R., Oleaga, A., Encinas-Grandes, A., Perez-Sanchez, R., 2013a. Cloning and characterization of a plasminogen-binding enolase from the saliva of the argasid tick *Ornithodoros moubata*. Veterinary Parasitology 191, 301–314.

Diaz-Martin, V., Manzano-Roman, R., Valero, L., Oleaga, A., Encinas-Grandes, A., Perez-Sanchez, R., 2013b. An insight into the proteome of the saliva of the argasid tick *Ornithodoros moubata* reveals important differences in saliva protein composition between the sexes. Journal of Proteomics 80, 216–235.

Fawcett, D.W., Doxsey, S.J., Buscher, G., 1981a. Salivary gland of the tick vector (*R. appendiculaius*) of East Coast Fever. I. Ultrastructure of the type III acinus. Tissue and Cell 13, 209–230.

Fawcett, D.W., Doxsey, S.J., Buscher, G., 1981b. Salivary gland of the tick vector (*R. appediculatus*) of East Coast Fever. II. Cellular basis for fluid secretion in the type III acinus. Tissue and Cell 13, 231–253.

Fawcett, D.W., Binnington, K., Voigt, W.P., 1986. The Cell Biology of the Ixodid Tick Salivary Gland. Ellis Horwood, Chichester, UK (Report).

Feldman-Muhsam, B., Borut, S., Saliternik-Givant, S., 1970. Salivary secretion of the male tick during copulation. Journal of Insect Physiology 16, 1945–1949.

Ferreira, B.R., Silva, J.S., 1999. Successive tick infestations selectively promote a T-helper 2 cytokine profile in mice. Immunology 96, 434–439.

Ferreira, C.A.S., Da Silva Vaz, I., da Silva, S.S., Haag, K.L., Valenzuela, J.G., Masuda, A., 2002. Cloning and partial characterization of a *Boophilus microplus* (Acari: Ixodidae) calreticulin. Experimental Parasitology 101, 25–34.

Fialová, A., Cimburek, Z., Iezzi, G., Kopecký, J., 2010. *Ixodes ricinus* tick saliva modulates tick-borne encephalitis virus infection of dendritic cells. Microbes and Infection 12, 580–585.

Fontaine, A., Diouf, I., Bakkali, N., Misse, D., Pages, F., Fusai, T., Rogier, C., Almeras, L., 2011. Implication of haematophagous arthropod salivary proteins in host-vector interactions. Parasites and Vectors 4, 187.

Francischetti, I.M., Valenzuela, J.G., Andersen, J.F., Mather, T.N., Ribeiro, J.M., 2002. Ixolaris, a novel recombinant tissue factor pathway inhibitor (TFPI) from the salivary gland of the tick, *Ixodes scapularis*: identification of factor X and factor Xa as scaffolds for the inhibition of factor VIIa/tissue factor complex. Blood 99, 3602–3612.

Francischetti, I.M., Mather, T.N., Ribeiro, J.M., 2003. Cloning of a salivary gland metalloprotease and characterization of gelatinase and fibrin(ogen)lytic activities in the saliva of the Lyme disease tick vector *Ixodes scapularis*. Biochemical and Biophysical Research Communications 305, 869–875.

Francischetti, I.M., Mather, T.N., Ribeiro, J.M., 2004. Penthalaris, a novel recombinant five-Kunitz tissue factor pathway inhibitor (TFPI) from the salivary gland of the tick vector of Lyme disease, *Ixodes scapularis*. Thrombosis and Haemostasis 91, 886–898.

Francischetti, I.M., Mather, T.N., Ribeiro, J.M., 2005a. Tick saliva is a potent inhibitor of endothelial cell proliferation and angiogenesis. Thrombosis and Haemostasis 94, 167–174.

Francischetti, I.M., My Pham, V., Mans, B.J., Andersen, J.F., Mather, T.N., Lane, R.S., Ribeiro, J.M., 2005b. The transcriptome of the salivary glands of the female western black-legged tick *Ixodes pacificus* (Acari: Ixodidae). Insect Biochemistry and Molecular Biology 35, 1142–1161.

Francischetti, I.M., Meng, Z., Mans, B.J., Gudderra, N., Hall, M., Veenstra, T.D., Pham, V.M., Kotsyfakis, M., Ribeiro, J.M., 2008a. An insight into the salivary transcriptome and proteome of the soft tick and vector of epizootic bovine abortion, *Ornithodoros coriaceus*. Journal of Proteomics 71, 493–512.

Francischetti, I.M., Mans, B.J., Meng, Z., Gudderra, N., Veenstra, T.D., Pham, V.M., Ribeiro, J.M., 2008b. An insight into the sialome of the soft tick, *Ornithodorus parkeri*. Insect Biochemistry and Molecular Biology 38, 1–21.

Francischetti, I.M., Sa-Nunes, A., Mans, B.J., Santos, I.M., Ribeiro, J.M., 2009. The role of saliva in tick feeding. Frontiers in Bioscience 14, 2051–2088.

Francischetti, I.M.B., Anderson, J.M., Manoukis, N., Pham, V.M., Ribeiro, J.M.C., 2011. An insight into the sialotranscriptome and proteome of the coarse bontlegged tick, *Hyalomma marginatum rufipes*. Journal of Proteomics 74, 2892–2908.

Francischetti, I.M., 2010. Platelet aggregation inhibitors from hematophagous animals. Toxicon 56, 1130–1144.

Franco, P.F., Silva, N.C., Fazito do Vale, V., Abreu, J.F., Santos, V.C., Gontijo, N.F., Valenzuela, J.G., Pereira, M.H., Sant'Anna, M.R., Gomes, A.P., et al., 2016. Inhibition of the classical pathway of the complement system by saliva of *Amblyomma cajennense* (Acari: Ixodidae). Experimental Parasitology 164, 91–96.

Frauenschuh, A., Power, C.A., Deruaz, M., Ferreira, B.R., Silva, J.S., Teixeira, M.M., Dias, J.M., Martin, T., Wells, T.N., Proudfoot, A.E., 2007. Molecular cloning and characterization of a highly selective chemokine-binding protein from the tick *Rhipicephalus sanguineus*. The Journal of Biological Chemistry 282, 27250–27258.

Fukumoto, S., Sakaguchi, T., You, M., Xuan, X., Fujisaki, K., 2006. Tick troponin I-like molecule is a potent inhibitor for angiogenesis. Microvascular Research 71, 218–221.

Gaede, K., Knülle, W., 1997. On the mechanism of water vapour sorption from unsaturated atmospheres by ticks. The Journal of Experimental Biology 200, 1491–1498.

Gao, X., Shi, L., Zhou, Y., Cao, J., Zhang, H., Zhou, J., 2011. Characterization of the anticoagulant protein Rhipilin-1 from the *Rhipicephalus haemaphysaloides* tick. Journal of Insect Physiology 57, 339–343.

Garcia, G.R., Gardinassi, L.G., Ribeiro, J.M., Anatriello, E., Ferreira, B.R., Moreira, H.N.S., Mafra, C., Martins, M.M., Szabo, M.P.J., de Miranda-Santos, I.K.F., et al., 2014. The sialotranscriptome of *Amblyomma triste*, *Amblyomma parvum* and *Amblyomma cajennense* ticks, uncovered by 454-based RNA-seq. Parasites and Vectors 7.

Garcia-Varas, S., Manzano-Roman, R., Fernandez-Soto, P., Encinas-Grandes, A., Oleaga, A., Perez-Sanchez, R., 2010. Purification and characterisation of a P-selectin-binding molecule from the salivary glands of *Ornithodoros moubata* that induces protective anti-tick immune responses in pigs. International Journal for Parasitology 40, 313–326.

Garg, R., Juncadella, I.J., Ramamoorthi, N., Ashish, N., Ananthanarayanan, S.K., Thomas, V., Rincón, M., Krueger, J.K., Fikrig, E., Yengo, C.M., et al., 2006. Cutting edge: CD4 is the receptor for the tick saliva immunosuppressor, Salp15. Journal of Immunology 177, 6579–6583.

Gern, L., Rais, O., 1996. Efficient transmission of *Borrelia burgdorferi* between cofeeding *Ixodes ricinus* ticks (Acari: Ixodidae). Journal of Medical Entomology 33, 189–192.

Gern, L., Schaible, U.E., Simon, M.M., 1993. Mode of inoculation of the Lyme disease agent *Borrelia burgdorferi* influences infection and immune responses in inbred strains of mice. The Journal of Infectious Diseases 167, 971–975.

Gillespie, R.D., Mbow, M.L., Titus, R.G., 2000. The immunomodulatory factors of bloodfeeding arthropod saliva. Parasite Immunology 22, 319–331.

Gillespie, R.D., Dolan, M.C., Piesman, J., Titus, R.G., 2001. Identification of an IL-2 binding protein in the saliva of the Lyme disease vector tick, *Ixodes scapularis*. Journal of Immunology 166, 4319–4326.

Gold, L.I., Rahman, M., Blechman, K.M., Greives, M.R., Churgin, S., Michaels, J., Callaghan, M.J., Cardwell, N.L., Pollins, A.C., Michalak, M., et al., 2006. Overview of the role for calreticulin in the enhancement of wound healing through multiple biological effects. The Journal of Investigative Dermatology. Symposium Proceedings 11, 57–65.

Gordon, J.R., Allen, J.R., 1991. Factors V and VII anticoagulant activities in the salivary glands of feeding *Dermacentor andersoni* ticks. The Journal of Parasitology 77, 167–170.

Gothe, R., 1999. Tick Toxicoses (Zeckentoxikosen). Hieronymusp, München.

Guo, X., Booth, C.J., Paley, M.A., Wang, X., DePonte, K., Fikrig, E., Narasimhan, S., Montgomery, R.R., 2009. Inhibition of neutrophil function by two tick salivary proteins. Infection and Immunity 77, 2320–2329.

Hai, V.V., Almeras, L., Socolovschi, C., Raoult, D., Parola, P., Pages, F., 2014. Monitoring human tick-borne disease risk and tick bite exposure in Europe: available tools and promising future methods. Ticks and Tick-Borne Diseases 5, 607–619.

Hajnická, V., Kocáková, P., Sláviková, M., Slovák, M., Gašperík, J., Fuchsberger, N., Nuttall, P.A., 2001. Anti-interleukin-8 activity of tick salivary gland extracts. Parasite Immunology 23, 483–489.

Hajnická, V., Vančová, I., Kocáková, P., Slovák, M., Gašperík, J., Sláviková, M., Hails, R.S., Labuda, M., Nuttall, P.A., 2005. Manipulation of host cytokine network by ticks: a potential gateway for pathogen transmission. Parasitology 130, 333–342.

Hajnická, V., Vančová-Štibrániová, I., Slovák, M., Kocáková, P., Nuttall, P.A., 2011. Ixodid tick salivary gland products target host wound healing growth factors. International Journal for Parasitology 41, 213–223.

Hall-Mendelin, S., Craig, S.B., Hall, R.A., O'Donoghue, P., Atwell, R.B., Tulsiani, S.M., Graham, G.C., 2011. Tick paralysis in Australia caused by *Ixodes holocyclus* Neumann. Annals of Tropical Medicine and Parasitology 105, 95–106.

Hannier, S., Liversidge, J., Sternberg, J.M., Bowman, A.S., 2003. *Ixodes ricinus* tick salivary gland extract inhibits IL-10 secretion and CD69 expression by mitogen-stimulated murine splenocytes and induces hyporesponsiveness in B lymphocytes. Parasite Immunology 25, 27–37.

Hannier, S., Liversidge, J., Sternberg, J.M., Bowman, A.S., 2004. Characterization of the B-cell inhibitory protein factor in *Ixodes ricinus* tick saliva: a potential role in enhanced *Borrelia burgdoferi* transmission. Immunology 113, 401–408.

Heinze, D.M., Carmical, J.R., Aronson, J.F., Thangamani, S., 2012a. Early immunologic events at the tick-host interface. PLoS One 7, e47301.

Heinze, D.M., Wikel, S.K., Thangamani, S., Alarcon-Chaidez, F.J., 2012b. Transcriptional profiling of the murine cutaneous response during initial and subsequent infestations with *Ixodes scapularis* nymphs. Parasites and Vectors 5, 26.

Heinze, D.M., Carmical, J.R., Aronson, J.F., Alarcon-Chaidez, F., Wikel, S., Thangamani, S., 2014. Murine cutaneous responses to the rocky mountain spotted fever vector, *Dermacentor andersoni*, feeding. Frontiers in Microbiology 5, 198.

Hermance, M.E., Thangamani, S., 2015. Tick saliva enhances Powassan virus transmission to the host, influencing its dissemination and the course of disease. Journal of Virology 89, 7852–7860.

Heyman, P., Cochez, C., Hofhuis, A., van der Giessen, J., Sprong, H., Porter, S.R., Losson, B., Saegerman, C., Donoso-Mantke, O., Niedrig, M., et al., 2010. A clear and present danger: tick-borne diseases in Europe. Expert Review of Anti-Infective Therapy 8, 33–50.

Hidano, A., Konnai, S., Yamada, S., Githaka, N., Isezaki, M., Higuchi, H., Nagahata, H., Ito, T., Takano, A., Ando, S., et al., 2014. Suppressive effects of neutrophil by Salp16-like salivary gland proteins from *Ixodes persulcatus* Schulze tick. Insect Molecular Biology 23, 466–474.

Hoffman, R., Benz, E.J., Shattil, S.J., 2009. Hematology: Basic Principles and Practice. Churchill Livingstone/Elsevier.

Hoffmann, A., Walsmann, P., Riesener, G., Paintz, M., Markwardt, F., 1991. Isolation and characterization of a thrombin inhibitor from the tick *Ixodes ricinus*. Die Pharmazie 46, 209–212.

Hojgaard, A., Biketov, S.F., Shtannikov, A.V., Zeidner, N.S., Piesman, J., 2009. Molecular identification of Salp15, a key salivary gland protein in the transmission of Lyme disease spirochetes, from *Ixodes persulcatus* and *Ixodes pacificus* (Acari: Ixodidae). Journal of Medical Entomology 46, 1458–1463.

Horká, H., Černá-Kýčková, K., Skallová, A., Kopecký, J., 2009. Tick saliva affects both prolifera-tion and distribution of *Borrelia burgdorferi* spirochetes in mouse organs and increases trans-mission of spirochetes to ticks. International Journal of Medical Microbiology 299, 373–380.

Horka, H., Staudt, V., Klein, M., Taube, C., Reuter, S., Dehzad, N., Andersen, J.F., Kopecky, J., Schild, H., Kotsyfakis, M., et al., 2012. The tick salivary protein sialostatin L inhibits the Th9-derived production of the asthma-promoting cytokine IL-9 and is effective in the prevention of experimental asthma. Journal of Immunology 188, 2669–2676.

Horn, F., dos Santos, P.C., Termignoni, C., 2000. *Boophilus microplus* anticoagulant protein: an antithrombin inhibitor isolated from the cattle tick saliva. Archives of Biochemistry and Biophysics 384, 68–73.

Hourcade, D.E., Akk, A.M., Mitchell, L.M., Zhou, H.F., Hauhart, R., Pham, C.T., 2016. Anti-complement activity of the *Ixodes scapularis* salivary protein Salp20. Molecular Immunology 69, 62–69.

Hovius, J.W., Ramamoorthi, N., Van't Veer, C., de Groot, K.A., Nijhof, A.M., Jongejan, F., van Dam, A.P., Fikrig, E., 2007. Identification of Salp15 homologues in *Ixodes ricinus* ticks. Vector Borne and Zoonotic Diseases 7, 296–303.

Hovius, J.W., de Jong, M.A., den Dunnen, J., Litjens, M., Fikrig, E., van der Poll, T., Gringhuis, S.I., Geijtenbeek, T.B., 2008a. Salp15 binding to DC-SIGN inhibits cytokine expression by impair-ing both nucleosome remodeling and mRNA stabilization. PLoS Pathogens 4, e31.

Hovius, J.W., Schuijt, T.J., de Groot, K.A., Roelofs, J.J., Oei, G.A., Marquart, J.A., de Beer, R., van 't Veer, C., van der Poll, T., Ramamoorthi, N., et al., 2008b. Preferential protection of *Borrelia burgdorferi* sensu stricto by a Salp15 homologue in *Ixodes ricinus* saliva. The Journal of Infectious Diseases 198, 1189–1197.

Hsu, M.-H., Sauer, J.R., 1975. Ion and water balance in the feeding lone star tick. Comparative Biochemistry and Physiology Part A: Physiology 52, 269–276.

Hynes, W.L., 2014. How ticks control microbes: innate immune responses. In: Sonnenshine, D., Roe, R.M. (Eds.), Biology of Ticks. Oxford University Press, Oxford, New York, pp. 129–146.

Ibelli, A.M., Kim, T.K., Hill, C.C., Lewis, L.A., Bakshi, M., Miller, S., Porter, L., Mulenga, A., 2014. A blood meal-induced *Ixodes scapularis* tick saliva serpin inhibits trypsin and throm-bin, and interferes with platelet aggregation and blood clotting. International Journal for Parasitology 44, 369–379.

Imamura, S., da Silva Vaz Junior, I., Sugino, M., Ohashi, K., Onuma, M., 2005. A serine protease inhib-itor (serpin) from *Haemaphysalis longicornis* as an anti-tick vaccine. Vaccine 23, 1301–1311.

Imamura, S., Namangala, B., Tajima, T., Tembo, M.E., Yasuda, J., Ohashi, K., Onuma, M., 2006. Two serine protease inhibitors (serpins) that induce a bovine protective immune response against *Rhipicephalus appendiculatus* ticks. Vaccine 24, 2230–2237.

Islam, M.K., Tsuji, N., Miyoshi, T., Alim, M.A., Huang, X., Hatta, T., Fujisaki, K., 2009. The Kunitz-like modulatory protein haemangin is vital for hard tick blood-feeding success. PLoS Pathogens 5, e1000497.

Iwanaga, S., Okada, M., Isawa, H., Morita, A., Yuda, M., Chinzei, Y., 2003. Identification and char-acterization of novel salivary thrombin inhibitors from the ixodidae tick, *Haemaphysalis longi-cornis*. European Journal of Biochemistry 270, 1926–1934.

Jablonka, W., Kotsyfakis, M., Mizurini, D.M., Monteiro, R.Q., Lukszo, J., Drake, S.K., Ribeiro, J.M., Andersen, J.F., 2015. Identification and mechanistic analysis of a novel tick-derived inhibitor of thrombin. PLoS One 10, e0133991.

Jacquet, M., Durand, J., Rais, O., Voordouw, M.J., 2016. Strain-specific antibodies reduce co-feed-ing transmission of the Lyme disease pathogen, *Borrelia afzelii*. Environmental Microbiology 18, 833–845.

Janeway, C.A., Travers, P., Walport, M., Capra, J.D., 1999. Immunobiology. In: The Immune System in Health and Disease. Elsevier Science Ltd/Garland Publishing, London-New York.

Jaworski, D.C., Rosell, R., Coons, L.B., Needham, G.R., 1992. Tick (Acari: Ixodidae) attachment cement and salivary gland cells contain similar immunoreactive polypeptides. Journal of Medical Entomology 29, 305–309.

Jaworski, D.C., Simmen, F.A., Lamoreaux, W., Coons, L.B., Muller, M.T., Needham, G.R., 1995. A secreted calreticulin protein in ixodid tick (*Amblyomma americanum*) saliva. Journal of Insect Physiology 41, 369–375.

Jaworski, D.C., Jasinskas, A., Metz, C.N., Bucala, R., Barbour, A.G., 2001. Identification and characterization of a homologue of the pro-inflammatory cytokine macrophage migration inhibitory factor in the tick, *Amblyomma americanum*. Insect Molecular Biology 10, 323–331.

Jelinski, J.W., 2016. Painless Hematophagy: The Functional Role of Novel Tick Metalloproteases in Pain Suppression. 401 p..

Jones, L.D., Davies, C.R., Steele, G.M., Nuttall, P.A., 1987. A novel mode of arbovirus transmission involving a nonviremic host. Science 237, 775–777.

Jones, L.D., Hodgson, E., Nuttall, P.A., 1989. Enhancement of virus transmission by tick salivary glands. The Journal of General Virology 70 (Pt 7), 1895–1898.

Joubert, A.M., Louw, A.I., Joubert, F., Neitz, A.W., 1998. Cloning, nucleotide sequence and expression of the gene encoding factor Xa inhibitor from the salivary glands of the tick, *Ornithodoros savignyi*. Experimental and Applied Acarology 22, 603–619.

Karczewski, J., Endris, R., Connolly, T.M., 1994. Disagregin is a fibrinogen receptor antagonist lacking the Arg-Gly-Asp sequence from the tick, *Ornithodoros moubata*. The Journal of Biological Chemistry 269, 6702–6708.

Karczewski, J., Waxman, L., Endris, R.G., Connolly, T.M., 1995. An inhibitor from the argasid tick *Ornithodoros moubata* of cell adhesion to collagen. Biochemical and Biophysical Research Communications 208, 532–541.

Karim, S., Ribeiro, J.M.C., 2015. An insight into the sialome of the lone star tick, *Amblyomma americanum*, with a glimpse on its time dependent gene expression. PLoS One 10.

Karim, S., Singh, P., Ribeiro, J.M., 2011. A deep insight into the sialotranscriptome of the gulf coast tick, *Amblyomma maculatum*. PLoS One 6, e28525.

Karpathy, S.E., Allerdice, M.E., Sheth, M., Dasch, G.A., Levin, M.L., 2016. Co-feeding transmission of the *Ehrlichia muris*-like agent to mice (*Mus musculus*). Vector-Borne and Zoonotic Diseases 16, 145–150.

Kato, N., Iwanaga, S., Okayama, T., Isawa, H., Yuda, M., Chinzei, Y., 2005. Identification and characterization of the plasma kallikrein-kinin system inhibitor, haemaphysalin, from hard tick, *Haemaphysalis longicornis*. Thrombosis and Haemostasis 93, 359–367.

Kaufman, W.R., Harris, R.A., 1983. Neural pathways mediating salivary fluid secretion in the ixodid tick *Amblyomma hebraeum*. Canadian Journal of Zoology 61, 1976–1980.

Kaufman, W.R., Phillips, J.E., 1973. Ion and water balance in the ixodid tick *Dermacentor andersoni*. I. Routes of ion and water excretion. Journal of Experimental Biology 58.

Kaufman, W.R., Wong, D.L., 1983. Evidence for multiple receptors mediating fluid secretion in salivary glands of ticks. European Journal of Pharmacology 87, 43–52.

Kaufman, W.R., Sloley, B.D., Tatchell, R.J., Zbitnew, G.L., Diefenbach, T.J., Goldberg, J.I., 1999. Quantification and cellular localization of dopamine in the salivary gland of the ixodid tick *Amblyomma hebraeum*. Experimental and Applied Acarology 23, 251–265.

Kaufman, W., 1976. The influence of various factors on fluid secretion by in vitro salivary glands of ixodid Ticks. Journal of Experimental Biology 64, 727–742.

Kaufman, W.R., 1978. Actions of some transmitters and their antagonists on salivary secretion in a tick. American Journal of Physiology 235, R76–R81.

Kazimírová, M., Štibrániová, I., 2013. Tick salivary compounds: their role in modulation of host defences and pathogen transmission. Frontiers in Cellular and Infection Microbiology 3, 43.

Kazimírová, M., Jančinová, V., Petríková, M., Takáč, P., Labuda, M., Nosál, R., 2002. An inhibitor of thrombin-stimulated blood platelet aggregation from the salivary glands of the hard tick *Amblyomma variegatum* (Acari: Ixodidae). Experimental and Applied Acarology 28, 97–105.

Kemp, D.H., Stone, B.F., Binnington, K.C., 1982. Tick attachment and feeding: role of the mouth-parts, feeding apparatus, salivary gland secretions and the host response. In: Obenchain, F.D., Galun, R. (Eds.), Physiology of Ticks.

Kern, A., Collin, E., Barthel, C., Michel, C., Jaulhac, B., Boulanger, N., 2011. Tick saliva represses innate immunity and cutaneous inflammation in a murine model of Lyme disease. Vector Borne and Zoonotic Diseases 11, 1343–1350.

Kim, D., Šimo, L., Park, Y., 2014. Orchestration of salivary secretion mediated by two different dopamine receptors in the blacklegged tick *Ixodes scapularis*. The Journal of Experimental Biology 217, 3656–3663.

Kim, T.K., Ibelli, A.M.G., Mulenga, A., 2015a. *Amblyomma americanum* tick calreticulin binds C1q but does not inhibit activation of the classical complement cascade. Ticks and Tick-Borne Diseases 6, 91–101.

Kim, T.K., Tirloni, L., Radulovic, Z., Lewis, L., Bakshi, M., Hill, C., Vaz, I.D., Logullo, C., Termignoni, C., Mulenga, A., 2015b. Conserved *Amblyomma americanum* tick Serpin19, an inhibitor of blood clotting factors Xa and XIa, trypsin and plasmin, has anti-haemostatic functions. International Journal for Parasitology 45, 613–627.

Kim, D., Urban, J., Boyle, D.L., Park, Y., 2016a. Multiple functions of Na/K-ATPase in dopamine-induced salivation of the blacklegged tick, *Ixodes scapularis*. Scientific Reports 6.

Kim, T.K., Tirloni, L., Pinto, A.F.M., Moresco, J., Yates, J.R., Vaz, I.D., Mulenga, A., 2016b. *Ixodes scapularis* tick saliva proteins sequentially secreted every 24 h during blood feeding. PLoS Neglected Tropical Diseases 10.

Knulle, W., Rudolph, D., 1982. Humidity Relationships and Water Balance of Ticks. Pergamon Press, Oxford, pp. 43–70.

Koči, J., Šimo, L., Park, Y., 2014. Autocrine/paracrine dopamine in the salivary glands of the black-legged tick *Ixodes scapularis*. Journal of Insect Physiology 62, 39–45.

Koh, C.Y., Kini, R.M., 2009. Molecular diversity of anticoagulants from haematophagous animals. Thrombosis and Haemostasis 102, 437–453.

Koh, C.Y., Kazimirova, M., Trimnell, A., Takac, P., Labuda, M., Nuttall, P.A., Kini, R.M., 2007. Variegin, a novel fast and tight binding thrombin inhibitor from the tropical bont tick. The Journal of Biological Chemistry 282, 29101–29113.

Kopecký, J., Kuthejlová, M., 1998. Suppressive effect of *Ixodes ricinus* salivary gland extract on mechanisms of natural immunity in vitro. Parasite Immunology 20, 169–174.

Kotál, J., Langhansová, H., Lieskovská, J., Andersen, J.F., Francischetti, I.M., Chavakis, T., Kopecký, J., Pedra, J.H., Kotsyfakis, M., Chmelař, J., 2015. Modulation of host immunity by tick saliva. Journal of Proteomics 128, 58–68.

Kotsyfakis, M., Sa-Nunes, A., Francischetti, I.M., Mather, T.N., Andersen, J.F., Ribeiro, J.M., 2006. Antiinflammatory and immunosuppressive activity of sialostatin L, a salivary cystatin from the tick *Ixodes scapularis*. The Journal of Biological Chemistry 281, 26298–26307.

Kotsyfakis, M., Horka, H., Salat, J., Andersen, J.F., 2010. The crystal structures of two salivary cystatins from the tick *Ixodes scapularis* and the effect of these inhibitors on the establishment of *Borrelia burgdorferi* infection in a murine model. Molecular Microbiology 77, 456–470.

Kotsyfakis, M., Schwarz, A., Erhart, J., Ribeiro, J.M., 2015. Tissue- and time-dependent transcription in *Ixodes ricinus* salivary glands and midguts when blood feeding on the vertebrate host. Scientific Reports 5, 9103.

Kramer, C., Nahmias, Z., Norman, D.D., Mulvihill, T.A., Coons, L.B., Cole, J.A., 2008. *Dermacentor variabilis*: regulation of fibroblast migration by tick salivary gland extract and saliva. Experimental Parasitology 119, 391–397.

Kramer, C.D., Poole, N.M., Coons, L.B., Cole, J.A., 2011. Tick saliva regulates migration, phagocytosis, and gene expression in the macrophage-like cell line, IC-21. Experimental Parasitology 127, 665–671.

Krause, P.J., Grant-Kels, J.M., Tahan, S.R., Dardick, K.R., Alarcon-Chaidez, F., Bouchard, K., Visini, C., Deriso, C., Foppa, I.M., Wikel, S., 2009. Dermatologic changes induced by repeated *Ixodes scapularis* bites and implications for prevention of tick-borne infection. Vector Borne and Zoonotic Diseases 9, 603–610.

Kročová, Z., Macela, A., Hernychová, L., Kroča, M., Pechová, J., Kopecký, J., 2003. Tick salivary gland extract accelerates proliferation of *Francisella tularensis* in the host. The Journal of Parasitology 89, 14–20.

Krolak, J.M., Ownby, C.L., Sauer, J.R., 1982. Alveolar structure of salivary glands of the lone star tick, *Amblyomma americanum* (L.): unfed females. The Journal of Parasitology 68, 61–82.

Kubeš, M., Fuchsberger, N., Labuda, M., Žuffová, E., Nuttall, P.A., 1994. Salivary gland extracts of partially fed *Dermacentor reticulatus* ticks decrease natural killer cell activity in vitro. Immunology 82, 113–116.

Kuthejlová, M., Kopecký, J., Štěpánová, G., Macela, A., 2001. Tick salivary gland extract inhibits killing of *Borrelia afzelii* spirochetes by mouse macrophages. Infection and Immunity 69, 575–578.

Labuda, M., Jones, L.D., Williams, T., Danielova, V., Nuttall, P.A., 1993a. Efficient transmission of tick-borne encephalitis virus between cofeeding ticks. Journal of Medical Entomology 30, 295–299.

Labuda, M., Jones, L.D., Williams, T., Nuttall, P.A., 1993b. Enhancement of tick-borne encephalitis virus transmission by tick salivary gland extracts. Medical and Veterinary Entomology 7, 193–196.

Labuda, M., Austyn, J.M., Zuffova, E., Kozuch, O., Fuchsberger, N., Lysy, J., Nuttall, P.A., 1996. Importance of localized skin infection in tick-borne encephalitis virus transmission. Virology 219, 357–366.

Labuda, M., Alves, M.J., Elečková, E., Kožuch, O., Filipe, A.R., 1997a. Transmission of tick-borne bunyaviruses by cofeeding ixodid ticks. Acta Virologica 41, 325–328.

Labuda, M., Kozuch, O., Zuffová, E., Eleckvá, E., Hails, R.S., Nuttall, P.A., 1997b. Tick-borne encephalitis virus transmission between ticks cofeeding on specific immune natural rodent hosts. Virology 235, 138–143.

Labuda, M., Trimnell, A.R., Ličková, M., Kazimírová, M., Davies, G.M., Lissina, O., Hails, R.S., Nuttall, P.A., 2006. An antivector vaccine protects against a lethal vector-borne pathogen. PLoS Pathogens 2, e27.

Lambson, B., Nene, V., Obura, M., Shah, T., Pandit, P., Ole-Moiyoi, O., Delroux, K., Welburn, S., Skilton, R., de Villiers, E., et al., 2005. Identification of candidate sialome components expressed in ixodid tick salivary glands using secretion signal complementation in mammalian cells. Insect Molecular Biology 14, 403–414.

Langhansová, H., Bopp, T., Schmitt, E., Kopecký, J., 2015. Tick saliva increases production of three chemokines including monocyte chemoattractant protein-1, a histamine-releasing cytokine. Parasite Immunology 37, 92–96.

Leboulle, G., Crippa, M., Decrem, Y., Mejri, N., Brossard, M., Bollen, A., Godfroid, E., 2002. Characterization of a novel salivary immunosuppressive protein from *Ixodes ricinus* ticks. The Journal of Biological Chemistry 277, 10083–10089.

Leger, E., Vourc'h, G., Vial, L., Chevillon, C., McCoy, K.D., 2013. Changing distributions of ticks: causes and consequences. Experimental and Applied Acarology 59, 219–244.

Lewis, L.A., Radulović, Z.M., Kim, T.K., Porter, L.M., Mulenga, A., 2015. Identification of 24 h *Ixodes scapularis* immunogenic tick saliva proteins. Ticks and Tick-Borne Diseases 6, 424–434.

Lew-Tabor, A.E., Moolhuijzen, P.M., Vance, M.E., Kurscheid, S., Valle, M.R., Jarrett, S., Minchin, C.M., Jackson, L.A., Jonsson, N.N., Bellgard, M.I., et al., 2010. Suppressive subtractive hybridization analysis of *Rhipicephalus* (*Boophilus*) *microplus* larval and adult transcript expression during attachment and feeding. Veterinary Parasitology 167, 304–320.

Lieskovská, J., Kopecký, J., 2012a. Effect of tick saliva on signalling pathways activated by TLR-2 ligand and *Borrelia afzelii* in dendritic cells. Parasite Immunology 34, 421–429.

Lieskovská, J., Kopecký, J., 2012b. Tick saliva suppresses IFN signalling in dendritic cells upon *Borrelia afzelii* infection. Parasite Immunology 34, 32–39.

Lieskovská, J., Páleníková, J., Langhansová, H., Chagas, A.C., Calvo, E., Kotsyfakis, M., Kopecký, J., 2015. Tick sialostatins L and L2 differentially influence dendritic cell responses to *Borrelia* spirochetes. Parasites and Vectors 8.

Lieskovská, J., Páleníková, J., Širmarová, J., Elsterová, J., Kotsyfakis, M., Chagas, A.C., Calvo, E., Růžek, D., Kopecký, J., 2015a. Tick salivary cystatin sialostatin L2 suppresses IFN responses in mouse dendritic cells. Parasite Immunology 37, 70–78.

Limo, M.K., Voigt, W.P., Tumbo-Oeri, A.G., Njogu, R.M., ole-MoiYoi, O.K., 1991. Purification and characterization of an anticoagulant from the salivary glands of the ixodid tick *Rhipicephalus appendiculatus*. Experimental Parasitology 72, 418–429.

Lindsay, P.J., Kaufman, W.R., 1986. Potentiation of salivary fluid secretion in ixodid ticks: a new receptor system for gamma-aminobutyric acid. Canadian Journal of Physiology and Pharmacology 64, 1119–1126.

Liu, X.Y., Bonnet, S.I., 2014. Hard tick factors implicated in pathogen transmission. PLoS Neglected Tropical Diseases 8, e2566.

Liu, L., Dai, J., Zhao, Y.O., Narasimhan, S., Yang, Y., Zhang, L., Fikrig, E., 2012. *Ixodes scapularis* JAK-STAT pathway regulates tick antimicrobial peptides, thereby controlling the agent of human granulocytic anaplasmosis. The Journal of Infectious Diseases 206, 1233–1241.

Liu, J., Renneker, S., Beyer, D., Kullmann, B., Seitzer, U., Ahmed, J., Bakheit, M.A., 2014a. Identification and partial characterization of a Salp15 homolog from *Ixodes ricinus*. Ticks and Tick-Borne Diseases 5, 318–322.

Liu, X.Y., de la Fuente, J., Cote, M., Galindo, R.C., Moutailler, S., Vayssier-Taussat, M., Bonnet, S.I., 2014b. IrSPI, a tick serine protease inhibitor involved in tick feeding and *Bartonella henselae* infection. PLoS Neglected Tropical Diseases 8, e2993.

Liyou, N., Hamilton, S., Elvin, C., Willadsen, P., 1999. Cloning and expression of ecto 5-nucleotidase from the cattle tick *Boophilus microplus*. Insect Molecular Biology 8, 257–266.

Macaluso, K.R., Mulenga, A., Simser, J.A., Azad, A.F., 2003. Differential expression of genes in uninfected and rickettsia-infected *Dermacentor variabilis* ticks as assessed by differential-display PCR. Infection and Immunity 71, 6165–6170.

Macháčková, M., Oborník, M., Kopecký, J., 2006. Effect of salivary gland extract from *Ixodes ricinus* ticks on the proliferation of *Borrelia burgdorferi* sensu stricto in vivo. Folia Parasitologica (Praha) 53, 153–158.

Madden, R.D., Sauer, J.R., Dillwith, J.W., 2004. A proteomics approach to characterizing tick salivary secretions. Experimental and Applied Acarology 32, 77–87.

Malouin, R., Winch, P., Leontsini, E., Glass, G., Simon, D., Hayes, E.B., Schwartz, B.S., 2003. Longitudinal evaluation of an educational intervention for preventing tick bites in an area with endemic Lyme disease in Baltimore County, Maryland. American Journal of Epidemiology 157, 1039–1051.

Mans, B.J., Neitz, A.W., 2004. Adaptation of ticks to a blood-feeding environment: evolution from a functional perspective. Insect Biochemistry and Molecular Biology 34, 1–17.

Mans, B.J., Ribeiro, J.M., 2008. Function, mechanism and evolution of the moubatin-clade of soft tick lipocalins. Insect Biochemistry and Molecular Biology 38, 841–852.

Mans, B.J., Gaspar, A.R., Louw, A.I., Neitz, A.W., 1998. Apyrase activity and platelet aggregation inhibitors in the tick *Ornithodoros savignyi* (Acari: Argasidae). Experimental and Applied Acarology 22, 353–366.

Mans, B.J., Louw, A.I., Neitz, A.W., 2002a. Evolution of hematophagy in ticks: common origins for blood coagulation and platelet aggregation inhibitors from soft ticks of the genus *Ornithodoros*. Molecular Biology and Evolution 19, 1695–1705.

Mans, B.J., Louw, A.I., Neitz, A.W., 2002b. Savignygrin, a platelet aggregation inhibitor from the soft tick *Ornithodoros savignyi*, presents the RGD integrin recognition motif on the Kunitz-BPTI fold. The Journal of Biological Chemistry 277, 21371–21378.

Mans, B.J., Louw, A.I., Neitz, A.W., 2002c. Disaggregation of aggregated platelets by savignygrin, a alphaIIbeta3 antagonist from *Ornithodoros savignyi*. Experimental and Applied Acarology 27, 231–239.

Mans, B.J., Louw, A.I., Neitz, A.W.H., 2003. The major tick salivary gland proteins and toxins from the soft tick, *Ornithodoros savignyi*, are part of the tick Lipocalin family: implications for the origins of tick toxicoses. Molecular Biology and Evolution 20, 1158–1167.

Mans, B.J., Andersen, J.F., Schwan, T.G., Ribeiro, J.M.C., 2008a. Characterization of anti-hemostatic factors in the argasid, Argas monolakensis: Implications for the evolution of blood-feeding in the soft tick family. Insect Biochemistry and Molecular Biology 38, 22–41.

Mans, B.J., Andersen, J.F., Francischetti, I.M., Valenzuela, J.G., Schwan, T.G., Pham, V.M., Garfield, M.K., Hammer, C.H., Ribeiro, J.M., 2008b. Comparative sialomics between hard and soft ticks: implications for the evolution of blood-feeding behavior. Insect Biochemistry and Molecular Biology 38, 42–58.

Mans, B.J., 2011. Evolution of vertebrate hemostatic and inflammatory control mechanisms in blood-feeding arthropods. Journal of Innate Immunity 3, 41–51.

Mans, B.J., 2014. Heme processing and the evolution of hematophagy. In: Sonenshine, D.E., Roe, M.R. (Eds.), Biology of Ticks. Oxford University Press, pp. 220–239.

Marchal, C., Schramm, F., Kern, A., Luft, B.J., Yang, X., Schuijt, T.J., Hovius, J.W., Jaulhac, B., Boulanger, N., 2011. Antialarmin effect of tick saliva during the transmission of Lyme disease. Infection and Immunity 79, 774–785.

Maruyama, S.R., Anatriello, E., Anderson, J.M., Ribeiro, J.M., Brandao, L.G., Valenzuela, J.G., Ferreira, B.R., Garcia, G.R., Szabo, M.P., Patel, S., et al., 2010. The expression of genes coding for distinct types of glycine-rich proteins varies according to the biology of three metastriate ticks, *Rhipicephalus (Boophilus) microplus*, *Rhipicephalus sanguineus* and *Amblyomma cajennense*. BMC Genomics 11, 363.

McNally, K.L., Mitzel, D.N., Anderson, J.M., Ribeiro, J.M., Valenzuela, J.G., Myers, T.G., Godinez, A., Wolfinbarger, J.B., Best, S.M., Bloom, M.E., 2012. Differential salivary gland transcript expression profile in *Ixodes scapularis* nymphs upon feeding or flavivirus infection. Ticks and Tick-Borne Diseases 3, 18–26.

McSwain, J.L., Essenberg, R.C., Sauer, J.R., 1992. Oral secretion elicited by effectors of signal transduction pathways in the salivary glands of *Amblyomma americanum* (Acari: Ixodidae). Journal of Medical Entomology 29, 41–48.

Medlock, J.M., Hansford, K.M., Bormane, A., Derdakova, M., Estrada-Pena, A., George, J.C., Golovljova, I., Jaenson, T.G., Jensen, J.K., Jensen, P.M., et al., 2013. Driving forces for changes in geographical distribution of *Ixodes ricinus* ticks in Europe. Parasites and Vectors 6, 1.

Mejri, N., Franscini, N., Rutti, B., Brossard, M., 2001. Th2 polarization of the immune response of BALB/c mice to *Ixodes ricinus* instars, importance of several antigens in activation of specific Th2 subpopulations. Parasite Immunology 23, 61–69.

Mercado-Curiel, R.F., Palmer, G.H., Guerrero, F.D., Brayton, K.A., 2011. Temporal characterisation of the organ-specific *Rhipicephalus microplus* transcriptional response to *Anaplasma marginale* infection. International Journal for Parasitology 41, 851–860.

Meredith, J., Kaufman, W.R., 1973. A proposed site of fluid secretion in the salivary gland of the ixodid tick, *Dermacentor andersoni*. Parasitology 67, 205–217.

Merino, O., Antunes, S., Mosqueda, J., Moreno-Cid, J.A., Perez de la Lastra, J.M., Rosario-Cruz, R., Rodriguez, S., Domingos, A., de la Fuente, J., 2013. Vaccination with proteins involved in tick-pathogen interactions reduces vector infestations and pathogen infection. Vaccine 31, 5889–5896.

Milhano, N., Saito, T.B., Bechelli, J., Fang, R., Vilhena, M., De Sousa, R., Walker, D.H., 2015. The role of *Rhipicephalus sanguineus* sensu lato saliva in the dissemination of *Rickettsia conorii* in C3H/HeJ mice. Medical and Veterinary Entomology 29, 225–229.

Montgomery, R.R., Lusitani, D., De Boisfleury Chevance, A., Malawista, S.E., 2004. Tick saliva reduces adherence and area of human neutrophils. Infection and Immunity 72, 2989–2994.

Moorhouse, D.E., Tatchell, R.J., 1966. The feeding processes of the cattle-tick *Boophilus microplus* (Canestrini): a study in host-parasite relations: Part I. Attachment to the host. Parasitology 56, 623–631.

Mori, A., Konnai, S., Yamada, S., Hidano, A., Murase, Y., Ito, T., Takano, A., Kawabata, H., Onuma, M., Ohashi, K., 2010. Two novel Salp15-like immunosuppressant genes from salivary glands of *Ixodes persulcatus* Schulze tick. Insect Molecular Biology 19, 359–365.

Motoyashiki, T., Tu, A.T., Azimov, D.A., Ibragim, K., 2003. Isolation of anticoagulant from the venom of tick, *Boophilus calcaratus*, from Uzbekistan. Thrombosis Research 110, 235–241.

Mudenda, L., Pierle, S.A., Turse, J.E., Scoles, G.A., Purvine, S.O., Nicora, C.D., Clauss, T.R.W., Ueti, M.W., Brown, W.C., Brayton, K.A., 2014. Proteomics informed by transcriptomics identifies novel secreted proteins in *Dermacentor andersoni* saliva. International Journal for Parasitology 44, 1029–1037.

Mulenga, A., Sugimoto, C., Sako, Y., Ohashi, K., Musoke, A., Shubash, M., Onuma, M., 1999. Molecular characterization of a *Haemaphysalis longicornis* tick salivary gland-associated 29-kilodalton protein and its effect as a vaccine against tick infestation in rabbits. Infection and Immunity 67, 1652–1658.

Mulenga, A., Macaluso, K.R., Simser, J.A., Azad, A.F., 2003a. Dynamics of Rickettsia-tick interactions: identification and characterization of differentially expressed mRNAs in uninfected and infected *Dermacentor variabilis*. Insect Molecular Biology 12, 185–193.

Mulenga, A., Macaluso, K.R., Simser, J.A., Azad, A.F., 2003b. The American dog tick, *Dermacentor variabilis*, encodes a functional histamine release factor homolog. Insect Biochemistry and Molecular Biology 33, 911–919.

Mulenga, A., Kim, T., Ibelli, A.M.G., 2013a. *Amblyomma americanum* tick saliva serine protease inhibitor 6 is a cross-class inhibitor of serine proteases and papain-like cysteine proteases that delays plasma clotting and inhibits platelet aggregation. Insect Molecular Biology 22, 306–319.

Mulenga, A., Kim, T.K., Ibelli, A.M.G., 2013b. Deorphanization and target validation of cross-tick species conserved novel *Amblyomma americanum* tick saliva protein. International Journal for Parasitology 43, 439–451.

Nakajima, C., da Silva Vaz Jr., I., Imamura, S., Konnai, S., Ohashi, K., Onuma, M., 2005. Random sequencing of cDNA library derived from partially-fed adult female *Haemaphysalis longicornis* salivary gland. The Journal of Veterinary Medical Science 67, 1127–1131.

Nakajima, C., Imamura, S., Konnai, S., Yamada, S., Nishikado, H., Ohashi, K., Onuma, M., 2006. A novel gene encoding a thrombin inhibitory protein in a cDNA library from *Haemaphysalis longicornis* salivary gland. The Journal of Veterinary Medical Science 68, 447–452.

Narasimhan, S., Koski, R.A., Beaulieu, B., Anderson, J.F., Ramamoorthi, N., Kantor, F., Cappello, M., Fikrig, E., 2002. A novel family of anticoagulants from the saliva of *Ixodes scapularis*. Insect Molecular Biology 11, 641–650.

Narasimhan, S., Deponte, K., Marcantonio, N., Liang, X., Royce, T.E., Nelson, K.F., Booth, C.J., Koski, B., Anderson, J.F., Kantor, F., et al., 2007. Immunity against *Ixodes scapularis* salivary proteins expressed within 24 hours of attachment thwarts tick feeding and impairs *Borrelia* transmission. PLoS One 2, e451.

Nazario, S., Das, S., de Silva, A.M., Deponte, K., Marcantonio, N., Anderson, J.F., Fish, D., Fikrig, E., Kantor, F.S., 1998. Prevention of *Borrelia burgdorferi* transmission in guinea pigs by tick immunity. The American Journal of Tropical Medicine and Hygiene 58, 780–785.

Needham, G.R., Pannabecker, T.L., 1983. Effects of octopamine, chlordimeform, and demethyl-chlordimeform on amine-controlled tick salivary glands isolated from feeding *Amblyomma americanum* (L.). Pesticide Biochemistry and Physiology 19, 133–140.

Needham, G.R., Teel, P.D., 1986. Water balance by ticks between bloodmeals. In: Sauer, J.R., Hair, J.A. (Eds.), Morphology, Physiology, and Behavioral Biology of Ticks.

Needham, G.R., Rosell, R., Greenwald, L., 1990. Ultrastructure of type-I salivary-gland acini in four species of ticks and the influence of hydration states on the type-I acini of *Amblyomma americanum*. Experimental and Applied Acarology 10, 83–104.

Nene, V., Lee, D., Quackenbush, J., Skilton, R., Mwaura, S., Gardner, M.J., Bishop, R., 2002. AvGI, an index of genes transcribed in the salivary glands of the ixodid tick *Amblyomma variegatum*. International Journal for Parasitology 32, 1447–1456.

Nene, V., Lee, D., Kang'a, S., Skilton, R., Shah, T., de Villiers, E., Mwaura, S., Taylor, D., Quackenbush, J., Bishop, R., 2004. Genes transcribed in the salivary glands of female *Rhipicephalus appendiculatus* ticks infected with *Theileria parva*. Insect Biochemistry and Molecular Biology 34, 1117–1128.

Nienaber, J., Gaspar, A.R., Neitz, A.W., 1999. Savignin, a potent thrombin inhibitor isolated from the salivary glands of the tick *Ornithodoros savignyi* (Acari: Argasidae). Experimental Parasitology 93, 82–91.

Nunn, M.A., Sharma, A., Paesen, G.C., Adamson, S., Lissina, O., Willis, A.C., Nuttall, P.A., 2005. Complement inhibitor of C5 activation from the soft tick *Ornithodoros moubata*. Journal of Immunology 174, 2084–2091.

Nuttall, P.A., Labuda, M., 2004. Tick-host interactions: saliva-activated transmission. Parasitology (129 Suppl.), S177–S189.

Nuttall, P.A., Trimnell, A.R., Kazimirova, M., Labuda, M., 2006. Exposed and concealed antigens as vaccine targets for controlling ticks and tick-borne diseases. Parasite Immunology 28, 155–163.

Oleaga, A., Escudero-Poblacion, A., Camafeita, E., Perez-Sanchez, R., 2007. A proteomic approach to the identification of salivary proteins from the argasid ticks *Ornithodoros moubata* and *Ornithodoros erraticus*. Insect Biochemistry and Molecular Biology 37, 1149–1159.

Oliveira, C.J., Sa-Nunes, A., Francischetti, I.M., Carregaro, V., Anatriello, E., Silva, J.S., Santos, I.K., Ribeiro, J.M., Ferreira, B.R., 2011. Deconstructing tick saliva: non-protein molecules with potent immunomodulatory properties. The Journal of Biological Chemistry 286, 10960–10969.

Oliveira, C.J., Anatriello, E., de Miranda-Santos, I.K., Francischetti, I.M., Sa-Nunes, A., Ferreira, B.R., Ribeiro, J.M.C., 2013. Proteome of *Rhipicephalus sanguineus* tick saliva induced by the secretagogues pilocarpine and dopamine. Ticks and Tick-Borne Diseases 4, 469–477.

Ouhelli, H., Pandey, V.S., Aboughal, A., 1987. Effect of infection by *Babesia* spp. on the development and survival of free-living stages of *Boophilus annulatus*. Veterinary Parasitology 23, 147–154.

Paesen, G.C., Adams, P.L., Harlos, K., Nuttall, P.A., Stuart, D.I., 1999. Tick histamine-binding proteins: isolation, cloning, and three-dimensional structure. Molecular Cell 3, 661–671.

Paesen, G.C., Siebold, C., Harlos, K., Peacey, M.F., Nuttall, P.A., Stuart, D.I., 2007. A tick protein with a modified Kunitz fold inhibits human tryptase. Journal of Molecular Biology 368, 1172–1186.

Pal, U., Li, X., Wang, T., Montgomery, R.R., Ramamoorthi, N., Desilva, A.M., Bao, F., Yang, X., Pypaert, M., Pradhan, D., et al., 2004. TROSPA, an *Ixodes scapularis* receptor for *Borrelia burgdorferi*. Cell 119, 457–468.

Páleníková, J., Lieskovská, J., Langhansová, H., Kotsyfakis, M., Chmelař, J., Kopecký, J., 2015. *Ixodes ricinus* salivary serpin IRS-2 affects Th17 differentiation via inhibition of the interleukin-6/STAT-3 signaling pathway. Infection and Immunity 83, 1949–1956.

Pannabecker, T., Needham, G.R., 1985. Effects of octopamine on fluid secretion by isolated salivary glands of a feeding ixodid tick. Archives of Insect Biochemistry and Physiology 2, 217–226.

Pechová, J., Štěpánová, G., Kovář, L., Kopecký, J., 2002. Tick salivary gland extract-activated transmission of *Borrelia afzelii* spirochaetes. Folia Parasitologica (Praha) 49, 153–159.

Pekáriková, D., Rajská, P., Kazimírová, M., Pecháňová, O., Takáč, P., Nuttall, P.A., 2015. Vasoconstriction induced by salivary gland extracts from ixodid ticks. International Journal for Parasitology 45, 879–883.

Pichu, S., Ribeiro, J.M.C., Mather, T.N., 2009. Purification and characterization of a novel salivary antimicrobial peptide from the tick, *Ixodes scapularis*. Biochemical and Biophysical Research Communications 390, 511–515.

Piesman, J., Happ, C.M., 2001. The efficacy of co-feeding as a means of maintaining *Borrelia burgdorferi*: a North American model system. Journal of Vector Ecology 26, 216–220.

Poole, N.M., Mamidanna, G., Smith, R.A., Coons, L., Cole, J., 2013a. Prostaglandin E2 in tick saliva regulates host cell migration and cytokine profile. FASEB Journal 27.

Poole, N.M., Nyinodo-Ogari, L., Kramer, C., Coons, L.B., Cole, J.A., 2013b. Effects of tick saliva on the migratory and invasive activity of Saos-2 osteosarcoma and MDA-MB-231 breast cancer cells. Ticks and Tick-Borne Diseases 4, 120–127.

Preston, S.G., Majtan, J., Kouremenou, C., Rysnik, O., Burger, L.F., Cabezas Cruz, A., Chiong Guzman, M., Nunn, M.A., Paesen, G.C., Nuttall, P.A., et al., 2013. Novel immunomodulators from hard ticks selectively reprogramme human dendritic cell responses. PLoS Pathogens 9, e1003450.

Prevot, P.P., Adam, B., Boudjeltia, K.Z., Brossard, M., Lins, L., Cauchie, P., Brasseur, R., Vanhaeverbeek, M., Vanhamme, L., Godfroid, E., 2006. Anti-hemostatic effects of a serpin from the saliva of the tick *Ixodes ricinus*. The Journal of Biological Chemistry 281, 26361–26369.

Prevot, P.P., Couvreur, B., Denis, V., Brossard, M., Vanhamme, L., Godfroid, E., 2007. Protective immunity against *Ixodes ricinus* induced by a salivary serpin. Vaccine 25, 3284–3292.

Qian, Y., Yuan, J., Essenberg, R.C., Bowman, A.S., Shook, A.L., Dillwith, J.W., Sauer, J.R., 1998. Prostaglandin E2 in the salivary glands of the female tick, *Amblyomma americanum* (L.): calcium mobilization and exocytosis. Insect Biochemistry and Molecular Biology 28, 221–228.

Radolf, J.D., Caimano, M.J., Stevenson, B., Hu, L.T., 2012. Of ticks, mice and men: understanding the dual-host lifestyle of Lyme disease spirochaetes. Nature Reviews Microbiology 10, 87–99.

Radulović, Z.M., Kim, T.K., Porter, L.M., Sze, S.H., Lewis, L., Mulenga, A., 2014. A 24-48 h fed *Amblyomma americanum* tick saliva immuno-proteome. BMC Genomics 15.

Ramachandra, R.N., Wikel, S.K., 1992. Modulation of host-immune responses by ticks (Acari: Ixodidae): effect of salivary gland extracts on host macrophages and lymphocyte cytokine production. Journal of Medical Entomology 29, 818–826.

Ramamoorthi, N., Narasimhan, S., Pal, U., Bao, F., Yang, X.F., Fish, D., Anguita, J., Norgard, M.V., Kantor, F.S., Anderson, J.F., et al., 2005. The Lyme disease agent exploits a tick protein to infect the mammalian host. Nature 436, 573–577.

Randolph, S.E., 2011. Transmission of tick-borne pathogens between co-feeding ticks: Milan Labuda's enduring paradigm. Ticks and Tick-Borne Diseases 2, 179–182.

Ribeiro, J.M., Mather, T.N., 1998. *Ixodes scapularis*: salivary kininase activity is a metallo dipeptidyl carboxypeptidase. Experimental Parasitology 89, 213–221.

Ribeiro, J.M., Makoul, G.T., Levine, J., Robinson, D.R., Spielman, A., 1985. Antihemostatic, anti-inflammatory, and immunosuppressive properties of the saliva of a tick, *Ixodes dammini*. The Journal of Experimental Medicine 161, 332–344.

Ribeiro, J.M., Makoul, G.T., Robinson, D.R., 1988. *Ixodes dammini*: evidence for salivary prostacyclin secretion. The Journal of Parasitology 74, 1068–1069.

Ribeiro, J.M., Weis, J.J., Telford 3rd, S.R., 1990. Saliva of the tick *Ixodes dammini* inhibits neutrophil function. Experimental Parasitology 70, 382–388.

Ribeiro, J.M., Endris, T.M., Endris, R., 1991. Saliva of the soft tick, *Ornithodoros moubata*, contains anti-platelet and apyrase activities. Comparative Biochemistry and Physiology A, Comparative Physiology 100, 109–112.

Ribeiro, J.M., Evans, P.M., MacSwain, J.L., Sauer, J., 1992. *Amblyomma americanum*: characterization of salivary prostaglandins E2 and F2 alpha by RP-HPLC/bioassay and gas chromatography-mass spectrometry. Experimental Parasitology 74, 112–116.

Ribeiro, J.M., Alarcon-Chaidez, F., Francischetti, I.M., Mans, B.J., Mather, T.N., Valenzuela, J.G., Wikel, S.K., 2006. An annotated catalog of salivary gland transcripts from *Ixodes scapularis* ticks. Insect Biochemistry and Molecular Biology 36, 111–129.

Ribeiro, J.M., Anderson, J.M., Manoukis, N.C., Meng, Z., Francischetti, I.M., 2011. A further insight into the sialome of the tropical bont tick, *Amblyomma variegatum*. BMC Genomics 12, 136.

Ribeiro, J.M., 1995. Blood-feeding arthropods: live syringes or invertebrate pharmacologists? Infectious Agents and Disease 4, 143–152.

Richter, D., Allgower, R., Matuschka, F.R., 2002. Co-feeding transmission and its contribution to the perpetuation of the Lyme disease spirochete *Borrelia afzelii*. Emerging Infectious Diseases 8, 1421–1425.

Rizzoli, A., Silaghi, C., Obiegala, A., Rudolf, I., Hubálek, Z., Földvári, G., Plantard, O., Vayssier-Taussat, M., Bonnet, S., Špitalská, E., et al., 2014. *Ixodes ricinus* and its transmitted pathogens in Urban and Peri-Urban areas in Europe: new hazards and relevance for public health. Frontiers in Public Health 2, 251.

Roller, L., Šimo, L., Mizoguchi, L., Slovák, M., Park, Y., Žitňan, D., 2015. Orcokinin-like immunoreactivity in central neurons innervating the salivary glands and hindgut of ixodid ticks. Cell and Tissue Research 360, 209–222.

Rosa, R.D., Capelli-Peixoto, J., Mesquita, R.D., Kalil, S.P., Pohl, P.C., Braz, G.R., Fogaça, A.C., Daffre, S., 2016. Exploring the immune signalling pathway-related genes of the cattle tick *Rhipicephalus microplus*: from molecular characterization to transcriptional profile upon microbial challenge. Developmental and Comparative Immunology 59, 1–14.

Sanders, M.L., Glass, G.E., Scott, A.L., Schwartz, B.S., 1998a. Kinetics and cross-species comparisons of host antibody responses to lone star ticks and American dog ticks (Acari: Ixodidae). Journal of Medical Entomology 35, 849–856.

Sanders, M.L., Jaworski, D.C., Sanchez, J.L., DeFraites, R.F., Glass, G.E., Scott, A.L., Raha, S., Ritchie, B.C., Needham, G.R., Schwartz, B.S., 1998b. Antibody to a cDNA-derived calreticulin protein from *Amblyomma americanum* as a biomarker of tick exposure in humans. The American Journal of Tropical Medicine and Hygiene 59, 279–285.

Sanders, M.L., Glass, G.E., Nadelman, R.B., Wormser, G.P., Scott, A.L., Raha, S., Ritchie, B.C., Jaworski, D.C., Schwartz, B.S., 1999. Antibody levels to recombinant tick calreticulin increase in humans after exposure to *Ixodes scapularis* (Say) and are correlated with tick engorgement indices. American Journal of Epidemiology 149, 777–784.

Sangamnatdej, S., Paesen, G.C., Slovak, M., Nuttall, P.A., 2002. A high affinity serotonin- and histamine-binding lipocalin from tick saliva. Insect Molecular Biology 11, 79–86.

Santos, I.K., Valenzuela, J.G., Ribeiro, J.M., de Castro, M., Costa, J.N., Costa, A.M., da Silva, E.R., Neto, O.B., Rocha, C., Daffre, S., et al., 2004. Gene discovery in *Boophilus microplus*, the cattle tick: the transcriptomes of ovaries, salivary glands, and hemocytes. Annals of the New York Academy of Sciences 1026, 242–246.

Sa-Nunes, A., Bafica, A., Antonelli, L.R., Choi, E.Y., Francischetti, I.M., Andersen, J.F., Shi, G.P., Chavakis, T., Ribeiro, J.M., Kotsyfakis, M., 2009. The immunomodulatory action of sialostatin L on dendritic cells reveals its potential to interfere with autoimmunity. Journal of Immunology 182, 7422–7429.

Sauer, J.R., McSwain, J.L., Bowman, A.S., Essenberg, R.C., 1995. Tick salivary gland physiology. Annual Review of Entomology 40, 245–267.

Sauer, J.R., Essenberg, R.C., Bowman, A.C., 2000. Salivary glands in ixodid ticks: control and mechanism of secretion. Journal of Insect Physiology 46, 1069–1078.

Schoeler, G.B., Wikel, S.K., 2001. Modulation of host immunity by haematophagous arthropods. Annals of Tropical Medicine and Parasitology 95, 755–771.

Schoeler, G.B., Manweiler, S.A., Wikel, S.K., 1999. *Ixodes scapularis*: effects of repeated infestations with pathogen-free nymphs on macrophage and T lymphocyte cytokine responses of BALB/c and C3H/HeN mice. Experimental Parasitology 92, 239–248.

Schuijt, T.J., Hovius, J.W.R., van Burgel, N.D., Ramamoorthi, N., Fikrig, E., van Dam, A.P., 2008. The tick salivary protein Salp15 inhibits the killing of serum-sensitive *Borrelia burgdorferi* sensu lato isolates. Infection and Immunity 76, 2888–2894.

Schuijt, T.J., Coumou, J., Narasimhan, S., Dai, J., Deponte, K., Wouters, D., Brouwer, M., Oei, A., Roelofs, J.J., van Dam, A.P., et al., 2011. A tick mannose-binding lectin inhibitor interferes with the vertebrate complement cascade to enhance transmission of the Lyme disease agent. Cell Host and Microbe 10, 136–146.

Schuijt, T.J., Bakhtiari, K., Daffre, S., Deponte, K., Wielders, S.J., Marquart, J.A., Hovius, J.W., van der Poll, T., Fikrig, E., Bunce, M.W., et al., 2013. Factor Xa activation of factor V is of paramount importance in initiating the coagulation system: lessons from a tick salivary protein. Circulation 128, 254–266.

Schwartz, B.S., Ribeiro, J.M., Goldstein, M.D., 1990. Anti-tick antibodies: an epidemiologic tool in Lyme disease research. American Journal of Epidemiology 132, 58–66.

Schwarz, A., Valdes, J.J., Kotsyfakis, M., 2012. The role of cystatins in tick physiology and blood feeding. Ticks and Tick-Borne Diseases 3, 117–127.

Schwarz, A., von Reumont, B.M., Erhart, J., Chagas, A.C., Ribeiro, J.M., Kotsyfakis, M., 2013. De novo *Ixodes ricinus* salivary gland transcriptome analysis using two next-generation sequencing methodologies. FASEB Journal 27, 4745–4756.

Schwarz, A., Tenzer, S., Hackenberg, M., Erhart, J., Gerhold-Ay, A., Mazur, J., Kuharev, J., Ribeiro, J.M.C., Kotsyfakis, M., 2014. A systems level analysis reveals transcriptomic and proteomic complexity in *Ixodes ricinus* midgut and salivary glands during early attachment and feeding. Molecular and Cellular Proteomics 13, 2725–2735.

Severinová, J., Salát, J., Kročová, Z., Řezníčková, J., Demová, H., Horká, H., Kopecký, J., 2005. Co-inoculation of *Borrelia afzelii* with tick salivary gland extract influences distribution of immunocompetent cells in the skin and lymph nodes of mice. Folia Microbiologica (Praha) 50, 457–463.

Shapiro, S.Z., Buscher, G., Dobbelaere, D.A., 1987. Acquired resistance to *Rhipicephalus appendiculatus* (Acari: Ixodidae): identification of an antigen eliciting resistance in rabbits. Journal of Medical Entomology 24, 147–154.

Silva, N.C., Vale, V.F., Franco, P.F., Gontijo, N.F., Valenzuela, J.G., Pereira, M.H., Sant'Anna, M.R., Rodrigues, D.S., Lima, W.S., Fux, B., et al., 2016. Saliva of *Rhipicephalus (Boophilus) microplus* (Acari: Ixodidae) inhibits classical and alternative complement pathways. Parasites and Vectors 9, 445.

Šimo, L., Slovák, M., Park, Y., Žitňan, D., 2008. Identification of a complex peptidergic neuroendocrine network in the hard tick, *Rhipicephalus appendiculatus*. Cell and Tissue Research 335, 639–655.

Šimo, L., Žitňan, D., Park, Y., 2009. Two novel neuropeptides in innervation of the salivary glands of the black-legged tick, *Ixodes scapularis*: myoinhibitory peptide and SIFamide. Journal of Comparative Neurology 517, 551–563.

Šimo, L., Koči, J., Žitňan, D., Park, Y., 2011. Evidence for D1 dopamine receptor activation by a paracrine signal of dopamine in tick salivary glands. PLoS One 6.

Šimo, L., Zitňan, D., Park, Y., 2012. Neural control of salivary glands in ixodid ticks. Journal of Insect Physiology 58, 459–466.

Šimo, L., Koči, J., Park, Y., 2013. Receptors for the neuropeptides, myoinhibitory peptide and SIFamide, in control of the salivary glands of the blacklegged tick *Ixodes scapularis*. Insect Biochemistry and Molecular Biology 43, 376–387.

Šimo, L., Daniel, S.E., Park, Y., Žitňan, D., 2014a. The nervous and sensory systems: structure, function, proteomics and genomics. In: Sonenshine, D.E., Roe, M.R. (Eds.), Biology of Ticks. Oxford University Press, pp. 309–367.

Šimo, L., Koči, J., Kim, D., Park, Y., 2014b. Invertebrate specific D1-like dopamine receptor in control of salivary glands in the black-legged tick *Ixodes scapularis*. Journal of Comparative Neurology 522, 2038–2052.

Skallová, A., Iezzi, G., Ampenberger, F., Kopf, M., Kopecký, J., 2008. Tick saliva inhibits dendritic cell migration, maturation, and function while promoting development of Th2 responses. Journal of Immunology 180, 6186–6192.

Slámová, M., Skallová, A., Páleniková, J., Kopecký, J., 2011. Effect of tick saliva on immune interactions between *Borrelia afzelii* and murine dendritic cells. Parasite Immunology 33, 654–660.

Slovák, M., Štibrániová, I., Hajnická, V., Nuttal, P.A., , 2014. Antiplatelet-derived growth factor (PDGF) activity in the saliva of ixodid ticks is linked with their long mouthparts. Parasite Immunology 36, 32–42.

Smith, A.A., Pal, U., 2014. Immunity-related genes in *Ixodes scapularis*. perspectives from genome information. Frontiers in Cellular and Infection Microbiology 4, 116.

Sonenshine, D.A., Roe, R.M., Anderson, J., 2014. Mouthparts and digestive system. In: Sonenshine, D.E., Roe, R.M. (Eds.), Biology of Ticks. Oxford University Press, New York, pp. 122–162.

Sonenshine, E.D., 1991. Biology of Ticks. Oxford University Press, New York, Oxford.

Steen, N.A., Barker, S.C., Alewood, P.F., 2006. Proteins in the saliva of the Ixodida (ticks): pharmacological features and biological significance. Toxicon 47, 1–20.

Steere, A.C., Sikand, V.K., Meurice, F., Parenti, D.L., Fikrig, E., Schoen, R.T., Nowakowski, J., Schmid, C.H., Laukamp, S., Buscarino, C., Krause, D.S., 1998. Vaccination against Lyme disease with recombinant Borrelia burgdorferi outer-surface lipoprotein A with adjuvant. Lyme Disease Vaccine Study Group. N Engl J Med. 23;339(4), 209–215.

Štibrániová, I., Lahová, M., Bartíková, P., 2013. Immunomodulators in tick saliva and their benefits. Acta Virologica 57, 200–216.

Stone, B.F., Binnington, K.C., Gauci, M., Aylward, J.H., 1989. Tick/host interactions for *Ixodes holocyclus*: role, effects, biosynthesis and nature of its toxic and allergenic oral secretions. Experimental and Applied Acarology 7, 59–69.

Sukumaran, B., Narasimhan, S., Anderson, J.F., DePonte, K., Marcantonio, N., Krishnan, M.N., Fish, D., Telford, S.R., Kantor, F.S., Fikrig, E., 2006. An *Ixodes scapularis* protein required for survival of *Anaplasma phagocytophilum* in tick salivary glands. The Journal of Experimental Medicine 203, 1507–1517.

Tan, A.W.L., Francischetti, I.M.B., Slovak, M., Kini, R.M., Ribeiro, J.M.C., 2015. Sexual differences in the sialomes of the zebra tick, *Rhipicephalus pulchellus*. Journal of Proteomics 117, 120–144.

Tian, Y., Chen, W., Mo, G., Chen, R., Fang, M., Yedid, G., Yan, X., 2016. An immunosuppressant peptide from the hard tick *Amblyomma variegatum*. Toxins 8.

Tirloni, L., Reck, J., Terra, R.M.S., Martins, J.R., Mulenga, A., Sherman, N.E., Fox, J.W., Yates, J.R., Termignoni, C., Pinto, A.F.M., et al., 2014. Proteomic analysis of cattle tick *Rhipicephalus (Boophilus) microplus* saliva: a comparison between partially and fully engorged females. PLoS One 9.

Tirloni, L., Islam, M.S., Kim, T.K., Diedrich, J.K., Yates III, J.R., Pinto, A.F.M., Mulenga, A., You, M.-J., Vaz Jr., I.D.S., 2015. Saliva from nymph and adult females of *Haemaphysalis longicornis*: a proteomic study. Parasites and Vectors 8.

Tonetti, N., Voordouw, M.J., Durand, J., Monnier, S., Gern, L., 2015. Genetic variation in transmission success of the Lyme borreliosis pathogen *Borrelia afzelii*. Ticks and Tick-Borne Diseases 6, 334–343.

Trager, W., 1939. Acquired immunity to ticks. The Journal of Parasitology 25, 57–81.

Tyson, K., Elkins, C., Patterson, H., Fikrig, E., de Silva, A., 2007. Biochemical and functional characterization of Salp20, an *Ixodes scapularis* tick salivary protein that inhibits the complement pathway. Insect Molecular Biology 16, 469–479.

Valdés, J.J., Cabezas-Cruz, A., Sima, R., Butterill, P.T., Růžek, D., Nuttall, P.A., 2016. Substrate prediction of *Ixodes ricinus* salivary lipocalins differentially expressed during *Borrelia afzelii* infection. Scientific Reports 6, 32372.

Valdés, J.J., 2014. Antihistamine response: a dynamically refined function at the host-tick interface. Parasites and Vectors 7, 491.

Valenzuela, J.G., Charlab, R., Mather, T.N., Ribeiro, J.M., 2000. Purification, cloning, and expression of a novel salivary anticomplement protein from the tick, *Ixodes scapularis*. The Journal of Biological Chemistry 275, 18717–18723.

Valenzuela, J.G., Francischetti, I.M., Pham, V.M., Garfield, M.K., Mather, T.N., Ribeiro, J.M., 2002. Exploring the sialome of the tick *Ixodes scapularis*. The Journal of Experimental Biology 205, 2843–2864.

Valenzuela, J.G., 2004. Exploring tick saliva: from biochemistry to 'sialomes' and functional genomics. Parasitology (129 Suppl.), S83–S94.

van de Locht, A., Stubbs, M.T., Bode, W., Friedrich, T., Bollschweiler, C., Hoffken, W., Huber, R., 1996. The ornithodorin-thrombin crystal structure, a key to the TAP enigma? The EMBO Journal 15, 6011–6017.

Vančová, I., Hajnická, V., Slovák, M., Kocáková, P., Paesen, G.C., Nuttall, P.A., 2010a. Evasin-3-like anti-chemokine activity in salivary gland extracts of ixodid ticks during blood-feeding: a new target for tick control. Parasite Immunology 32, 460–463.

Vančová, I., Hajnická, V., Slovák, M., Nuttall, P.A., 2010b. Anti-chemokine activities of ixodid ticks depend on tick species, developmental stage, and duration of feeding. Veterinary Parasitology 167, 274–278.

Villar, M., Popara, M., Mangold, A.J., de la Fuente, J., 2014. Comparative proteomics for the characterization of the most relevant *Amblyomma* tick species as vectors of zoonotic pathogens worldwide. Journal of Proteomics 105, 204–216.

Villar, M., Lopez, V., Ayllón, N., Cabezas-Cruz, A., Lopez, J.A., Vazquez, J., Alberdi, P., de la Fuente, J., 2016. The intracellular bacterium *Anaplasma phagocytophilum* selectively manipulates the levels of vertebrate host proteins in the tick vector *Ixodes scapularis*. Parasites and Vectors 9, 467.

Voordouw, M.J., 2015. Co-feeding transmission in Lyme disease pathogens. Parasitology 142, 290–302.

Vu Hai, V., Almeras, L., Audebert, S., Pophillat, M., Boulanger, N., Parola, P., Raoult, D., Pages, F., 2013a. Identification of salivary antigenic markers discriminating host exposition between two European ticks: *Rhipicephalus sanguineus* and *Dermacentor reticulatus*. Comparative Immunology, Microbiology and Infectious Diseases 36, 39–53.

Vu Hai, V., Pages, F., Boulanger, N., Audebert, S., Parola, P., Almeras, L., 2013b. Immunoproteomic identification of antigenic salivary biomarkers detected by *Ixodes ricinus*-exposed rabbit sera. Ticks and Tick-Borne Diseases 4, 459–468.

Wagemakers, A., Coumou, J., Schuijt, T.J., Oei, A., Nijhof, A.M., van 't Veer, C., van der Poll, T., Bins, A.D., Hovius, J.W., 2016. An *Ixodes ricinus* tick salivary lectin pathway inhibitor protects *Borrelia burgdorferi* sensu lato from human complement. Vector Borne and Zoonotic Diseases 16, 223–228.

Walker, A.R., Fletcher, J.D., Gill, H.S., 1985. Structural and histochemical changes in the salivary glands in *Rhipicephalus appendiculatus* during feeding. International Journal for Parasitology 15, 80–100.

Wang, H., Nuttall, P.A., 1999. Immunoglobulin-binding proteins in ticks: new target for vaccine development against a blood-feeding parasite. Cellular and Molecular Life Sciences 56, 286–295.

Wang, X., Coons, L.B., Taylor, D.B., Stevens Jr., S.E., Gartner, T.K., 1996. Variabilin, a novel RGD-containing antagonist of glycoprotein IIb-IIIa and platelet aggregation inhibitor from the hard tick *Dermacentor variabilis*. The Journal of Biological Chemistry 271, 17785–17790.

Waxman, L., Connolly, T.M., 1993. Isolation of an inhibitor selective for collagen-stimulated platelet aggregation from the soft tick *Ornithodoros moubata*. The Journal of Biological Chemistry 268, 5445–5449.

Waxman, L., Smith, D.E., Arcuri, K.E., Vlasuk, G.P., 1990. Tick anticoagulant peptide (TAP) is a novel inhibitor of blood coagulation factor Xa. Science 248, 593–596.

Wikel, S.K., Alarcon-Chaidez, F.J., 2001. Progress toward molecular characterization of ectoparasite modulation of host immunity. Veterinary Parasitology 101, 275–287.

Wikel, S.K., Ramachandra, R.N., Bergman, D.K., Burkot, T.R., Piesman, J., 1997. Infestation with pathogen-free nymphs of the tick *Ixodes scapularis* induces host resistance to transmission of *Borrelia burgdorferi* by ticks. Infection and Immunity 65, 335–338.

Wikel, S.K., 1979. Acquired resistance to ticks. Expression of resistance by C4-deficient guinea pigs. The American Journal of Tropical Medicine and Hygiene 28, 586–590.

Wikel, S.K., 1996. Host immunity to ticks. Annual Review of Entomology 41, 1–22.

Wikel, S.K., 1999. Tick modulation of host immunity: an important factor in pathogen transmission. International Journal for Parasitology 29, 851–859.

Wikel, S., 2013. Ticks and tick-borne pathogens at the cutaneous interface: host defenses, tick countermeasures, and a suitable environment for pathogen establishment. Frontiers in Microbiology 4, 337.

Willadsen, P., Bird, P., Cobon, G.S., Hungerford, J., 1995. Commercialisation of a recombinant vaccine against *Boophilus microplus*. Parasitology (110 Suppl.), S43–S50.

Willadsen, P., 2004. Anti-tick vaccines. Parasitology (129 Suppl.), S367–S387.

Wright, C.L., Sonenshine, D.E., Gaff, H.D., Hynes, W.L., 2015. *Rickettsia parkeri* transmission to *Amblyomma americanum* by cofeeding with *Amblyomma maculatum* (Acari: Ixodidae) and potential for spillover. Journal of Medical Entomology 52, 1090–1095.

Wu, J., Wang, Y., Liu, H., Yang, H., Ma, D., Li, J., Li, D., Lai, R., Yu, H., 2010. Two immunoregulatory peptides with antioxidant activity from tick salivary glands. The Journal of Biological Chemistry 285, 16606–16613.

Xiang, F-y., Zhou, Y-z., Zhou, J-l, 2012. Identification of differentially expressed genes in the salivary gland of *Rhipicephalus haemaphysaloides* by the suppression subtractive hybridization approach. Journal of Integrative Agriculture 11, 1528–1536.

Xu, X.-L., Cheng, T.-Y., Yang, H., Yan, F., Yang, Y., 2015. De novo sequencing, assembly and analysis of salivary gland transcriptome of *Haemaphysalis flava* and identification of sialoprotein genes. Infection Genetics and Evolution 32, 135–142.

Xu, T., Lew-Tabor, A., Rodriguez-Valle, M., 2016. Effective inhibition of thrombin by *Rhipicephalus microplus* serpin-15 (RmS-15) obtained in the yeast *Pichia pastoris*. Ticks and Tick-Borne Diseases 7, 180–187.

Yoder, J.A., Benoit, J.B., Rellinger, E.J., Tank, J.L., 2006. Developmental profiles in tick water balance with a focus on the new Rocky Mountain spotted fever vector, *Rhipicephalus sanguineus*. Medical and Veterinary Entomology 20, 365–372.

Yu, D., Liang, J., Yu, H., Wu, H., Xu, C., Liu, J., Lai, R., 2006. A tick B-cell inhibitory protein from salivary glands of the hard tick, *Hyalomma asiaticum* asiaticum. Biochemical and Biophysical Research Communications 343, 585–590.

Yu, X., Gong, H., Zhou, Y., Zhang, H., Cao, J., Zhou, J., 2015. Differential sialotranscriptomes of unfed and *fed Rhipicephalus haemaphysaloides*, with particular regard to differentially expressed genes of cysteine proteases. Parasites and Vectors 8, 597.

Zeidner, N.S., Schneider, B.S., Nuncio, M.S., Gern, L., Piesman, J., 2002. Coinoculation of *Borrelia* spp. with tick salivary gland lysate enhances spirochete load in mice and is tick species-specific. The Journal of Parasitology 88, 1276–1278.

Zemtsova, G., Killmaster, L.F., Mumcuoglu, K.Y., Levin, M.L., 2010. Co-feeding as a route for transmission of *Rickettsia conorii* israelensis between *Rhipicephalus sanguineus* ticks. Experimental and Applied Acarology 52, 383–392.

Zhang, P., Tian, Z., Liu, G., Xie, J., Luo, J., Zhang, L., Shen, H., 2011. Characterization of acid phosphatase from the tick *Haemaphysalis longicornis*. Veterinary Parasitology 182, 287–296.

Zhu, K., Bowman, A.S., Brigham, D.L., Essenberg, R.C., Dillwith, J.W., Sauer, J.R., 1997. Isolation and characterization of americanin, a specific inhibitor of thrombin, from the salivary glands of the lone star tick *Amblyomma americanum* (L. Experimental Parasitology 87, 30–38.

Zivkovic, Z., Esteves, E., Almazan, C., Daffre, S., Nijhof, A.M., Kocan, K.M., Jongejan, F., de la Fuente, J., 2010. Differential expression of genes in salivary glands of male *Rhipicephalus* (*Boophilus*) *microplus* in response to infection with *Anaplasma marginale*. BMC Genomics 11, 186.

Chapter 6

Insect-Borne Pathogens and Skin Interface: Flagellate Parasites and Skin Interface

INTRODUCTION

Among arthropod-borne diseases, infections induced by flagellate parasites occupy a significant place in public health, affecting humans and other mammals. The most worldwide distributed are leishmaniases transmitted by the sand fly (Fig. 6.1). It is also the most studied, especially the clinical manifestations occurring at the skin interface where *Leishmania* parasites are inoculated together with sand fly saliva. This saliva has been also particularly well investigated, and several molecules have been identified affecting the host immunity and facilitating pathogen transmission. Less studied are *Trypanosoma* spp., encompassing in humans, *T. cruzi* (American trypanosomiasis or Chagas disease) transmitted by the reduviid bugs and *Trypanosoma brucei gambiense* and *Trypanosoma brucei rhodesiense* (African trypanosomiasis or sleeping sickness) transmitted by the tsetse fly (Fig. 6.1). The interaction of *Trypanosoma* with the host skin and the role of arthropod saliva have been much less explored. For this chapter, we have chosen to describe the physiopathology of leishmaniases and sleeping sickness at the skin interface.

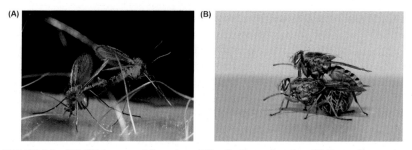

FIGURE 6.1   (A) *Phlebotomus dubosqui*, sand flies feeding and mating. (B) Tsetse flies mating. *((A) Copyright Ed Rowton. (B) Photo: L. De Vooght & L. Verhelst.)*

Skin and Arthropod Vectors. https://doi.org/10.1016/B978-0-12-811436-0.00006-X
193

LEISHMANIASIS

Natalia Tavares[1], Léa Castellucci[2], Camila Indiani de Oliveira[1] and Cláudia Brodskyn[1]
[1]Gonçalo Moniz Institute- Osvaldo cruz Foundation (IG-FIOCRUZ-Ba), Salvador, Brazil;
[2]Serviço de Imunologia do Hospital Universitário Edgar Santos-Federal University of Bahia, Salvador, Brazil

INTRODUCTION

Parasites from the genus *Leishmania* (Trypanosomatida: Trypanosomatidae) are the etiological agent of leishmaniasis. 53 *Leishmania* species have been described and, of these, 20 species are pathogenic for humans. This complex of diseases presents four main clinical manifestations, referred to visceral leishmaniasis (VL), localized cutaneous leishmaniasis (LCL), diffuse cutaneous leishmaniasis (DCL), and mucocutaneous leishmaniasis (MCL) (Alvar et al., 2012).

The first descriptions about the occurrence of these diseases date to 2500 BC in ancient writings of lesions similar to current LCL. However, only in the 18th century, the reports clinically detailed the cutaneous manifestation of leishmaniasis (Hoare, 1938). Later, Leishman identified the etiological microorganism as "trypanosomes" in the beginning of the 20th century, leading to *in vitro* cultivation of promastigotes and description of its life cycle (Leishman, 1903; Donovan, 1903). Successive studies, then, reported different clinical manifestations and their causative species. In 1923, an experimental model of infection by *Leishmania* and the intradermal test (Montenegro skin test) were first described. Currently, the Montenegro skin test is still in use for the diagnosis of leishmaniasis (Alvar et al., 2012; Montenegro, 1926). In the following decades, there was an intense effort to define the taxonomy of the genus *Leishmania*. Only in the end of 20th century, *Leishmania* species were categorized into three subgenera: *Leishmania*, *Viannia*, and *Sauroleishmania*. Recently, this classification was reviewed and the *Leishmania* genus is composed of two phylogenetic lineages: *Euleishmania* and *Paraleishmania* (Schonian et al., 2010; Schonian et al., 2010). Table 6.1 summarizes the classification of *Leishmania* species that are pathogenic for humans, their vector, and geographic distribution.

These diseases are endemic in 98 countries, where 350 million people live at risk of infection. Based on estimates, approximately 300,000 and 950,000 new cases of VL and LCL, respectively, occur each year. Only six countries concentrate more than 90% of global VL cases in the world: India, Bangladesh, Sudan, South Sudan, Ethiopia, and Brazil. The distribution of cutaneous leishmaniasis (CL) is wider, occurring mainly in three epidemiological regions: the Americas, the Mediterranean basin, and from the Middle East to Central Asia (Akhoundi et al., 2016). In these regions, 10 countries have the highest estimated case counts and account for 75% of global CL cases: Afghanistan, Algeria, Colombia, Brazil, Iran, Syria, Ethiopia, North Sudan, Costa Rica, and

Peru. Regarding mortality data, an estimate of 10% is based on case-fatality rate, suggesting that, on average, 30,000 leishmaniasis deaths occur per year. Then, leishmaniasis are currently considered a major global public health problem due to the increased burden over the lasts decades (Alvar et al., 2012).

The majority of VL infection is asymptomatic, but malnutrition and immune suppression predispose to clinical disease, which tends to be chronic, affecting especially children. The period of incubation ranges from 10 days to 1 year and the most common symptoms are fever, chills, weight loss, and anorexia. Clinically, these patients present splenomegaly, hepatomegaly may not be present, wasting, and pallor of mucous membranes. As the VL progresses, signs of malnutrition develop, and intercurrent infection is frequent (Alvar et al., 2012).

CL is the most frequent form of Leishmaniasis, and clinical features vary between regions. A classical lesion can occur anywhere on the body but generally originate at the site of inoculation. It starts as a macule followed by a papule or nodule that grows slowly to a typical round to oval lesion. A crust develops centrally, which may fall and expose an ulcer with raised edge, severely inflamed. The lesion spontaneously heals gradually over months or years, leaving a depressed scar with altered pigmentation and apparently conferring immunity. Lymphatic involvement, such as lymphadenitis or lymphadenopathy, is frequent in lesions caused by species of the *Viannia* subgenus (Table 6.1) (Barral et al., 1992, 1995).

On the other hand, widely disseminated cutaneous nodules, macules, papules, or plaques, without ulceration of the lesion, characterize DCL. Mucosal tissues, usually, are not affected by this condition that does not heal spontaneously. Standard treatments are effective in the beginning, but the disease relapses and becomes unresponsive to further treatment (Handler et al., 2015).

The MCL clinical form is mainly caused by *Leishmania braziliensis* and *L. panamensis*, which can metastasize to the mucosal tissues of the mouth and upper respiratory tract by lymphatic or hematogenous dissemination. This disease manifests from several months to decades after a cutaneous lesion and it does not heal spontaneously. The nose is always involved with nodules and infiltration of the septum, leading to obstruction of the nostril (Davies et al., 2003). Then, this inflammation perforates the septum with collapse and broadening of the nose. Other sites can be involved in MCL, such as pharynx, palate, larynx, trachea, and upper lip. In most severe cases, there is mutilation with obstruction and destruction of the nose, pharynx, and larynx. The most common cause of death related to MCL is pneumonia due to secondary infections, which are frequent in the affected tissues (Llambrich et al., 2009; Yucel et al., 2013).

EARLY EVENTS DURING *LEISHMANIA* TRANSMISSION

The skin provides a protective barrier against pathogens and consists of two layers: epidermis and dermis. The keratinocytes are the major cellular constituent of the epidermis, which are constantly replaced by basal layer stem cells and represent the first line of defense against pathogens. Besides,

TABLE 6.1 Species of *Leishmania* Clinically Relevant for Human Diseases, Their Respective Vector and Geographic Distribution

Groups	*Leishmania* Subgenera	Species	Confirmed Vector	Geographic Distribution
Euleishmania	Leishmania	L. donovani	Phlebotomus alexandri, Phlebotomus celiae, Phlebotomus argentipes, Phlebotomus orientalis, Phlebotomus martini	Central Africa, South Asia, Middle East, India, China Ethiopia, Kenya
		L. aethiopica	Phlebotomus longipes, Phlebotomus pedifer, Phlebotomus sergenti	
		L. amazonensis	Lutzomyia flaviscutellata, Lutzomyia longipalpis, Lutzomyia novica, Lutzomyia reducta	Bolivia, Brazil, Venezuela
		L. infantum	Lutzomyia almeriol, Lutzomyia cruzi, Lu. longipalpis, Lutzomyia evansi, Phlebotomus kandelakii, Phlebotomus ariasi, Phlebotomus major, Phlebotomus balcanicus, Phlebotomus wui, Phlebotomus chinensis, Phlebotomus tobbi, Phlebotomus smirnovi, Phlebotomus turanicus, Phlebotomus langeroni, Phlebotomus perfiliewi, Phlebotomus longiductus, Phlebotomus perniciosus, Phlebotomus sichuanensis	North Africa, Mediterranean countries, Middle, East, Central Asia, Brazil, Venezuela, Bolivia, Mexico
		L. major	Phlebotomus caucasicus, Phlebotomus duboscqi, Phlebotomus papatasi, Phlebotomus salehi	Central and North Africa, Middle East, Central Asia
		L. mexicana	Lutzomyia ayacuchensis, Lutzomyia olmeca, Lutzomyia ovallesi	USA, Ecuador, Peru, Venezuela

Subgenus	Species	Vector	Distribution
	L. tropica	*Phlebotomus arabicus, P. sergenti, Phlebotomus guggisbergi, Phlebotomus rossi, Phlebotomus saevus,*	Central and North Africa, Middle East, Central Asia, India
	L. venezuelensis		Venezuela
Viannia	*L. braziliensis*	*Lutzomyia carrerai, Lutzomyia complexa, Lutzomyia fischeri, Lutzomyia gomezi, Lutzomyia llanosmartins, Lutzomyia migonei, Lutzomyia neivai, Lutzomyia ovailesi, Lutzomyia shawi, Lutzomyia panamensis, Lutzomyia spinicrassa, Lutzomyia whitmani, Lutzomyia wellcomei, Lutzomyia ylephiletor, Lutzomyia yucumensis*	Amazon basin, Brazil, Bolivia, Peru, Guatemala, Venezuela
	L. guyanensis	*Lutzomyia anduzei, Lu. shawi, Lutzomyia ayacuchensis, Lutzomyia umbratilis, Lu. whitmani*	Brazil, French Guiana, Suriname
	L. lainsoni	*Lutzomyia ubiquitalis*	Brazil, Bolivia, Peru
	L. lindenbergi		Brazil
	L. naiffi	*Lutzomyia ayrozai, Lutzomyia squamiventris*	
	L. panamensis	*Lutzomyia gomezi, Lutzomyia hartmanni, Lutzomyia panamensis, Lutzomyia trapidoi, Lutzomyia yuilli*	Brazil, Panama, Venezuela, Colombia
	L. peruviana	*Lu. ayacuchensis, Lutzomyia peruensis, Lu. verrucarum*	Peru, Bolivia
Paraleishmania	*L. shawi*	*Lu. whitmani*	Brazil
	L. colombiensis	*Lu. hartmani*	Colombia

keratinocytes are an important part of the innate immune system due to their ability to produce cytokines, chemokines, and antimicrobial peptides on stimulation (Nicholas et al., 2016a; Pastar et al., 2014). Other cell types are also present in the epidermis, including melanocytes and specialized dendritic cells (DCs), called Langerhans cells. Melanocytes are located in the lower part of the epidermis and are producers of melanin, the pigment responsible for skin color (Nicholas et al., 2016a; Nicoletti et al., 2015). Langerhans cells represent a subset of DCs, the professional antigen-presenting cell type, which play an essential role in cutaneous immunity (Zhang et al., 2016).

The dermis provides the major structural support of the skin since it is a connective tissue underlying the epidermis. Collagen and elastic fibers compose the dermal layer of the skin, while fibroblasts are the predominant cell type. They secrete the compounds of the extracellular matrix (ECM), predominantly collagen and fibronectin (Nicholas et al., 2016a, 2016b) (Fig. 6.2).

Skin Laceration

To access blood vessels and feed, the sand fly must overcome this entire skin structure by puncturing a passage on it. The sand fly needs to lacerate these skin

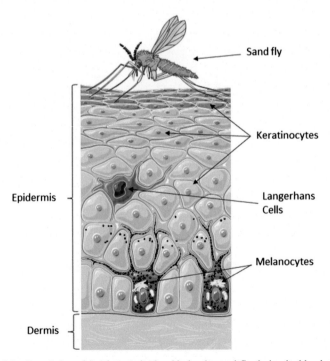

FIGURE 6.2 Inoculation of *Leishmania* in the skin by the sand fly during its blood meal. Two layers compose the skin: epidermis and dermis. The first one does not have blood vessels. Then, the vector needs to dilacerate the epidermis to access the dermic blood vessels.

tissues due to their small cutting mouthparts, causing hemorrhage of surface capillaries. Besides the mechanical action of sand fly proboscis, the formation of the blood pool also depends on the enzymatic action of insect saliva (Charlab et al., 1999). The sand fly saliva contains a large number of immunomodulatory substances and with the ability to inhibit hemostasis. These effects of saliva benefit the sand fly by keeping the blood flowing while it minimizes the host responses. In addition, the sand fly delivers *Leishmania* parasites into a host skin site profoundly altered by the effects of sand fly saliva (Titus et al., 2006).

Leishmania Antigens

A second component of *Leishmania* transmission is the promastigote secretory gel (PSG). During the metacyclogenesis, as the parasites acquire infectivity in the sand fly gut, they secrete a unique class of serine-rich proteophosphoglycans (PPGs), which condense to form a gel, in which the parasites are embedded (Bates, 2007; Rogers et al., 2004). Besides the role of sand fly saliva in the infection of *Leishmania*, the PSG also facilitates the infection of mammalian tissues (Ready, 2013; Rogers and Bates, 2007). In that way, the PSG plug in the sand fly midgut blocks, impairs the ingestion of the blood, and increases the biting persistence or feeding on multiple hosts, which leads to the transmission of *Leishmania* by regurgitation (Ready, 2013).

Besides PPGs, *Leishmania* parasites secrete or express on their surface a broad range of glycoconjugates containing phosphoglycan, such as membrane-bound lipophosphoglycan (LPG), and secreted acid phosphatase (sAP). The surface of *Leishmania* also contains molecules attached to glycosylphosphatidylinositol (GPI) anchors, including glycosylinositolphospholipids (GIPL) (Beverley and Turco, 1998; Ilg et al., 1999; McConville and Ferguson, 1993). It has been shown that the interaction between *Leishmania* and host cells could be mediated by PGs. Parasites deficient in LPG are not able to survive in the midgut of the vector after blood meal excretion, revealing an important function of LPG for the parasite life cycle (Sacks et al., 2000). The ability of *Leishmania* to invade host cells is also impaired when the surface expression of GIPL is inhibited (Suzuki et al., 2002). On the other hand, the function of *Leishmania* sAP is still unclear. *Leishmania mexicana* parasites deficient in this enzyme maintain their infectivity, but this seems to be different for other species (Doyle and Dwyer, 1993; Wiese, 1998). The role of PPGs in the interaction between different *Leishmania* species and the host remains to be answered.

Infective Dose

The third and last component of *Leishmania* transmission is the dose of parasites delivered into the skin. The quantification of *Leishmania* promastigotes delivered by a single sand fly to the skin of the mammalian host is a difficult task to achieve due to the lack of experimental models for natural transmission. Only in the early 2000, a reproducible animal model was described for *Leishmania* infection by a sand fly bite of BALB/c and C57BL/6 mice (Kamhawi et al., 2000). This is a much superior approach than the traditional needle injection since it takes in consideration all features involved in natural transmission. After that, several groups have studied the biology of *Leishmania* transmission between different species and their respective vector. These recent studies revealed a remarkable range (10–100,000) in the dose of parasites. For Old World species, such as *Leishmania major* transmitted by *Phlebotomus duboscqui*, most of the transmitting flies (approximately 75%) deliver 600 or fewer promastigotes (Kimblin et al., 2008). Considering the New World species of *Leishmania*, the sand fly *Lutzomyia longipalpis* is the vector for *Leishmania infantum* and the average of parasites expelled from this sand fly is 1200 (Secundino et al., 2012). Besides, *Lu. longipalpis* is permissive to infection with other *Leishmania* species in the laboratory, but the amount of *L. mexicana* released is also 1000 on average (Rogers et al., 2002). Such great variability in the number of transmitted parasites raises the potential role of dose in determining the clinical outcome of *Leishmania* infection and the average number of expelled promastigotes seems specific for each sand fly.

To feed from the blood of a vertebrate host, the sand fly must lacerate dermic vessels and feed in the blood-feeding pool formed (telmophagous bite). In this environment, the promastigotes are exposed to different blood leukocytes, mostly neutrophils and monocytes. However, as an obligate and promiscuous intracellular parasite, *Leishmania* has a broad range of possible host cells (Kaye and Scott, 2011).

SKIN DISRUPTION BY THE SAND FLY AND ACCESS TO BLOOD (INTERACTION WITH LEUKOCYTES)

Neutrophils

The successful establishment of *Leishmania* infection depends on getting residence in long-lived macrophages without triggering their defense mechanisms. However, neutrophils are the first cell type recruited to the site of infection and readily phagocytose parasites (Peters et al., 2008; Ribeiro-Gomes et al., 2012). They are able to alter the microenvironment due to the release of inflammatory mediators, directing the adaptive immune response. The consequence of neutrophil activities in the early moments of infection for the disease outcome (parasite clearance or disease exacerbation) depends on *Leishmania* species and host susceptibility or resistance (Ribeiro-Gomes et al., 2004).

Several elegant reports associated the presence of neutrophils with disease development in the resistant mouse model for LCL caused by *L. major* (Charmoy et al., 2016; Gonzalez-Lombana et al., 2013). The cross talk between *L. major*-infected apoptotic neutrophils with DCs in the skin of resistant C57BL/6 mice inhibited the specific CD4+ T cells response (Ribeiro-Gomes et al., 2012, 2015). Besides, studies with human neutrophils *in vitro* have shown that the infection by *L. major* modulates apoptotic pathways to prolong the life span of the neutrophil, favoring parasite survival (Aga et al., 2002; Sarkar et al., 2013). These studies highlight the importance of apoptosis, independent of its induction or inhibition, to the early establishment of *L. major* infection in neutrophils.

Similar results are observed when neutrophils from susceptible BALB/c mice are infected *in vitro* with *L. braziliensis* and express markers of apoptosis. However, these cells also express activation markers, produce reactive oxygen species (ROS), release tumor necrosis factor, and neutrophil elastase, leading to an efficient parasite killing (Carlsen et al., 2015; Falcao et al., 2015). In agreement with that, the depletion of neutrophils *in vivo* enhances *L. braziliensis* multiplication (Novais et al., 2009). Regarding neutrophils obtained from LCL patients infected with *L. braziliensis*, their phenotype is activated, similar to neutrophils from BALB/c, after exposure to parasites *in vitro* (Conceicao et al., 2016). Although the authors did not evaluate the cell viability, the results indicate that neutrophils are not apoptotic. Together, these reports suggest that neutrophils play a significant role during the first moments of infection by *L. braziliensis* (Fig. 6.3).

The consequences for *Leishmania amazonensis* infection in neutrophils are still controversial and dependent on the host. Several studies with human cells indicate that neutrophils play a protective role against *L. amazonensis* since they express an activated phenotype and this is related with a leishmanicidal function (Guimaraes-Costa et al., 2009; Rochael et al., 2015; Tavares et al., 2014). The mechanisms behind *L. amazonensis* killing by human neutrophils range from the extrusion of NETs (neutrophil extracellular traps) (Guimaraes-Costa et al., 2009; Rochael et al., 2015) to the degranulation dependent on LTB$_4$ (leukotriene B$_4$)

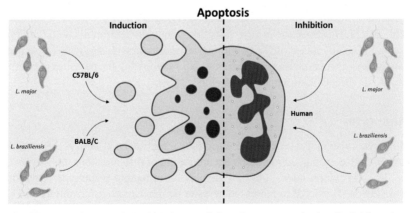

FIGURE 6.3 Inhibition or induction of apoptosis is an important mechanism for *Leishmania* to evade the immune response and maintain the infection. *Leishmania* species are able to lead to different apoptosis outcome depending on the cell type and the host.

production (Tavares et al., 2014, 2016). Besides, these studies reinforce the production of ROS as an important issue for parasite elimination (Afonso et al., 2008).

The role of neutrophils in the killing of *L. amazonensis* was also demonstrated *in vivo* with the experimental model of infection with BALB/c mice. The depletion of neutrophils led to a faster lesion development and increased *L. amazonensis* burden, accompanied by augmented production of interleukin (IL)-10 and IL-17 (Sousa et al., 2014). However, in the C57BL/6 mice, the development of larger lesions corresponds to a larger infiltrate of neutrophils (Roma et al., 2016). The authors also report a significant production of ROS by C57BL/6 mice but without any effect on *L. amazonensis* killing. The results suggest that ROS could regulate the inflammatory response, controlling the recruitment of neutrophils to the lesion (Roma et al., 2016). In this way, the presence of neutrophils is important for parasite killing, but they could also account for tissue injury.

Few studies address the role of neutrophils in the infection by *L. mexicana*, but a recent report showed the induction of NETs without impairing parasite survival (Hurrell et al., 2015). Furthermore, neutropenic- or neutrophil-depleted mice are able to control the infection, leading to a complete healing of the lesion. Together, these data point out a negative role of neutrophils for host defense in the mouse model against certain forms of CL.

Regarding VL, there is a lack of information about the role of neutrophils during the infection by *L. infantum*. The protective role of these cells was already reported for two different experimental models. The impaired recruitment of neutrophils in C57BL/6 mice is associated with increased parasite load in the liver and spleen (Sacramento et al., 2015). These findings suggest a critical role of neutrophils for the protective response against *L. infantum*. Besides, the depletion of neutrophils in BALB/c mice leads to similar results since it slowed the clearance of extracellular promastigotes, suggesting that neutrophils participate in the initial parasite clearance (Rousseau et al., 2001; Thalhofer et al., 2011). However, the *in vitro* infection of human neutrophils with *L. infantum* induces the release of NETs without affecting parasite survival. These promastigotes express nucleases that disrupt NETs, allowing the parasites to escape their toxic effects (Guimaraes-Costa et al., 2014).

Monocytes

Monocytes also play a crucial role during the infection by *L. major*. Susceptible BALB/c mice present an increased accumulation of inflammatory monocytes in the draining lymph nodes, but they display an immature phenotype (Oghumu et al., 2015). This leads to increased parasite burden caused by an amplified Th2 host immune response, favoring parasite establishment. Besides that, monocytes also play a significant role in the natural transmission of *L. major* to C57BL/6 mice. The inflammatory infiltrate at the feeding site presents an earlier and significant larger recruitment of leukocytes, predominantly monocytes, in protected mice (Teixeira et al., 2014). This creates an inhospitable environment for parasite growth. In agreement with that, the reduction of monocytes at the lesion infiltrate turns C57BL/6-resistant mice unable to resolve the lesion and they develop a chronic disease (Goncalves et al., 2011; Kling et al., 2013).

On the other hand, the immune response can contribute to the pathology associated with the disease. The severity observed in leishmaniasis depends of both parasite growth and immune response. Then, increased monocyte infiltrate could be important for control parasite replication, but it could lead to larger and progressive lesions (Gonzalez-Lombana et al., 2013). Together, these results indicate that changes in the proportion of monocytes present in the lesion could be important for parasite killing, but it could also account for extensive pathology in leishmaniasis (Campos et al., 2014; Passos et al., 2015).

There is a lack of information about the role of monocytes during *Leishmania* infection in the human context. However, recent studies have addressed the function of monocytes purified from the blood of patients and stimulated *in vitro* with *L. braziliensis*. Cells from individuals with subclinical infection are less permissive to parasite entry and have a greater ability to kill *L. braziliensis* when compared to cells from patients (Muniz et al., 2016). This could be related to the balance between the production of ROS and reactive nitrogen species by monocytes. It is known that nitric oxide alone is not sufficient to control *L. braziliensis* infection and it may contribute to tissue damage, while ROS is involved in parasite killing (Carneiro et al., 2016).

LESION DEVELOPMENT AND IMMUNOPATHOGENESIS

CD4 T Cells: Th1 Versus Th2

The pathogenesis of LCL depends on host, parasite, and environmental factors including comorbidities and vector components. Among these elements, the host's immune response has been extensively investigated and a large body of evidence has accumulated aimed at explaining the immunopathogenesis of lesion development. Even though LCL is caused by a variety of species, most research has been focused on mouse models employing *L. major* (Sacks and Noben-Trauth, 2002). The conventional model employs a high dose of parasites injected into a subcutaneous site. BALB/c mice are highly susceptible: they develop large skin ulcers, which expand and metastasize, whereas C57BL/6 mice develop small lesions, which cure in 10–12 weeks and are considered resistant. Resistance is linked to a Th1 response, with production of IL-12 and IFN-γ and inhibition of Th2 cytokine production. IFN-γ induces macrophage activation and parasite killing with the production of reactive oxygen and nitrogen intermediates. Susceptibility, on the contrary, is related to a predominant Th2 response, determined by the presence of IL-4. Th2 cytokines lead to the development of severe lesions in mice infected with *L. major*, probably by deactivating infected cells. The production of reactive nitrogen intermediates by IFN-γ-activated macrophages is inhibited by transforming growth factor beta (TGF-β), IL-4, IL-13, and IL-10 (Bogdan et al., 2000) indicating that the down-regulation of Th2 cytokines is crucial for the development of acquired resistance. Although this model explains experimental *L. major* infection, it cannot be translated to all LCL-causing *Leishmania* parasites. For example, both BALB/c and C57BL/6 mice are susceptible to *L. amazonensis* infection (Afonso and Scott, 1993) and develop chronic nonhealing lesions. In experimental *L. amazonensis* infection, resistance and susceptibility are not entirely associated with the Th1/Th2 paradigm. For example, this parasite inhibits DC maturation and subsequent IL-12 production, which is necessary to drive Th1 expansion (Boggiatto et al., 2009 ; Ji et al., 2002). In fact, it has been shown that T cells play a pathogenic role in *L. amazonensis* infection (Terabe et al., 1999) and, para-doxically, IFN-γ can induce parasite proliferation as well as parasite death (Qi et al., 2004). In the host macrophage, *L. amazonensis* parasites can repress NF-kB, inhibiting iNOS expression and NO production, thus favoring sur-vival (Calegari-Silva et al., 2009). For parasite killing, it requires both NO and superoxide (Mukbel et al., 2007).

In contrast to *L. amazonensis*, both BALB/c and C57BL/6 mice are resis-tant to infection with *L. braziliensis*; another species that causes LCL and that is prevalent in the Americas. BALB/c mice were ranked as most suscep-tible since *L. braziliensis* does not induce severe or lasting cutaneous lesions in this mouse strain. The analysis of the immune response has shown that *L. braziliensis*-infected BALB/c mice produce less IL-4, when compared with *L. major*-infected mice. Besides, treating *L. braziliensis*-infected BALB/c mice with anti-IFN-γ significantly enhanced lesion size and prevented mice from resolving the infection (DeKrey et al., 1998). These data suggest that an IFN-γ-dependent mechanism is responsible for the killing of *L. braziliensis* in BALB/c mice and that its weak infectivity in this mouse strain may be due to the inability of the parasite to elicit a strong and sustained IL-4 production. BALB/c mice inoculated in the ear dermis with *L. braziliensis* develop self-healing lesions (de Moura et al., 2005). In this model, a mixed Th1/Th2 type immune response was observed, with the presence of IFN-γ, IL-4, and IL-10-secreting cells. In a comparative study, BALB/c mice infected with *L. braziliensis* developed small, nodular lesions that self-healed, differently from *L. major*-infected BALB/c mice, which displayed progressive ulcers (Rocha et al., 2007). Similar phe-notypes were also observed in C57BL/6 mice. The "resistance" observed in *L. braziliensis*-infected mice was associated with a significantly lower IL-4 and IL-13 production paralleled by a higher presence of IFN-γ and iNOS. Previous works showed that IFN-γ$^{-/-}$ mice infected with *L. braziliensis* develop uncon-trolled lesions (DeKrey et al., 1998), whereas IL-12p40$^{-/-}$ mice (lacking both IL-12 and IL-23) develop chronic lesions (de Souza-Neto et al., 2004). Since lymph node cells from the latter produced low levels of IFN-γ, this cytokine was again implicated in control of *L. braziliensis* infection. These results were con-firmed on infection of IL-12p35$^{-/-}$ mice (which lack IL-12 only) that displayed uncontrolled lesions as did IL-12p35p40$^{-/-}$ mice (Rocha et al., 2007). A similar phenotype was observed on infection of STAT4$^{-/-}$ mice, suggesting that IL-12 is implicated in the response to *L. braziliensis* as observed in *L. major*.

CD8$^+$ T Cells

Although CD4$^+$ Th1 cells secreting IFN-γ are associated with macrophage activation and parasite killing, CD8$^+$ T cells have been shown to play a deleterious role in both experimental and human CL caused by *L. braziliensis* (Scott and Novais, 2016). Using a murine model, it was demonstrated that the pathology associated with unregulated CD8 function is not due to IFN-γ-production, but rather to the excessive perforin-dependent and antigen-specific cytolytic activity by CD8$^+$ T cells. Although CD8$^+$ T cells provide macrophage activation protection by releasing IFN-γ and lead to differentiation of CD4 Th1 response (Uzonna et al., 2004), in experimental LCL caused by *L. braziliensis* infection, increased pathology is promoted by the cytolytic function of CD8$^+$ T cells. Again, these results are different from those observed for *L. major* in which CD8$^+$ T cells contribute to resistance as seen by increased pathology in mice that lack this subset (Belkaid et al., 2002; Uzonna et al., 2004). Similarly, CD8$^+$ T cells have also been implicated in a protective role in studies addressing vaccine development (Gurunathan et al., 1997; Jayakumar et al., 2011; Kronenberg et al., 2010; Maroof et al., 2012; Stager and Rafati, 2012).

Importantly, the role of CD8$^+$ T cells in the pathogenesis of LCL caused by *L. braziliensis* has been also demonstrated in human infection. In these patients, an increase in CD8$^+$ T cells-expressing granzyme has been observed (Faria et al., 2009). CD8$^+$ T cells present at the lesion site exhibit cytolytic properties, and the frequency of granzyme-secreting cells is positively correlated with lesion size (Santos Cda et al., 2013). Transcriptional profiling of CL lesions caused by *L. braziliensis* showed an upregulation of pathways associated with cytotoxicity and inflammasome activation (Novais et al., 2014). Authors hinted at the participation of NK cells and danger-associated molecular patterns resulting from cytotoxic activity in driving this extensive in situ inflammatory reaction. These results corroborate the highly inflammatory nature of *L. braziliensis* infection in which strong cellular responses are present paralleled by a scarcity of parasites (Carvalho et al., 2012). Of note, in mucosal lesions, where there is an excessive Th1 response (Bacellar et al., 2002; Gaze et al., 2006), an important role has been attributed to Th17 cells and infiltrating neutrophils (Boaventura et al., 2010), suggesting that different subsets may be associated with immunopathogenesis of leishmaniasis caused by *L. braziliensis*. Achieving a balanced immune response in which IFN-γ-production induces macrophage activation and parasite killing without overt pathology is key in resolving infection (Dutra et al., 2011).

WOUND AND HEALING IN LEISHMANIASIS

CL is characterized by an intense inflammatory response orchestrated by different types of cells, including, neutrophils, macrophages, CD4, and CD8[+] T cells. Actually, it is well established that the pathology of disease is caused by host immune response, rather than by the parasite. It is well known that the clinical outcome of leishmaniasis infections is the result of a complex interaction of parasite strain, host genetics, and environmental factor.

Inflammatory Genes

The tissue repair begins immediately after a skin injury as an attempt to provide a protective barrier against other stimuli (Clark et al., 1996). In response to an infection, the host cells remodel their tissues through the production of collagen and ECM (Majka et al., 2002; Vaalamo et al., 1999). Fibroblasts migrate to the destroyed tissue and connective tissue matrices are deposited (Izzo et al., 2004; Price et al., 2003; Singh et al., 2004). The severity of the disease generally depends on the host's ability to form granulomas; in its absence, the parasite spreads resulting in diffuse CL (Sakthianandeswaren et al., 2005). Moreover, several parasites produce proteases that degrade matrix proteins, breaking down the cutaneous barrier to disseminate the infection (Lira et al., 1997). After being inoculated into the skin, *Leishmania* promastigotes are exposed to the dermis microenvironment, remaining there until the first contact with the macrophages or other host cell. During this process, there is a loosening of connective tissue matrix of the dermis, allowing the establishment of infection (Ghosh et al., 1999; Lira et al., 1997; McGwire et al., 2003). Furthermore, data demonstrated that *L. amazonensis* promastigotes interact with type 1 collagen modifying the organization of collagen fibers. The proteases released in this process reshape the environment and increase the chances of promastigotes actively interacting with the matrix while "looking" for their host cell (Petropolis et al., 2014).

It is known that the genetic control of wound healing in leishmaniasis involves several families of growth and transcription factors. Studies have shown that *TGFB* gene signaling (which encodes the cytokine TGF-β) is involved in the healing of lesions by regulating cell migration, proliferation, and differentiation (Yang et al., 2007). TGF-β is a cytokine involved in inflammatory processes, fibrosis, and has been implicated in tissue repair and remodeling due to its ability to stimulate fibronectin, proteoglycans, collagen synthesis, fibroblast proliferation, and angiogenesis (Smith et al., 2007). Data have shown that the application of exogenous TGF-β accelerates the healing of acute lesions (Shah et al., 1995; Wu et al., 1997), and the application of recombinant TGF-β improves healing in animal models (Beck et al., 1993). TGF-β also induces connective tissue growth factor (CTGF), which modulates fibroblast growth, collagen synthesis, fibronectin secretion, and ECM (Xiao et al., 2006). A TGF-β ligand initiates signaling by binding and assembling type I (TGFBR1) and type II (TGFBR2) receptors on the cell surface. This allows the type II receptor to phosphorylate the kinase domain of the type I receptor, which then propagates the signal through the phosphorylation of the SMAD proteins (Shi and Massague, 2003). The SMAD proteins are homologs of both the *Drosophila* protein, mothers against decapentaplegic, and the *Caenorhabditis elegans* protein SMA (from gene *sma* for small body size) (Budi et al., 2017). Irregularity in this signaling pathway has been implicated in several developmental disorders as well as some human diseases, including cancer, fibrosis, and autoimmune diseases (Blobe et al., 2000). Eight SMAD proteins are encoded in the human genome (Massague, 1998), but only five act as substrates for receptors in the TGF-β family, including SMAD2. Once phosphorylated by TGF-β receptors, SMAD2 forms a heteromeric complex with a second class of SMAD proteins, known as Co-SMADs. This complex is then translocated to the nucleus to regulate gene expression (Wrana and Attisano, 2000). The biological function of SMAD2 in wound healing is not known because SMAD2-deficient mice are lethal for the embryo (Yagi et al., 1999). However, studies have suggested that SMAD2 is closely involved in the healing process (Ishida et al., 2004) and overexpression of SMAD2 affects the proliferation and differentiation of epidermal keratinocytes.

Gene polymorphisms in the TGF-β signaling pathway, in particular *TGFB1*, *TGFBR2*, *CTGF (Connection Tissue Growth Factor)*, and *SMAD2/3/7*, were described as risk factors for human CL in multicase families from an endemic area of tegumentary leishmaniasis. TGF-β upregulates *CTGF* mRNA (Klass et al., 2009). According to interlocus analysis, the wound healing processes important in CL versus mucosal disease (ML) caused by *L. braziliensis* indicate that *CTGF* regulated via the *SMAD2* arm of the TGF-β signaling pathway is required for wound healing in CL disease. In contrast, ML disease was associated with polymorphism in *SMAD3*, suggesting that alternative regulation of gene expression via the TGF-β signaling pathway may lead to ML disease. However, further functional data would be required to determine the downstream events following signaling via SMAD3 in ML compared to signaling via SMAD2 for CL disease might be. Both forms of disease were influenced by polymorphisms in the negative regulator *SMAD7* that blocks the TGF-β pathway upstream of both *SMAD2* and *SMAD3* (Fig. 6.4) (Castellucci et al., 2012).

The importance of the wound healing processes in cutaneous forms of leishmaniasis has also been demonstrated from studies mapping murine susceptibility genes (Sakthianandeswaren et al., 2005, 2009, 2010). In particular, fine mapping in the region of chromosome 9 in mice (chromosome 11q24 in humans) identified Fli1 (Friend leukemia virus integration 1; FLI1 in humans) as a novel candidate influencing both resistance to *L. major* and an enhanced wound healing response (Sakthianandeswaren et al., 2010). Subsequently, it was demonstrated that polymorphisms at *FLI1* are associated with CL caused by *L. braziliensis* in humans (Castellucci et al., 2011).

Member of the Ets transcription factors family, *FLI1* interacts with the TGF-β pathway and plays an important role in hematopoiesis, embryonic

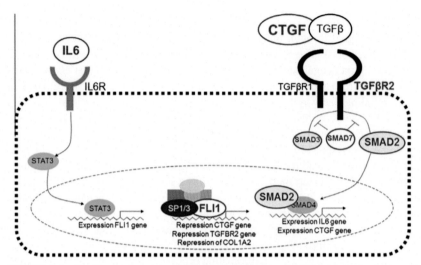

FIGURE 6.4 Signaling pathways important for the activation of SMAD proteins in the wound healing in Leishmaniasis. *CTGF*, connective tissue growth factor; *TGF-β*, transforming growth factor beta.

development, and vasculogenesis (Truong and Ben-David, 2000). Studies have shown that *FLI1* controls the expression of collagen by inhibiting the expression of *CTGF*, and therefore can act on the wound healing pathway (Nakerakanti et al., 2006). In congenital mouse strains, the FLil gene is expressed approximately three times less in animals with the resistant phenotype, suggesting this negative regulation as an important factor for the normal function of macrophages against pathogens and increased lesion healing, correlated with the organization and abundant collagen deposition and a faster inflammatory process compared to the susceptible strains (Sakthianandeswaren et al., 2010).

Additionally there is an association documented between ML and the polymorphism −174 G/C at *IL6* (Castellucci et al., 2006). It has recently been demonstrated that homocysteine-dependent IL6 stimulation positively regulates genes essential for epigenetic DNA methylation via *FLI1* expression (Thaler et al., 2011). In the other direction, it was also seen that Flil is an IL-6 regulator in a murine model of systemic lupus erythematosus (Sato and Zhang, 2014), indicating the existence of a possible feedback loop mechanisms between these two factors. This suggests that although there are many immunological functions for both IL-6 and FLI1 that could be responsible for the association with tegumentary leishmaniasis caused by *L. braziliensis*, there may be a direct functional link between these two genes that mediate susceptibility to disease. In the wound healing response, both *FLI1* (Nakerakanti et al., 2006) and *IL6* (Gressner et al., 2011) repress *CTGF*, and all three genes interact with the TGF-β signaling pathway (Fig. 6.3).

MMP-1 (metalloproteinase type I) is a collagenase that degrades type 1 collagen (Visse and Nagase, 2003). In response to injury, MMP-1 expression occurs rapidly, followed by a gradual decrease, and is undetectable at the time of complete reepithelization (Saarialho-Kere et al., 1993). Transcriptome analysis of human CL lesions caused by *L. braziliensis* showed that *MMP1* was among the 10 most expressed genes in infected tissues indicating that along with other inflammatory genes, *MMP1* might have an important role in the tissue damage observed in CL lesions (Novais et al., 2014).

IL-22 and Tissue Repair

More recently, Gimblet et al. (2015) showed the importance of IL-22 in the tissue repair in the C57BL/6 mice infected by *L. major* and *L. braziliensis*. Member of IL-10 subfamily can regulate wound healing and IL-22 is one of this cytokine that can induce proliferation, migration of epithelial cells, and fibroblasts and also can reconstitute the ECM (Boniface et al., 2005). IL-22 inhibits epidermal differentiation and induces proinflammatory gene expression and migration of human keratinocytes (McGee et al., 2013). This cytokine is important for the wound healing of skin, gut, and lungs (Aujla et al., 2008; Pickert et al., 2009) and sometimes has a pathologic role, such as in psoriasis (Van Belle et al., 2012).

The authors demonstrated that in IL22$^{-/-}$ mice, the infection by these two species of *Leishmania* induced more severe disease with higher score of pathology when compared to wild-type mice. Besides that, it was also observed that there was no significant difference in the number of parasites between these two strains of mice, but KO mice displayed a higher level of IL-1alpha and beta expression at the site of lesion. Importantly, there is a threshold of pathology induced by a high number of parasites that allows IL-22 to limit the pathology induced by *Leishmania*.

Resident Memory T Cells

Another important point to keep the immunity in the skin concerns the presence of skin-resident memory T cells (Trm). Resident memory T cells are nonrecirculating memory T cells that stay for a long time in epithelial barrier tissues, including the gastrointestinal tract, lung, skin, and reproductive tract. These cells remain in the absence of antigens, present intense effector functions, and are responsible for rapid on-site immune protection against some pathogens in peripheral tissues (Fan and Rudensky, 2016).

It was discovered that the healthy skin surface of an adult human displayed nearly 20 billion T cells, approximately twice, which are present in the entire blood volume (Clark et al., 2006). Human skin T cells were all CD45RO+ memory T cells, expressed the skin homing addressins CLA and CCR4 and presented potent effector functions and a diverse T cell repertoire (Clark, 2015).

Studies observed the location of skin tropic memory T cells and found that 98% were located in human skin under noninflamed conditions and only 2% were in the circulation (Clark, 2015). These findings changed the idea that T cells must be recruited from peripheral blood during infectious challenges and seemed that at least some subsets of tissue tropic T cells spend the majority of their time in peripheral tissues.

Several studies have described Trm cells as responsible for the immunity against acute viral infections, such as vaccinia, herpes simplex, influenza, and lymphocytic choriomeningitis virus (Gebhardt et al., 2009; Jiang et al., 2012; Schenkel and Masopust, 2014; Teijaro et al., 2011). These Trm cells are in the gut, brain, lung, and skin (Clark et al., 2006; Teijaro et al., 2011; Wakim et al., 2010), and their location allows them to immediately control a challenge infection without the mobilization of circulating T cells. Additionally, Trm cells can promote recruitment of effector cells from the circulation (Schenkel and Masopust, 2014) and induce antigen-independent innate immunity (Ariotti et al., 2014; Schenkel and Masopust, 2014). The majority of the studies has focused the CD8+ T cells as Trm, but few reports have related the role of CD4 T cells acting as Trm.

Recently, Glennie et al. (2015) identified for the first time a population of skin-resident CD4 T cells that form in response to a chronic parasitic infection. These CD4 T cells remain in the skin even at distal sites of the primary infection until 1 year after the infection, produce IFN-γ, and enhance the recruitment of circulating memory cells to the site of *Leishmania* challenge. These cells by themselves were not able to confer protection, as observed in experiments of transplantation, but they recruited specific circulating memory T cells in chemokine-dependent mechanism enhancing the level of protection against a challenge with *L. major* (Glennie et al., 2015).

There is no available vaccine to *Leishmania* in humans, and the findings about these cells are important for the design of new formulations or strategies that induce the presence of skin-specific *Leishmania*-Trm, providing an appropriate protection. In addition, it is important to recall that some salivary proteins from the sand flies, shown to protect against an infection with different species of *Leishmania*, possibly, might induce the appearance of these cells, favoring an inhospitable environment to the establishment of infection by *Leishmania*.

TRYPANOSOMIASIS

Dorien Mabille, Louis Maes and Guy Caljon
University of Antwerp, Wilrijk, Belgium

HUMAN AFRICAN TRYPANOSOMIASIS

Human African trypanosomiasis (HAT), also known as sleeping sickness, is a neglected tropical disease occurring exclusively in the African continent in a region confined by the Sahara and the Kalahari Desert and is caused by the *rhodesiense* and *gambiense* subspecies of *Trypanosoma brucei*. Millions of people in 36 countries in sub-Saharan Africa are at risk of infection. Untreated infections almost invariably result in a fatal outcome (Checchi and Barrett, 2008). HAT is characterized by two distinct disease stages. During the hemo-lymphatic stage of the disease the patients show nonspecific symptoms such as fever, headaches, joint pains, and itching due to irritation of the sensory nerve endings. In the meningoencephalitic stage, the parasite has invaded the central nervous system causing changes in behavior, confusion, sensory disturbances, poor coordination, and the typical disturbances of the sleep cycle (Malvy and Chappuis, 2011). *Rhodesiense* HAT (rHAT) is characterized by an acute disease progression leading to coma and death within months of infection, whereas *gambiense* HAT (gHAT) has a more chronic presentation taking months to years to progress into the fatal stage of the disease (Checchi and Barrett, 2008). *T. b. gambiense* has an anthroponotic transmission cycle and is responsible for almost 98% of human cases, whereas for *T. b. rhodesiense*, the transmission is zoonotic with a high prevalence in both domestic and wild animals (Malvy and Chappuis, 2011).

HAT is a vector-borne disease solely transmitted by tsetse flies of the genus *Glossina*. More than 30 tsetse fly species and subspecies exist, which prefer specific biotopes, display specific host feeding preferences, and play differential roles in parasite transmission (Caljon et al., 2014; Leak, 1999). The *Glossina* genus is divided into three subgenera: (1) *Palpalis*; a main vector for the transmission of *T. b. gambiense* infections in West and Central Africa. They occur mainly along rivers and streams and are therefore named the "riverine species"; (2) *Morsitans*; prevalent in the East African savannah woodlands and lowland forest and responsible for the transmission of *T. b.rhodesiense*; (3) *Fusca* species; occurring in moist West African forest and savannah (Leak, 1999; Wamwiri and Changasi, 2016). The tsetse fly is a biological vector of try-panosomiasis (Fig. 6.5) meaning that the parasites need to undergo a complex development in the tsetse fly before natural transmission can occur (Van Den Abbeele et al., 2010).

Although the annual number of reported HAT cases is steadily declining (less than 3000 gHAT cases in 2015; Aksoy et al., 2017) due to control programs led by WHO, nongovernmental organizations, and public–private partnerships, history tells that sleeping sickness can rapidly recrudesce if control efforts are discontinued. It is also important to note that despite a historically tremendous effort in vaccination research, not a single effective vaccine that protects humans from infection is yet available (Magez et al., 2010). Nevertheless, the current trend of reducing sleeping sickness cases holds promise of moving into the eradication phase for this disease (Aksoy, 2011; Simarro et al., 2015).

A range of trypanosome species (*T. congolense*, *T. vivax*, and *T. brucei brucei*) is also responsible for animal trypanosomiasis (AAT) infections in livestock and wildlife causing enormous socioeconomic losses (Cherenet et al., 2004).

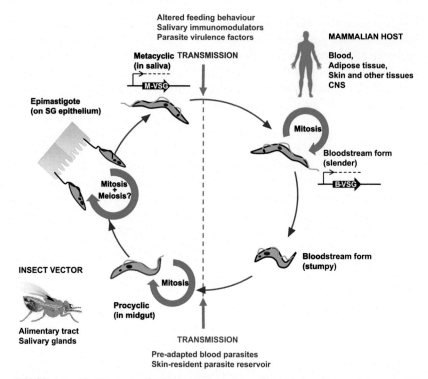

FIGURE 6.5 **The life cycle of African trypanosomes.** Simplified overview of the main life cycle stages of *Trypanosoma brucei* encountered in the vertebrate and invertebrate hosts. Indicated in red are the strategies the parasite takes advantage of to ensure efficient transmission. *(Figure adapted from Caljon, G., De Vooght, L., Van Den Abbeele, J., 2014. The biology of tsetse-trypanosome interactions. In: Magez, S., Radwanska, M. (Eds.), Trypanosomes and Trypanosomiasis, Springer-Verlag, Wien.)*

In cattle, the main cause of animal mortality following a trypanosome infection is anemia and secondary infections (Naessens, 2006). To the advantage of scientific research, *T. brucei brucei* is used as a main and safe nonhuman pathogenic model organism closely resembling *T. b. rhodesiense*, which allows gaining insights into the immunology of sleeping sickness on the basis of experimental rodent infections. To date, very limited trypanosome infection studies have included the tsetse fly and therefore our main understanding of trypanosome infection immunology excludes important aspects related to natural transmission by the insect vector.

EARLY EVENTS DURING *TRYPANOSOMA* TRANSMISSION

Fly Feeding Behavior

Tsetse flies are obligate blood-feeding insects meaning that both male and female flies require a blood meal every 3–4 days (Caljon et al., 2014; Leak, 1999).

Similar to sand flies, the vector of *Leishmania*, tsetse flies rely on a pool feeding (telmophagous) strategy, which involves the laceration of the skin with their proboscis and blood ingestion from a superficial skin lesion (Lehane, 2005). During this feeding process, parasite transmission can occur between vector and mammalian host. The skin of the vertebrate host is a first barrier to invading pathogens, which is breached by the tsetse mouthparts. The actual infection of the mammalian host readily occurs during the probing phase, where the fly tries to generate a local blood pool from which it can feed. During this process, the vector introduces parasites and salivary factors into the skin for which actual blood feeding is not a prerequisite (Lehane, 2005; Van Den Abbeele et al., 2007). Given that the feeding behavior and parasite transmission are linked, this is an interphase for adaptive modification by the parasite and an opportunity for intervention (e.g., through vector control) (Fig. 6.6).

The presence of *T. brucei* parasites in the tsetse salivary glands was shown to have an extensive impact on the tsetse behavior and on gene transcription and protein translation in the salivary glands (Matetovici et al., 2016; Telleria et al., 2014; Van Den Abbeele et al., 2010). Differential RNAseq analyses revealed a prominent tissue damage response as a result of infection with a concomitant upregulation of a number of tissue repair genes. Surprisingly, also immunity-related genes are induced indicating the involvement of the immunodeficiency and JAK–STAT pathways in antitrypanosome responses of this tissue (Matetovici et al., 2016). A major impact is noted on the genes encoding the major saliva factors, indicating a compromised salivary function. Assessment of the anticoagulant and antiplatelet activity of saliva from infected flies revealed a severely compromised antithrombin activity and a reduced capacity to inhibit platelet aggregation. Biochemical assays even showed the presence of a thrombin-enhancing activity in trypanosome-infected saliva from which the identity remains to be identified. These overall physiological changes are compatible with the observed extended probing time required for infected flies to locate a suitable blood vessel (Van Den Abbeele et al., 2010). These changes in the feeding behavior of the tsetse fly are anticipated to increase the likelihood of interrupted feeding, enhancing vector–host contact and therefore the chance of parasite transmission to multiple hosts (Caljon et al., 2016a).

Trypanosomes responsible for AAT such as *T. congolense* and *T. vivax* have a much less complex life cycle than *T. brucei* and do not colonize the salivary glands but instead infect the proboscis and the cibarium primarily around sensillae that are playing a role in the detection of an incoming blood flux[.17]. Combination of the reduced diameter of the proboscis due to the presence of parasites and the interference with the sensory structures was also found to negatively affect the blood-feeding performance in favor of increased probability of parasite transmission. These combined observations show that tsetse transmitted trypanosomes have developed a variety of strategies to modify the insect vector behavior to their advantage.

SALIVARY FACTORS
Anti-hemostasis (serine protease inhibitors, apyrase, ADA)
Immunomodulators (Gloss2, apyrase, ADA, nuclease)
Immunodominance/allergenicity (Tsal, TAg5)
PAMP masking (CTL, galectin, ficolin)

TRYPANOSOME FACTORS
Antigenic variation (M-VSG /B-VSG)
Polyclonal B cell activation (VSG, CpG)
B cell homeostasis and memory depletion
T cell suppression (TSIF)
Immunomodulation (KHC, Adc)
Complement inactivation (M-VSG / B-VSG)

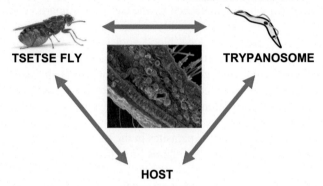

TSETSE FLY   **TRYPANOSOME**

HOST
IMMUNE FACTORS
Immune cells (neutrophils,eosinophils, mast cells, CD4+ and CD8+T cells, B cells, ILCs)
Somatic cells (adipocytes, keratinocytes)
PRRs (SR-A, TLR9)
Antiparasitic effectors (TNF-α , NO)
Antibodies (IgG, IgM, IgE)
Complement factors (C1 – C9)

FIGURE 6.6 **Interactions between the vertebrate host and tsetse fly-transmitted trypano-somes.** Overview of various known salivary, parasite-derived and host-derived immune factors that modulate the tripartite interaction between vector, parasite, and host. *TNF-α*, tumor necrosis factor; *VSG*, variant-specific surface glycoprotein. *(Figure adapted from Caljon, G., Stijlemans, B., De Trez, C., Van Den Abbeele, J., 2017. Early immunological responses upon tsetse fly mediated trypanosome inoculation. Arthropod Vector. Controller of Disease Transmission. Volume 2. Vector Saliva-Host-Pathogen Interactions. Wikel, S.K., Aksoy, S., Dimopoulos, D. (Eds.), Elsevier, Academic Press.)*

Salivary Components

Tsetse flies inoculate *Trypanosoma* parasites in the skin together with a complex mixture of salivary components. The tsetse fly saliva consists of more than 250 different proteins of which some are known to interfere with the host hemostatic responses (Alves-Silva et al., 2010; Caljon et al., 2010; Cappello et al., 1996, 1998; Li and Aksoy, 2000; Mant and Parker, 1981; Zhao et al., 2015). Suppression of vasoconstriction, platelet aggregation, and blood coagulation is important for hematophagous insects to efficiently acquire a blood meal. However, information on the tsetse salivary factors and their role in the regulation of physiological and immunological processes is still very scarce.

Based on proteomic and transcriptomic data, important salivary components of the tsetse fly were identified, including the tsetse thrombin inhibitor, a potent anticoagulant peptide important during the blood-feeding process (Cappello et al., 1998). A salivary apyrase belonging to the 5′nucleotidase-related gene family was found to be a key thromboregulator in tsetse saliva relying on the combined inhibition of ADP-induced aggregation and antagonizing the fibrinogen receptor (Caljon et al., 2010). This enzyme also uses ATP as a substrate. Given that ATP can bind to the inflammatory $P2X_7$ receptor on macrophages, DCs, and neutrophils, this enzymatic activity could also inhibit the downstream processing and secretion of mature IL-1β (Gombault et al., 2012; Karmakar et al., 2016). The tsetse salivary gland growth factors (TSGF-1, TSGF-2) are putative adenosine deaminases that share evolutionary conserved amino acid residues with enzymes responsible for the deamination of adenosine (Li and Aksoy, 2000). Given that adenosine enhances IgE-dependent mast cell degranulation, its enzymatic conversion to inosine might contribute to reducing mast cell degranulation (Tilley et al., 2000). Inosine was moreover shown to display some antiinflammatory properties (Hasko et al., 2000). Tsetse antigen 5, related to vespid toxin allergens, has prominent immunogenic/allergenic properties with the capacity to induce anaphylactic reactions (Caljon et al., 2009). Also the tsetse salivary proteins 1 and 2 (Tsal1 and Tsal 2), representing approximately 40% of the total saliva protein content of *Glossina morsitans*, are highly immunogenic and have been used in the development of serological tests to detect tsetse fly exposure (Caljon et al., 2006a, 2015). These proteins are high-affinity nucleic acid binding proteins with residual nuclease activity (Caljon et al., 2012). Their role at the vector–host interphase remains to be elucidated, although the nucleolytic activity could be anticipated to help in cleaving nucleic acids released from lysed cells thereby reducing viscosity of the blood meal (Chagas et al., 2014). These enzymes might also degrade chromatin released from apoptotic neutrophils, which represents an innate immune defense strategy of the mammalian host to limit the spread of pathogens (Chagas et al., 2014; Kolaczkowska and Kubes, 2013). Additionally, it has been described for telmophagous vectors, e.g., black flies, that salivary serine protease inhibitors are not only involved in the inhibition of thrombin but can also affect enzymes such as elastase and cathepsin G, which can influence the immunological conditions in the bite site microenvironment (Tsujimoto et al., 2012).

In general, the tsetse salivary factors are believed to transform the skin barrier into an immune-tolerant organ to support parasite development. Tsetse saliva induces a prominent Th2 response and was found to suppress T and B cell responses against an intradermally inoculated heterologous antigen (i.e., ovalbumin) (Caljon et al., 2006a). By performing experimental infections using purified *T. brucei* parasites and extracted tsetse fly saliva, a role for saliva in the infection initiation process has also been demonstrated. The saliva is able to accelerate the infection onset, which is associated with a reduced expression of certain inflammatory genes (*il6, il12*) at the initial inoculation site in the skin (Caljon et al., 2006b). The mechanism of this inhibition of dermal parasite-induced inflammation remains to be further explored. Recently, an immunoregulatory peptide, Gloss 2, was shown to inhibit mitogen-activated protein kinase (MAPK) signaling and the secretion of tumor necrosis factor α (TNF-α), IFN-γ,

IL-6, and IL-10 by splenocytes in response to a toll-like receptors (TLR) trigger (i.e., lipopolysaccharide). This immune suppression might possibly also apply to responses against trypanosomal pathogen-associated molecular patterns and contribute to facilitating the early survival of trypanosomes at the initial inoculation site (Bai et al., 2015).

Other salivary factors that could play a role at the dermal interphase are a soluble C-type lectin, galectins, and ficolins (Alves-Silva et al., 2010) that might interfere with parasite recognition by the host immune system. No nonprotein immunomodulatory components have yet been identified in saliva although the presence of two transcripts encoding putative prostaglandin E2 synthases (Alves-Silva et al., 2010) might point to the possible presence of PGE2 with inhibitory functions on DC maturation.

Infective Dose

Experimental infections with purified bloodstream form (BSF) *T. brucei* and *T. congolense* have revealed that the intradermal route of infection is very stringent and represents an effective barrier against low numbers of parasites (Wei et al., 2011). B cell-deficient ($\mu MT^{-/-}$) mice and T/B cell-deficient ($RAG\text{-}2^{-/-}$) mice showed the same innate resistance, revealing that the dermal barrier capacity relates to innate immune cell functions independent of B and T lymphocytes and antibodies. $CD1d^{-/-}$ mice are even more resistant, suggesting that CD1d-restricted natural killer T cells are induced during intradermal infection and suppress the innate resistance mechanism(s). Using gene-deficient mice and by pharmacological inhibition, respectively, inducible nitric oxide synthase (iNOS) and TNF-α were shown to contribute to the innate resistance to intradermal infection with low numbers of BSF *T. congolense* parasites (Wei et al., 2011). Experimental intradermal injections revealed that more than 200 BSF parasites are required to establish an infection through the intradermal route (Caljon et al., 2016b). In contrast, the median infective dose (which causes infections in 50% of the injected mice) for metacyclic (MCF) parasites was calculated to be as low as seven MCF parasites. This observation clearly indicates that MCF parasites are particularly adapted to establishing infections through the intradermal route. This could, for instance, relate to differences in the induction of innate responses or in higher resistance to NO and TNF-α. Infection studies conducted during the colonial period in humans indicate that the skin of man is a more stringent barrier, requiring 300–450 MCF *T. b. rhodesiense* parasites (Fairbairn and Godfrey, 1957). This study indicated a high variability among the various tested individuals with a lowest infective dose of 170 and the highest noninfective dose of 1067 parasites. The combined observations from the human and mouse infections indicate that MCF parasites are highly infective but that host/parasite-dependent variations apply.

CHANGES IN THE SKIN (HOST) FOLLOWING AN INFECTIVE TSETSE FLY BITE

Early Immunological Events

Following inoculation of MCF trypomastigotes by the tsetse fly in the skin microenvironment, the parasites transform into proliferative long-slender bloodstream form (LS-BSF) trypomastigotes (Caljon et al., 2016b). In 20% of the rHAT cases and rarely in gHAT, a chancre can occur at the initial bite site approximately 1 week after parasite inoculation (Aksoy, 2011; Cochran and Rosen, 1983; Hedley et al., 2016; Malvy and Chappuis, 2011). The chancre is characterized by local erythema, edema, heat, tenderness, and lack of suppuration and is the result of a local immune response. Trypanosomal drugs have been shown to alleviate the symptoms associated with this local immune response suggesting that it is mainly parasite-induced (Akol and Murray, 1985). The kinetics and size of the chancre were found to be determined by the number of MCF trypomastigotes that are inoculated in the skin (Barry and Emergy, 1984). This typical ulceration could even be induced by the inoculation of a single MCF *T. brucei* parasite in goats (Dwinger et al., 1987). Observations of high percentages (50%–75%) of degenerating parasites in some host species (e.g., goat and sheep) displaying a strong chancre response (Dwinger et al., 1988; Mwangi et al., 1995) and the inverse in mice and rabbits with a milder dermal response (Caljon et al., 2016b; Luckins and Gray, 1978) suggests that also a host-dependent degree of parasite degradation at the primary inoculation site could determine the ulceration progression. This skin ulceration is not unique to *T. brucei* and has also been documented following *T. congolense* infections in a range of animals such as rabbits, goats, calves, and sheep (Akol and Murray, 1982; Dwinger et al., 1987, 1988; Mwangi et al., 1990). In calves, chancres can be very large (up to 100 mm) and characterized by a heavy recruitment of neutrophils. About 24 h after infection initiation other immune cells are recognized in the dermal infiltrates, including lymphocytes, plasma cells, and macrophages (Akol and Murray, 1982; Roberts et al., 1969). Depletion of CD4+ T lymphocytes in calves was shown to significantly reduce this ulcerative response, showing the importance of these lymphocytes in orchestrating the onset of the dermal reaction. Chancre formation did not impact the infection process (Naessens et al., 2003), which is consistent with the observations in the T/B cell-deficient RAG2$^{-/-}$ mice where the dermal barrier remained unaffected (Wei et al., 2011).

Prior intradermal exposure to noninfective doses of parasites or to bites of uninfected flies could play an important role in the infection process. It was shown that noninfective doses of *T. congolense* primes mice to become more susceptible to a subsequent intradermal dose of another *T. congolense* variant. A similar exacerbating effect was obtained by preexposure of mice to *T. brucei* lysate and a challenge 6 months later with a different *T. brucei* strain. Illustrating that an impact of a priming response is not confined to parasite antigens, immunization of mice against tsetse saliva was also found to have the adverse effect of enhancing tsetse-mediated parasite infection onset (Caljon et al., 2006b). It is possible that hypersensitivity reactions modulate parasite extravasation to the blood circulation of sensitized hosts (Caljon et al., 2009). Strong immediate and delayed-type hypersensitivity responses have been described (Caljon et al., 2009; Ellis et al., 1986) and could shape the immunological composition of the dermal infection site. Experiments that compared the immunological cell composition of the skin of naive and saliva-preexposed rabbits indeed showed that hypersensitivity reactions to saliva resulted in a much more pronounced edema and cellular infiltration of eosinophils and mast cells (Ellis et al., 1986).

Establishment of Infection

From the initial inoculation site in the skin, the majority of the parasites will rapidly gain access to the draining lymph nodes. Parasites that enter the draining lymph nodes cause a local immunological reaction called Winterbottom sign, which is characterized by posterior cervical lymphadenopathy in humans (Malvy and Chappuis, 2011). Microscopic analysis of lymph node aspirates is therefore often used for the early diagnosis of sleeping sickness (Büscher, 2014). However, it has been described that a small subpopulation of parasites remains in the skin and multiplies close to the initial inoculation site, supported by the observation of multiple parasites with double flagella in the skin (Caljon et al., 2016b). Histological and electron microscopy studies have reported a locally expanding trypanosome population at the bite site, which was shown to be derived directly from the tsetse fly inoculum and not from reinvasion of the skin by parasites from the bloodstream (Caljon et al., 2016b). This local parasite population in the interstitial spaces interacts with adipocytes and the surrounding collagen fibers (peri-adipocyte basket) (Fig. 6.7), which could be responsible for retention of the parasites in the dermis or play a beneficial role in local parasite expansion (Caljon et al., 2016b; Luckins and Gray, 1978; Mwangi et al., 1995; Wolbach and Binger, 1912). Intriguingly, the entanglement left free the flagellar pocket, which is a primary site required for endocytosis and essential in the variant-specific surface glycoprotein (VSG) recycling process (Caljon et al., 2016b). The role of the dermal trypanosome population as a reservoir for parasite transmission and recrudescence remains to be further established. However, it is increasingly recognized that the dermal trypanosomes in chronic, low parasitemic patients that escape parasitological diagnosis could be a major bottleneck in the HAT elimination program (Caljon et al., 2016b; Casas-Sanchez and Acosta-Serrano, 2016; Trindade et al., 2016). It also remains to be further explored whether the dermal parasite population resembles metabolically the life cycle stages in the bloodstream and those described to occur in adipose tissue, the latter being a tissue where large amounts of parasites were described to accumulate (Trindade et al., 2016).

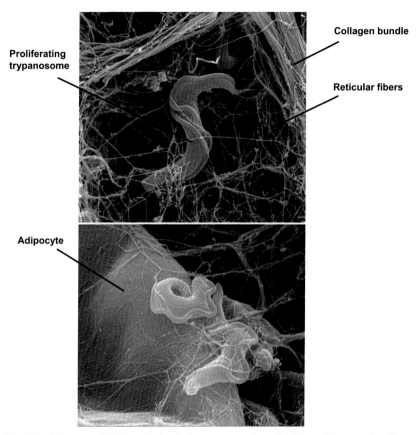

FIGURE 6.7 Dermal interaction of *Trypanosoma brucei* with collagen and adipocytes. Scanning electron microscopy (SEM) images of dermal *T. brucei* populations interacting with dermal adipocytes and collagen fibrous structures (bundles and reticular fibers). Images were made following an infective tsetse fly bite on mouse ears.

From the skin and the lymph, the parasite enters the bloodstream. In the blood, different forms of the *T. brucei* parasite can be recognized; (1) proliferating LS-BSF parasites and (2) a quiescent, short-lived short stumpy BSF (SS-BSF). The differentiation process to stumpy forms depends on quorum-sensing mechanism (Mony et al., 2014; Mony and Matthews, 2015) and is also linked to the process of antigenic variation (see Immune Escape Mechanisms) (Zimmermann et al., 2017). Differentiation into the short-lived SS-BSF is an important self-regulation mechanism to control parasite densities and preadapts the parasite to uptake by the tsetse fly vector (MacGregor et al., 2012; Rico et al., 2013).

Immune Escape Mechanisms

As trypanosomes are extracellular parasites, they are obliged to use various mechanisms to escape elimination by the innate and adaptive immune system, where especially the antiparasitic effects of host antibodies are to be avoided. A major mechanism relies on antigenic variation of the major surface glycoprotein, the VSG. VSG switching is an intrinsically programmed genetic process that occurs independent of the presence of B cells or antibody pressure. The VSG coat consists of approximately 5×10^6 glycoprotein homodimers of 50–60 kDa subunits anchored in the plasma membrane by a GPI anchor (Mehlert et al., 2002). In addition to a pivotal role in antigenic variation, the VSG coat can occlude invariant proteins on the parasite surface from host antibodies (Ferrante and Allison, 1983; Stijlemans et al., 2004). VSGs are already expressed at the metacyclic stage (M-VSGs) in the tsetse fly salivary glands in anticipation of exposure to the vertebrate immune system. The M-VSG surface coat repertoire was shown to be encoded by five canonical MCF genes in *T. brucei* (Cross et al., 2014; Ramey-Butler et al., 2015). At the individual cell level, this results in a largely homogenous M-VSG expression but with an overall diverse antigenic repertoire in the inoculated MCF population, which might contribute to enhancing the success of infection establishment. On entry of trypanosomes in the bloodstream they are described to switch from M-VSG to the expression of bloodstream VSG (B-VSG) genes (Barry and Emergy, 1984; Barry et al., 1998). These *b-vsg* genes are organized in polycistronic transcription units at the telomeres. A combination of an extensive potential for rearrangements from a large repertoire of silent *vsg*-genes and pseudogenes and regulation of transcription from a single fully active telemetric expression site yields a nearly inexhaustible capacity to remodel the homogenous B-VSG cell surface repertoire to continuously escape elimination by host antibodies (Hall et al., 2013; Mugnier et al., 2015; Shimogawa et al., 2015). *T. brucei* is very efficient in turning over its surface-exposed VSG molecules by a highly active membrane recycling from the flagellar pocket (Engstler et al., 2007). *T. brucei* can also activate an endogenous phospholipase C that cleaves off the dimyristoyl glycerol (DMG) from the GPI anchor resulting in the release of soluble VSG (sVSG) that includes a glycosylinositolphosphate (GIP) moiety. These evasion mechanisms prevent activation of the classical pathway of complement activation as these enable removal of IgGs from the parasite surface (Engstler et al., 2007). VSG also inhibits the alternative pathway of complement activation, which occurs in the absence of specific antibodies, by preventing the complement lysis process to proceed beyond the establishment of the C3 convertase (Devine et al., 1986; Ferrante and Allison, 1983). Beside VSG, *T. brucei* also abundantly expresses calflagins that are present in lipid raft microdomains in the flagellar membrane and play a yet unclear role in conferring resistance to early elimination by antibodies (Emmer et al., 2010; Rotureau et al., 2012).

Beside the capacity to undergo antigenic variation to escape antibody responses, trypanosome infection can induce several B cell dysfunctionalities

(Assoku et al., 1977). The concept that trypanosomes can induce a functional B cell-deficient state and can escape antibody responses, might be the reason why, as described above, B cell-deficient mice are not more susceptible to infection than wild-type littermates (Magez et al., 2008; Wei et al., 2011). CpG motifs in the trypanosomal genomic DNA trigger TLR9-signaling events and induce polyclonal B cell activation (Drennan et al., 2005; Shoda et al., 2001). This results in the generation of poly-specific and auto-reactive antibodies, mainly of the IgM isotype (Buza and Naessens, 1999; Hudson et al., 1976; Radwanska et al., 2000). Driving unselective differentiation of B cells is anticipated to negatively affect antibody-mediated effector functions against the parasite. Trypanosome infections were also shown in mice to avoid the buildup of protective humoral responses by ablation of B cell lymphopoiesis in primary and secondary lymphoid organs. Already after the first parasitemic peak of a virulent *T. brucei* and *T. congolense* infection, both the bone marrow and splenic B cell compartment were documented to become severely compromised (Bockstal et al., 2011; Cnops et al., 2015a; Obishakin et al., 2014; Radwanska et al., 2008). Massive cell death of immature transitional B cells underlies the impaired levels of mature marginal zone and follicular B cells (Bockstal et al., 2011; Radwanska et al., 2008). *T. brucei* infection was also shown to abrogate vaccine-induced protective memory responses against unrelated antigens via its impact on effector B cells, such as plasma cells (De Trez et al., 2015; Radwanska et al., 2008). The degree of follicular B cell ablation seems to relate to parasite virulence and the level of IFN-γ induction, with low virulent *T. b. gambiense* infections in mice being associated with a normal follicular B cell composition of the spleen (Cnops et al., 2016). In *T. b.* gHAT patients, only a moderate reduction in historical vaccine-induced antibody levels is observed (Lejon et al., 2014).

The trypanosomal VSG also plays a key role in regulating the parasitemia and the induction of pathogenicity. GIP-sVSG is recognized by scavenger receptor type A (SR-A) expressed on macrophages and DCs (Hereld et al., 1986; Leppert et al., 2007). SR-A triggering leads to the activation of NF-κB and MAPK pathways and downstream expression of the proinflammatory cytokines TNF-α, IL-6, and IL-12 (Paulnock et al., 2010). In conditions where macrophages are already primed with IFN-γ or exposed to DMG, the recognition of GIP-sVSG evokes hyperactivation of the cells with an excessive proinflammatory cytokine response, which can lead to the development of pathogenicity features such as cachexia and anemia. Only animals able to dampen this inflammatory cytokine storm by tissue protective IL-10 can exhibit an alleviated pathogenicity (Stijlemans et al., 2007). Importantly, the balance of these different activation/deactivation signals may determine the outcome of infection (Namangala et al., 2001; Paulnock and Coller, 2001). Recently, it was shown that different lymphocyte populations play a role in IFN-γ production, whereby NK and NKT cells are the earliest IFN-γ producers, followed by CD8+ and CD4+ T cells (Cnops et al., 2015b). Production of the trypanotoxic molecules TNF-α and nitric oxide (NO) controls the early peak parasitemia and is important in the dermal immune barrier function as described above (Magez et al., 1997, 1998; Wei et al., 2011). NO and prostaglandins, raised as part of the antiparasitic response, on the other hand also contribute to the severe T cell proliferation impairment (Schleifer and Mansfield, 1993) that is observed very early on infection. These suppressed T cell proliferative responses are associated with decreased interleukin-2 (IL-2) production and IL-2 receptor expression (Sileghem et al., 1986, 1987). A parasite membrane protein, trypanosome-suppressive immunomodulating factor, was shown to be capable of inducing macrophages with a T cell suppressive character (Gomez-Rodriguez et al., 2009). In addition, these macrophages were poorly capable of activating specific T cells due to a hampered presentation of antigenic peptides in major histocompatibility complex class II (Namangala et al., 2000).

Trypanosoma parasites have developed several mechanisms to regulate the inflammatory host responses. *T. brucei* was shown to activate transmembrane receptor-like adenylate cyclases on phagocytosis by mononuclear phagocytic cells. The induced adenylate cyclase activity results in the generation of elevated concentrations of cyclic adenosine monophosphate (cAMP) inside these phagocytic cells. The downstream effect of cAMP is the activation of protein kinase A and a subsequent reduction in the synthesis of the antitrypanosomal TNF-α (Lucas et al., 1994; Magez et al., 1997; Rolin et al., 1996; Salmon et al., 2012). *T. brucei* also secretes kinase heavy chain 1 (TbKHC1) though a noncanonical pathway, which is responsible for reducing the production of NO. By binding onto SIGN-R1, TbKHC1 modifies the balance of iNOS and arginase in an IL-10 dependent manner. By reducing iNOS in favor of increased arginase activity, this induces lower NO levels and a higher availability of L-ornithine, which is a main source for the synthesis of polyamines by trypanosomes (De Muylder et al., 2013). Corroborating the importance of TbKHC1, deficient parasites were hampered in infection onset following a tsetse-mediated parasite inoculation.

CONCLUDING REMARKS

Trypanosomes have evolved various mechanisms to cope with the highly differential conditions encountered in the vertebrate and invertebrate hosts. On inoculation in the skin, the natural infective parasite forms efficiently cope with antiparasitic effectors and shape the hosts' innate/adaptive immune response to establish an infection. Not only the parasite has to survive the hostile host environment but also it has to control its own growth and occupy suitable tissue niches to ensure its transmission. The skin has in that respect been highly underexplored as infection niche. The immune responses evoked in this tissue and the specific parasite adaptations and interactions with immune and somatic cells remain to be unraveled. This will be essential in the quest for novel windows of opportunity for intervention.

REFERENCES

Afonso, L.C., Scott, P., 1993. Immune responses associated with susceptibility of C57BL/10 mice to *Leishmania amazonensis*. Infection and Immunity 61, 2952–2959.

Afonso, L., Borges, V.M., Cruz, H., Ribeiro-Gomes, F.L., DosReis, G.A., Dutra, A.N., Clarencio, J., de Oliveira, C.I., Barral, A., Barral-Netto, M., Brodskyn, C.I., 2008. Interactions with apoptotic but not with necrotic neutrophils increase parasite burden in human macrophages infected with *Leishmania amazonensis*. Journal of Leukocyte Biology 84, 389–396.

Aga, E., Katschinski, D.M., van Zandbergen, G., Laufs, H., Hansen, B., Muller, K., Solbach, W., Laskay, T., 2002. Inhibition of the spontaneous apoptosis of neutrophil granulocytes by the intracellular parasite *Leishmania* major. Journal of Immunology 169, 898–905.

Akhoundi, M., Kuhls, K., Cannet, A., Votypka, J., Marty, P., Delaunay, P., Sereno, D., 2016. A historical overview of the classification, evolution, and dispersion of leishmania parasites and sandflies. PLoS Neglected Tropical Diseases 10, e0004349.

Akol, G.W., Murray, M., 1982. Early events following challenge of cattle with tsetse infected with *Trypanosoma congolense*: development of the local skin reaction. The Veterinary Record 110, 295–302.

Akol, G.W., Murray, M., 1985. Induction of protective immunity in cattle by tsetse-transmitted cloned isolates of *Trypanosoma congolense*. Annals of Tropical Medicine and Parasitology 79, 617–627.

Aksoy, S., Buscher, P., Lehane, M., Solano, P., Van Den Abbeele, J., 2017. Human African trypanosomiasis control: achievements and challenges. PLoS Neglected Tropical Diseases 11, e0005454.

Aksoy, S., 2011. Sleeping sickness elimination in sight: time to celebrate and reflect, but not relax. PLoS Neglected Tropical Diseases 5, e1008.

Alvar, J., Velez, I.D., Bern, C., Herrero, M., Desjeux, P., Cano, J., Jannin, J., den Boer, M., WHO Leishmaniasis Control Team, 2012. Leishmaniasis worldwide and global estimates of its incidence. PLoS One 7, e35671.

Alves-Silva, J., Ribeiro, J.M., Van Den Abbeele, J., Attardo, G., Hao, Z., Haines, L.R., Soares, M.B., Berriman, M., Aksoy, S., Lehane, M.J., 2010. An insight into the sialome of *Glossina morsitans* morsitans. BMC Genomics 11, 213.

Ariotti, S., Hogenbirk, M.A., Dijkgraaf, F.E., Visser, L.L., Hoekstra, M.E., Song, J.Y., Jacobs, H., Haanen, J.B., Schumacher, T.N., 2014. T cell memory. Skin-resident memory CD8(+) T cells trigger a state of tissue-wide pathogen alert. Science 346, 101–105.

Assoku, R.K., Tizard, I.R., Neilsen, K.H., 1977. Free fatty acids, complement activation, and polyclonal B-cell stimulation as factors in the immunopathogenesis of African trypanosomiasis. Lancet 2, 956–959.

Aujla, S.J., Chan, Y.R., Zheng, M., Fei, M., Askew, D.J., Pociask, D.A., Reinhart, T.A., McAllister, F., Edeal, J., Gaus, K., Husain, S., Kreindler, J.L., Dubin, P.J., Pilewski, J.M., Myerburg, M.M., Mason, C.A., Iwakura, Y., Kolls, J.K., 2008. IL-22 mediates mucosal host defense against gram-negative bacterial pneumonia. Nature Medicine 14, 275–281.

Bacellar, O., Lessa, H., Schriefer, A., Machado, P., Ribeiro de Jesus, A., Dutra, W.O., Gollob, K.J., Carvalho, E.M., 2002. Up-regulation of Th1-type responses in mucosal leishmaniasis patients. Infection and Immunity 70, 6734–6740.

Bai, X., Yao, H., Du, C., Chen, Y., Lai, R., Rong, M., 2015. An immunoregulatory peptide from tsetse fly salivary glands of *Glossina morsitans* morsitans. Biochimie 118, 123–128.

Barral, A., Barral-Netto, M., Almeida, R., de Jesus, A.R., Grimaldi Junior, G., Netto, E.M., Santos, I., Bacellar, O., Carvalho, E.M., 1992. Lymphadenopathy associated with *Leishmania braziliensis* cutaneous infection. The American Journal of Tropical Medicine and Hygiene 47, 587–592.

Barral, A., Teixeira, M., Reis, P., Vinhas, V., Costa, J., Lessa, H., Bittencourt, A.L., Reed, S., Carvalho, E.M., Barral-Netto, M., 1995. Transforming growth factor-beta in human cutaneous leishmaniasis. The American Journal of Pathology 147, 947–954.

Barry, J.D., Emergy, D.L., 1984. Parasite development and host responses during the establishment of *Trypanosoma brucei* infection transmitted by tsetse fly. Parasitology 88 (Pt 1), 67–84.

Barry, J.D., Graham, S.V., Fotheringham, M., Graham, V.S., Kobryn, K., Wymer, B., 1998. VSG gene control and infectivity strategy of metacyclic stage *Trypanosoma brucei*. Molecular and Biochemical Parasitology 91, 93–105.

Bates, P.A., 2007. Transmission of Leishmania metacyclic promastigotes by phlebotomine sand flies. International Journal for Parasitology 37, 1097–1106.

Beck, L.S., Amento, E.P., Xu, Y., Deguzman, L., Lee, W.P., Nguyen, T., Gillett, N.A., 1993. TGF-beta 1 induces bone closure of skull defects: temporal dynamics of bone formation in defects exposed to rhTGF-beta 1. Journal of Bone and Mineral Research 8, 753–761.

Belkaid, Y., Von Stebut, E., Mendez, S., Lira, R., Caler, E., Bertholet, S., Udey, M.C., Sacks, D., 2002. CD8+ T cells are required for primary immunity in C57BL/6 mice following low-dose, intradermal challenge with Leishmania major. Journal of Immunology 168, 3992–4000.

Beverley, S.M., Turco, S.J., 1998. Lipophosphoglycan (LPG) and the identification of virulence genes in the protozoan parasite Leishmania. Trends in Microbiology 6, 35–40.

Blobe, G.C., Schiemann, W.P., Lodish, H.F., 2000. Role of transforming growth factor beta in human disease. The New England Journal of Medicine 342, 1350–1358.

Boaventura, V.S., Santos, C.S., Cardoso, C.R., de Andrade, J., Dos Santos, W.L., Clarencio, J., Silva, J.S., Borges, V.M., Barral-Netto, M., Brodskyn, C.I., Barral, A., 2010. Human mucosal leishmaniasis: neutrophils infiltrate areas of tissue damage that express high levels of Th17-related cytokines. European Journal of Immunology 40, 2830–2836.

Bockstal, V., Guirnalda, P., Caljon, G., Goenka, R., Telfer, J.C., Frenkel, D., Radwanska, M., Magez, S., Black, S.J., 2011. *T. brucei* infection reduces B lymphopoiesis in bone marrow and truncates compensatory splenic lymphopoiesis through transitional B-cell apoptosis. PLoS Pathogens 7, e1002089.

Bogdan, C., Donhauser, N., Doring, R., Rollinghoff, M., Diefenbach, A., Rittig, M.G., 2000. Fibroblasts as host cells in latent leishmaniosis. The Journal of Experimental Medicine 191, 2121–2130.

Boggiatto, P.M., Jie, F., Ghosh, M., Gibson-Corley, K.N., Ramer-Tait, A.E., Jones, D.E., Petersen, C.A., 2009. Altered dendritic cell phenotype in response to *Leishmania amazonensis* amastigote infection is mediated by MAP kinase, ERK. The American Journal of Pathology 174 (5), 1818–1826.

Boniface, K., Bernard, F.X., Garcia, M., Gurney, A.L., Lecron, J.C., Morel, F., 2005. IL-22 inhibits epidermal differentiation and induces proinflammatory gene expression and migration of human keratinocytes. Journal of Immunology 174, 3695–3702.

Budi, E.H., Duan, D., Derynck, R., 2017. Transforming growth factor-β receptors and smads: regulatory complexity and functional versatility. Trends in Cell Biology (17), 30063–30066 pii: S0962-8924.

Büscher, P., 2014. Diagnosis of African trypanosomiasis. In: Magez, S., Radwanska, M. (Eds.), Trypanosomes and Trypanosomiasis. Springer-Verlag, Wien.

Buza, J., Naessens, J., 1999. Trypanosome non-specific IgM antibodies detected in serum of *Trypanosoma congolense*-infected cattle are polyreactive. Veterinary Immunology and Immunopathology 69, 1–9.

Calegari-Silva, T.C., Pereira, R.M., De-Melo, L.D., Saraiva, E.M., Soares, D.C., Bellio, M., Lopes, U.G., 2009. NF-kappaB-mediated repression of iNOS expression in *Leishmania amazonensis* macrophage infection. Immunology Letters 127, 19–26.

Caljon, G., Van Den Abbeele, J., Sternberg, J.M., Coosemans, M., De Baetselier, P., Magez, S., 2006a. Tsetse fly saliva biases the immune response to Th2 and induces anti-vector antibodies that are a useful tool for exposure assessment. International Journal for Parasitology 36, 1025–1035.

Caljon, G., Van Den Abbeele, J., Stijlemans, B., Coosemans, M., De Baetselier, P., Magez, S., 2006b. Tsetse fly saliva accelerates the onset of *Trypanosoma brucei* infection in a mouse model associated with a reduced host inflammatory response. Infection and Immunity 74, 6324–6330.

Caljon, G., Broos, K., De Goeyse, I., De Ridder, K., Sternberg, J.M., Coosemans, M., De Baetselier, P., Guisez, Y., Den Abbeele, J.V., 2009. Identification of a functional Antigen5-related allergen in the saliva of a blood feeding insect, the tsetse fly. Insect Biochemistry and Molecular Biology 39, 332–341.

Caljon, G., De Ridder, K., De Baetselier, P., Coosemans, M., Van Den Abbeele, J., 2010. Identification of a tsetse fly salivary protein with dual inhibitory action on human platelet aggregation. PLoS One 5, e9671.

Caljon, G., De Ridder, K., Stijlemans, B., Coosemans, M., Magez, S., De Baetselier, P., Van Den Abbeele, J., 2012. Tsetse salivary gland proteins 1 and 2 are high affinity nucleic acid binding proteins with residual nuclease activity. PLoS One 7, e47233.

Caljon, G., De Vooght, L., Van Den Abbeele, J., 2014. The biology of tsetse-trypanosome interactions. In: Magez, S., Radwanska, M. (Eds.), Trypanosomes and Trypanosomiasis. Springer-Verlag, Wien.

Caljon, G., Hussain, S., Vermeiren, L., Van Den Abbeele, J., 2015. Description of a nanobody-based competitive immunoassay to detect tsetse fly exposure. PLoS Neglected Tropical Diseases 9, e0003456.

Caljon, G., De Muylder, G., Durnez, L., Jennes, W., Vanaerschot, M., Dujardin, J.C., 2016a. Alice in microbes' land: adaptations and counter-adaptations of vector-borne parasitic protozoa and their hosts. FEMS Microbiology Reviews 40, 664–685.

Caljon, G., Van Reet, N., De Trez, C., Vermeersch, M., Pérez-Morga, D., Van Den Abbeele, J., 2016b. The dermis as a delivery site of *Trypanosoma brucei* for tsetse flies. PLoS Pathogens 12, e1005744.

Caljon, G., Stijlemans, B., De Trez, C., Van Den Abbeele, J., 2017. Early immunological responses upon tsetse fly mediated trypanosome inoculation. In: Wikel, S.K., Aksoy, S., Dimopoulos, D. (Eds.), Arthropod Vector. Controller of Disease Transmission. Volume 2. Vector Saliva-Host-Pathogen Interactions. Elsevier, Academic Press.

Campos, T.M., Passos, S.T., Novais, F.O., Beiting, D.P., Costa, R.S., Queiroz, A., Mosser, D., Scott, P., Carvalho, E.M., Carvalho, L.P., 2014. Matrix metalloproteinase 9 production by monocytes is enhanced by TNF and participates in the pathology of human cutaneous Leishmaniasis. PLoS Neglected Tropical Diseases 8, e3282.

Cappello, M., Bergum, P.W., Vlasuk, G.P., Furmidge, B.A., Pritchard, D.I., Aksoy, S., 1996. Isolation and characterization of the tsetse thrombin inhibitor: a potent antithrombotic peptide from the saliva of *Glossina morsitans* morsitans. The American Journal of Tropical Medicine and Hygiene 54, 475–480.

Cappello, M., Li, S., Chen, X., Li, C.B., Harrison, L., Narashimhan, S., Beard, C.B., Aksoy, S., 1998. Tsetse thrombin inhibitor: bloodmeal-induced expression of an anticoagulant in salivary glands and gut tissue of *Glossina morsitans* morsitans. Proceedings of the National Academy of Sciences of the United States of America 95, 14290–14295.

Carlsen, E.D., Jie, Z., Liang, Y., Henard, C.A., Hay, C., Sun, J., de Matos Guedes, H., Soong, L., 2015. Interactions between neutrophils and *Leishmania braziliensis* amastigotes facilitate cell activation and parasite clearance. Journal of Innate Immunity 7, 354–363.

Carneiro, P.P., Conceicao, J., Macedo, M., Magalhaes, V., Carvalho, E.M., Bacellar, O., 2016. The role of nitric oxide and reactive oxygen species in the killing of *Leishmania braziliensis* by monocytes from patients with cutaneous leishmaniasis. PLoS One 11, e0148084.

Carvalho, L.P., Passos, S., Schriefer, A., Carvalho, E.M., 2012. Protective and pathologic immune responses in human tegumentary leishmaniasis. Frontiers in Immunology 3, 301.

Casas-Sanchez, A., Acosta-Serrano, A., 2016. Skin deep. Elife 5.

Castellucci, L., Menezes, E., Oliveira, J., Magalhaes, A., Guimaraes, L.H., Lessa, M., Ribeiro, S., Reale, J., Noronha, E.F., Wilson, M.E., Duggal, P., Beaty, T.H., Jeronimo, S., Jamieson, S.E., Bales, A., Blackwell, J.M., de Jesus, A.R., Carvalho, E.M., 2006. IL6 -174 G/C promoter poly-morphism influences susceptibility to mucosal but not localized cutaneous leishmaniasis in Brazil. The Journal of Infectious Diseases 194, 519–527.

Castellucci, L., Jamieson, S.E., Miller, E.N., de Almeida, L.F., Oliveira, J., Magalhaes, A., Guimaraes, L.H., Lessa, M., Lago, E., de Jesus, A.R., Carvalho, E.M., Blackwell, J.M., 2011. FLI1 polymorphism affects susceptibility to cutaneous leishmaniasis in Brazil. Genes and Immunity 12, 589–594.

Castellucci, L., Jamieson, S.E., Almeida, L., Oliveira, J., Guimaraes, L.H., Lessa, M., Fakiola, M., Jesus, A.R., Nancy Miller, E., Carvalho, E.M., Blackwell, J.M., 2012. Wound healing genes and suscep-tibility to cutaneous leishmaniasis in Brazil. Infection, Genetics and Evolution 12, 1102–1110.

Chagas, A.C., Oliveira, F., Debrabant, A., Valenzuela, J.G., Ribeiro, J.M., Calvo, E., 2014. Lundep, a sand fly salivary endonuclease increases *Leishmania* parasite survival in neutrophils and inhibits XIIa contact activation in human plasma. PLoS Pathogens 10, e1003923.

Charlab, R., Valenzuela, J.G., Rowton, E.D., Ribeiro, J.M., 1999. Toward an understanding of the biochemical and pharmacological complexity of the saliva of a hematophagous sand fly *Lutzomyia longipalpis*. Proceedings of the National Academy of Sciences of the United States of America 96, 15155–15160.

Charmoy, M., Hurrell, B.P., Romano, A., Lee, S.H., Ribeiro-Gomes, F., Riteau, N., Mayer-Barber, K., Tacchini-Cottier, F., Sacks, D.L., 2016. The Nlrp3 inflammasome, IL-1beta, and neutro-phil recruitment are required for susceptibility to a nonhealing strain of Leishmania major in C57BL/6 mice. European Journal of Immunology 46, 897–911.

Checchi, F., Barrett, M.P., 2008. African sleeping sickness. British Medical Journal 336, 679–680.

Cherenet, T., Sani, R.A., Panandam, J.M., Nadzr, S., Speybroeck, N., van den Bossche, P., 2004. Seasonal prevalence of bovine trypanosomosis in a tsetse-infested zone and a tsetse-free zone of the Amhara Region, North-West Ethiopia. The Onderstepoort Journal of Veterinary Research 71, 307–312.

Clark, R.A., Ashcroft, G.S., Spencer, M.J., Larjava, H., Ferguson, M.W., 1996. Re-epithelialization of normal human excisional wounds is associated with a switch from alpha v beta 5 to alpha v beta 6 integrins. The British Journal of Dermatology 135, 46–51.

Clark, R.A., Chong, B., Mirchandani, N., Brinster, N.K., Yamanaka, K., Dowgiert, R.K., Kupper, T.S., 2006. The vast majority of CLA+ T cells are resident in normal skin. Journal of Immunology 176, 4431–4439.

Clark, R.A., 2015. Resident memory T cells in human health and disease. Science Translational Medicine 7, 269rv261.

Cnops, J., De Trez, C., Bulte, D., Radwanska, M., Ryffel, B., Magez, S., 2015a. IFN-gamma mediates early B-cell loss in experimental African trypanosomosis. Parasite Immunology 37, 479–484.

Cnops, J., De Trez, C., Stijlemans, B., Keirsse, J., Kauffmann, F., Barkhuizen, M., Keeton, R., Boon, L., Brombacher, F., Magez, S., 2015b. NK-, NKT- and CD8-derived IFNgamma drives myeloid cell activation and erythrophagocytosis, resulting in trypanosomosis-associated acute anemia. PLoS Pathogens 11, e1004964.

Cnops, J., Kauffmann, F., De Trez, C., Baltz, T., Keirsse, J., Radwanska, M., Muraille, E., Magez, S., 2016. Maintenance of B cells during chronic murine *T. b. gambiense* infection. Parasite Immunology 38, 642–647.

Cochran, R., Rosen, T., 1983. African trypanosomiasis in the United States. Archives of Dermatology 119, 670–674.

Conceicao, J., Davis, R., Carneiro, P.P., Giudice, A., Muniz, A.C., Wilson, M.E., Carvalho, E.M., Bacellar, O., 2016. Characterization of neutrophil function in human cutaneous leishmaniasis caused by *Leishmania braziliensis*. PLoS Neglected Tropical Diseases 10, e0004715.

Cross, G.A., Kim, H.S., Wickstead, B., 2014. Capturing the variant surface glycoprotein repertoire (the VSGnome) of *Trypanosoma brucei* Lister 427. Molecular and Biochemical Parasitology 195, 59–73.

Davies, C.R., Kaye, P., Croft, S.L., Sundar, S., 2003. Leishmaniasis: new approaches to disease control. BMJ 326(7385), 377–382.

de Moura, T.R., Novais, F.O., Oliveira, F., Clarencio, J., Noronha, A., Barral, A., Brodskyn, C., de Oliveira, C.I., 2005. Toward a novel experimental model of infection to study American cutaneous leishmaniasis caused by *Leishmania braziliensis*. Infection and Immunity 73, 5827–5834.

De Muylder, G., Daulouede, S., Lecordier, L., Uzureau, P., Morias, Y., Van Den Abbeele, J., Caljon, G., Herin, M., Holzmuller, P., Semballa, S., Courtois, P., Vanhamme, L., Stijlemans, B., De Baetselier, P., Barrett, M.P., Barlow, J.L., McKenzie, A.N., Barron, L., Wynn, T.A., Beschin, A., Vincendeau, P., Pays, E., 2013. A *Trypanosoma brucei* kinesin heavy chain promotes parasite growth by triggering host arginase activity. PLoS Pathogens 9, e1003731.

de Souza-Neto, S.M., Carneiro, C.M., Vieira, L.Q., Afonso, L.C., 2004. *Leishmania braziliensis*: partial control of experimental infection by interleukin-12 p40 deficient mice. Memorias do Instituto Oswaldo Cruz 99, 289–294.

De Trez, C., Katsandegwaza, B., Caljon, G., Magez, S., 2015. Experimental african trypanosome infection by needle passage or natural tsetse fly challenge Thwarts the development of collagen-induced arthritis in DBA/1 prone mice via an impairment of antigen specific B cell autoantibody titers. PLoS One 10, e0130431.

DeKrey, G.K., Lima, H.C., Titus, R.G., 1998. Analysis of the immune responses of mice to infection with *Leishmania braziliensis*. Infection and Immunity 66, 827–829.

Devine, D.V., Falk, R.J., Balber, A.E., 1986. Restriction of the alternative pathway of human complement by intact *Trypanosoma brucei* subsp. *gambiense*. Infection and Immunity 52, 223–229.

Donovan, C., 1903. On the possibility of the occurrence of trypanosomiasis in India. British Medical Journal 79.

Doyle, P.S., Dwyer, D.M., 1993. Leishmania: immunochemical comparison of the secretory (extracellular) acid phosphatases from various species. Experimental Parasitology 77, 435–444.

Drennan, M.B., Stijlemans, B., Van den Abbeele, J., Quesniaux, V.J., Barkhuizen, M., Brombacher, F., De Baetselier, P., Ryffel, B., Magez, S., 2005. The induction of a type 1 immune response following a *Trypanosoma brucei* infection is MyD88 dependent. Journal of Immunology 175, 2501–2509.

Dutra, W.O., de Faria, D.R., Lima Machado, P.R., Guimaraes, L.H., Schriefer, A., Carvalho, E., Gollob, K.J., 2011. Immunoregulatory and effector activities in human cutaneous and mucosal leishmaniasis: understanding mechanisms of pathology. Drug Development Research 72, 430–436.

Dwinger, R.H., Lamb, G., Murray, M., Hirumi, H., 1987. Dose and stage dependency for the development of local skin reactions caused by *Trypanosoma congolense* in goats. Acta Tropica 44, 303–314.

Dwinger, R.H., Rudin, W., Moloo, S.K., Murray, M., 1988. Development of *Trypanosoma congolense*, *T vivax* and *T brucei* in the skin reaction induced in goats by infected *Glossina morsitans centralis*: a light and electron microscopical study. Research in Veterinary Science 44, 154–163.

Ellis, J.A., Shapiro, S.Z., ole Moi-Yoi, O., Moloo, S.K., 1986. Lesions and saliva-specific antibody responses in rabbits with immediate and delayed hypersensitivity reactions to the bites of *Glossina morsitans centralis*. Veterinary Pathology 23, 661–667.

Emmer, B.T., Daniels, M.D., Taylor, J.M., Epting, C.L., Engman, D.M., 2010. Calflagin inhibition prolongs host survival and suppresses parasitemia in *Trypanosoma brucei* infection. Eukaryotic Cell 9, 934–942.

Engstler, M., Pfohl, T., Herminghaus, S., Boshart, M., Wiegertjes, G., Heddergott, N., Overath, P., 2007. Hydrodynamic flow-mediated protein sorting on the cell surface of trypanosomes. Cell 131, 505–515.

Fairbairn, H., Godfrey, D.G., 1957. The local reaction in man at the site of infection with *Trypanosoma rhodesiense*. Annals of Tropical Medicine and Parasitology 51, 464–470.

Falcao, S.A., Weinkopff, T., Hurrell, B.P., Celes, F.S., Curvelo, R.P., Prates, D.B., Barral, A., Borges, V.M., Tacchini-Cottier, F., de Oliveira, C.I., 2015. Exposure to *Leishmania braziliensis* triggers neutrophil activation and apoptosis. PLoS Neglected Tropical Diseases 9, e0003601.

Fan, X., Rudensky, A.Y., 2016. Hallmarks of tissue-resident lymphocytes. Cell 164, 1198–1211.

Faria, D.R., Souza, P.E., Duraes, F.V., Carvalho, E.M., Gollob, K.J., Machado, P.R., Dutra, W.O., 2009. Recruitment of CD8(+) T cells expressing granzyme A is associated with lesion progression in human cutaneous leishmaniasis. Parasite Immunology 31, 432–439.

Ferrante, A., Allison, A.C., 1983. Alternative pathway activation of complement by African trypanosomes lacking a glycoprotein coat. Parasite Immunology 5, 491–498.

Gaze, S.T., Dutra, W.O., Lessa, M., Lessa, H., Guimaraes, L.H., Jesus, A.R., Carvalho, L.P., Machado, P., Carvalho, E.M., Gollob, K.J., 2006. Mucosal leishmaniasis patients display an activated inflammatory T-cell phenotype associated with a nonbalanced monocyte population. Scandinavian Journal of Immunology 63, 70–78.

Gebhardt, T., Wakim, L.M., Eidsmo, L., Reading, P.C., Heath, W.R., Carbone, F.R., 2009. Memory T cells in nonlymphoid tissue that provide enhanced local immunity during infection with herpes simplex virus. Nature Immunology 10, 524–530.

Ghosh, A., Bandyopadhyay, K., Kole, L., Das, P.K., 1999. Isolation of a laminin-binding protein from the protozoan parasite *Leishmania donovani* that may mediate cell adhesion. The Biochemical Journal 337 (Pt 3), 551–558.

Gimblet, C., Loesche, M.A., Carvalho, L., Carvalho, E.M., Grice, E.A., Artis, D., Scott, P., 2015. IL-22 protects against tissue damage during cutaneous leishmaniasis. PLoS One 10, e0134698.

Glennie, N.D., Yeramilli, V.A., Beiting, D.P., Volk, S.W., Weaver, C.T., Scott, P., 2015. Skin-resident memory CD4+ T cells enhance protection against Leishmania major infection. The Journal of Experimental Medicine 212, 1405–1414.

Gombault, A., Baron, L., Couillin, I., 2012. ATP release and purinergic signaling in NLRP3 inflammasome activation. Frontiers in Immunology 3, 414.

Gomez-Rodriguez, J., Stijlemans, B., De Muylder, G., Korf, H., Brys, L., Berberof, M., Darji, A., Pays, E., De Baetselier, P., Beschin, A., 2009. Identification of a parasitic immunomodulatory protein triggering the development of suppressive M1 macrophages during African trypanosomiasis. The Journal of Infectious Diseases 200, 1849–1860.

Goncalves, R., Zhang, X., Cohen, H., Debrabant, A., Mosser, D.M., 2011. Platelet activation attracts a subpopulation of effector monocytes to sites of Leishmania major infection. The Journal of Experimental Medicine 208, 1253–1265.

Gonzalez-Lombana, C., Gimblet, C., Bacellar, O., Oliveira, W.W., Passos, S., Carvalho, L.P., Goldschmidt, M., Carvalho, E.M., Scott, P., 2013. IL-17 mediates immunopathology in the absence of IL-10 following Leishmania major infection. PLoS Pathogens 9, e1003243.

Gressner, O.A., Peredniene, I., Gressner, A.M., 2011. Connective tissue growth factor reacts as an IL-6/STAT3-regulated hepatic negative acute phase protein. World Journal of Gastroenterology 17, 151–163.

Guimaraes-Costa, A.B., Nascimento, M.T., Froment, G.S., Soares, R.P., Morgado, F.N., Conceicao-Silva, F., Saraiva, E.M., 2009. Leishmania amazonensis promastigotes induce and are killed by neutrophil extracellular traps. Proceedings of the National Academy of Sciences of the United States of America 106, 6748–6753.

Guimaraes-Costa, A.B., DeSouza-Vieira, T.S., Paletta-Silva, R., Freitas-Mesquita, A.L., Meyer-Fernandes, J.R., Saraiva, E.M., 2014. 3'-nucleotidase/nuclease activity allows *Leishmania* parasites to escape killing by neutrophil extracellular traps. Infection and Immunity 82, 1732–1740.

Gurunathan, S., Sacks, D.L., Brown, D.R., Reiner, S.L., Charest, H., Glaichenhaus, N., Seder, R.A., 1997. Vaccination with DNA encoding the immunodominant LACK parasite antigen confers protective immunity to mice infected with Leishmania major. The Journal of Experimental Medicine 186, 1137–1147.

Hall, J.P., Wang, H., Barry, J.D., 2013. Mosaic VSGs and the scale of *Trypanosoma brucei* antigenic variation. PLoS Pathogens 9, e1003502.

Handler, M.Z., Patel, P.A., Kapila, R., Al-Qubati, Y., Schwartz, R.A., 2015. Cutaneous and mucocutaneous leishmaniasis: differential diagnosis, diagnosis, histopathology, and management. Journal of the American Academy of Dermatology 73, 911–926 927–918.

Hasko, G., Kuhel, D.G., Nemeth, Z.H., Mabley, J.G., Stachlewitz, R.F., Virag, L., Lohinai, Z., Southan, G.J., Salzman, A.L., Szabo, C., 2000. Inosine inhibits inflammatory cytokine production by a posttranscriptional mechanism and protects against endotoxin-induced shock. Journal of Immunology 164, 1013–1019.

Hedley, L., Fink, D., Sparkes, D., Chiodini, P.L., 2016. African sleeping sickness. British Journal of Hospital Medicine (London) 77, C157–C160.

Hereld, D., Krakow, J.L., Bangs, J.D., Hart, G.W., Englund, P.T., 1986. A phospholipase C from *Trypanosoma brucei* which selectively cleaves the glycolipid on the variant surface glycoprotein. The Journal of Biological Chemistry 261, 13813–13819.

Hoare, C.A., 1938. Early discoveries regarding the parasite of oriental sore. Transactions of the Royal Society of Tropical Medicine and Hygiene 32, 67–92.

Hudson, K.M., Byner, C., Freeman, J., Terry, R.J., 1976. Immunodepression, high IgM levels and evasion of the immune response in murine trypanosomiasis. Nature 264, 256–258.

Hurrell, B.P., Schuster, S., Grun, E., Coutaz, M., Williams, R.A., Held, W., Malissen, B., Malissen, M., Yousefi, S., Simon, H.U., Muller, A.J., Tacchini-Cottier, F., 2015. Rapid sequestration of *Leishmania mexicana* by neutrophils contributes to the development of chronic lesion. PLoS Pathogens 11, e1004929.

Ilg, T., Montgomery, J., Stierhof, Y.D., Handman, E., 1999. Molecular cloning and characterization of a novel repeat-containing Leishmania major gene, ppg1, that encodes a membrane-associated form of proteophosphoglycan with a putative glycosylphosphatidylinositol anchor. The Journal of Biological Chemistry 274, 31410–31420.

Ishida, Y., Kondo, T., Takayasu, T., Iwakura, Y., Mukaida, N., 2004. The essential involvement of cross-talk between IFN-gamma and TGF-beta in the skin wound-healing process. Journal of Immunology 172, 1848–1855.

Izzo, A.A., Izzo, L.S., Kasimos, J., Majka, S., 2004. A matrix metalloproteinase inhibitor promotes granuloma formation during the early phase of *Mycobacterium tuberculosis* pulmonary infection. Tuberculosis 84, 387–396.

Jayakumar, A., Castilho, T.M., Park, E., Goldsmith-Pestana, K., Blackwell, J.M., McMahon-Pratt, D., 2011. TLR1/2 activation during heterologous prime-boost vaccination (DNA-MVA) enhances CD8+ T Cell responses providing protection against *Leishmania* (*Viannia*). PLoS Neglected Tropical Diseases 5, e1204.

Ji, J., Sun, J., Qi, H., Soong, L., 2002. Analysis of T helper cell responses during infection with Leishmania amazonensis. Am J Trop Med Hyg 66(4), 338–345.

Jiang, X., Clark, R.A., Liu, L., Wagers, A.J., Fuhlbrigge, R.C., Kupper, T.S., 2012. Skin infection generates non-migratory memory CD8+ T(RM) cells providing global skin immunity. Nature 483, 227–231.

Kamhawi, S., Belkaid, Y., Modi, G., Rowton, E., Sacks, D., 2000. Protection against cutaneous leishmaniasis resulting from bites of uninfected sand flies. Science 290, 1351–1354.

Karmakar, M., Katsnelson, M.A., Dubyak, G.R., Pearlman, E., 2016. Neutrophil P2X7 receptors mediate NLRP3 inflammasome-dependent IL-1beta secretion in response to ATP. Nature Communications 7, 10555.

Kaye, P., Scott, P., 2011. Leishmaniasis: complexity at the host-pathogen interface. Nature Reviews Microbiology 9, 604–615.

Kimblin, N., Peters, N., Debrabant, A., Secundino, N., Egen, J., Lawyer, P., Fay, M.P., Kamhawi, S., Sacks, D., 2008. Quantification of the infectious dose of Leishmania major transmitted to the skin by single sand flies. Proceedings of the National Academy of Sciences of the United States of America 105, 10125–10130.

Klass, B.R., Grobbelaar, A.O., Rolfe, K.J., 2009. Transforming growth factor beta1 signalling, wound healing and repair: a multifunctional cytokine with clinical implications for wound repair, a delicate balance. Postgraduate Medical Journal 85, 9–14.

Kling, J.C., Mack, M., Korner, H., 2013. The absence of CCR7 results in dysregulated monocyte migration and immunosuppression facilitating chronic cutaneous leishmaniasis. PLoS One 8, e79098.

Kolaczkowska, E., Kubes, P., 2013. Neutrophil recruitment and function in health and inflammation. Nature Reviews Immunology 13, 159–175.

Kronenberg, K., Brosch, S., Butsch, F., Tada, Y., Shibagaki, N., Udey, M.C., von Stebut, E., 2010. Vaccination with TAT-antigen fusion protein induces protective, CD8(+) T cell-mediated immunity against Leishmania major. The Journal of Investigative Dermatology 130, 2602–2610.

Leak, 1999. Tsetse Biology and Ecology: Their Role in the Epidemiology and Control of Trypanosomosis. Cabi Publishing, Wallingford, UK.

Lehane, M.J., 2005. The Biology of Blood-Sucking in Insects. Cambridge University Press.

Leishman, W.B., 1903. On the possibility of the occurrence of trypanosomiasis in India. British Medical of Journal 57, 1252.

Lejon, V., Mumba Ngoyi, D., Kestens, L., Boel, L., Barbe, B., Kande Betu, V., van Griensven, J., Bottieau, E., Muyembe Tamfum, J.J., Jacobs, J., Buscher, P., 2014. Gambiense human african trypanosomiasis and immunological memory: effect on phenotypic lymphocyte profiles and humoral immunity. PLoS Pathogens 10, e1003947.

Leppert, B.J., Mansfield, J.M., Paulnock, D.M., 2007. The soluble variant surface glycoprotein of African trypanosomes is recognized by a macrophage scavenger receptor and induces I kappa B alpha degradation independently of TRAF6-mediated TLR signaling. Journal of Immunology 179, 548–556.

Li, S., Aksoy, S., 2000. A family of genes with growth factor and adenosine deaminase similarity are preferentially expressed in the salivary glands of *Glossina m. morsitans*. Gene 252, 83–93.

Lira, R., Rosales-Encina, J.L., Arguello, C., 1997. Leishmania mexicana: binding of promastigotes to type I collagen. Experimental Parasitology 85, 149–157.

Llambrich, A., Zaballos, P., Terrasa, F., Torne, I., Puig, S., Malvehy, J., 2009. Dermoscopy of cutaneous leishmaniasis. The British Journal of Dermatology 160, 756–761.

Lucas, R., Magez, S., De Leys, R., Fransen, L., Scheerlinck, J.P., Rampelberg, M., Sablon, E., De Baetselier, P., 1994. Mapping the lectin-like activity of tumor necrosis factor. Science 263, 814–817.

Luckins, A.G., Gray, A.R., 1978. An extravascular site of development of *Trypanosoma congolense*. Nature 272, 613–614.

MacGregor, P., Szoor, B., Savill, N.J., Matthews, K.R., 2012. Trypanosomal immune evasion, chronicity and transmission: an elegant balancing act. Nature Reviews Microbiology 10, 431–438.

Magez, S., Geuskens, M., Beschin, A., del Favero, H., Verschueren, H., Lucas, R., Pays, E., de Baetselier, P., 1997. Specific uptake of tumor necrosis factor-alpha is involved in growth control of *Trypanosoma brucei*. The Journal of Cell Biology 137, 715–727.

Magez, S., Stijlemans, B., Radwanska, M., Pays, E., Ferguson, M.A., De Baetselier, P., 1998. The glycosyl-inositol-phosphate and dimyristoylglycerol moieties of the glycosylphosphatidylinositol anchor of the trypanosome variant-specific surface glycoprotein are distinct macrophage-activating factors. Journal of Immunology 160, 1949–1956.

Magez, S., Schwegmann, A., Atkinson, R., Claes, F., Drennan, M., De Baetselier, P., Brombacher, F., 2008. The role of B-cells and IgM antibodies in parasitemia, anemia, and VSG switching in *Trypanosoma brucei*-infected mice. PLoS Pathogens 4, e1000122.

Magez, S., Caljon, G., Tran, T., Stijlemans, B., Radwanska, M., 2010. Current status of vaccination against African trypanosomiasis. Parasitology 137, 2017–2027.

Majka, S.M., Kasimos, J., Izzo, L., Izzo, A.A., 2002. Cryptococcus neoformans pulmonary granuloma formation is associated with matrix metalloproteinase-2 expression. Medical Mycology 40, 323–328.

Malvy, D., Chappuis, F., 2011. Sleeping sickness. Clinical Microbiology and Infection 17, 986–995.

Mant, M.J., Parker, K.R., 1981. Two platelet aggregation inhibitors in tsetse (*Glossina*) saliva with studies of roles of thrombin and citrate in in vitro platelet aggregation. British Journal of Haematology 48, 601–608.

Maroof, A., Brown, N., Smith, B., Hodgkinson, M.R., Maxwell, A., Losch, F.O., Fritz, U., Walden, P., Lacey, C.N., Smith, D.F., Aebischer, T., Kaye, P.M., 2012. Therapeutic vaccination with recombinant adenovirus reduces splenic parasite burden in experimental visceral leishmaniasis. The Journal of Infectious Diseases 205, 853–863.

Massague, J., 1998. TGF-beta signal transduction. Annual Review of Biochemistry 67, 753–791.

Matetovici, I., Caljon, G., Van Den Abbeele, J., 2016. Tsetse fly tolerance to *T. brucei* infection: transcriptome analysis of trypanosome-associated changes in the tsetse fly salivary gland. BMC Genomics 17, 971.

McConville, M.J., Ferguson, M.A., 1993. The structure, biosynthesis and function of glycosylated phosphatidylinositols in the parasitic protozoa and higher eukaryotes. The Biochemical Journal 294 (Pt 2), 305–324.

McGee, H.M., Schmidt, B.A., Booth, C.J., Yancopoulos, G.D., Valenzuela, D.M., Murphy, A.J., Stevens, S., Flavell, R.A., Horsley, V., 2013. IL-22 promotes fibroblast-mediated wound repair in the skin. The Journal of Investigative Dermatology 133, 1321–1329.

McGwire, B.S., Chang, K.P., Engman, D.M., 2003. Migration through the extracellular matrix by the parasitic protozoan Leishmania is enhanced by surface metalloprotease gp63. Infection and Immunity 71, 1008–1010.

Mehlert, A., Bond, C.S., Ferguson, M.A., 2002. The glycoforms of a *Trypanosoma brucei* variant surface glycoprotein and molecular modeling of a glycosylated surface coat. Glycobiology 12, 607–612.

Montenegro, J., 1926. Cutaneous reaction in leishmaniasis. Archives of Dermatology and Syphilology 13, 187–194.

Mony, B.M., Matthews, K.R., 2015. Assembling the components of the quorum sensing pathway in African trypanosomes. Molecular Microbiology 96, 220–232.

Mony, B.M., MacGregor, P., Ivens, A., Rojas, F., Cowton, A., Young, J., Horn, D., Matthews, K., 2014. Genome-wide dissection of the quorum sensing signalling pathway in *Trypanosoma brucei*. Nature 505, 681–685.

Mugnier, M.R., Cross, G.A., Papavasiliou, F.N., 2015. The in vivo dynamics of antigenic variation in *Trypanosoma brucei*. Science 347, 1470–1473.

Mukbel, R.M., Patten Jr., C., Gibson, K., Ghosh, M., Petersen, C., Jones, D.E., 2007. Macrophage killing of *Leishmania amazonensis* amastigotes requires both nitric oxide and superoxide. The American Journal of Tropical Medicine and Hygiene 76, 669–675.

Muniz, A.C., Bacellar, O., Lago, E.L., Carvalho, A.M., Carneiro, P.P., Guimaraes, L.H., Rocha, P.N., Carvalho, L.P., Glesby, M., Carvalho, E.M., 2016. Immunologic markers of protection in *Leishmania* (*Viannia*) *braziliensis* infection: a 5-year cohort study. The Journal of Infectious Diseases 214, 570–576.

Mwangi, D.M., Hopkins, J., Luckins, A.G., 1990. Cellular phenotypes in *Trypanosoma congolense* infected sheep: the local skin reaction. Parasite Immunology 12, 647–658.

Mwangi, D.M., Hopkins, J., Luckins, A.G., 1995. *Trypanosoma congolense* infection in sheep: ultrastructural changes in the skin prior to development of local skin reactions. Veterinary Parasitology 60, 45–52.

Naessens, J., Mwangi, D.M., Buza, J., Moloo, S.K., 2003. Local skin reaction (chancre) induced following inoculation of metacyclic trypanosomes in cattle by tsetse flies is dependent on CD4 T lymphocytes. Parasite Immunology 25, 413–419.

Naessens, J., 2006. Bovine trypanotolerance: a natural ability to prevent severe anaemia and haemophagocytic syndrome? International Journal for Parasitology 36, 521–528.

Nakerakanti, S.S., Kapanadze, B., Yamasaki, M., Markiewicz, M., Trojanowska, M., 2006. Fli1 and Ets1 have distinct roles in connective tissue growth factor/CCN2 gene regulation and induction of the profibrotic gene program. The Journal of Biological Chemistry 281, 25259–25269.

Namangala, B., Brys, L., Magez, S., De Baetselier, P., Beschin, A., 2000. *Trypanosoma brucei brucei* infection impairs MHC class II antigen presentation capacity of macrophages. Parasite Immunology 22, 361–370.

Namangala, B., De Baetselier, P., Noel, W., Brys, L., Beschin, A., 2001. Alternative versus classical macrophage activation during experimental African trypanosomosis. Journal of Leukocyte Biology 69, 387–396.

Nicholas, M.N., Jeschke, M.G., Amini-Nik, S., 2016a. Cellularized bilayer pullulan-gelatin hydrogel for skin regeneration. Tissue Engineering Part A 22, 754–764.

Nicholas, M.N., Jeschke, M.G., Amini-Nik, S., 2016b. Methodologies in creating skin substitutes. Cellular and Molecular Life Sciences 73, 3453–3472.

Nicoletti, G., Brenta, F., Bleve, M., Pellegatta, T., Malovini, A., Faga, A., Perugini, P., 2015. Long-term in vivo assessment of bioengineered skin substitutes: a clinical study. Journal of Tissue Engineering and Regenerative Medicine 9, 460–468.

Novais, F.O., Santiago, R.C., Bafica, A., Khouri, R., Afonso, L., Borges, V.M., Brodskyn, C., Barral-Netto, M., Barral, A., de Oliveira, C.I., 2009. Neutrophils and macrophages cooperate in host resistance against *Leishmania braziliensis* infection. Journal of Immunology 183, 8088–8098.

Novais, F.O., Nguyen, B.T., Beiting, D.P., Carvalho, L.P., Glennie, N.D., Passos, S., Carvalho, E.M., Scott, P., 2014. Human classical monocytes control the intracellular stage of *Leishmania braziliensis* by reactive oxygen species. The Journal of Infectious Diseases 209, 1288–1296.

Obishakin, E., de Trez, C., Magez, S., 2014. Chronic *Trypanosoma congolense* infections in mice cause a sustained disruption of the B-cell homeostasis in the bone marrow and spleen. Parasite Immunology 36, 187–198.

Oghumu, S., Stock, J.C., Varikuti, S., Dong, R., Terrazas, C., Edwards, J.A., Rappleye, C.A., Holovatyk, A., Sharpe, A., Satoskar, A.R., 2015. Transgenic expression of CXCR3 on T cells enhances susceptibility to cutaneous Leishmania major infection by inhibiting monocyte maturation and promoting a Th2 response. Infection and Immunity 83, 67–76.

Passos, S., Carvalho, L.P., Costa, R.S., Campos, T.M., Novais, F.O., Magalhaes, A., Machado, P.R., Beiting, D., Mosser, D., Carvalho, E.M., Scott, P., 2015. Intermediate monocytes contribute to pathologic immune response in *Leishmania braziliensis* infections. The Journal of Infectious Diseases 211, 274–282.

Pastar, I., Stojadinovic, O., Yin, N.C., Ramirez, H., Nusbaum, A.G., Sawaya, A., Patel, S.B., Khalid, L., Isseroff, R.R., Tomic-Canic, M., 2014. Epithelialization in wound healing: a comprehensive review. Advances in Wound Care 3, 445–464.

Paulnock, D.M., Coller, S.P., 2001. Analysis of macrophage activation in African trypanosomiasis. Journal of Leukocyte Biology 69, 685–690.

Paulnock, D.M., Freeman, B.E., Mansfield, J.M., 2010. Modulation of innate immunity by African trypanosomes. Parasitology 137, 2051–2063.

Peters, N.C., Egen, J.G., Secundino, N., Debrabant, A., Kimblin, N., Kamhawi, S., Lawyer, P., Fay, M.P., Germain, R.N., Sacks, D., 2008. In vivo imaging reveals an essential role for neutrophils in leishmaniasis transmitted by sand flies. Science 321, 970–974.

Petropolis, D.B., Rodrigues, J.C., Viana, N.B., Pontes, B., Pereira, C.F., Silva-Filho, F.C., 2014. Leishmania amazonensis promastigotes in 3D Collagen I culture: an in vitro physiological environment for the study of extracellular matrix and host cell interactions. PeerJ 2, e317.

Pickert, G., Neufert, C., Leppkes, M., Zheng, Y., Wittkopf, N., Warntjen, M., Lehr, H.A., Hirth, S., Weigmann, B., Wirtz, S., Ouyang, W., Neurath, M.F., Becker, C., 2009. STAT3 links IL-22 signaling in intestinal epithelial cells to mucosal wound healing. The Journal of Experimental Medicine 206, 1465–1472.

Price, N.M., Gilman, R.H., Uddin, J., Recavarren, S., Friedland, J.S., 2003. Unopposed matrix metalloproteinase-9 expression in human tuberculous granuloma and the role of TNF-alpha-dependent monocyte networks. Journal of Immunology 171, 5579–5586.

Qi, H., Ji, J., Wanasen, N., Soong, L., 2004. Enhanced replication of *Leishmania amazonensis* amastigotes in gamma interferon-stimulated murine macrophages: implications for the pathogenesis of cutaneous leishmaniasis. Infection and Immunity 72, 988–995.

Radwanska, M., Magez, S., Michel, A., Stijlemans, B., Geuskens, M., Pays, E., 2000. Comparative analysis of antibody responses against HSP60, invariant surface glycoprotein 70, and variant surface glycoprotein reveals a complex antigen-specific pattern of immunoglobulin isotype switching during infection by *Trypanosoma brucei*. Infection and Immunity 68, 848–860.

Radwanska, M., Guirnalda, P., De Trez, C., Ryffel, B., Black, S., Magez, S., 2008. Trypanosomiasis-induced B cell apoptosis results in loss of protective anti-parasite antibody responses and abolishment of vaccine-induced memory responses. PLoS Pathogens 4, e1000078.

Ramey-Butler, K., Ullu, E., Kolev, N.G., Tschudi, C., 2015. Synchronous expression of individual metacyclic variant surface glycoprotein genes in *Trypanosoma brucei*. Molecular and Biochemical Parasitology 200, 1–4.

Ready, P.D., 2013. Biology of phlebotomine sand flies as vectors of disease agents. Annual Review of Entomology 58, 227–250.

Ribeiro-Gomes, F.L., Otero, A.C., Gomes, N.A., Moniz-De-Souza, M.C., Cysne-Finkelstein, L., Arnholdt, A.C., Calich, V.L., Coutinho, S.G., Lopes, M.F., DosReis, G.A., 2004. Macrophage interactions with neutrophils regulate Leishmania major infection. Journal of Immunology 172, 4454–4462.

Ribeiro-Gomes, F.L., Peters, N.C., Debrabant, A., Sacks, D.L., 2012. Efficient capture of infected neutrophils by dendritic cells in the skin inhibits the early anti-leishmania response. PLoS Pathogens 8, e1002536.

Ribeiro-Gomes, F.L., Romano, A., Lee, S., Roffe, E., Peters, N.C., Debrabant, A., Sacks, D., 2015. Apoptotic cell clearance of Leishmania major-infected neutrophils by dendritic cells inhibits CD8(+) T-cell priming in vitro by Mer tyrosine kinase-dependent signaling. Cell Death and Disease 6, e2018.

Rico, E., Rojas, F., Mony, B.M., Szoor, B., Macgregor, P., Matthews, K.R., 2013. Bloodstream form pre-adaptation to the tsetse fly in *Trypanosoma brucei*. Frontiers in Cellular and Infection Microbiology 3, 78.

Roberts, C.J., Gray, M.A., Gray, A.R., 1969. Local skin reactions in cattle at the site of infection with *Trypanosoma congolense* by *Glossina morsitans* and *G. tachinoides*. Transactions of the Royal Society of Tropical Medicine and Hygiene 63, 620–624.

Rocha, F.J., Schleicher, U., Mattner, J., Alber, G., Bogdan, C., 2007. Cytokines, signaling pathways, and effector molecules required for the control of *Leishmania (Viannia) braziliensis* in mice. Infection and Immunity 75, 3823–3832.

Rochael, N.C., Guimaraes-Costa, A.B., Nascimento, M.T., DeSouza-Vieira, T.S., Oliveira, M.P., Garcia e Souza, L.F., Oliveira, M.F., Saraiva, E.M., 2015. Classical ROS-dependent and early/rapid ROS-independent release of neutrophil extracellular traps triggered by Leishmania parasites. Scientific Reports 5, 18302.

Rogers, M.E., Bates, P.A., 2007. Leishmania manipulation of sand fly feeding behavior results in enhanced transmission. PLoS Pathogens 3, e91.

Rogers, M.E., Chance, M.L., Bates, P.A., 2002. The role of promastigote secretory gel in the origin and transmission of the infective stage of *Leishmania mexicana* by the sandfly *Lutzomyia longipalpis*. Parasitology 124, 495–507.

Rogers, M.E., Ilg, T., Nikolaev, A.V., Ferguson, M.A., Bates, P.A., 2004. Transmission of cutaneous leishmaniasis by sand flies is enhanced by regurgitation of fPPG. Nature 430, 463–467.

Rolin, S., Hanocq-Quertier, J., Paturiaux-Hanocq, F., Nolan, D., Salmon, D., Webb, H., Carrington, M., Voorheis, P., Pays, E., 1996. Simultaneous but independent activation of adenylate cyclase and glycosylphosphatidylinositol-phospholipase C under stress conditions in *Trypanosoma brucei*. The Journal of Biological Chemistry 271, 10844–10852.

Roma, E.H., Macedo, J.P., Goes, G.R., Goncalves, J.L., Castro, W., Cisalpino, D., Vieira, L.Q., 2016. Impact of reactive oxygen species (ROS) on the control of parasite loads and inflammation in *Leishmania amazonensis* infection. Parasites and Vectors 9, 193.

Rotureau, B., Subota, I., Buisson, J., Bastin, P., 2012. A new asymmetric division contributes to the continuous production of infective trypanosomes in the tsetse fly. Development (Cambridge, England) 139, 1842–1850.

Rousseau, D., Demartino, S., Ferrua, B., Michiels, J.F., Anjuere, F., Fragaki, K., Le Fichoux, Y., Kubar, J., 2001. In vivo involvement of polymorphonuclear neutrophils in *Leishmania infantum* infection. BMC Microbiology 1, 17.

Saarialho-Kere, U.K., Kovacs, S.O., Pentland, A.P., Olerud, J.E., Welgus, H.G., Parks, W.C., 1993. Cell-matrix interactions modulate interstitial collagenase expression by human keratinocytes actively involved in wound healing. The Journal of Clinical Investigation 92, 2858–2866.

Sacks, D., Noben-Trauth, N., 2002. The immunology of susceptibility and resistance to Leishmania major in mice. Nature Reviews Immunology 2, 845–858.

Sacks, D.L., Modi, G., Rowton, E., Spath, G., Epstein, L., Turco, S.J., Beverley, S.M., 2000. The role of phosphoglycans in Leishmania-sand fly interactions. Proceedings of the National Academy of Sciences of the United States of America 97, 406–411.

Sacramento, L., Trevelin, S.C., Nascimento, M.S., Lima-Junior, D.S., Costa, D.L., Almeida, R.P., Cunha, F.Q., Silva, J.S., Carregaro, V., 2015. Toll-like receptor 9 signaling in dendritic cells regulates neutrophil recruitment to inflammatory foci following *Leishmania infantum* infection. Infection and Immunity 83, 4604–4616.

Sakthianandeswaren, A., Elso, C.M., Simpson, K., Curtis, J.M., Kumar, B., Speed, T.P., Handman, E., Foote, S.J., 2005. The wound repair response controls outcome to cutaneous leishmaniasis. Proceedings of the National Academy of Sciences of the United States of America 102, 15551–15556.

Sakthianandeswaren, A., Foote, S.J., Handman, E., 2009. The role of host genetics in leishmaniasis. Trends in Parasitology 25, 383–391.

Sakthianandeswaren, A., Curtis, J.M., Elso, C., Kumar, B., Baldwin, T.M., Lopaticki, S., Kedzierski, L., Smyth, G.K., Foote, S.J., Handman, E., 2010. Fine mapping of Leishmania major susceptibility Locus lmr2 and evidence of a role for Fli1 in disease and wound healing. Infection and Immunity 78, 2734–2744.

Salmon, D., Vanwalleghem, G., Morias, Y., Denoeud, J., Krumbholz, C., Lhomme, F., Bachmaier, S., Kador, M., Gossmann, J., Dias, F.B., De Muylder, G., Uzureau, P., Magez, S., Moser, M., De Baetselier, P., Van Den Abbeele, J., Beschin, A., Boshart, M., Pays, E., 2012. Adenylate cyclases of *Trypanosoma brucei* inhibit the innate immune response of the host. Science 337, 463–466.

Santos Cda, S., Boaventura, V., Ribeiro Cardoso, C., Tavares, N., Lordelo, M.J., Noronha, A., Costa, J., Borges, V.M., de Oliveira, C.I., Van Weyenbergh, J., Barral, A., Barral-Netto, M., Brodskyn, C.I., 2013. CD8(+) granzyme B(+)-mediated tissue injury vs. CD4(+)IFNgamma(+)-mediated parasite killing in human cutaneous leishmaniasis. The Journal of Investigative Dermatology 133, 1533–1540.

Sarkar, A., Aga, E., Bussmeyer, U., Bhattacharyya, A., Moller, S., Hellberg, L., Behnen, M., Solbach, W., Laskay, T., 2013. Infection of neutrophil granulocytes with *Leishmania major* activates ERK 1/2 and modulates multiple apoptotic pathways to inhibit apoptosis. Medical Microbiology and Immunology 202, 25–35.

Sato, S., Zhang, X.K., 2014. The friend leukaemia virus integration 1 (Fli-1) transcription factor affects lupus nephritis development by regulating inflammatory cell infiltration into the kidney. Clinical and Experimental Immunology 177, 102–109.

Schenkel, J.M., Masopust, D., 2014. Tissue-resident memory T cells. Immunity 41, 886–897.

Schleifer, K.W., Mansfield, J.M., 1993. Suppressor macrophages in African trypanosomiasis inhibit T cell proliferative responses by nitric oxide and prostaglandins. Journal of Immunology 151, 5492–5503.

Schonian, G., Mauricio, I., Cupolillo, E., 2010. Is it time to revise the nomenclature of Leishmania? Trends in Parasitology 26, 466–469.

Scott, P., Novais, F.O., 2016. Cutaneous leishmaniasis: immune responses in protection and pathogenesis. Nature Reviews Immunology 16, 581–592.

Secundino, N.F., de Freitas, V.C., Monteiro, C.C., Pires, A.C., David, B.A., Pimenta, P.F., 2012. The transmission of *Leishmania infantum* chagasi by the bite of the *Lutzomyia longipalpis* to two different vertebrates. Parasites and Vectors 5, 20.

Shah, M., Foreman, D.M., Ferguson, M.W., 1995. Neutralisation of TGF-beta 1 and TGF-beta 2 or exogenous addition of TGF-beta 3 to cutaneous rat wounds reduces scarring. Journal of Cell Science 108 (Pt 3), 985–1002.

Shi, Y., Massague, J., 2003. Mechanisms of TGF-beta signaling from cell membrane to the nucleus. Cell 113, 685–700.

Shimogawa, M.M., Saada, E.A., Vashisht, A.A., Barshop, W.D., Wohlschlegel, J.A., Hill, K.L., 2015. Cell surface proteomics provides insight into stage-specific remodeling of the host-parasite interface in *Trypanosoma Brucei*. Molecular and Cellular Proteomics 14, 1977–1988.

Shoda, L.K., Kegerreis, K.A., Suarez, C.E., Roditi, I., Corral, R.S., Bertot, G.M., Norimine, J., Brown, W.C., 2001. DNA from protozoan parasites *Babesia bovis, Trypanosoma cruzi*, and *T. brucei* is mitogenic for B lymphocytes and stimulates macrophage expression of interleukin-12, tumor necrosis factor alpha, and nitric oxide. Infection and Immunity 69, 2162–2171.

Sileghem, M., Hamers, R., De Baetselier, P., 1986. Active suppression of interleukin 2 secretion in mice infected with *Trypanosoma brucei* AnTat 1.1.E. Parasite Immunology 8, 641–649.

Sileghem, M., Hamers, R., De Baetselier, P., 1987. Experimental *Trypanosoma brucei* infections selectively suppress both interleukin 2 production and interleukin 2 receptor expression. European Journal of Immunology 17, 1417–1421.

Simarro, P.P., Cecchi, G., Franco, J.R., Paone, M., Diarra, A., Priotto, G., Mattioli, R.C., Jannin, J.G., 2015. Monitoring the progress towards the elimination of gambiense human african trypanosomiasis. PLoS Neglected Tropical Diseases 9, e0003785.

Singh, K.P., Gerard, H.C., Hudson, A.P., Boros, D.L., 2004. Dynamics of collagen, MMP and TIMP gene expression during the granulomatous, fibrotic process induced by *Schistosoma mansoni* eggs. Annals of Tropical Medicine and Parasitology 98, 581–593.

Smith, C.A., Stauber, F., Waters, C., Alway, S.E., Stauber, W.T., 2007. Transforming growth factor-beta following skeletal muscle strain injury in rats. Journal of Applied Physiology 102, 755–761.

Sousa, L.M., Carneiro, M.B., Resende, M.E., Martins, L.S., Dos Santos, L.M., Vaz, L.G., Mello, P.S., Mosser, D.M., Oliveira, M.A., Vieira, L.Q., 2014. Neutrophils have a protective role during early stages of *Leishmania amazonensis* infection in BALB/c mice. Parasite Immunology 36, 13–31.

Stager, S., Rafati, S., 2012. CD8(+) T cells in leishmania infections: friends or foes? Frontiers in Immunology 3, 5.

Stijlemans, B., Conrath, K., Cortez-Retamozo, V., Van Xong, H., Wyns, L., Senter, P., Revets, H., De Baetselier, P., Muyldermans, S., Magez, S., 2004. Efficient targeting of conserved cryptic epitopes of infectious agents by single domain antibodies. African trypanosomes as paradigm. The Journal of Biological Chemistry 279, 1256–1261.

Stijlemans, B., Guilliams, M., Raes, G., Beschin, A., Magez, S., De Baetselier, P., 2007. African trypanosomosis: from immune escape and immunopathology to immune intervention. Veterinary Parasitology 148, 3–13.

Suzuki, E., Tanaka, A.K., Toledo, M.S., Takahashi, H.K., Straus, A.H., 2002. Role of beta-D-galactofuranose in Leishmania major macrophage invasion. Infection and Immunity 70, 6592–6596.

Tavares, N.M., Araujo-Santos, T., Afonso, L., Nogueira, P.M., Lopes, U.G., Soares, R.P., Bozza, P.T., Bandeira-Melo, C., Borges, V.M., Brodskyn, C., 2014. Understanding the mechanisms controlling *Leishmania amazonensis* infection in vitro: the role of LTB4 derived from human neutrophils. The Journal of Infectious Diseases 210, 656–666.

Tavares, N., Afonso, L., Suarez, M., Ampuero, M., Prates, D.B., Araujo-Santos, T., Barral-Netto, M., DosReis, G.A., Borges, V.M., Brodskyn, C., 2016. Degranulating neutrophils promote leukotriene B4 production by infected macrophages to kill *Leishmania amazonensis* parasites. Journal of Immunology 196, 1865–1873.

Teijaro, J.R., Turner, D., Pham, Q., Wherry, E.J., Lefrancois, L., Farber, D.L., 2011. Cutting edge: tissue-retentive lung memory CD4 T cells mediate optimal protection to respiratory virus infection. Journal of Immunology 187, 5510–5514.

Teixeira, C., Gomes, R., Oliveira, F., Meneses, C., Gilmore, D.C., Elnaiem, D.E., Valenzuela, J.G., Kamhawi, S., 2014. Characterization of the early inflammatory infiltrate at the feeding site of infected sand flies in mice protected from vector-transmitted Leishmania major by exposure to uninfected bites. PLoS Neglected Tropical Diseases 8, e2781.

Telleria, E.L., Benoit, J.B., Zhao, X., Savage, A.F., Regmi, S., Alves e Silva, T.L., O'Neill, M., Aksoy, S., 2014. Insights into the trypanosome-host interactions revealed through transcriptomic analysis of parasitized tsetse fly salivary glands. PLoS Neglected Tropical Diseases 8, e2649.

Terabe, M., Kuramochi, T., Hatabu, T., Ito, M., Ueyama, Y., Katakura, K., Kawazu, S., Onodera, T., Matsumoto, Y., 1999. Non-ulcerative cutaneous lesion in immunodeficient mice with *Leishmania amazonensis* infection. Parasitology International 48, 47–53.

Thaler, R., Agsten, M., Spitzer, S., Paschalis, E.P., Karlic, H., Klaushofer, K., Varga, F., 2011. Homocysteine suppresses the expression of the collagen cross-linker lysyl oxidase involving IL-6, Fli1, and epigenetic DNA methylation. The Journal of Biological Chemistry 286, 5578–5588.

Thalhofer, C.J., Chen, Y., Sudan, B., Love-Homan, L., Wilson, M.E., 2011. Leukocytes infiltrate the skin and draining lymph nodes in response to the protozoan *Leishmania infantum* chagasi. Infection and Immunity 79, 108–117.

Tilley, S.L., Wagoner, V.A., Salvatore, C.A., Jacobson, M.A., Koller, B.H., 2000. Adenosine and inosine increase cutaneous vasopermeability by activating A(3) receptors on mast cells. The Journal of Clinical Investigation 105, 361–367.

Titus, R.G., Bishop, J.V., Mejia, J.S., 2006. The immunomodulatory factors of arthropod saliva and the potential for these factors to serve as vaccine targets to prevent pathogen transmission. Parasite Immunology 28, 131–141.

Trindade, S., Rijo-Ferreira, F., Carvalho, T., Pinto-Neves, D., Guegan, F., Aresta-Branco, F., Bento, F., Young, S.A., Pinto, A., Van Den Abbeele, J., Ribeiro, R.M., Dias, S., Smith, T.K., Figueiredo, L.M., 2016. *Trypanosoma brucei* parasites occupy and functionally adapt to the adipose tissue in mice. Cell Host and Microbe 19, 837–848.

Truong, A.H., Ben-David, Y., 2000. The role of Fli-1 in normal cell function and malignant transformation. Oncogene 19, 6482–6489.

Tsujimoto, H., Kotsyfakis, M., Francischetti, I.M., Eum, J.H., Strand, M.R., Champagne, D.E., 2012. Simukunin from the salivary glands of the black fly *Simulium vittatum* inhibits enzymes that regulate clotting and inflammatory responses. PLoS One 7, e29964.

Uzonna, J.E., Joyce, K.L., Scott, P., 2004. Low dose Leishmania major promotes a transient T helper cell type 2 response that is down-regulated by interferon gamma-producing CD8+ T cells. The Journal of Experimental Medicine 199, 1559–1566.

Vaalamo, M., Kariniemi, A.L., Shapiro, S.D., Saarialho-Kere, U., 1999. Enhanced expression of human metalloelastase (MMP-12) in cutaneous granulomas and macrophage migration. The Journal of Investigative Dermatology 112, 499–505.

Van Belle, A.B., de Heusch, M., Lemaire, M.M., Hendrickx, E., Warnier, G., Dunussi-Joannopoulos, K., Fouser, L.A., Renauld, J.C., Dumoutier, L., 2012. IL-22 is required for imiquimod-induced psoriasiform skin inflammation in mice. Journal of Immunology 188, 462–469.

Van Den Abbeele, J., Caljon, G., Dierick, J.F., Moens, L., De Ridder, K., Coosemans, M., 2007. The *Glossina morsitans* tsetse fly saliva: general characteristics and identification of novel salivary proteins. Insect Biochemistry and Molecular Biology 37, 1075–1085.

Van Den Abbeele, J., Caljon, G., De Ridder, K., De Baetselier, P., Coosemans, M., 2010. *Trypanosoma brucei* modifies the tsetse salivary composition, altering the fly feeding behavior that favors parasite transmission. PLoS Pathogens 6, e1000926.

Visse, R., Nagase, H., 2003. Matrix metalloproteinases and tissue inhibitors of metalloproteinases: structure, function, and biochemistry. Circulation Research 92, 827–839.

Wakim, L.M., Woodward-Davis, A., Bevan, M.J., 2010. Memory T cells persisting within the brain after local infection show functional adaptations to their tissue of residence. Proceedings of the National Academy of Sciences of the United States of America 107, 17872–17879.

Wamwiri, F.N., Changasi, R.E., 2016. Tsetse flies (*Glossina*) as vectors of human african trypanosomiasis: a review. International Journal of Biomedical Research 2016, 6201350.

Wei, G., Bull, H., Zhou, X., Tabel, H., 2011. Intradermal infections of mice by low numbers of African trypanosomes are controlled by innate resistance but enhance susceptibility to reinfection. The Journal of Infectious Diseases 203, 418–429.

Wiese, M., 1998. A mitogen-activated protein (MAP) kinase homologue of *Leishmania mexicana* is essential for parasite survival in the infected host. The EMBO Journal 17, 2619–2628.

Wolbach, S.B., Binger, C.A., 1912. A contribution to the parasitology of trypanosomiasis. The Journal of Medical Research 27, 83–108 113.

Wrana, J.L., Attisano, L., 2000. The Smad pathway. Cytokine and Growth Factor Reviews 11, 5–13.

Wu, L.L., Cox, A., Roe, C.J., Dziadek, M., Cooper, M.E., Gilbert, R.E., 1997. Transforming growth factor beta 1 and renal injury following subtotal nephrectomy in the rat: role of the renin-angiotensin system. Kidney International 51, 1553–1567.

Xiao, R., Liu, F.Y., Luo, J.Y., Yang, X.J., Wen, H.Q., Su, Y.W., Yan, K.L., Li, Y.P., Liang, Y.S., 2006. Effect of small interfering RNA on the expression of connective tissue growth factor and type I and III collagen in skin fibroblasts of patients with systemic sclerosis. The British Journal of Dermatology 155, 1145–1153.

Yagi, K., Goto, D., Hamamoto, T., Takenoshita, S., Kato, M., Miyazono, K., 1999. Alternatively spliced variant of Smad2 lacking exon 3. Comparison with wild-type Smad2 and Smad3. The Journal of Biological Chemistry 274, 703–709.

Yang, X.O., Panopoulos, A.D., Nurieva, R., Chang, S.H., Wang, D., Watowich, S.S., Dong, C., 2007. STAT3 regulates cytokine-mediated generation of inflammatory helper T cells. The Journal of Biological Chemistry 282, 9358–9363.

Yucel, S., Ozcan, D., Seckin, D., Allahverdiyev, A.M., Kayaselcuk, F., Haberal, M., 2013. Visceral leishmaniasis with cutaneous dissemination in a renal transplant recipient. European Journal of Dermatology 23, 892–893.

Zhang, X., Liu, Q., Wang, J., Li, G., Weiland, M., Yu, F.S., Mi, Q.S., Gu, J., Zhou, L., 2016. TIM-4 is differentially expressed in the distinct subsets of dendritic cells in skin and skin-draining lymph nodes and controls skin Langerhans cell homeostasis. Oncotarget 7, 37498–37512.

Zhao, X., Alves e Silva, T.L., Cronin, L., Savage, A.F., O'Neill, M., Nerima, B., Okedi, L.M., Aksoy, S., 2015. Immunogenicity and serological cross-reactivity of saliva proteins among different tsetse species. PLoS Neglected Tropical Diseases 9, e0004038.

Zimmermann, H., Subota, I., Batram, C., Kramer, S., Janzen, C.J., Jones, N.G., Engstler, M., 2017. A quorum sensing-independent path to stumpy development in *Trypanosoma brucei*. PLoS Pathogens 13, e1006324.

Chapter 7

Skin and Other Pathogens: Malaria and Plague

INTRODUCTION

Insects are vectors of parasites, viruses, and bacteria. While numerous parasites and viruses are transmitted by insects, relatively few bacterial pathogens are insect-borne. In Chapter 6, we focused on flagellate parasites and in Chapter 8 on vector-borne viruses. In this chapter, we describe the latest advances in research on malaria, caused by the apicomplexan protist *Plasmodium* and transmitted by anopheline mosquitoes, and plague, caused by the gram-negative bacterium *Yersinia pestis* and transmitted by fleas, with an emphasis on the role of skin innate immunity.

Because of the high impact of malaria on human health, with it causing ~half a million deaths annually worldwide, it has been the focus of a tremendous amount of research since its discovery. Despite the progress that has been made, currently there is no effective vaccine available. Recent studies of the skin stages of malaria have improved our understanding of the immune response to the parasite and will help guide future vaccine development. Plague has been known to mankind for thousands of years and is responsible for hundreds of millions of deaths over that time. However, compared to malaria, relatively little is known about the fate of *Y. pestis* in the skin and potential effect of flea transmission plague pathogenesis. Improvements in public health and the advent of antibiotics have greatly reduced the threat of plague, but *Y. pestis* causes sporadic cases in countries around the world where it remains endemic in wild rodent populations, with significant outbreaks occurring in Central Africa and Madagascar. Additionally, because *Y. pestis* is highly virulent and has been developed and used as a biological weapon in the past, it continues to be a pathogen of great concern from a biodefense perspective.

In this chapter, we present the current status of malaria and plague research, with an emphasis on the roles of the arthropod vectors in transmission and pathogenesis. For both, the role of the skin in the persistence of the pathogen constitutes a key element in the process of transmission and development of host immunity.

Skin and Arthropod Vectors. https://doi.org/10.1016/B978-0-12-811436-0.00007-1

EMERGENCE OF A SKIN PHASE IN MAMMALIAN MALARIA

Pauline Formaglio[1], Laura Mac-Daniel[2,3], Rogerio Amino[2] and Robert Ménard[2]

[1]*Otto-von-Guericke University, Magdeburg, Germany;* [2]*Institut Pasteur, Paris, France;* [3]*Loyola University Chicago, Maywood, IL, United States*

INTRODUCTION

Malaria has plagued humans since the dawn of time. The earliest records of its typical symptoms date as far back as Chinese texts from about 2700 BC, and plasmodial DNA was detected in Egyptian mummies from ~2000 BC (Nerlich et al., 2008). Throughout human history, malaria has been largely distributed over the globe, reaching up to the Arctic Circle, concerning, for example, during the 19th century an estimated 90% of the world population (Mendis et al., 2009). During the 20th century, the implementation of mosquito control measures and the use of systematic antimalarial treatments have significantly alleviated the malaria burden. Malaria was progressively eradicated from several areas in the world, including Europe and the United States. During the 1950–60s, the global malaria eradication campaign led by the World Health Organization (WHO), despite failing to globally eradicate the disease, did reduce malaria-associated morbidity and mortality across much of the tropical and subtropical world. Sub-Saharan Africa, however, has remained highly infested; this region continues to pay the heaviest toll, concentrating over 90% of the malaria deaths (Carter and Mendis, 2002).

Into the 21st century, renewed political will and financial commitment have allowed unprecedented efforts to limit the malarial pandemics (Alonso and Tanner, 2013). The reinforcement of antivectorial measures, including the massive distribution of insecticide-treated mosquito nets and indoor residual spraying, the use of rapid diagnostic tests and of more effective drug treatments, particularly the artemisinin-based combination therapies, have combined to improve the malaria condition, including in Africa. According to WHO, in 2015 there were roughly 200 million clinical cases and 450,000 deaths due to malaria, which represents a 60% reduction in global mortality rates and a 40% decrease in the incidence of clinical disease in Africa between 2000 and 2015 (Bhatt et al., 2015). Nonetheless, malaria is still endemic in about 100 countries and roughly 3 billion people—almost half of the world population—live in areas where they are at risk of contracting the disease.

Malaria is due to the *Plasmodium* parasite. There are more than 250 different species of *Plasmodium* that infect mammals, birds, and reptiles, and five that infect humans (Martinsen et al., 2008). *Plasmodium falciparum* is widespread, causes roughly 50% of all malarial infections and over 75% of those occurring in sub-Saharan Africa. It is the most dangerous species, being responsible for almost all malarial deaths. Its virulence mainly stems from its distinctive property to induce the so-called cytoadherence, i.e., the binding of infected red blood cells (RBCs) to endothelia in the microcirculation—particularly in the brain where it results in the most feared malaria complication, cerebral malaria. *Plasmodium vivax* is also largely distributed, except in Africa where it is mostly present in Madagascar, and accounts for 65% of the malaria cases in Asia and South America (Howes et al., 2016). *Plasmodium ovale*, closely related to *P. vivax*, has a relatively low prevalence and primarily concentrates in sub-Saharan Africa, while *Plasmodium malariae* is globally distributed but causes milder symptoms. The fifth species, *Plasmodium knowlesi*, is a primate parasite that was reported to also infect humans in 1965 for the first time in Malaysia and is now increasingly present in South East Asia. The duration of a reproduction cycle in the blood, which causes the characteristic periodic fever with chills, rigors, and sweating, differs between species: 3 days for *P. malariae,* causing the so-called "quartan" malaria (the febrile episode returning every third day), 2 days for *P. falciparum*, *P. vivax*, and *P. ovale*, causing "tertian" malaria, and only 1 day for *P. knowlesi.*

Human-infecting *Plasmodium* parasites are transmitted by *Anopheles* mosquitoes, which act as the definitive host in the parasite life cycle. About 500 *Anopheles* species are known, but only 40 are capable of transmitting *Plasmodium* (Sinka et al., 2012). Notably, *Anopheles* is not restricted to malaria endemic areas and many species live in colder latitudes, as mentioned with the past occurrence of malaria in cold places of the globe. *Anopheles gambiae* mosquitoes are the most efficient *Plasmodium* transmitters in sub-Saharan Africa and the main vector of *P. falciparum*; they are strongly anthropophilic (i.e., feeds mostly on humans), typically bite at dusk and dawn, and can feed indoors as well as outdoors. More worryingly, mosquitoes may adapt to the challenge of the large deployment of bed nets by biting in broad daylight (Sougoufara et al., 2014).

The parasite life cycle is typically divided into three main phases (Fig. 7.1). All symptoms and complications of the disease take place during the erythrocytic phase, when parasites undergo cycles of multiplication inside RBCs. One erythrocytic cycle starts by the penetration of the parasite form called merozoite inside an RBC, proceeds with the asexual multiplication of the merozoite into ~8–32 new merozoites, and ends by the egress of daughter merozoites into the bloodstream. The second, "sexual/mosquito" phase starts when newly invaded merozoites differentiate into sexual stages, called gametocytes, which remain quiescent inside RBCs. Once taken up by a mosquito and inside its gut, gametocytes activate and egress RBCs. Male gametes "exflagellate" and fertilize female macrogametes, and the ensuing zygote transforms in the gut wall into an oocyst stage. The oocyst then generates thousands of sporozoites, which eventually invade salivary glands and await transmission. The third, preerythrocytic (PE) phase of the parasite's life is initiated by the transmission of sporozoites to a susceptible host. It essentially consists, in mammals, of a single round of parasite multiplication transformation inside hepatocytes, where one sporozoite transforms into thousands of merozoites.

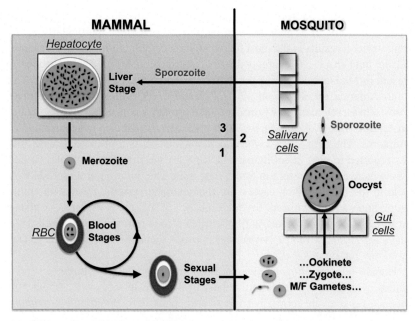

FIGURE 7.1 Simplified view of the *Plasmodium* life cycle. The *Plasmodium* life cycle consists of three phases. The **erythrocytic phase** (**1**) starts when the red blood cell (RBC)-infecting forms (called merozoites) are released from the liver. It consists of cycles of asexual parasite multiplication inside RBCs that generate, at each cycle, 8–32 new erythrocytic merozoites (erythrocytic merogony), resulting in an exponential growth of the parasite. The **sexual/mosquito phase** (**2**) starts when some newly internalized merozoites do not multiply but differentiate into male or female gametocytes (gametocytogenesis). Once in the mosquito midgut, gametocytes activate into gametes. The male gametocyte replicates its genome three times, forming eight flagellated male microgametes. Male microgametes and female macrogametes fertilize in the midgut lumen, forming a zygote, and then a motile ookinete that crosses the midgut epithelium to transform into an oocyst. The oocyst undergoes multiplication and generates sporozoites (sporogony), which are released into the mosquito body cavity (hemocoel) and are passively transported by the hemolymph to the salivary glands. The **preerythrocytic phase** (**3**) follows sporozoite inoculation by the infected mosquito. Sporozoites are injected into the host, passively reach the liver by the blood circulation, arrest in the liver, and invade hepatocytes, where they asexually multiply and transform into thousands of hepatic merozoites (hepatic merogony).

Below, we focus on the PE phase. After presenting the structural and functional features of the sporozoite that are relevant to appreciate its infectivity in vivo, we provide a brief historical description of the early discoveries that have defined the main steps of the *Plasmodium* life cycle. We then concentrate on how a skin component of the PE phase has been recognized and how the biology of this increasingly complex skin phase is currently being dissected using intravital imaging and other tools.

STRUCTURAL AND FUNCTIONAL FEATURES OF THE *PLASMODIUM* SPOROZOITE

Plasmodium is a unicellular eukaryote that belongs to the phylum Apicomplexa, formerly known as Sporozoa, probably the largest taxon of protists (Morrison, 2009). Almost all known apicomplexans (more than 5000 species have been described) are parasites of vertebrate or invertebrate hosts. Several apicomplexans are of medical importance beyond *Plasmodium*, including *Toxoplasma*, *Cryptosporidium*, and *Isospora*. Apicomplexans of veterinary relevance include *Eimeria*, *Theileria*, and *Babesia* that infect cattle and poultry.

Plasmodium displays structural and functional features that are conserved in the phylum (Fig. 7.2). It is an intracellular obligatory parasite, i.e., multiplies only inside host cells, and thus alternates between extra- and intracellular forms. Extracellular forms are called zoites (Morrissette and Sibley, 2002); they are elongated and polarized cells, shaped by longitudinal microtubules anchored anteriorly at a polar ring, and secrete at their tip proteins stored in internal vesicles called rhoptries (club-shaped) and micronemes (rodlike). Immediately underneath the zoite external membrane is an actin-myosin motor, also called glideosome.

The motor endows zoites with the ability to move using a substrate-dependent type of locomotion called gliding, as the moving cell does not overtly deform its shape during movement (Boucher and Bosch, 2015). The motor also powers invasion of host cells, a process that occurs by the formation of a ring-like tight junction between the zoite tip and the cell membrane, through which the zoite actively slides into an invagination of the cell membrane that becomes the parasitophorous vacuole (PV). Inside the PV the parasite typically multiplies by a process of multiple fission called schizogony.

The *Plasmodium* sporozoite is an aggressive apicomplexan zoite. It is needle-shaped, 10-µm long, and 1-µm wide and exhibits vigorous gliding at speeds up to 2 µm/s, about 10 times faster than the fastest mammalian cells such as leukocytes (Vanderberg, 1974). It also has the ability to "traverse" host cells, a behavior that is not displayed by all apicomplexan zoites (Mota et al., 2001). During host cell traversal, the sporozoite enters the cell by breaching its plasma membrane, without making a PV, glides through its cytosol and exits the cell by again rupturing its membrane. These morphological and functional traits make the sporozoite the ultimate infectious weapon, tailored for the rough journey from the skin to the safe intrahepatocytic home.

UNRAVELING THE *PLASMODIUM* LIFE CYCLE

Early Discoveries—A Bit(e) of History

A Parasite as the Causing Agent of Malaria

At the end of the 1870s, after Louis Pasteur and Robert Koch had proposed the germ theory of infection, malaria was described as caused by a bacillus,

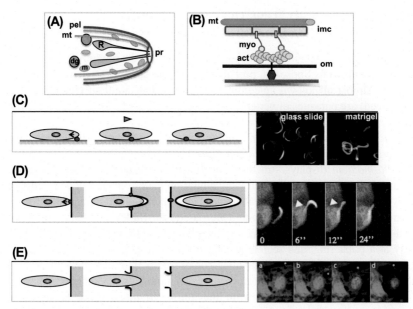

FIGURE 7.2 Structure and functional capacities of the *Plasmodium* sporozoite. (A) The apical pole of an apicomplexan zoite. The zoite apical tip contains a polar ring (pr), a microtubule-organizing center, from which microtubules (mt) run posteriorly underneath a triple-membraned pellicle (pel). The zoite anterior pole contains vesicles called rhopries (R) and micronemes (m), which discharge their content at the time of motility (m) and invasion (m+R), and dense granules (dg) that are discharged when the zoite is intracellular. (B) The zoite motor. Actin (act) and an apicomplexan-specific single-headed unconventional myosin (myo) are located between the outer membrane (om) and the double-membraned inner membrane complex (imc) made up of adjacent flattened vesicles (alveoli). Actin is linked to the cytoplasmic tail of transmembrane proteins that bind to the substrate, while myosin is bound to the imc, and also likely to the underlying mt, via proteins called gliding-associated proteins. (C) Gliding motility. *Left*, Gliding is viewed as being powered by the actomyosin motor arranged in linear fashion, whereby myosin-generated force posteriorly cap F-actin/force-transducing protein (*red dot*) complexes, moving the zoite forward if complexes are fixed to the substrate. *Right*, *Plasmodium berghei* sporozoites glide in a circular pattern when on glass slides or in a helical pattern when in matrigel. (D) Host cell invasion. *Left*, The sporozoite anterior tip forms with the host cell membrane a tight junction (*red dot*) that is used by the parasite as traction point to propel itself inside the cell. The entering sporozoite invaginates the host cell plasma membrane, which becomes a PV that seals off at the zoite posterior tip when entry is complete. *Right*, a P. berghei sporozoite entering a host cell, with a presumed entry junction/constriction. Numbers indicate seconds. (E) Host cell traversal. *Left*, the sporozoite migrates inside the host cell cytosol and breaches its plasma membrane twice to enter and exit the cell. *Right*, a red sporozoite glides through a green fluorescent protein (GFP)-expressing cell in the presence of propidium iodide (PI) in the medium, leading to GFP efflux and PI influx and the observed color switch of the traversed cell. ((D) *Reproduced from Amino, R., Giovannini, D., Thiberge, S., Gueirard, P., Boisson, B., Dubremetz, J.-F., Prévost, M.-C., Ishino, T., Yuda, M., Ménard, R., 2008. Host cell traversal is important for progression of the malaria parasite through the dermis to the liver. Cell Host and Microbe 3, 88–96. (E) Reproduced from Tavares, J., Formaglio, P., Thiberge, S., Mordelet, E., van Rooijen, N., Medvinsky, A., Ménard, R., Amino, R., 2013. Role of host cell traversal by the malaria sporozoite during liver infection. Journal of Experimental Medicine 210, 905–915.)*

Bacillus malariae. In 1880, Alphonse Laveran, a French army doctor working in Algeria, saw in a wet film of fresh blood from a malaria patient *"on the edges of spherical pigmented bodies, thin elements resembling flagella that were moving with great agility,"* the first observation of exflagellation (Laveran, 1880). As Laveran had performed autopsies of malaria patients and confirmed the presence of a black pigment (now known as hemozoin) associated with lesions of the liver, spleen and cerebral vessels (he called *melanemia*), movement from pigmented bodies suggested to him that the agent of malaria was a parasitic animal, which he called *Oscillaria malariae.*

Although initially met with skepticism, Laveran's discovery was confirmed during the 1880s by parasitologists in Italy, where the disease was endemic and responsible for 15,000 deaths annually. Ettore Marchiafava and Angelo Celli described the developmental cycle of the parasite in RBCs and gave the organism its generic name, *Plasmodium* (Marchiafava and Celli, 1883). Camillo Golgi, who would reach fame for his pioneering work on the central nervous system, linked the cyclic and synchronous bursting of parasitized RBCs to the classical febrile peaks of malaria and differentiated tertian and quartan malaria. By the early 1890s, the Italian scientists had defined the *Plasmodium* species responsible for benign tertian (*P. vivax*), malignant tertian (*P. falciparum*), and quartan (*P. malariae*) malaria (Golgi, 1889).

Mosquitoes as Vectors (and Hosts)

Much research that followed addressed the mode of transmission of *Plasmodium* to humans. The association of intermittent fevers and marshes had long been recognized, and the hypothesis that mosquitoes transmitted malaria was commonplace, although the way they might transfer the parasite was a matter of debate. In the late 1890s, British and Italian scientists identified *Plasmodium* in mosquitoes and made the crucial observations that incriminated the bites of mosquitoes as the mode of transmission.

In London, Patrick Manson, who in 1877 had linked mosquitoes to transmission of human filarial parasites, instructed his student working in India, Ronald Ross, to *"follow the flagellum"* in mosquitoes. Ross did not find the flagellum, but in 1897 he observed pigmented bodies (oocysts) within the gut wall of a mosquito that had fed on a malaria patient (Ross, 1897). Unable to complete his studies on human malaria, Ross turned to an avian (sparrow) model of malaria combining *Plasmodium relictum* and *Culex* mosquitoes. In 1898 he saw that mosquito gut pigmented bodies were filled with *"rods"* (sporozoites) that eventually gained access to the mosquito salivary glands, which led him to imagine that these rods would be *"poured out in vast numbers under the skin of the man or bird"* (Ross, 1898). Ross then demonstrated parasite passage between birds via the bites of culicine mosquitoes, against his mentor and others' speculation that transmission occurred by ingesting water contaminated by parasites released by dead or discharging mosquitoes.

Also in 1898, scientists of the Italian school of malariology led by Battista Grassi identified *Anopheles claviger* as a local transmitter of human malaria and revealed the specific dependence of human malaria on *Anopheles* mosquitoes (Grassi, 1898). Amico Bignami, along with Giuseppe Bastianelli, carried out in 1898 the decisive experiments that demonstrated that malaria was transmitted to humans by mosquito bites, by allowing mosquitoes collected in a highly malarious region of Italy to feed on a malaria-free hospitalized patient in controlled conditions (Bignami, 1898). In 1900 *A. claviger* mosquitoes that had fed on a malaria patient in Rome were sent to London and allowed to bite Manson's son, who developed tertian malaria 15 days later.

By 1900 the link between malaria transmission and mosquito bites was accepted, which meant that the disease was controllable by reducing mosquito numbers and limiting contacts between humans and mosquitoes. Over the next decades, various methods of mosquito and mosquito bite control contributed to eradicate malaria from Italy, a milestone achieved by 1950.

The Liver as the First Site of Multiplication in the Mammalian Host

After mosquitoes had been linked to malaria transmission via the sporozoite stage in the mosquito salivary glands, a gap remained in the cycle of events between sporozoite inoculation and RBC infection. In 1903, the renowned German protozoologist Fritz Schaudinn appeared to have solved the problem when he reported that *P. vivax* sporozoites directly invaded RBCs (Schaudinn, 1903). Although this observation could not be repeated, Schaudinn's model was adopted and became dogma for more than 40 years.

However, important findings were later made using avian malaria, particularly *Plasmodium gallinaceum*, which had become the dominant study model. Work from several authors showed that the infective bite was followed by an incubation period during which the parasite was not found in the blood, which suggested the existence of a "tissue phase." By the late 1930s "exo-erythrocytic" parasites developing in macrophages at the inoculation site in the skin, as well as in reticuloendothelial cells in the liver and spleen, were found in *P. gallinaceum* sporozoite-infected chickens (Huff, 1947). These stages give rise to successive PE cycles of infection, a situation specific to avian malaria.

In mammals, the site of initial parasite multiplication has remained a puzzle until 1948. That year, Henry Shortt et al. showed that following *Plasmodium cynomolgi* sporozoite delivery by over 500 *Anopheles maculipennis* mosquitoes into rhesus monkey and systematic histological analysis of virtually all organs 7 days after transmission, exo-erythrocytic parasites (schizonts) were found only inside parenchymal cells in the liver (Shortt and Garnham, 1948; Shortt et al., 1948). In the years that followed, the same authors found exo-erythrocytic forms (EEFs) of *P. vivax* and of *P. falciparum* in the liver of humans. Mammalian malaria parasites typically undergo a single schizogonic cycle inside hepatocytes, which can be arrested for extended periods of time (up to tens of years) in some

Plasmodium species as persistent forms, called hypnozoites (Krotoski et al., 1982).

The main pieces of the malaria life cycle were thus in place almost 70 years after the discovery of the parasite. Two crucial parasitological breakthroughs that would have profound effect on molecular research were later achieved. One was the development of a rodent model of malaria, which started in 1948 with the isolation from tree rats in Central Africa of the first rodent-infecting malaria parasite, *Plasmodium berghei*, by Ignace Vincke and Marcel Lips (Vincke and Lipps, 1948). *P. berghei,* along with other rodent-infecting species such as *Plasmodium yoelii*, are now adapted to mice and rats, easily maintained in laboratory-bred *Anopheles* mosquitoes and have become powerful models for investigating PE malaria infection in vivo. The second breakthrough was the ability to grow erythrocytic stages of *P. falciparum* inside RBCs in continuous in vitro cultures (Trager and Jensen, 1976), and therefore to collect unlimited amounts of the most pathogenic human parasite.

Evidence of a Skin Phase in the *Plasmodium* Life Cycle

Initial Evidence

The idea that blood-feeding mosquitoes must inject sporozoites in the blood has long been the common assumption, favored by the influential report that in humans sporozoites appear in the circulation as early as during the biting act by the mosquito (Fairley, 1947). Nonetheless, mosquito physiology and studies on the biting process clearly indicate otherwise (Griffiths and Gordon, 1952; Kashin, 1966; Ribeiro et al., 1985). A bite starts by a probing phase, during which the mosquito, searching for blood, thrusts its mouthparts repeatedly through the host skin. This phase frequently lasts less than 30 s with *Anopheles*, being successful and followed by blood suction or unsuccessful and followed by the search for another site or host (Li and Rossignol, 1992). When a blood vessel is sensed, the proboscis might cannulate or rupture the vessel; consequently, the mosquito might ingest circulating or leaked blood in a process named "vessel feeding" or "pool feeding," respectively. Salivation, and thus potential release of sporozoites, occurs during the probing phase of the bite, when the antihemostatic activity of saliva is needed to optimize blood vessel finding and blood ingestion, as well as during withdrawal of the mouthparts. Clearly, some saliva may also be ejected during blood ingestion; however, the use of *Plasmodium* sporozoite-harboring mosquitoes showed that parasites ejected during blood suction were mostly ingurgitated with the blood into the midgut (Beier et al., 1992; Kebaier and Vanderberg, 2006).

The notion that sporozoites are mostly deposited in the extravascular areas of the skin during the probing phase, rather than directly into the bloodstream, was shown by two studies. The first (Sidjanski and Vanderberg, 1997) used *P. yoelii* and a bite site removal approach: the authors surgically ablated the bite site at different times after completion of feeding and showed that fewer mice became infected (i.e., displayed blood-stage parasites) when the bite site was removed before 5 min postfeeding compared to 15 min postfeeding. The second study (Matsuoka et al., 2002) used *P. berghei* and a probing interruption approach: the authors allowed mosquitoes to probe (4 times 10 s), but not blood feed, and showed this was sufficient to induce parasitemia, while infection was inhibited if the site probed by mosquitoes was subsequently heated. Both studies thus suggested that sporozoites were injected by mosquitoes into the skin and needed minutes to get access to the blood circulation. The existence of a skin step in the PE phase of malaria would soon be confirmed and extended by the use of more sophisticated techniques.

New Tools, New Proofs

The last two decades have witnessed a flurry of technical breakthroughs that have changed the face of malaria research. Relevant to the study of PE stages is the combination of powerful intravital imaging techniques of mice and the possibility to genetically modify, and thus appropriately tag, rodent-infecting *Plasmodium* parasites. Imaging techniques have been revolutionized by advancements in both microscopy instrumentation and data collection/processing software, which are presented in another chapter of this book (Chapter 12). In our own studies, we preferentially used a high-speed spinning-disk confocal microscope, which collects data faster than traditional, slow laser scanning confocal microscopy, making it particularly suited for tracking in real-time fast-moving objects such as a gliding *Plasmodium* sporozoite (Amino et al., 2007).

All in vivo imaging studies on PE stages performed so far have used rodent-infecting *P. berghei* or *P. yoelii* parasites. Following the first success in site-directed genetic modification of *P. berghei* (van Dijk et al., 1996; Ménard et al., 1997), several transgenic parasites expressing a fluorescent marker through a variety of stage-specific or constitutive promoters were generated, including *P. berghei* expressing green fluorescent protein (GFP) (Natarajan et al., 2001; Franke-Fayard et al., 2004; Ishino et al., 2006), the improved red fluorescent protein RedStar (Frevert et al., 2005) or mCherry (Hopp et al., 2015), and *P. yoelii* expressing GFP (Tarun et al., 2006; Ono et al., 2007) or RedStar (Gueirard et al., 2010). While fluorescence microscopy is most suited for tracking individual sporozoites and analyzing host-parasite interactions, other transgenic parasites have been generated for quantitative studies, i.e., measuring parasite loads, such as *P. berghei* expressing β-galactosidase (Engelmann et al., 2006) or expressing luciferase (Franke-Fayard et al., 2008).

The first applications of intravital imaging to the *Plasmodium* sporozoite have allowed to see *P. berghei* sporozoites gliding in their natural context, in the mosquito (Frischknecht et al., 2004) and the mouse (Vanderberg and Frevert, 2004; Amino et al., 2006). The latter two studies showed sporozoite transmission in the extravascular tissue of the skin, rapid gliding exhibited by the majority of

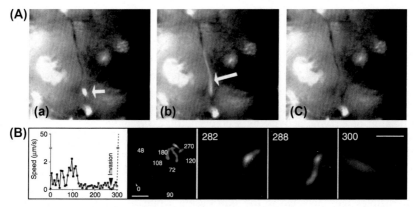

FIGURE 7.3 Initial imaging evidence of sporozoite invasion of blood vessels in the skin. (A) The three individual frames from a time-lapse sequence show invasion of a dermal blood vessel (*dark line*) by a sporozoite. The *white arrow* indicates the sporozoite before (a) or during (b) egress. In (b), the sporozoite is seen as an elongate streak during a 100 ms exposure. In (c), the sporozoite had left the observation field. (B) The panel on the left shows the velocity profile of the sporozoite analyzed by time-lapse microscopy on the panels on the right. The first microscopy panel shows the maximum-intensity projection (numbers, sporozoite position at various times in seconds) of a sporozoite gliding in the dermis (red), decreasing its speed as it glides along the blood vessel wall (green) and moving through the vessel wall (yellow). The single-time frames on the right (time in seconds indicated) show the sporozoite invading the vessel wall (*arrowheads*, site of sporozoite invasion). The blood vessel is in blue. Scale bar, 10 μm. ((A) *Evidence from Vanderberg, J.P., Frevert, U., 2004. Intravital microscopy demonstrating antibody-mediated immobilisation of* Plasmodium berghei *sporozoites injected into skin by mosquitoes. International Journal for Parasitology 34, 991–996. (B) Evidence from Amino, R., Thiberge, S., Martin, B., Celli, S., Shorte, S., Frischknecht, F., Ménard, R., 2006. Quantitative imaging of* Plasmodium *transmission from mosquito to mammal. Nature Medicine 12, 220–224.)*

dermal sporozoites, and sporozoite translocation into the bloodstream through the walls of undamaged blood vessels (Fig. 7.3), therefore bringing visual evidence for a skin phase in the sporozoite's life.

Sporozoites in the Skin

Dynamics of Sporozoite Transmission Into the Skin

The number of sporozoites ejected during a mosquito bite is a recurrent and important question. The first attempts to answer this, made at a time when sporozoites were assumed to be injected into the bloodstream, used ex vivo settings such as mosquitoes salivating into fluid-filled capillary tubes (Rosenberg et al., 1990; Beier et al., 1991), onto glass slides (Golenda et al., 1992) or through mouse skin stretched over blood-filled chambers (Ponnudurai et al., 1991). Results showed significant variations, as expected, from the distinct experimental conditions that may not reflect salivation during a natural bite.

After it was recognized that sporozoites are injected into the skin, three quantification studies were performed using natural biting of mice by *Anopheles stephensi* mosquitoes and sporozoite counting in biopsies or homogenates from the bite site. Using wild-type *P. yoelii* and a 3 min-bite time in mice, qRT-PCR of sporozoite 18S rRNA extracted from the bite site reported an estimated mean of 123 sporozoites injected per mosquito (Medica and Sinnis, 2005). Using *P. berghei* sporozoites expressing β-galactosidase and a 10 min-bite time, quantification of enzyme activity by a colorimetric assay indicated 120–280 sporozoites deposited per mosquito (Engelmann et al., 2006). Using *P. berghei* expressing a red fluorescent protein, direct fluorescence microscopy reported means of 281 versus 452 sporozoites for 3 and 15 min bite times, respectively (Jin et al., 2007). These studies essentially agree on the fact that relatively few sporozoites are deposited in the skin during a bite, at least compared to the sporozoite load in the salivary glands (which frequently exceeds 10,000 in laboratory-reared mosquitoes). Moreover, the numbers of sporozoites deposited in the skin only weakly correlated with the numbers of sporozoites present in the salivary glands. Therefore the main factor influencing the number of sporozoites ejected appears to be the duration of probing, which in turn depends on the degree of vascularization of the bitten tissue and the likelihood of rapidly finding a blood source.

The relatively small number of sporozoites ejected into the skin is at least partly the consequence of the relatively low rate of sporozoite release through the mosquito proboscis, which is surprisingly constant at about 1–2 per second during ex vivo salivation (Frischknecht et al., 2004) or transmission to mice (Jin et al., 2007). This is due to the physical constraints of the salivary system, as the lumen of the salivary duct maximally narrows to about 1 μm in diameter, barely larger than the sporozoite section (Fig. 7.4). Another factor that may account for the reduced number of sporozoites ejected was revealed by dynamic sporozoite imaging in the mosquito salivary system (Frischknecht et al., 2004). This study showed that, in resting mosquitoes, while sporozoites inside salivary cavities are immotile, some glide from the cavities into primary ducts and beyond, at the slow speed of ~0.1 μm/s (less than a 10th its maximal speed), and can travel long distances (up to ~1 μm/5 h in the 2 mm-long salivary duct system; see Fig. 7.4B). It also showed that only those sporozoites that have colonized ducts by gliding prior to salivation can be ejected during salivation, while those sitting inside cavities are not displaced even during prolonged salivation, making a nonreleasable pool. It might be speculated that downmodulation of sporozoite motility in the salivary glands may limit the number of sporozoites transmitted to the minimum necessary for host infection. It may also ensure selection of the fittest sporozoites for transmission into a hostile environment where rapid escape is crucial for survival.

Sporozoite Motility in the Skin

Sporozoite gliding motility in the skin is activated within seconds after transmission (Vanderberg and Frevert, 2004), switching from the sluggish motility of

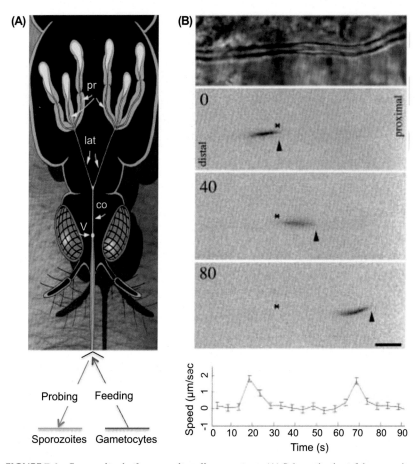

FIGURE 7.4   Sporozoites in the mosquito salivary system. (A) Schematization of the mosquito salivary system. *Anopheles* has two tri-lobed salivary glands, each drained by three primary (pr) ducts. The primary ducts fuse into lateral (lat) ducts, which fuse into the common (co) duct that runs into the proboscis at the salivary valve (V). Saliva, along with potential sporozoites, is ejected by the salivary canal, whereas blood, along with potential gametocytes, is ingested by the food canal. (B) A sporozoite moves down a primary duct in a stop-and-go fashion. The upper panel is a transmission light image of a salivary duct in an isolated gland. The three center panels, taken 40 s apart, show a sporozoite progressing through the duct of the upper panel; scale bar: 10 μm. The lower panel shows the velocity of the sporozoite and its stop-and-go motion pattern. *(Reproduced from Frischknecht, F., Baldacci, P., Martin, B., Zimmer, C., Thiberge, S., Olivo-Marin, J.C., Shorte, S., Ménard, R., 2004. Imaging movement of malaria parasites during transmission by* Anopheles *mosquitoes. Cellular Microbiology 6, 687–694.)*

sporozoites inside mosquito salivary ducts to the robust motility (~1–2 μm/s) of most (>75%) sporozoites in the skin (Amino et al., 2006). What activates sporozoite motility in the skin is unknown. In vitro, the motility of sporozoites dissected out from mosquito salivary glands is activated in the presence of serum (Vanderberg, 1974) or albumin alone (Kebaier and Vanderberg, 2010), as well

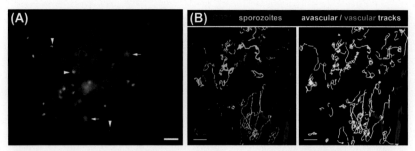

FIGURE 7.5   Sporozoite motility in the skin. (A) Maximum-intensity projections of the fluorescent signal (red) over 200 s showing the motility of sporozoites 2 min after injection by a mosquito into the skin of a mouse ear. *Arrowheads* indicate sporozoites (green) and *arrows* autofluorescent patches on the skin surface. Scale bar, 50 μm. (B) Maximum-intensity projection over 240 s showing trajectories of sporozoites (green) 10 min after intradermal injection into a mouse ear. Blood vessels are labeled with antibodies to CD31 (pan-endothelial junction molecule; far red). On the right panel, sporozoites were manually tracked using Imaris software, and tracks overlapping with CD31+ vessels were colored in orange and tracks in between CD31+ vessels in white. Scale bar, 25 μm. *((A) Reproduced from Amino, R., Thiberge, S., Martin, B., Celli, S., Shorte, S., Frischknecht, F., Ménard, R., 2006. Quantitative imaging of* Plasmodium *transmission from mosquito to mammal. Nature Medicine 12, 220–224. (B) Reproduced from Hopp, C.S., Chiou, K., Ragheb, D.R.T., Salman, A.M., Khan, S.M., Liu, A.J., Sinnis, P., 2015. Longitudinal analysis of* Plasmodium *sporozoite motility in the dermis reveals component of blood vessel recognition. eLife 4, e07789.)*

as by small, immobilized RGD (Arg-Gly-Asp) tripeptides (Perschmann et al., 2011). It is likely that both albumin (or other biochemical) and adhesive signals are also involved in vivo in sparking sporozoite motility. However, despite gliding at similar speeds, sporozoites on 2D glass slides and in the skin display very different motility paths; while the former glide in almost perfect circles, the latter display tortuous, nonlinear paths with abrupt changes in direction (Fig. 7.5). Motility is not continuous, as sporozoites frequently pause during their journey in the skin.

What guides this complex sporozoite movement is not understood. There is currently no evidence for sporozoite chemotaxis. One hypothesis is that moving sporozoites respond to structural cues in their environment. In favor of this, a study found that the *P. berghei* sporozoite path in the skin varies in different locations such as the ear (confined) and tail (linear) in mice, and could be somewhat reproduced *in vitro* in micropatterned arrays containing obstacles of defined spacing and sizes (Hellmann et al., 2011). However, although physical obstacles can alter the sporozoite migration pattern, it is unclear whether they are sufficient to guide sporozoite progression and optimize vessel finding. Another study proposed that sporozoites undergo a switch in motility pattern, occurring at around 20 min posttransmission, from a high speed/linear path away from blood vessels allowing maximal dispersal from the initial location, to a more constrained pattern around blood vessels (Hopp et al., 2015; Fig. 7.5B). It remains to be established whether the switch is predetermined or the consequence of specific sporozoite interactions with ligands associated with blood vessel walls. In any case, both studies (Hellmann et al., 2011; Hopp et al., 2015) agree that the sporozoite path is not explained by a Brownian random walk, leaving open the question of directed sporozoite migration in host tissues.

Another issue still unresolved is the time sporozoites stay in the skin before gaining access to the blood circulation, or may stay in the skin without losing their capacity to reach and invade a hepatocyte. Imaging studies of *P. berghei* sporozoites transmitted either by bite (Amino et al., 2006) or syringe inoculation (Hopp et al., 2015) both reported an overall decrease in sporozoite speed during the first hour and limited residual motility after 2 h (Hopp et al., 2015). A subsequent fluorescence microscopy analysis of biopsy specimens removed at various times posttransmission in the ear pinna, or the abdomen indicated that sporozoite emigration from the skin was largely restricted to a 2-h period (Kebaier et al., 2009). A third work (Yamauchi et al., 2007), however, using *P. yoelii* sporozoite injection into the ear and qPCR analysis, concluded that the majority of liver-infective sporozoites remained in the skin for (at least) 3 h, and that the pattern of sporozoite exit from the bite site resembled a slow trickle. Although these differences, which may be due to the different parasites, mouse species, or experimental conditions used, must be explained, it remains that sporozoite sojourn in the skin lasts far longer than anticipated. This opens an interesting window of opportunity for immune attack, particularly by neutralizing and immobilizing antibodies.

Sporozoite Traversal of Host Cells

One striking feature of sporozoite gliding in the skin is its speed, which after transmission equals that of a sporozoite gliding on a glass slide, showing that sporozoites are well adapted to migrate in the constrained cellular environment of the host skin. As mentioned above, the *Plasmodium* sporozoite expresses the capacity to traverse host cells. A number of proteins have been implicated in rupturing the host cell membrane, including a protein named sporozoite protein essential for cell traversal 2 (SPECT2) that harbors a typical membrane-attack/perforin–like domain (Ishino et al., 2005; Tavares et al., 2014) found in pore-forming proteins such as components of the mammalian complement system and perforin.

Sporozoite traversal of host cells was first observed as *P. berghei* sporozoites interacting with mouse peritoneal macrophages in vitro in a "*needling manner and inducing an outward flow of host cell cytoplasm at the point of egress*" (Vanderberg et al., 1990). Sporozoites were then found to traverse hepatocytes both in vitro (Mota et al., 2001) and in vivo (Frevert et al., 2005), and initial work suggested that hepatocyte traversal progressively increased the sporozoite capacity to invade hepatocytes inside a PV and to differentiate. However, this hypothesis was not confirmed by the use of SPECT2/cell traversal-deficient (traversal[KO]) sporozoite mutants, which invade and develop inside hepatocytes normally (Ishino et al., 2004, 2005).

Intravital imaging later revealed that cell traversal was important at many steps of the sporozoite's life from the inoculation site to the liver parenchyma. Imaging was used to objectivate traversal events in vivo, as the leakage of an intracellular, genetically encoded fluorescence marker or the incorporation of an extracellular, cell-impermeant fluorescent dye, and to characterize the phenotype of traversal[KO] sporozoites and thus the contribution of cell traversal to sporozoite infectivity. In the dermis, sporozoites traverse host cells, as evidenced by the incorporation of extracellular propidium iodide into the nuclei of cells located on the trajectories of gliding sporozoites (Formaglio et al., 2014; see Fig. 7.6A). Cell traversal in the dermis is important, as shown by the rapid immobilization of most (>80%) traversal[KO] sporozoites by dermal cells and their internalization by resident professional phagocytes (Amino et al., 2008; see Fig. 7.6B). A congruent view of sporozoite traversal was obtained when imaging sporozoites inside liver sinusoids. Gliding sporozoites were frequently seen traversing Kupffer cells, the macrophages in the liver that reside in the lumen of the sinusoids, and traversal[KO] sporozoites injected intravenously were cleared by Kupffer cells (Tavares et al., 2013). Moreover, most sporozoite translocation events (~75%) across the liver sinusoidal barrier occur by traversal of one of the two cell types that compose the barrier, endothelial cells, or Kupffer cells, although traversal[KO] sporozoites can still translocate by a paracellular pathway or via cell traversal–independent transcellular migration (Tavares et al., 2013). Therefore, it seems that sporozoites can traverse any cell type they encounter during their journey to hepatocytes, which maximizes their survival (when traversing phagocytes and preventing phagocytosis) and their progression (when traversing isolated cells or cellular barriers). How sporozoites cross the endothelium of capillaries in the skin has not been reported.

What causes sporozoites to switch from "cell traversal" to "cell invasion" when they finally reach hepatocytes is unclear. One study suggested that the switch may be induced by the level of sulfation of cell-associated heparan sulfate proteoglycans (HSPGs) (Coppi et al., 2007). Indeed, it is known that the circumsporozoite protein, the major surface protein of the sporozoite, specifically binds the glycosaminoglycan chains of HSPGs synthesized by hepatocytes (Sinnis et al., 1996), which are more highly sulfated than HSPGs in other organs. The HSPG sulfation levels on different cell types may thus be used as a global positioning system by the sporozoite, capable of both arresting sporozoites in the liver and activating their invasive capacity. Another important question is the fate of the cells traversed by sporozoites, which has immunological implications. Intravital imaging using two independent cell-impermeant nucleic acid fluorescent dyes, the first to reveal the initial traversal event and the second to test the transient or permanent loss of integrity of the wounded membrane (Formaglio et al., 2014), showed that only 20%–30% of the traversed cells resealed their breached membrane while the majority did not. Likewise, most Kupffer cells in the liver sinusoids exhibited permanent loss of membrane integrity and died after being traversed by a sporozoite (Tavares et al., 2013).

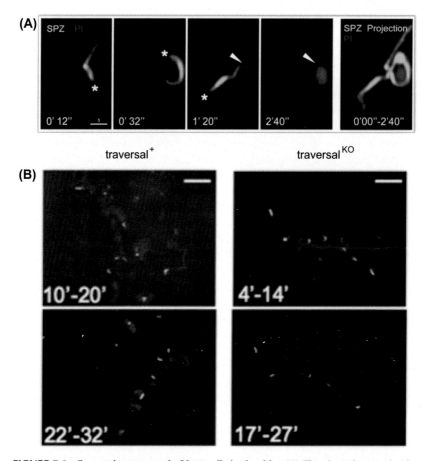

FIGURE 7.6 Sporozoites traversal of host cells in the skin. (A) Time-lapse images showing a skin cell traversed by a sporozoite (green) in a mouse previously injected with propidium iodide (PI; red). Frames 1 to 4 are the maximum-intensity projections of three consecutive focal planes and frame 5 is the time projection of the corresponding sequence. *Asterisks* point to the parasite apical end and *white arrowheads* indicate the PI+ traversed cell. Scale bar, 5 μm. (B) Maximum-intensity projections of the fluorescent signals (red) of wild-type sporozoites (traversal+, left) or SPECT2-deficient sporozoites (traversalKO, right) gliding in mouse skin during the indicated period of time. Scale bar, 50 μm. *((A) Reproduced from Formaglio, P., Tavares, J., Ménard, R., Amino, R., 2014. Loss of host cell plasma membrane integrity following cell traversal by Plasmodium sporozoites in the skin. Parasitology International 63, 237–244. (B) Reproduced from Amino, R., Giovannini, D., Thiberge, S., Gueirard, P., Boisson, B., Dubremetz, J.-F., Prévost, M.-C., Ishino, T., Yuda, M., Ménard, R., 2008. Host cell traversal is important for progression of the malaria parasite through the dermis to the liver. Cell Host and Microbe 3, 88–96.)*

Fate of Skin Sporozoites

The Skin as a Crossroad—Tripartite Fate of Skin Sporozoites

Initial quantitative studies on the fate of *P. berghei* sporozoites injected by mosquitoes into the ear pinna indicated that only ~25% of the sporozoites followed the expected path by invading blood vessels (Amino et al., 2006). Consecutive to (frequently prolonged) sporozoite contact with blood vessel walls, these events were authenticated by the rapid disappearance of the parasites from the observation field at speeds similar to those of erythrocytes in the dermal capillaries and postcapillary venules, i.e., ~40–50 μm/s. Another 15% of the sporozoites left the skin by invading lymphatic vessels, scored as sporozoites drifting at low velocity (<0.5 μm/s) by sideward movement, contrary to sporozoites that glide with their anterior pole leading. The remainder ~60% of the inoculated sporozoites lingered in the skin with a lowered average motility (Amino et al., 2006). This "blood–lymph–skin" tripartite fate of skin sporozoites was confirmed by subsequent studies. After injection of *P. yoelii* sporozoites by syringe in the ear pinna and qPCR analysis of parasite DNA, approximately 20% of the initial sporozoite inoculum was found to be drained by lymphatics to the auricular lymph node (LN) (Yamauchi et al., 2007). When *P. berghei* sporozoites were inoculated by mosquitoes into the ear pinna or ventral abdomen and detected by fluorescence microscopy analysis of biopsies, 57% of the sporozoites from the ear pinna and 72% from abdominal tissues had departed the skin within 2 h after transmission (Kebaier et al., 2009). Although the tripartite fate of skin sporozoites appears to be common to infection of laboratory rodents by rodent parasite species, the proportions of each of the three paths are likely to vary with many factors, including the mode of injection, the vessel density at the bite site, the parasite species, and the host–parasite adaptation, as further discussed below. Importantly, access to the LN requires sporozoite active motility (Amino et al., 2006; Radtke et al., 2015), and a higher proportion of sporozoites reaches the LN after inoculation by a mosquito than by a syringe.

Parasites in the Lymphatic System

After inoculation of sporozoites by mosquito or syringe into the footpads of mice, sporozoites were indeed found in the proximal (popliteal) draining lymph node (dLN) (Amino et al., 2006). Sporozoites were located in the subcapsular zone of LNs, where they were initially (1–2 h) motile but were increasingly found inside CD11c+ dendritic cells (DCs) over time. At about 8-h posttransmission, sporozoite-associated fluorescence was no longer detectable in LNs. Moreover, sporozoites were rarely detected in the distal (iliac) dLN and were not found in the lymph collected from the thoracic duct, indicating that the lymphatic route unlikely leads sporozoites back to the blood, as earlier speculated (Vaughan et al., 1999; Krettli and Dantas, 2000), and is therefore not productive for the parasite life cycle. Interestingly, rare parasites in the LNs appeared to develop. Imaging analysis detected round-shaped, exo-erythrocytic fluorescent parasites until 24-h posttransmission, albeit of smaller sizes than those of their liver counterparts, which disappeared at 50 h. RT-PCR analysis showed that antigens normally expressed exclusively by liver stages, but not sporozoites, were detected in infected LNs until 24 h after transmission (Amino et al., 2006). Therefore, although most LN sporozoites are rapidly taken up by DCs, a few can initiate, but not complete, development, thereby expressing early liver-stage proteins in the dLN.

Parasite Development in the Skin

More decisive parasite development was later found to occur in the skin of mice. One day after mosquito inoculation of *P. berghei* sporozoites into the ear, around 10% of the injected sporozoites were detected as round, developing EEFs (Gueirard et al., 2010; Voza et al., 2012). Parasite growth was demonstrated by their increase in size measured by fluorescence microscopy, and by a sevenfold increase in the bioluminescence signals produced by luciferase-producing parasites in the ear skin between day 1 and day 3 posttransmission (Gueirard et al., 2010). Immunohistological analysis showed that parasites developed inside various cell types of the skin, including keratinocytes in the epidermis, dermal cells (presumably fibroblasts), and cells associated with sebaceous glands of hair follicles (Fig. 7.7A–C). In hair follicles, an immunoprivileged site of the

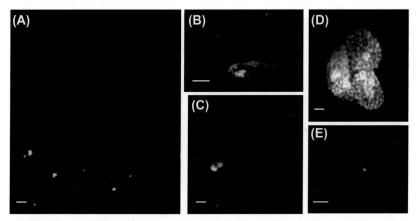

FIGURE 7.7 **Parasite development in the skin.** (A–C) Confocal images of skin thin sections showing *Plasmodium berghei* parasites in the dermis (A; scale bar, 10 μm) or epidermis (B; scale bar, 10 μm), or associated with a hair follicle (C; scale bar, 5 μm), of a mouse. Parasites are in green, nuclei in blue (DAPI), and keratinocytes (stained with K5 antibodies) in red. (D) Intravital imaging of a merozoite-filled, green fluorescent *P. berghei* exo-erythrocytic form in a mouse ear at day 3 postinoculation. Scale bar, 5 μm. (E) Red fluorescent *Plasmodium yoelii* merozoites inside a superficial keratinocyte in a mouse ear at day 2 postinoculation. Scale bar, 10 μm. *(Reproduced from Gueirard, P., Tavares, J., Thiberge, S., Bernex, F., Ishino, T., Milon, G., Franke-Fayard, B., Janse, C.J., Ménard, R., Amino, R., 2010. Development of the malaria parasite in the skin of the mammalian host. Proceedings of the National Academy of Sciences of the United States of America 107, 18640–18645.)*

mammalian body characterized by a strongly immunosuppressive environment (Bertolini et al., 2013), parasites were detected for over 2 weeks (Gueirard et al., 2010). Most surprisingly, schizonts filled with tens or hundreds of merozoites were occasionally detected by fluorescence microscopy; they were detected using both *P. berghei* and *P. yoelii* and typically in dermal cells, where parasite development appeared to be more efficient than in other sites in the skin (Fig. 7.7D and E). Ex vivo infections, using cells infected in the skin, grown in vitro and reinjected intravenously in mice, demonstrated that merozoites generated inside skin cells were infective (Gueirard et al., 2010). Whether skin merozoites can initiate an infection in situ could not be decisively addressed (Gueirard et al., 2010; Voza et al., 2012), although in any event skin merozoites would constitute only a very small contingent produced with delay compared to hepatic merozoites. Notably, both skin and hepatic merozoites egress from host cells by inducing the formation of merosomes, which are merozoite-filled vesicles that bud from the host cell plasma membrane and separate from the cell into the environment (Sturm et al., 2006), emphasizing the similarity of parasite development inside skin and liver cells.

CONCLUSION: A NEW PREERYTHROCYTIC PHASE IN MAMMALIAN MALARIA

The studies reported above in rodents have revealed important insights into the biology of the PE stages, such as parasite development inside skin cells, which shows its independence of any hepatocyte-specific factor as has long been assumed, or the apparent parasite persistence in association with hair follicles. Together, they also indicate that up to the 2/3 of the sporozoites inoculated into the skin eventually remain in the skin–LN area. In the skin, although many sporozoites are cleared, some invade host cells and start developing into EEFs, of which a few fully mature and generate merozoites.

It should be kept in mind that *P. berghei* and *P. yoelii* sporozoites do not naturally infect laboratory mice but rather wild thicket rats in Africa, and the time of adaptation in laboratory systems may not have been enough to optimize the rodent species/mouse combination. The apparently poor efficiency in leaving the skin may be due to inadequate guidance to blood vessels in mice, resulting in an increase in the occurrence of secondary pathways. Alternatively, some plasticity in sporozoite behavior may exist across mammalian *Plasmodium* species.

The most pressing question is to what extent the rodent picture reflects the situation in humans. It can be assumed that human-infecting sporozoites are also injected into the skin by the respective *Anopheles* mosquito species. The presence of sporozoites in human skin has been reported once in skin histological sections of human volunteers bitten by *P. vivax*-infected mosquitoes (Boyd and Kitchen, 1939) and does not appear to have been investigated since. Skin histological studies following injection by syringe of *P. cynomolgi* sporozoites into rhesus monkeys showed that sporozoites remained in the skin from 2 h (Lloyd, 1949) to more than 4 h (Hawking, 1948).

Whether human-infecting sporozoites injected into human skin may reach dLNs or develop inside skin cells is also unknown. A mosquito-transmitted *P. vivax* sporozoite was detected in a dLN of a human volunteer 24 h after the mosquito bite (Boyd and Kitchen, 1939). *P. falciparum* sporozoites are known to invade only primary human hepatocytes in vitro, but so were *P. yoelii* sporozoites before the demonstration of their development in the skin of mice. Interestingly, the molecular basis of host cell invasion by *P. yoelii* and *P. falciparum* sporozoites appears to be conserved, in particular the host cell CD81 dependence of sporozoite invasion of hepatocytes (Silvie et al., 2003).

The most important implications of this newly discovered skin phase may be immunological. Sporozoites appear to spend more time than anticipated in the skin, and thus expand the window of opportunity for neutralizing antibodies. The skin, where the first antiplasmodial responses are initiated, contains a rich array of immunologically competent cells. The sporozoites leave trails of antigens in the skin, in the matrix during gliding, and inside dead cells after traversal, reach the local LN, and invade skin cells where they release liver-stage material. Moreover, the new biology of PE stages may impact the vaccine approach based on live-attenuated sporozoites, which is increasingly considered for use in humans. These immunological and vaccine aspects are the subject of the associated Chapter 11.

INNATE IMMUNE RESPONSES TO FLEA-TRANSMITTED *YERSINIA PESTIS* IN THE SKIN

Jeffrey G. Shannon
National Institute of Allergy and Infectious Diseases (NIAID), National Institutes of Health (NIH), Hamilton, MT, United States

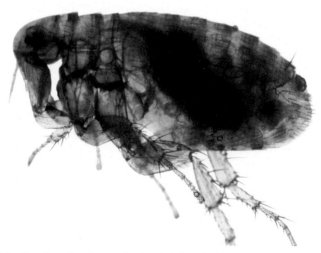

FIGURE 7.8   Female rat flea, *Xenopsylla cheopis*, infected with *Yersinia pestis*. *(Photo: B. Joseph Hinnebusch.)*

YERSINIA PESTIS AND PLAGUE

Plague is caused by the gram-negative bacterial pathogen *Y. pestis* and presents clinically as three distinct forms, pneumonic, bubonic, and primary septicemic. Pneumonic plague results from inhalation of aerosolized *Y. pestis*. The bubonic and primary septicemic forms of plague are typically transmitted by fleas and will be the focus of this section. Bubonic plague is the most common form in humans and results from deposition of *Y. pestis* in the skin via the bite of an infected flea. The bacteria survive in the dermis and quickly transit through the lymph to the regional dLN. There, *Y. pestis* replicates forming an enlarged, painful, swollen LN termed a bubo. If the infection is left untreated, the bacteria can replicate to large numbers, eventually destroy the normal LN architecture and spread systemically through the blood. The primary septicemic form of plague is also transmitted by fleas and occurs when infected fleas introduce *Y. pestis* directly into the bloodstream during feeding, thus bypassing the LN stage of disease. Humans with disseminated *Y. pestis* infection display general flulike symptoms such as fever, chills, lethargy, and myalgia. In the absence of antibiotic treatment, plague is often fatal. The rat flea species *Xenopsylla cheopis* (Fig. 7.8) is most commonly associated with maintenance of plague epizootic cycles in endemic regions and has been the most thoroughly studied, but a variety of flea species are capable of transmitting *Y. pestis* (Bitam et al., 2010).

Y. pestis may have had a greater impact on human civilization than any other pathogen. There have been three plague pandemics in recorded history, each believed to have spread from Asia along trade routes. The first pandemic affected the eastern Roman Empire beginning in the 6th century and was called the Justinian plague. The second pandemic, now known as the Black Death, had devastating effects on the European population in the 14th century and may have caused hundreds of millions of deaths worldwide. The third pandemic began in the late 18th century and is ongoing. This pandemic spread *Y. pestis* to every continent except Antarctica and has caused millions of deaths. Today, plague is endemic in countries in Asia, Africa, North America, and South America where *Y. pestis* is maintained by rodent-flea enzootic cycles. Improved living conditions, sanitation, and the availability of antibiotics have greatly reduced the public health impact of plague; however, sporadic outbreaks of plague occur in endemic areas, with hundreds of human plague cases being reported to the WHO each year.

Y. pestis evolved from its closest ancestor *Yersinia pseudotuberculosis* within the last 3000–6000 years (Hinnebusch et al., 2016; Rasmussen et al., 2015). Despite their genetic similarity, the two species differ dramatically in transmission mechanisms and pathogenesis. *Y. pseudotuberculosis* is orally transmitted and typically causes relatively mild, self-limiting gastrointestinal disease. In contrast, *Y. pestis* is transmitted by a flea vector or the aerosol route, is extremely virulent (LD50 < 10 organisms), and can spread systemically from peripheral routes of inoculation. The evolutionary steps leading from *Y. pseudotuberculosis* to *Y. pestis* involve both gene acquisition and gene loss culminating in the ability of *Y. pestis* to survive in the flea midgut while avoiding acute toxicity to the flea, replicate, form a cohesive biofilm within the flea, and disseminate from peripheral routes of infection, all traits absent in *Y. pseudotuberculosis* (Hinnebusch et al., 2016). One key evolutionary step was the acquisition of a plasmid encoding Ymt, a phospholipase D protein essential for survival of the bacteria within the flea midgut (Hinnebusch et al., 2002). Another was the mutation of the *ureD* gene encoding a subunit of urease. Urease activity of *Y. pseudotuberculosis* is toxic to fleas, and loss of this activity facilitated flea colonization (Chouikha and Hinnebusch, 2014). Additionally, conversion of the transcriptional regulator *rcs* and two phosphodiesterase genes to pseudogenes in *Y. pestis* increases cyclic-di-GMP levels, a bacterial second messenger molecule, resulting in greatly increased colonization and biofilm formation in the flea gut (Hinnebusch et al., 2016; Sun ct al., 2014).

TRANSMISSION OF *YERSINIA PESTIS* BY FLEAS

Fleas are flightless, blood-feeding insects of the order Siphonaptera. They are ectoparasites of warm-blooded vertebrates, mostly mammals and to a lesser extent birds. Two skin-piercing lacinae combine with an epipharynx to form the proboscis that is used for both delivery of saliva into host tissue and acquiring a blood meal. The flea proboscis is highly flexible, allowing it to bend laterally after penetrating the skin as it probes for a small blood vessel. Fleas are considered capillary feeders, in that they typically draw blood directly from blood vessels they have cannulated with their proboscis; however, occasional feeding on pooled blood in the tissue has been observed (Deoras and Prasad, 1967).

In nature, fleas become infected with *Y. pestis* by feeding on a host, typically a rodent, with high-level bacteremia. Studies of flea feeding behavior and infection of fleas with *Y. pestis* have been facilitated by the development of artificial flea feeding systems. These systems typically employ a heated chamber covered with mouse skin that contains blood loaded with a known concentration of *Y. pestis*. They have permitted a variety of experiments characterizing *Y. pestis*–flea interactions including identification of genetic determinants responsible for flea colonization by *Y. pestis* (Hinnebusch et al., 2016), the dose of *Y. pestis* needed in the blood meal to establish infection in the flea 10^7–10^8 bacteria/mL blood (Lorange et al., 2005), the transmission efficiency of *Y. pestis* by infected fleas, and the precise mechanisms by which *Y. pestis* is transmitted by fleas. Two distinct flea transmission mechanisms have been described.

Regurgitative Transmission

The most fully elucidated mechanism is referred to as regurgitative transmission and is dependent on biofilm formation by *Y. pestis* in the flea gut (Hinnebusch, 2012; Hinnebusch et al., 1996). The bacterial biofilm obstructs the proventricular valve of the flea digestive tract, thereby interfering with blood meal acquisition. This phenomenon is referred to as flea blockage. Blocked fleas make repeated attempts to feed during which bacteria from the biofilm can be regurgitated into the skin. The number of bacteria transmitted by a blocked flea is highly variable, with one study of individual infected fleas feeding on an artificial feeding system showing a range from fewer than ten to several thousand bacteria, with a median of 82 bacteria transmitted. Transmission occurred in only ~50% of the feeding events (Lorange et al., 2005). Bacteria are most commonly deposited in the dermis, after which they traffic to the dLN and subsequently cause the bubonic form of disease. However, fleas cannulate small blood vessels in the skin with their mouthparts during feeding (Deoras and Prasad, 1967; Lavoipierre, 1965) and thus can deposit bacteria intravenously resulting in primary septicemic plague (Sebbane et al., 2006).

Early-Phase Transmission

The second flea transmission mechanism is called early-phase transmission (EPT). EPT occurs during flea feeding on a naïve animal within 1–4 days after acquisition of the pathogen by feeding on infected blood. Less is known about the exact mechanisms responsible for EPT, but it appears to be biofilm-independent (Eisen et al., 2006, 2015). EPT may be more mechanical in nature, but the mechanism has not been described. The relative importance of biofilm-dependent transmission versus EPT in maintaining plague in nature is a subject of ongoing debate.

INNATE IMMUNE RESPONSE TO *YERSINIA PESTIS* IN THE SKIN
Role of Neutrophils Versus Macrophages

Once in the skin, bacteria quickly confront the mammalian host innate immune system. Neutrophils predominate the early host response (<12h postinfection [hpi]) to *Y. pestis* in the dermis, with a large influx of neutrophils seen by 2 hpi (Bosio et al., 2012; Shannon et al., 2013). Transgenic mice expressing high levels of GFP in neutrophils infected with red fluorescent *Y. pestis* have been used in intravital microscopy experiments to image *Y. pestis*–host cell interactions in vivo (Shannon et al., 2015). In these studies, large numbers of neutrophils were seen swarming around the injected bacteria in the dermis, with some of the bacteria associating with neutrophils and moved out of the field of view (Fig. 7.9, leftmost panels). Flow cytometry of cells isolated from infected mouse ears revealed that >80% of the cell-associated *Y. pestis* at 4 hpi were associated with neutrophils, with the remainder predominantly associated with macrophages (Shannon et al., 2013). The ultimate fate of these intracellular bacteria in vivo is not known. *Y. pestis* can survive and replicate within mouse and human macrophages in vitro (Pujol and Bliska, 2005). In contrast, most *Y. pestis* phagocytosed by neutrophils are killed (Spinner et al., 2008, 2013b). It has

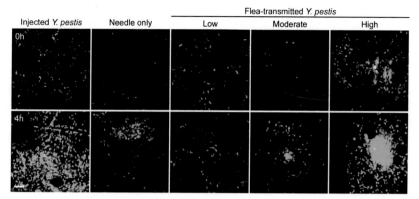

FIGURE 7.9   Imaging neutrophil and macrophage response to flea-transmitted *Yersinia pestis* in the dermis in vivo. Confocal images of Lys-eGFP mouse ears after being fed on for 50 min by fleas infected and blocked with *Y. pestis* pMcherry (Low = 2 fleas, Moderate = 5 fleas, High = 2 fleas). Images of needle-inoculated *Y. pestis* expressing dsRed (~1000 CFU) and a sterile needle-stick site are shown for comparison. Mice were injected with SytoxBlue i.p. prior to flea feeding. Upper panels show t = 0 h and lower panels show t = 4 h. SytoxBlue staining was used to identify the flea bite sites (blue, shown in upper panels only for all experiments except the needle-inoculated *Y. pestis*). GFP^{dim} cells are macrophages, GFP^{bright} cells are neutrophils, and red is *Y. pestis* pMcherry (near the center of each image, in the vicinity of intense SytoxBlue staining). Left, center, and right panels show low (~10 or fewer bacteria), moderate (~100s of bacteria), and high (~1000s of bacteria) levels of transmission, respectively. Scale bar represents 100 μm. (*Figure reproduced from Shannon, J.G., Bosio, C.F., Hinnebusch, B.J., 2015. Dermal neutrophil, macrophage and dendritic cell responses to* Yersinia pestis *transmitted by fleas. PLoS Pathogens 11, e1004734.*)

been proposed that macrophages provide a protective niche for *Y. pestis* early in infection, giving the pathogen time to upregulate virulence factors essential for evasion of the more microbicidal neutrophils; however, it remains possible that a proportion of inoculated bacteria remains extracellular and it is this population that goes on to disseminate systemically.

In a recent study, Gonzalez et al. infected mice intradermally with an equal mixture of 10 genetically tagged, but otherwise identical *Y. pestis* strains and found that, on average, just over two of the strains could be isolated from the dLN (Gonzalez et al., 2015). This suggests that the dermis represents a strong bottleneck that allows very few of the injected bacteria through. The authors also showed that neutrophil depletion dramatically increased the number of *Y. pestis* at the injection site, indicating that neutrophils can at least partially control the proliferation of *Y. pestis* remaining in the dermis (Gonzalez et al., 2015).

Role of Dendritic Cells

Y. pestis dissemination from the skin to the dLN is a key step in bubonic plague pathogenesis. DCs were long assumed to play a role in this dissemination due to their ability to phagocytose particulate antigens or pathogens and traffic to lymphoid tissue via the afferent lymph. However, studies using mice deficient in CCR7, a chemokine receptor required for homing of DCs from periphery to LNs, revealed no role for DCs in dissemination of *Y. pestis* from the skin (Gonzalez et al., 2015; Shannon et al., 2013). Additionally, these studies showed that neutrophil depletion had no effect on *Y. pestis* dissemination kinetics. Gonzalez et al. also observed cell-free *Y. pestis* in the lymphatic vessels in fixed tissue samples early after intradermal infection (Gonzalez et al., 2015). These data, combined with the fact that *Y. pestis* can be found in the dLN within 10 min to 1 h after injection or flea transmission, respectively, suggest that *Y. pestis* dissemination to the dLN can occur passively without the aid of host cells or motility of the pathogen.

POTENTIAL EFFECTS OF FLEA TRANSMISSION ON PLAGUE PATHOGENESIS

Much of the work on bubonic plague pathogenesis to date has focused on models using intradermal or subcutaneous needle inoculation of animals with *Y. pestis* grown in traditional bacteriologic medium. Obviously, infection in the context of flea feeding has the potential to alter the host–*Y. pestis* interactions and disease pathogenesis. As mentioned previously, fleas are primarily capillary feeders that probe the skin with their mouthparts until they cannulate a small blood vessel and draw a blood meal. Like other hematophagous arthropods, fleas deposit saliva in the skin during feeding and salivary components likely disrupt the normal hemostatic mechanisms present in the mammalian host, facilitating blood meal acquisition. Transcriptomic and proteomic analyses of cat flea (*Ctenocephalides felis*) and rat flea (*X. cheopis*) salivary glands revealed a number of molecules with potential to affect blood clotting, vasodilation, and inflammation (Andersen et al., 2007; Ribeiro et al., 2012). These include apyrase (hydrolyzes ADP and ATP, molecules responsible for platelet aggregation and inflammation, respectively), adenosine deaminases (convert adenosine to inosine, possibly to reduce mast cell degranulation (Tilley et al., 2000)), and esterases (may affect platelet aggregation (Cheeseman et al., 2001)). Interestingly, the most abundant proteins in the flea saliva are members of a family of closely related acid phosphatases that have lost their enzymatic activity (Andersen et al., 2007). The function of these inactive phosphatases is unknown.

In a recent study, Bosio et al. examined the innate host cell response of mice to uninfected rat flea feeding by flow cytometry and histology (Bosio et al., 2014). The authors found transient, nonpapular, nonedematous erythematous lesions present at some fleabite sites and that flea feeding induced a mild inflammatory response characterized by a small increase in neutrophils and macrophages in mouse skin. They detected no clear monocyte, eosinophil, basophil, or mast cell response to the fleabites up to 48-h postfeeding. Similarly mild inflammation was reported in a study of rat (*Rattus norvegicus*) responses to *X. cheopis* feeding (Vaughan et al., 1989). Bosio et al. also found that feeding of 11–22 uninfected fleas on mouse ears prior to injection with *Y. pestis* had no effect on bacterial survival in the skin or dLN up to 24 hpi (Bosio et al., 2014). This was the case in naïve animals and animals that had been exposed to fleas repeatedly over the course of 10 weeks to simulate the constant flea exposure wild rodents experience. The repeated flea exposure induced an immune response to flea salivary proteins but did not alter susceptibility to plague following challenge by blocked fleas. Thus, in this study, *Y. pestis* infection in the context of flea feeding did not appear to affect plague pathogenesis.

Another study used intravital microscopy to image the cellular response to both uninfected and *Y. pestis*-infected fleabites (Shannon et al., 2015). One obstacle encountered in this study is that while flea feeding sometimes leaves a visible area of erythema at the feeding site (Fig. 7.10A), frequently fleas leave no visible indication of where they have fed. Injection of mice with the non–cell-permeant fluorescent DNA stain SytoxBlue facilitated identification of fleabite sites by allowing the detection of minute amounts of cell damage inflicted during insertion of flea mouthparts into the skin (Fig. 7.10B and C). Imaging up to 4 h after flea feeding revealed surprisingly few neutrophils recruited to the bite sites despite the presence of tissue damage (Fig. 7.10D). Bite sites containing variable numbers of flea-transmitted *Y. pestis* after feeding by infected blocked fleas were also imaged. Neutrophil recruitment to bite sites containing bacteria was increased compared to uninfected flea bites and correlated with the number of bacteria deposited in the skin, which can vary from <10 to >1000 bacteria (Fig. 7.9) (Shannon et al., 2015). Overall, less *Y. pestis*–neutrophil interaction was observed compared to needle-inoculated bacteria. Additionally, interactions between flea-transmitted bacteria and macrophages may be more common than was observed after needle inoculation of *Y. pestis* (Fig. 7.11). This may have significant implications for pathogenesis as macrophages are much more permissive for *Y. pestis* intracellular survival and growth than neutrophils. Tissue-resident DCs near the bite site were also seen migrating toward flea-transmitted *Y. pestis* in the skin, but not uninfected flea bites; however, no direct *Y. pestis*–DC interactions were observed (Shannon et al., 2015).

Another important consideration is that flea-transmitted *Y. pestis* are coming from the flea digestive tract and have a very different transcriptional profile compared to broth-grown cultures (Vadyvaloo et al., 2010). *Y. pestis* relies on the type III secretion system (T3SS) and associated effectors with potent antiinflammatory and antiphagocytic effects on mammalian cells (Viboud and Bliska, 2005); however, this system is temperature regulated and is not expressed during growth in the flea. The *Y. pestis* PhoPQ two-component regulatory system is important for defense against host defenses such as antimicrobial peptides, low pH, and osmotic stress (O'Loughlin et al., 2009). The *Y. pestis* toxin complex (Tc) proteins have antiphagocytic effects (Spinner et al., 2013a; Vadyvaloo et al., 2010). The genes encoding the PhoPQ system and the Tc proteins are downregulated in broth cultures but induced in the flea midgut (Vadyvaloo et al., 2010).

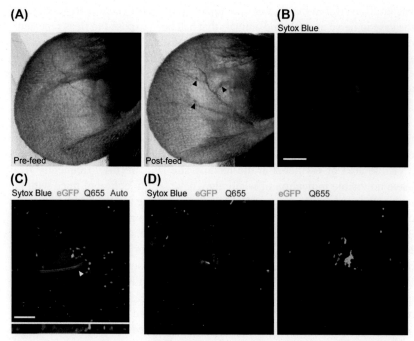

FIGURE 7.10 Characterization of flea bites in mouse skin. (A) Dissecting microscope images of a mouse ear before and after being fed on by three blocked fleas for 50 min. *Arrowheads* indicate small spots of erythema that can sometimes be found on the ear after flea feeding. Often there is no obvious indication of where fleas have fed, which necessitates the use of a fluorescent reporter of cell damage at the bite site. (B) Confocal image of the ear of a mouse injected i.p. with SytoxBlue prior to being fed on by an uninfected flea. The blue color indicates an area where cells have been damaged by flea feeding. (C) Confocal image of the ear of a LysM-eGFP mouse that was injected with SytoxBlue i.p. and Qtracker655 vascular dye i.v. prior to being fed on by a blocked flea. During feeding the flea was anesthetized with isoflurane and its embedded mouthparts cut with microscissors. GFP^{dim} cells are macrophages, GFP^{bright} cells are neutrophils, red is the Q655 vascular dye, blue is the SytoxBlue, and magenta is autofluorescence of the flea mouthparts. The *arrowhead* indicates where the flea mouthparts pierce the skin. The lower panel is an x-z cross section through the bite site. The intense red coloration in the area is blood leakage that occurred during flea feeding. SytoxBlue stains the cells damaged by flea mouthparts probing of the skin. (D) Confocal images of a mouse ear prepared exactly as in (C) after being fed on by two uninfected fleas. The 0 h and 4 h time points are shown. Some GFP^{bright} neutrophils are recruited to uninfected fleabites over the first 4 h. Scale bars represent 100 μm. *((D) Figure reproduced from Shannon, J.G., Bosio, C.F., Hinnebusch, B.J., 2015. Dermal neutrophil, macrophage and dendritic cell responses to* Yersinia pestis *transmitted by fleas. PLoS Pathogens 11, e1004734.)*

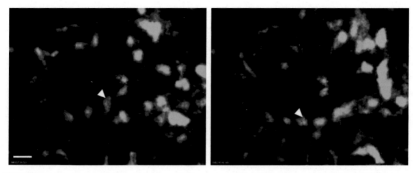

FIGURE 7.11 **Increased association of flea-transmitted *Yersinia pestis* with eGFP^{dim} macrophages in the dermis in vivo.** A Lys-eGFP mouse ear after being fed on for 50 min by five *Y. pestis* pMcherry blocked fleas. Confocal images of ~1 h 17 min and 2 h time points (~1.5- and 2.25-h postfeeding, respectively) of a 4 h movie are shown. The flea bite site was identified by SytoxBlue staining prior to acquisition of time series. Images were digitally magnified 6× from original. *Arrowhead* in each panel indicates *Y. pestis* (red) associated with the same GFP^{dim} macrophage (green) at each time point. Scale bar represents 20 μm. *(Figure reproduced from Shannon, J.G., Bosio, C.F., Hinnebusch, B.J., 2015. Dermal neutrophil, macrophage and dendritic cell responses to* Yersinia pestis *transmitted by fleas. PLoS Pathogens 11, e1004734.)*

Additionally, *Y. pestis* originating from the flea gut will be associated with biofilm extracellular matrix (ECM) polysaccharide. The effects of *Y. pestis* biofilm ECM on innate immune cells are unknown, but these components structurally similar to ECM of *Staphylococcus epidermidis* shown to inhibit innate immune effectors (Vuong et al., 2004). Thus, *Y. pestis* growing within the flea may be preadapted for survival after transmission into the mammalian host, at least for a short time until the T3SS is induced.

CONCLUSIONS

Overall, data from microscopy and flow cytometry studies demonstrate that uninfected flea feeding induces minimal skin inflammation in rodent models (Bosio et al., 2014; Shannon et al., 2015). Whether flea saliva plays an active role in dampening the inflammatory response is unknown. It is possible that the extremely small size of the flea mouthparts minimizes tissue damage done during feeding, thus reducing inflammatory cell recruitment. If saliva is playing a role, it does not appear to suppress the neutrophil recruitment in response to bacteria present at the bite site. There does appear to be less neutrophil interaction with flea-transmitted *Y. pestis* compared to needle-inoculated bacteria. Whether this difference affects plague pathogenesis is unknown. Due to the highly variable number of bacteria transmitted by individual fleas and the extremely low LD$_{50}$ of *Y. pestis* by the intradermal route, at present it is difficult to quantitatively assess any potential differences in LD$_{50}$ between flea transmission and needle inoculation. It should also be noted that laboratory mice and rats appear to become desensitized to flea saliva after repeated flea exposure over time and exhibit less inflammation at fleabite sites (Bosio et al., 2014). In contrast, dogs, cats, and humans often become allergically sensitized to flea salivary components and subsequently develop a hypersensitivity response to fleabites. It is possible that the more inflammatory environment at fleabite sites in these sensitized animals might affect *Y. pestis* pathogenesis, but this has not yet been studied. In the future, it will be important to consider the potential impact of flea saliva and bacterial gene expression changes within the flea when designing experiments and interpreting results from in vivo models of *Y. pestis* infection. Clearly, more work is needed to improve our understanding of the role of the flea vector on bubonic plague pathogenesis.

ACKNOWLEDGMENTS

This research was supported by the Intramural Research Program of the NIH, National Institute of Allergy and Infectious Diseases.

REFERENCES

Alonso, P.L., Tanner, M., 2013. Public health challenges and prospects for malaria control and elimination. Nature Medicine 19, 150–155.

Amino, R., Thiberge, S., Martin, B., Celli, S., Shorte, S., Frischknecht, F., Ménard, R., 2006. Quantitative imaging of *Plasmodium* transmission from mosquito to mammal. Nature Medicine 12, 220–224.

Amino, R., Thiberge, S., Blazquez, S., Baldacci, P., Renaud, O., Shorte, S., Ménard, R., 2007. Imaging malaria sporozoites in the dermis of the mammalian host. Nature Protocols 2, 1705–1712.

Amino, R., Giovannini, D., Thiberge, S., Gueirard, P., Boisson, B., Dubremetz, J.-F., Prévost, M.-C., Ishino, T., Yuda, M., Ménard, R., 2008. Host cell traversal is important for progression of the malaria parasite through the dermis to the liver. Cell Host and Microbe 3, 88–96.

Andersen, J.F., Hinnebusch, B.J., Lucas, D.A., Conrads, T.P., Veenstra, T.D., Pham, V.M., Ribeiro, J.M.C., 2007. An insight into the sialome of the oriental rat flea, *Xenopsylla cheopis* (Rots). BMC Genomics 8, 102.

Beier, J.C., Davis, J.R., Vaughan, J.A., Noden, B.H., Beier, M.S., 1991. Quantitation of *Plasmodium falciparum* sporozoites transmitted in vitro by experimentally infected *Anopheles gambiae* and *Anopheles stephensi*. American Journal of Tropical Medicine and Hygiene 44, 564–570.

Beier, M.S., Davis, J.R., Pumpuni, C.B., Noden, B.H., Beier, J.C., 1992. Ingestion of *Plasmodium falciparum* sporozoites during transmission by anopheline mosquitoes. American Journal of Tropical Medicine and Hygiene 47, 195–200.

Bertolini, M., Meyer, K.C., Slominski, R., Kobayashi, K., Ludwig, R.J., Paus, R., 2013. The immune system of mouse vibrissae follicles: cellular composition and indications of immune privilege. Experimental Dermatology 22, 593–598.

Bhatt, S., Weiss, D.J., Cameron, E., Bisanzio, D., Mappin, B., Dalrymple, U., Battle, K.E., Moyes, C.L., Henry, A., Eckhoff, P.A., Wenger, E.A., Briët, O., Penny, M.A., Smith, T.A., Bennett, A., Yukich, J., Eisele, T.P., Griffin, J.T., Fergus, C.A., Lynch, M., Lindgren, F., Cohen, J.M., Murray, C.L.J., Smith, D.L., Hay, S.I., Cibulskis, R.E., Gething, P.W., 2015. The effect of malaria control on *Plasmodium falciparum* in Africa between 2000 and 2015. Nature 526, 207–211.

Bignami, A., 1898. The inoculation theory of malarial infection: account of a successful experiment with mosquitoes. Lancet ii, 1461–1463.

Bitam, I., Dittmar, K., Parola, P., Whiting, M.F., Raoult, D., 2010. Fleas and flea-borne diseases. International Journal of Infectious Diseases 14, e667–676.

Bosio, C.F., Jarrett, C.O., Gardner, D., Hinnebusch, B.J., 2012. Kinetics of the innate immune response to *Yersinia pestis* after intradermal infection in a mouse model. Infection and Immunity.

Bosio, C.F., Viall, A.K., Jarrett, C.O., Gardner, D., Rood, M.P., Hinnebusch, B.J., 2014. Evaluation of the murine immune response to *Xenopsylla cheopis* flea saliva and its effect on transmission of *Yersinia pestis*. PLoS Neglected Tropical Diseases 8, e3196.

Boucher, L.E., Bosch, J., 2015. The apicomplexan glideosome and adhesins – structure and function. Journal of Structural Biology 190, 93–114.

Boyd, M.F., Kitchen, S.F., 1939. The demonstration of sporozoites in human tissues. The American Journal of Tropical Medicine and Hygiene 19, 27–31.

Carter, R., Mendis, K.N., 2002. Evolutionary and historical aspects of the burden of malaria. Clinical Microbiology Reviews 15, 564–594.

Cheeseman, M.T., Bates, P.A., Crampton, J.M., 2001. Preliminary characterisation of esterase and platelet-activating factor (PAF)-acetylhydrolase activities from cat flea (*Ctenocephalides felis*) salivary glands. Insect Biochemistry and Molecular Biology 31, 157–164.

Chouikha, I., Hinnebusch, B.J., 2014. Silencing urease: a key evolutionary step that facilitated the adaptation of *Yersinia pestis* to the flea-borne transmission route. Proceedings of the National Academy of Sciences of the United States of America 111, 18709–18714.

Coppi, A., Tewari, R., Bishop, J.R., Bennett, B.L., Lawrence, R., Esko, J.D., Billker, O., Sinnis, P., 2007. Heparan sulfate proteoglycans provide a signal to *Plasmodium* sporozoites to stop migrating and productively invade host cells. Cell Host and Microbe 2, 316–327.

Deoras, P.J., Prasad, R.S., 1967. Feeding mechanism of Indian fleas *X. cheopis* (Roths) and *X. astia* (Roths). Indian Journal of Medical Research 55, 1041–1050.

Eisen, R.J., Bearden, S.W., Wilder, A.P., Montenieri, J.A., Antolin, M.F., Gage, K.L., 2006. Early-phase transmission of *Yersinia pestis* by unblocked fleas as a mechanism explaining rapidly spreading plague epizootics. Proceedings of the National Academy of Sciences of the United States of America 103, 15380–15385.

Eisen, R.J., Dennis, D.T., Gage, K.L., 2015. The role of early-phase transmission in the spread of *Yersinia pestis*. Journal of Medical Entomology 52, 1183–1192.

Engelmann, S., Sinnis, P., Matuschewski, K., 2006. Transgenic *Plasmodium berghei* sporozoites expressing β-galactosidase for quantification of sporozoite transmission. Molecular and Biochemical Parasitology 146, 30–37.

Fairley, N.H., 1947. Sidelights on malaria in man obtained by subinoculation experiments. Transactions of the Royal Society of Tropical Medicine and Hygiene 40, 621–676.

Formaglio, P., Tavares, J., Ménard, R., Amino, R., 2014. Loss of host cell plasma membrane integrity following cell traversal by *Plasmodium* sporozoites in the skin. Parasitology International 63, 237–244.

Franke-Fayard, B., Trueman, H., Ramesar, J., Mendoza, J., van der Keur, M., van der Linden, R., Sinden, R.E., Waters, A.P., Janse, C.J., 2004. A *Plasmodium berghei* reference line that constitutively expresses GFP at a high level throughout the complete life cycle. Molecular and Biochemical Parasitology 137, 23–33.

Franke-Fayard, B., Djokovic, D., Dooren, M.W., Ramesar, J., Waters, A.P., Falade, M.O., Kranendonk, M., Martinelli, A., Cravo, P., Janse, C.J., 2008. Simple and sensitive antimalarial drug screening in vitro and in vivo using transgenic luciferase expressing *Plasmodium berghei* parasites. International Journal for Parasitology 38, 1651–1662.

Frevert, U., Engelmann, S., Zougbédé, S., Stange, J., Ng, B., Matuschewski, K., Liebes, L., Yee, H., 2005. Intravital observation of *Plasmodium berghei* sporozoite infection of the liver. PLoS Biology 3, e192.

Frischknecht, F., Baldacci, P., Martin, B., Zimmer, C., Thiberge, S., Olivo-Marin, J.C., Shorte, S., Ménard, R., 2004. Imaging movement of malaria parasites during transmission by *Anopheles* mosquitoes. Cellular Microbiology 6, 687–694.

Golenda, C.F., Burge, R., Schneider, I., 1992. *Plasmodium falciparum* and *P. berghei*: detection of sporozoites and the circumsporozoite proteins in the saliva of *Anopheles stephensi* mosquitoes. Parasitology Research 78, 563–569.

Golgi, C., 1889. Sul ciclo evolutivo dei parassiti malarici nella febbre terzana: diagnosi differenziale tra i parassiti endoglobulari malarici della terzana e quelli della quartana. Archivio per le Scienze Mediche (Torino) 13, 173–196.

Gonzalez, R.J., Lane, M.C., Wagner, N.J., Weening, E.H., Miller, V.L., 2015. Dissemination of a highly virulent pathogen: tracking the early events that define infection. PLoS Pathogens 11, e1004587.

Grassi, B., 1898. La malaria propagata per mezzo di peculiari insetti. Nota preliminare. Accademia dei Lincei 7, 234–240.

Griffiths, R.B., Gordon, R.M., 1952. An apparatus which enables the process of feeding by mosquitoes to be observed in the tissues of a live rodent, together with an account of the ejection of saliva and its significance in malaria. Annals of Tropical Medicine and Parasitology 46, 311–319.

Gueirard, P., Tavares, J., Thiberge, S., Bernex, F., Ishino, T., Milon, G., Franke-Fayard, B., Janse, C.J., Ménard, R., Amino, R., 2010. Development of the malaria parasite in the skin of the mammalian host. Proceedings of the National Academy of Sciences of the United States of America 107, 18640–18645.

Hawking, F., 1948. Pre-erythrocytic stage in mammalian malaria parasites. Nature 161, 175.

Hellmann, J.K., Münter, S., Kudryashev, M., Schulz, S., Heiss, K., Müller, A.-K., Matuschewski, K., Spatz, J.P., Schwarz, U.S., Frischknecht, F., 2011. Environmental constraints guide migration of malaria parasites during transmission. PloS Pathogens 7, e1002080.

Hinnebusch, B.J., Perry, R.D., Schwan, T.G., 1996. Role of the *Yersinia pestis* hemin storage (*hms*) locus in the transmission of plague by fleas. Science 273, 367–370.

Hinnebusch, B.J., Rudolph, A.E., Cherepanov, P., Dixon, J.E., Schwan, T.G., Forsberg, Å., 2002. Role of Yersinia murine toxin in survival of *Yersinia pestis* in the midgut of the flea vector. Science 296, 733–735.

Hinnebusch, B.J., Chouikha, I., Sun, Y.C., 2016. Ecological opportunity, evolution, and the emergence of flea-borne plague. Infection and Immunity 84, 1932–1940.

Hinnebusch, B.J., 2012. Biofilm-dependent and biofilm-independent mechanisms of transmission of *Yersinia pestis* by fleas. Advances in Experimental Medicine and Biology 954, 237–243.

Hopp, C.S., Chiou, K., Ragheb, D.R.T., Salman, A.M., Khan, S.M., Liu, A.J., Sinnis, P., 2015. Longitudinal analysis of *Plasmodium* sporozoite motility in the dermis reveals component of blood vessel recognition. eLife 4, e07789.

Howes, R.E., Battle, K.E., Mendis, K.N., Smith, D.L., Cibulskis, R.E., Baird, J.K., Hay, S.I., 2016. Global epidemiology of *Plasmodium vivax*. American Journal of Tropical Medicine and Hygiene pii: 16–0141.

Huff, C.G., 1947. Life cycle of malarial parasites. Annual Review of Microbiology 1, 43–60.

Ishino, T., Yano, K., Chinzei, Y., Yuda, M., 2004. Cell-passage activity is required for the malarial parasite to cross the liver sinusoidal cell layer. PLoS Biology 2, e4.

Ishino, T., Chinzei, Y., Yuda, M., 2005. A *Plasmodium* sporozoite protein with a membrane attack complex domain is required for breaching the liver sinusoidal cell layer prior to hepatocyte infection. Cellular Microbiology 7, 199–208.

Ishino, T., Orito, Y., Chinzei, Y., Yuda, M., 2006. A calcium-dependent protein kinase regulates *Plasmodium* ookinete access to the midgut epithelial cell. Molecular Microbiology 59, 1175–1184.

Jin, Y., Kebaier, C., Vanderberg, J.P., 2007. Direct microscopic quantification of dynamics of *Plasmodium berghei* sporozoite transmission from mosquitoes to mice. Infection and Immunity 75, 5532–5539.

Kashin, P., 1966. Electronic recording of the mosquito bite. Journal of Insect Physiology 12, 281–286.

Kebaier, C., Vanderberg, J.P., 2006. Reingestion of *Plasmodium berghei* sporozoites after delivery into the host by mosquitoes. American Journal of Tropical Medicine and Hygiene 75, 1200–1204.

Kebaier, C., Vanderberg, J.P., 2010. Initiation of *Plasmodium* sporozoite motility by albumin is associated with induction of intracellular signaling. International Journal of Parasitology 40, 25–33.

Kebaier, C., Voza, T., Vanderberg, J., 2009. Kinetics of mosquito-injected *Plasmodium* sporozoites in mice: fewer sporozoites are injected into sporozoite-immunized mice. PLoS Pathogens 5, e1000399.

Krettli, A.U., Dantas, L.A., 2000. Which routes do *Plasmodium* sporozoites use for successful infections of vertebrates? Infection and Immunity 68, 3064–3065.

Krotoski, W.A., Collins, W.E., Bray, R.S., Garnhman, P.C., Cogswell, F.B., Gwadz, R.W., Killick-Kendrick, R., Wolf, R., Sinden, R., Koontz, L.C., Stanfill, P.S., 1982. Demonstration of hypnozoites in sporozoite-transmitted *Plasmodium* vivax infection. American Journal of Tropical Medicine and Hygiene 31, 1291–1293.

Laveran, A., 1880. Bulletin de l'Académie de Médecine, séance du 23 novembre 1880: note sur un nouveau parasite trouvé dans le sang de plusieurs malades atteints de fièvre palustre. 2ème Série IX, 1235–1236.

Lavoipierre, M.M.J., 1965. Feeding mechanism of blood-sucking arthropods. Nature 208, 302–303.

Li, X., Rossignol, J., 1992. Blood vessel location time by *Anopheles stephensi* (Diptera: Culicidae). Journal of Medical Entomology 29, 122–124.

Lloyd, O.C., 1949. The fate of sporozoites of *P. cynomolgi* injected into the skin of rhesus monkeys. The Journal of Pathology and Bacteriology 61, 144–146.

Lorange, E.A., Race, B.L., Sebbane, F., Hinnebusch, B.J., 2005. Poor vector competence of fleas and the evolution of hypervirulence in *Yersinia pestis*. The Journal of Infectious Diseases 191, 1907–1912.

Marchiafava, E., Celli, A., 1883. Sulle alterazioni dei globuli rossi nelle infezioni da malaria e sulla genesi della melanemia. Atto della Accademia Nazionale dei Lincei Series 3 (18), 381.

Martinsen, E.S., Perkins, S.L., Schall, J.J., 2008. A three-genome phylogeny of malaria parasites (*Plasmodium* and closely related genera): evolution of life-history traits and host switches. Molecular Phylogenetics and Evolution 47, 261–273.

Matsuoka, H., Yoshida, S., Hirai, M., Ishii, A., 2002. A rodent malaria, *Plasmodium berghei*, is experimentally transmitted to mice by merely probing of infective mosquito, *Anopheles stephensi*. Parasitology International 51, 17–23.

Medica, D.L., Sinnis, P., 2005. Quantitative dynamics of *Plasmodium yoelii* sporozoite transmission by infected anopheline mosquitoes. Infection and Immunity 73, 4363–4369.

Ménard, R., Sultan, A.A., Cortes, C., Altszuler, R., van Dijk, M.R., Janse, C.J., Waters, A.P., Nussenzweig, R.S., Nussenzweig, V., 1997. Circumsporozoite protein is required for development of malaria sporozoites in mosquitoes. Nature 385, 336–340.

Mendis, K., Rietveld, A., Warsame, M., Bosman, A., Greenwood, B., Wernsdorfer, W.H., 2009. From malaria control to eradication: the WHO perspective. Tropical Medicine and International Health 14, 1–7.

Morrison, D.A., 2009. Evolution of the Apicomplexa: where are we now? Trends in Parasitology 25, 375–382.

Morrissette, N.S., Sibley, L.D., 2002. Cytoskeleton of apicomplexan parasites. Microbiology and Molecular Biology Reviews 66, 21–38.

Mota, M.M., Pradel, G., Vanderberg, J.P., Hafalla, J.C., Frevert, U., Nussenzweig, R.S., Nussenzweig, V., Rodríguez, A., 2001. Migration of *Plasmodium* sporozoites through cells before infection. Science 291, 141–144.

Natarajan, R., Thathy, V., Mota, M.M., Hafalla, J.C., Ménard, R., Vernick, K.D., 2001. Fluorescent *Plasmodium berghei* sporozoites and pre-erythrocytic stages: a new tool to study mosquito and mammalian host interactions with malaria parasites. Cellular Microbiology 3, 371–379.

Nerlich, A.G., Schraut, B., Dittrich, S., Jelinek, T., Zink, A.R., 2008. *Plasmodium falciparum* in ancien Egypt. Emerging Infectious Diseases 14, 1317–1319.

O'Loughlin, J.L., Spinner, J.L., Minnich, S.A., Kobayashi, S.D., 2009. *Yersinia pestis* two-component gene regulatory systems promote survival in human neutrophils. Infection and Immunity 78, 773–782.

Ono, T., Tadakuma, T., Rodriguez, A., 2007. *Plasmodium yoelii yoelii* 17XNL constitutively expressing GFP throughout the life cycle. Experimental Parasitology 115, 310–313.

Perschmann, N., Hellmann, J.K., Frischknecht, F., Spatz, J.P., 2011. Induction of malaria parasite migration by synthetically tunable microenvironments. Nano Letters 11, 4468–4474.

Ponnudurai, T., Lensen, A.H.W., van Gemert, G.J.A., Bolmer, M.G., Meuwissen, J.H.E.T., 1991. Feeding behavior and sporozoite ejection by infected *Anopheles stephensi*. Transactions of the Royal Society of Tropical Medicine and Hygiene 85, 175–180.

Pujol, C., Bliska, J.B., 2005. Turning *Yersinia* pathogenesis outside in: subversion of macrophage function by intracellular yersiniae. Clinical Immunology 114, 216–226.

Radtke, A.J., Kastenmüller, W., Espinosa, D.A., Gerner, M.Y., Tse, S.-W., Sinnis, P., Germain, R.N., Zavala, F.P., Cockburn, I.A., 2015. Lymph-node resident CD8α+ dendritic cells capture antigens from migratory malaria sporozoites and induce CD8+ T cell responses. PLoS Pathogens 11, e1004637.

Rasmussen, S., Allentoft, M.E., Nielsen, K., Orlando, L., Sikora, M., Sjogren, K.G., Pedersen, A.G., Schubert, M., Van Dam, A., Kapel, C.M., et al., 2015. Early divergent strains of *Yersinia pestis* in Eurasia 5,000 years ago. Cell 163, 571–582.

Ribeiro, J.M., Rossignol, P.A., Spielman, A., 1985. *Aedes aegypti*: model for blood finding strategy and prediction of parasite manipulation. Experimental Parasitology 60, 118–132.

Ribeiro, J.M., Assumpcao, T.C., Ma, D., Alvarenga, P.H., Pham, V.M., Andersen, J.F., Francischetti, I.M., Macaluso, K.R., 2012. An insight into the sialotranscriptome of the cat flea, *Ctenocephalides felis*. PLoS One 7, e44612.

Rosenberg, R., Wirtz, R.A., Schneider, I., Burge, R., 1990. An estimation of the number of malaria sporozoites ejected by a feeding mosquito. Transactions of the Royal Society of Tropical Medicine and Hygiene 84, 209–212.

Ross, R., 1897. On some peculiar pigmented cells found in two mosquitoes fed on malaria blood. British Medical Journal ii, 1786–1788.

Ross, R., 1898. The role of the mosquito in the evolution of the malaria parasite. Lancet ii, 488–489.

Schaudinn, F., 1903. Studien über krankheitserregende Protozoen II. *Plasmodium vivax* (Grassi et Feletti) der Erreger des Tertianfiebers beim Menschen. Arbeit Kaiserlich Gesund 19, 169–250.

Sebbane, F., Jarrett, C.O., Gardner, D., Long, D., Hinnebusch, B.J., 2006. Role of the *Yersinia pestis* plasminogen activator in the incidence of distinct septicemic and bubonic forms of flea-borne plague. Proceedings of the National Academy of Sciences of the United States of America 103, 5526–5530.

Shannon, J.G., Hasenkrug, A.M., Dorward, D.W., Nair, V., Carmody, A.B., Hinnebusch, B.J., 2013. *Yersinia pestis* subverts the dermal neutrophil response in a mouse model of bubonic plague. mBio 4 e00170–13.

Shannon, J.G., Bosio, C.F., Hinnebusch, B.J., 2015. Dermal neutrophil, macrophage and dendritic cell responses to *Yersinia pestis* transmitted by fleas. PLoS Pathogens 11, e1004734.

Shortt, H.E., Garnham, P.C., 1948. Pre-erythrocytic stage in mammalian malaria parasites. Nature 161, 126.

Shortt, H.E., Garnham, P.C.C., Malamos, B., 1948. The pre-erythrocytic stage of mammalian malaria. British Medical Journal 1, 192–194.

Sidjanski, S., Vanderberg, J.P., 1997. Delayed migration of *Plasmodium* sporozoites from the mosquito bite site to the blood. American Journal of Tropical Medicine and Hygiene 57, 426–429.

Silvie, O., Rubinstein, E., Franetich, J.-F., Prenant, M., Belnoue, E., Rénia, L., Hannoun, L., Eling, W., Levy, S., Boucheix, C., Mazier, D., 2003. Hepatocyte CD81 is required for *Plasmodium falciparum* and *Plasmodium yoelii* sporozoite infectivity. Nature Medicine 9, 93–96.

Sinka, M.E., Bangs, M.J., Manguin, S., Rubio-Palis, Y., Chareonviriyaphap, T., Coetzee, M., Mbogo, C.M., Hemingway, J., Patil, A.P., Temperley, W.H., Gething, P.W., Kabaria, C.W., Burkot, T.R., Harbach, R.E., Hay, S.I., 2012. A global map of dominant malaria vectors. Parasites and Vectors 5, 69–80.

Sinnis, P., Willnow, T.E., Briones, M.R., Herz, J., Nussenzweig, V., 1996. Remnant lipoproteins inhibit malaria sporozoite invasion of hepatocytes. Journal of Experimental Medicine 184, 945–954.

Sougoufara, S., Diédhiou, S.M., Doucouré, S., Diagne, N., Sembène, P.M., Harry, M., Trape, J.-F., Sokhna, C., Ndiath, M.O., 2014. Biting by *Anopheles funestus* in broad daylight after use of long-lasting insecticidal nets: a new challenge to malaria elimination. Malaria Journal 13, 125.

Spinner, J.L., Cundiff, J.A., Kobayashi, S.D., 2008. *Yersinia pestis* type III secretion system-dependent inhibition of human polymorphonuclear leukocyte function. Infection and Immunity 76, 3754–3760.

Spinner, J.L., Carmody, A.B., Jarrett, C.O., Hinnebusch, B.J., 2013a. Role of *Yersinia pestis* toxin complex family proteins in resistance to phagocytosis by polymorphonuclear leukocytes. Infection and Immunity 81, 4041–4052.

Spinner, J.L., Winfree, S., Starr, T., Shannon, J.G., Nair, V., Steele-Mortimer, O., Hinnebusch, B.J., 2013b. *Yersinia pestis* survival and replication within human neutrophil phagosomes and uptake of infected neutrophils by macrophages. Journal of Leukocyte Biology.

Sturm, A., Amino, R., van de Sand, C., Regen, T., Retzlaff, S., Rennenberg, A., Krueger, A., Pollok, J.M., Ménard, R., Heussler, V.T., 2006. Manipulation of host hepatocytes by the malaria parasite for delivery into liver sinusoids. Science 313, 1287–1290.

Sun, Y.C., Jarrett, C.O., Bosio, C.F., Hinnebusch, B.J., 2014. Retracing the evolutionary path that led to flea-borne transmission of *Yersinia pestis*. Cell Host and Microbe 15, 578–586.

Tarun, A.S., Baer, K., Dumpit, R.F., Gray, S., Lejarcegui, N., Frevert, U., Kappe, S.H., 2006. Quantitative isolation and in vivo imaging of malaria parasite liver stages. International Journal of Parasitology 36, 1283–1293.

Tavares, J., Formaglio, P., Thiberge, S., Mordelet, E., van Rooijen, N., Medvinsky, A., Ménard, R., Amino, R., 2013. Role of host cell traversal by the malaria sporozoite during liver infection. Journal of Experimental Medicine 210, 905–915.

Tavares, J., Amino, R., Ménard, R., 2014. The role of MACPF proteins in the biology of malaria and other apicomplexan parasites. In: Anderluch, G., Gilbert, R. (Eds.), MACPF/CDC Proteins: Agents of Defence, Attack and Invasion, Subcellular Biochemistry, vol. 80. Springer Science, pp. 241–253.

Tilley, S.L., Wagoner, V.A., Salvatore, C.A., Jacobson, M.A., Koller, B.H., 2000. Adenosine and inosine increase cutaneous vasopermeability by activating A(3) receptors on mast cells. Journal of Clinical Investigation 105, 361–367.

Trager, W., Jensen, J.B., 1976. Human malaria parasites in continuous culture. Science 193, 673–675.

Vadyvaloo, V., Jarrett, C., Sturdevant, D.E., Sebbane, F., Hinnebusch, B.J., 2010. Transit through the flea vector induces a pretransmission innate immunity resistance phenotype in *Yersinia pestis*. PLoS Pathogens 6, e10000783.

Van Dijk, M.R., Janse, C.J., Waters, A.P., 1996. Expression of a *Plasmodium* gene introduced into subtelomeric regions of *Plasmodium berghei* chromosomes. Science 271, 662–665.

Vanderberg, J.P., Frevert, U., 2004. Intravital microscopy demonstrating antibody-mediated immobilisation of *Plasmodium berghei* sporozoites injected into skin by mosquitoes. International Journal for Parasitology 34, 991–996.

Vanderberg, J.P., Chew, S., Stewart, M.J., 1990. *Plasmodium* sporozoite interactions with macrophages in vitro: a videomicroscopic analysis. Journal of Protozoology 37, 528–536.

Vanderberg, J.P., 1974. Studies on the motility of *Plasmodium* sporozoites. Journal of Protozoology 21, 527–537.

Vaughan, J.A., Jerse, A.E., Azad, A.F., 1989. Rat leucocyte response to the bites of rat fleas (Siphonaptera: Pulicidae). Journal of Medical Entomology 26, 449–453.

Vaughan, J.A., Scheller, L.F., Wirtz, R.A., Azad, A.F., 1999. Infectivity of *Plasmodium berghei* sporozoites delivered by intravenous inoculation versus mosquito bite: implications for sporozoite vaccine trials. Infection and Immunity 67, 4285–4289.

Viboud, G.I., Bliska, J.B., 2005. *Yersinia* outer proteins: role in modulation of host cell signaling responses and pathogenesis. Annals Review of Microbiology 59, 69–89.

Vincke, I.H., Lips, M., 1948. Un nouveau *Plasmodium* d'un rongeur sauvage du Congo, *Plasmodium berghei* n.sp. Annales de la Société Belge de Médecine Tropicale 28, 97–104.

Voza, T., Miller, J.L., Kappe, S.H., Sinnis, P., 2012. Extrahepatic exoerythrocytic forms of rodent malaria parasites at the site of inoculation: clearance after immunization, susceptibility to primaquine, and contribution to blood-stage infection. Infection and Immunity 80, 2158–2164.

Vuong, C., Voyich, J.M., Fischer, E.R., Braughton, K.R., Whitney, A.R., DeLeo, F.R., Otto, M., 2004. Polysaccharide intercellular adhesin (PIA) protects *Staphylococcus epidermidis* against major components of the human innate immune system. Cellular Microbiology 6, 269–275.

Yamauchi, L.M., Coppi, A., Snounou, G., Sinnis, P., 2007. *Plasmodium* sporozoites trickle out of the injection site. Cellular Microbiology 9, 1215–1222.

FURTHER READING

Ménard, R., Tavares, J., Cockburn, I., Markus, M., Zavala, F., Amino, R., 2013. Looking under the skin: the first steps in malarial infection and immunity. Nature Reviews Microbiology 11, 701–712.

Chapter 8

Insect-Borne Viruses and Host Skin Interface

Christopher G. Mueller[1], Van-Mai Cao-Lormeau[2]

[1]*CNRS UPR3572, Université de Strasbourg, Strasbourg, France;* [2]*Pôle de recherche et de veille sur les maladies infectieuses émergentes, Institut Louis Malardé, Papeete, French Polynesia*

INTRODUCTION

The arthropod is the number one human killer by transmitting potentially deadly parasites (protozoans and helminths), bacteria, and viruses. Viruses transmitted by arthropods are named arboviruses, a term that refers to arthropod-borne virus. The arboviruses that are pathogenic to humans belong to five families: *Togaviridae, Flaviviridae, Bunyaviridae, Reoviridae, and Rhabdoviridae* and are mostly transmitted by mosquitoes, ticks, or sand flies (Table 8.1). During a blood feed, the infected arthropod injects virus-containing saliva into the host skin. The injected virus is then able to infect either skin-resident cells or inflammatory cells, recruited from the blood stream to the wound. Alternatively, the virus can directly enter the blood stream, infect leukocytes, and spread to all distant organs. Here, we will review the progress in our understanding of the molecular and cellular interactions associated with arbovirus transmission and infection at the skin interface. We will describe the skin immune structure, the immediate and delayed infection events, and relate these to viral pathogenicity. We will take the example of some mosquito-borne viruses currently recognized as global public health concerns: dengue and Zika (flavivirus), chikungunya (alphavirus).

Arboviruses

Classification and Structure

Arboviruses have evolved convergently and fall into several different viral taxa that also include nonarthropod-borne members. The common feature of arboviruses is that they have a biological cycle in both hematophagous arthropods and vertebrates. More than 534 arboviruses have been described and about 134 are known to be pathogenic to humans (Gubler, 2001). Most arboviruses are RNA viruses with either double- or single-stranded RNA (dsRNA, ssRNA) genomes

Skin and Arthropod Vectors. https://doi.org/10.1016/B978-0-12-811436-0.00008-3

275

TABLE 8.1 Main Human Disease-Causing Arboviruses; in Bold, Insect-Borne Viruses

Family	Genus		Sero-type	Vector	Symptoms
Togaviridae	Alphavirus	Chikungunya		**Mosquito (Aedes)**	Encephalitis Myalgia
		O'nyong-nyong		**Anopheles**	Arthralgia
		Ross River		**Culex, Aedes**	
		Sindbis		**?**	
Flaviviridae	Flavivirus	Venezuelan	1–6	**Mosquito (genus?)**	Encephalitis
		Semliki Forest		**Genus?**	
		Yellow fever	**1–4**	**Mosquito (Aedes)**	**Hemorrhagic fever,**
		Dengue		**Aedes**	**Encephalitis**
		Zika		**Aedes**	
		West Nile		**Mosquito (Culex)**	**Encephalitis**
		Japanese encephalitis		**Culex**	
		Murray Valley		**Culex**	
		Saint Louis		**Culex**	
		Omsk hemorrhagic fever		Tick	Hemorrhagic fever
		Kyasanur forest disease			
		Langat virus			
		Tick-borne encephalitis		Tick	Encephalitis
		Powassan encephalitis			

Family	Genus	Virus/Disease	No.	Vector	Clinical
Bunyaviridae	*Orthobunyavirus*	**La Crosse** **Batai** **Bwamba fever** **California encephalitis** **Jamestown Canyon** **Tahyna**		**Mosquito (Aedes)** *Culex* *Anopheles* **Genus?** **Genus?** **Genus?**	**Encephalitis**
	Phlebovirus *Orthobunyavirus*	**Rift Valley** **Bunyamwera** **Oropouche**	3	**Mosquito** **Aedes** **Aedes/Culex**	**Hemorrhagic fever**
	Nairovirus *Phlebovirus*	**Crimean–Congo** **Heartland** **SFTS**[a]	7	**Tick**	Hemorrhagic f. Myalgia
		Pappataci fever		**Sand fly**	**Fever**
Reoviridae	*Seadornavirus*	**Banna virus encephalitis**		**Mosquito (Culex)**	**Encephalitis**
	Coltivirus *Orbivirus*	**Colorado tick fever** **Kemerovo tick-borne viral fever**	23	**Tick**	Fever Myalgia Encephalitis
Rhabdoviridae		**Chandipura**		**Sand fly**	**Encephalitis**

[a]*Severe fever with thrombocytopenia syndrome.*

of positive (+) or negative (−) polarity. *Togaviridae, Flaviviridae, Bunyaviridae,* and *Rhabdoviridae* are enveloped viruses. The alphavirus genome consists of a single (+)ssRNA with two open reading frames that are translated into two polyproteins comprising structural and nonstructural proteins. The flavivirus genome is similarly structured with a single (+)ssRNA with a single open reading frame translated into a polyprotein that is cleaved into mature structural and nonstructural polypeptides by viral and host proteases. Bunyaviruses are comprised of tripartite (−)ssRNA genomes and the rhabdovirus genome is a single (−) ssRNA. Reoviruses have a nonenveloped capsid structure comprising 10 or more dsRNA segments.

After receptor-mediated endocytosis of togaviruses, flaviviruses, and phleboviruses, an acid-induced conformational change activates the viral envelope glycoproteins to enable the viral envelope to fuse with the membrane of endocytic vacuoles (Albornoz et al., 2016; Marsh et al., 1983; Stiasny and Heinz, 2006). After reovirus internalization, most likely via clathrin-dependent pits, the outer capsid protein is removed by expose of the μ1 viral membrane-penetration protein. Its processing and conformational alterations convey endosomal membrane rupture and cytoplasmic delivery of transcriptionally active viral particles (Danthi et al., 2010).

Arbovirus Transmission Cycles

Arboviruses are maintained in natural transmission cycles in which they replicate in hematophagous arthropods and are transmitted to vertebrate hosts during a blood meal. Most of the arboviruses that are pathogenic to humans are transmitted by mosquitoes or ticks. Such arboviruses are initially maintained in an enzootic sylvatic cycle involving rodents, birds, or nonhuman primates as reservoir hosts (Fig. 8.1). Spillover from enzootic cycle sometimes occurs through incidental biting of humans by enzootic and/or bridge vectors. Some arboviruses may achieve further amplification in domestic animals within a rural epizootic cycle. Others may alter their host range, from nonhuman primate to humans, leading to urban epidemic transmission cycle (Weaver and Barrett, 2004).

Vertical transmission from an infected vector to its progeny is a common feature of arboviral infections (Lequime et al., 2016; Lumley et al., 2017; Nuttall et al., 1994). For tick-borne viruses, transovarial, transstadial, and venereal transmissions also occur (Bente et al., 2013; Xia et al., 2016).

Clinical Symptoms of Arboviral Infection in Human

Arboviral infections are mostly asymptomatic. When pathogenic, clinical patterns can range from mild febrile illness to more severe disease (encephalitis, hemorrhagic signs). For the purpose of surveillance and reporting, based on their clinical presentation, arboviral infections are often categorized into two primary groups: neuroinvasive disease and nonneuroinvasive disease (Table 8.1). The neuroinvasive diseases are characterized by an onset of fever with headache, myalgia, stiff neck, and can develop further into states of altered mental status, seizures, limb weakness, aseptic meningitis, encephalitis, or acute flaccid paralysis. Acute flaccid paralysis may result from anterior myelitis, peripheral neuritis, or postinfectious peripheral demyelinating neuropathy (i.e., Guillain–Barre' syndrome). Some arboviruses can also cause more characteristic clinical manifestations, such as severe polyarthralgia or arthritis caused by the chikungunya virus or other alphaviruses (e.g., Mayaro, Ross River, O'nyong-nyong). Dengue virus infection mostly produces a febrile illness but severe hemorrhagic signs can also appear that sometimes lead to plasma leakage and fever, but forms of encephalitis have also been observed in dengue-infected patients (Carod-Artal et al., 2013).

The finding that all arboviruses except the African swine fever are RNA viruses suggests that their genetic plasticity has played an important role in the adaptation to the alternating vertebrate–invertebrate host cycle. For instance, two groups of flaviviruses could be distinguished by genetic phylogenetic analysis that showed an association with two different enzootic amplification or reservoir hosts. One group is transmitted by the *Culex* mosquito species and generally associates with bird reservoirs, while the other transmitted by *Aedes* mosquitoes is maintained in sylvatic primate cycles (Gaunt et al., 2001).

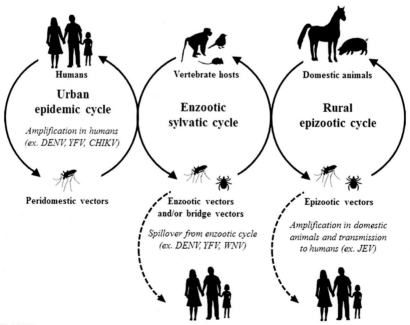

FIGURE 8.1 Arboviruses transmission cycles. *CHIKV*, chikungunya virus; *DENV*, dengue virus; *JEV*, Japanese encephalitis virus; *VEEV*, Venezuelan equine encephalitis virus; *WNV*, West Nile virus; *YFV*, yellow fever virus.

Interestingly, disease type also shows some relationship with this distribution as the neurotropic viruses are often associated with *Culex* mosquitoes, while the viscerotropic viruses (notably those causing hemorrhagic disease) are more frequently transmitted by *Aedes* mosquito species (Gaunt et al., 2001). In contrast, tick-borne flaviviruses produce encephalitic and hemorrhagic disease in humans, and this does not appear to correlate with either their phylogenetic or their geographical characteristics.

Epidemics

One of the earliest alphavirus epidemics reported was due to Semliki Forest virus, with infections having occurred among French soldiers in the Central African Republic (Mathiot et al., 1990). More recently, in 2006, chikungunya virus caused a large outbreak in the island of La Réunion, with about one-third of the population being affected (Pialoux et al., 2007). Since, chikungunya outbreaks have been increasingly reported in African, Asian, and Latin American countries, and autochthonous transmission also occurred in Europe. Among the flaviviruses, dengue virus is still recognized as a major public health concern with about 390 million people exposed to the risk of dengue every year (Bhatt et al., 2013). This virus has caused numerous outbreaks, such as those described for the Americas (San Martin et al., 2010). Recently, Zika virus, another flavivirus, unexpectedly emerged outside Africa and Asia, causing in 2007 a first outbreak in the Pacific area, in Yap, Federated States of Micronesia (Duffy et al., 2009), and 6 years later in French Polynesia (Cao-Lormeau et al., 2014). In 2015, Zika virus appeared in Brazil and quickly spread to other Latin American countries. In a few months Zika virus left the status of neglected tropical disease to be notified a Public Health Emergency of International Concern regarding the occurrence of clusters of severe neurological disorders in adults, notably Guillain–Barré syndrome, and congenital abnormalities in fetuses and babies, including microcephaly, in recently affected areas (Lessler et al., 2016).

Vertebrate Host Skin

Structure of the Skin

The skin is the body's largest organ that functions to protect it from harmful external insults (see Chapter 1). In addition to shielding to heat, cold, dryness, or UV light, the skin prevents pathogens from entering the body by forming a highly cornified and keratinized multicell-layered epidermis. Blood-feeding mosquitoes breach this protection by the aid of their rigid proboscis penetrating into the dermis. The human epidermis comprises keratinocytes, the sensatory Merkel cells, occasional CD8[+] T cells, and the Langerhans cells that are related to macrophages/dendritic cells and function as sentinels of the immune system (Tay et al., 2013). In the mouse epidermis reside additionally γδ-T cells. Melanocytes are localized in the basal epithelial layer in man and in hair follicles in rodents. The dermis comprises blood endothelial cells, lymphatic endothelial cells, fibroblasts, nerve cells, and different immune cells. These include dermal dendritic cells and macrophages, memory αβ-T cells, γδ-T cells, and mast cells. The antigen-presenting cells (dendritic cells and macrophages) together with T cells are more frequent around blood or lymphatic endothelial cells, which has generated the notion of the dermal microvascular unit (Braverman, 1993). While B cells, neutrophils, eosinophils, granulocytes, or plasmacytoid dendritic cells are rare in the normal dermis, their numbers can dramatically rise with inflammation. Indeed, the mosquito bite creates a wound that generates an inflammatory reaction that subsequently recruits these cells from the blood stream. In addition, arboviruses trigger innate and adaptive immune receptor signaling. Depending on the virus, the skin area or age of the host, the proportion of human infections with cutaneous reaction is either negligible or surpasses 50%. During the 1998 Malaysian epidemic of chikungunya, skin manifestations were seen in about half of the cases and consisted of pruriginous macular rash predominating on the thorax, facial edema, or, in children, a bullous rash with localized petechiae and gingivorrhagia (Pialoux et al., 2007). West Nile virus (5%–50% of patients) and dengue virus (over 50% of patients) also develop macular erythema (Saleem and Shaikh, 2008). Therefore, the skin is an important organ not only as arbovirus entry but also as reactive and diagnostic element. Better understanding of immediate and delayed infection events is likely to have important consequences on predicting and preventing viral pathogenicity.

Investigative Methods for Cutaneous Arbovirus Infections

It is important to review the means of scientific investigation employed in the context of skin arbovirus infection and physiopathology. Cells can be easily obtained from all species and virtually all organs thanks to improved isolation methods and optimized cell culture conditions (see Chapter 12). However, care must be taken not to underestimate the changes that occur in the gene expression profiles of the cells when they are liberated from their surroundings and placed into culture. Still, the basic characteristics of a cell type, i.e., fibroblasts versus endothelial cells or T cells versus macrophages, will remain valid, and thus allows to make comparisons between different cell types. To more accurately address the relative importance of one cell type in relation to another, the cell diversity and perhaps even its three-dimensional multicellular arrangement should be maintained. The use of animal models such as rodents or even primates (Clark et al., 2013) therefore presents advantages, although, as noted above, rodent and human skin show differences in cell composition. Also, to study the inflammatory skin reaction and to assess the physiological impacts, live animals are a requisite. In spite of obvious ethical limitations to perform infection studies in man, studies on human volunteers have been performed (Wu et al., 2000). The alternative is to resort to human skin explants, obtained in the form of punch biopsies, from plastic surgery or occasionally from cadavers. Animal studies present the advantage of a systemic follow-up of skin viral infection and the comparison between blood and skin entry or syringe versus insect inoculation (see Chapter 12). However, an important limitation is that some animals, in particular rodents, are not the natural host to many arboviruses that are pathogenic to man. Hence, neonatal animals or mice deficient for interferon type I signaling have frequently been used to overcome this problem, although it presents a major restriction in the innate immune response and greatly deforms the physiological impact of the viral infections. Another alternative is to use humanized mice, where human immune cell progenitors are introduced into immunodeficient mice to reconstitute its immune system (Cox et al., 2012). Infections are measured by production of viral progeny, by detection of viral transcripts, and by staining for viral proteins. The first two techniques are quantitative and informative on whether or not a productive viral life cycle is induced. However, under most conditions they require isolated cells. When cells cannot be purified, tissue sections are frequently stained for viral-encoded proteins, yet this only provides limited information regarding the productive life cycle. Similarly, the cell response is determined more accurately when cells are isolated, enabling quantified gene expression measures or cell morphology changes resulting of cell activation, cell senescence, or death.

Nonimmune Skin Cells

Among the nonimmune skin cells, fibroblasts, keratinocytes, and endothelial cells are those most studied.

Fibroblasts

Mouse and human fibroblasts were productively infected with West Nile virus (Hoover and Fredericksen, 2014; Lazear et al., 2011), dengue (Bustos-Arriaga et al., 2011), chikungunya (Couderc et al., 2008; Sourisseau et al., 2007), Rift Valley fever (do Valle et al., 2010), or Zika virus (Hamel et al., 2015). An active viral life cycle is best seen when fibroblasts are undergoing active cell growth in culture while viral progeny is less easily detectable in quiescent cells in culture or in skin tissues. Since primary fibroblasts grow well in culture from virtually any skin biopsy, this cell type facilitates investigations into the infection mechanism and the cell response. Viral infection of fibroblasts led to an interferon type I response (Arnold et al., 2004; do Valle et al., 2010; Hamel et al., 2015; Scherbik et al., 2007), but a low tumor necrosis factor alpha (TNFα) production could also be observed that may contribute to interferon type I to reduce viral yield (Arnold et al., 2004; Cheng et al., 2004).

Keratinocytes

Keratinocytes are less easy to grow, requiring enriched cell medium to select, amplify, and maintain proliferative cells. West Nile viral antigen was detected in mouse keratinocytes after in situ inoculation, and viral progeny was detected in the cell culture supernatant (Lim et al., 2011; Welte et al., 2009). In addition, epithelial cells of hair follicles or their adnexal glands expressed West Nile viral antigen (Lim et al., 2011). Primary human foreskin keratinocytes were permissive to dengue and Zika virus infection, as detected by viral nucleic acid and antigen measures (Hamel et al., 2015; Surasombatpattana et al., 2011). However, the chikungunya virus poorly infected human primary keratinocytes (Bernard et al., 2015). Administration of dengue virus directly into human cadaver skin revealed expression of viral antigen in basal keratinocytes (Limon-Flores et al., 2005) although dengue virus injection into mouse skin did not infect keratinocytes (Schmid and Harris, 2014).

Endothelial Cells

Dengue virus RNA and antigens were detected in sinusoidal endothelial cells of the liver (Jessie et al., 2004), and studies have observed endothelial cell primary or antigen-dependent enhancement infection in cell cultures (Srikiatkhachorn and Kelley, 2014). However, otherwise, endothelial cell infection in the skin is not well studied. More attention is given to the disruption of the endothelial brain barrier for neurotropic arbovirus (i.e., West Nile, Japanese encephalitis virus) than to whether or not skin endothelial cells are viral infectious targets. Recent studies have highlighted the importance of skin blood endothelial cells for extravasation of leukocytes at the site of infection through the action of insect salivary components (Pingen et al., 2016; Schmid et al., 2016).

Immune Cells

The skin immune system comprises a resident and an inflammatory compartment. Key resident immune cells are the Langerhans cells of the epidermis, the dendritic cells, and the macrophages of the dermis (see Chapter 1). However, T cells, as well as the dermal mast cells also play an important role in viral pathogenicity (Londono-Renteria et al., 2017; Troupin et al., 2016). The relevance of the immune cells as viral targets is supported by a number of studies where arboviruses were administered either into skin with a functional or an incomplete immune compartment or when immune cells were isolated and then exposed to virus. When analyzed in comparison with nonimmune skin cells, the immune cells release more viral progeny (Hamel et al., 2015). In addition, from a physiopathological point of view, they play a more important role by the capacity of some (dendritic cells and macrophages) to activate the adaptive immune response leading to virus-specific helper and cytotoxic T cells and the production of virus-specific immunoglobulins. Based on findings that carbohydrate-binding receptors such as dendritic cell–specific intercellular adhesion molecule 3–grabbing nonintegrin (DC-SIGN/CD209), mannose receptor (MR/CD206), or L-SIGN bound mosquito-derived glycosylated viral envelope proteins from dengue, West Nile, Zika, Japanese encephalitis virus, or the Uukuniemi phlebovirus (Davis et al., 2006; Goncalves et al., 2013; Hamel et al., 2015; Miller et al., 2008; Shimojima et al., 2014; Tassaneetrithep et al., 2003), the focus has been directed to those immune cells that express these C-type lectins. Indeed, evidence that DC-SIGN plays an important role in arbovirus immunopathology was provided by studies of gene promoter polymorphism, showing that higher DC-SIGN expression increased the risk of dengue fever and chikungunya infection (Chaaithanya et al., 2016; Sakuntabhai et al., 2005).

Macrophages

Although glycosylated and nonglycosylated West Nile virus infected murine macrophages equally, the glycosylated variant triggered more inflammatory cytokines and showed enhanced pathogenicity (Shirato et al., 2004, 2006). Among the skin-resident immune cells, dermal macrophages and CD14[+] dermal dendritic cells express most DC-SIGN and MR. To overcome ethical and technical limitations, cells resembling CD14[+] dendritic cells can be derived from human monocytes in culture with granulocyte-macrophage colony-stimulating factor and IL-4. The monocyte-derived dendritic cells were readily infected by dengue (Tassaneetrithep et al., 2003) or Zika virus (Hamel et al., 2015). Macrophages derived from monocytes in the presence of macrophage colony-stimulating factor were also generally susceptible to dengue viral infection although they became resistant in the presence of IL-10 that inhibits endosomal acidification (Kwan et al., 2008). While these cell culture experiments provide essential insights into the handling of arbovirus and the cell response elicited, it is necessary to extend these findings to genuine skin immune cells. Thus, although IL-10 was found to induce the expression of DC-SIGN on macrophages derived from monocytes (Kwan et al., 2008), ex vivo-isolated skin macrophages did not upregulate the C-type lectin in the presence of IL-10 (Schaeffer et al., 2015). Instead DC-SIGN expression by dermal macrophages was strongly induced by IL-4 rendering the cells highly permissive to dengue virus infection (Schaeffer et al., 2015).

Dendritic Cells

Human skin-isolated dermal CD14[+] dendritic cells were productively infected by dengue virus and viral progeny increased after stimulation with IL-4 (Schaeffer et al., 2015). Infection of dermal CD14[+] dendritic cells was confirmed in an independent study (Cerny et al., 2014). CD1a/c-expressing dendritic cells, a dendritic cell subset distinct from the CD14[+] type, were likewise permissive to dengue virus (Cerny et al., 2014; Schaeffer et al., 2015). Langerhans cells of the epidermis have been suspected to be an infectious target since CD1a[+] cells residing in the epidermis of human volunteers stained positive for dengue virus antigen (Wu et al., 2000). Indeed, ex vivo-purified human Langerhans cells were found to be infected (Cerny et al., 2014); however, an independent study did not observe productive Langerhans cell infection (Schaeffer et al., 2015). In the mouse, Langerhans cells were also poorly infected (Cerny et al., 2014; Schmid and Harris, 2014). It is possible that such discrepancies arise from the usage of dengue viral strains that differ in virulence. The infected dendritic cells responded to dengue virus by expression of interferon type I responsive genes but little TNFα, which instead was seen mostly produced by the dermal macrophages (Cerny et al., 2014; Schaeffer et al., 2015). Hence, in view of their potent T cell stimulatory activity, it is likely that an important impact of dermal dendritic cell infection in viral pathogenicity is the induction of an adaptive T cell or B cell immune response, while dermal macrophages would play a more important role in local inflammatory responses.

Mast Cells

Mast cells that are integral part of the skin immune compartment are also susceptible to dengue virus infection and react not only by production of interferon and interferon-regulated genes but also by the release of chemokines and mediators of vascular permeability (St John et al., 2011; Troupin et al., 2016). One effect of this is the recruitment of monocytes that can then differentiate locally into inflammatory dendritic cells that in turn are susceptible to viral infection (Cerny et al., 2014; Schmid and Harris, 2014).

Insect Vectors

Insects

To be maintained in nature arboviruses go through complex transmission cycles involving vertebrate hosts and hematophagous insect vectors of different orders (Diptera, Anoplura, Siphonaptera, and Hemiptera) and two families of ticks (Ixodidae and Argasidae). The *Aedes* genus transmits dengue, chikungunya, Zika virus, and yellow fever virus (Lambrechts et al., 2010). Chikungunya has undergone envelope glycoprotein substitutions that enhance transmission by *Aedes albopictus* by increasing oral infectivity for midgut epithelial cells (Tsetsarkin et al., 2016). Arboviruses must infect all major compartments of the insect from infection of the midgut to dissemination and invasion of the salivary glands to be transmitted to the vertebrate host. In general, arboviruses infect mosquito cells without causing significant damage, with little evidence of the cytopathic effects usually seen in vertebrate cells. This allows mosquitoes to remain infectious for their lifetime, compared to vertebrates that typically clear the infection from their systems or succumb to infection.

Impact of Insect Salivary Compounds

There are increasing numbers of studies reporting on the impact of mosquito salivary compounds on arbovirus infection (Briant et al., 2014). Indeed, mosquito saliva comprises a great variety of molecules with hemostatic, allergenic, and immunoregulatory activity, and even further complexity is introduced by the effect of the arbovirus itself on salivary gland gene expression. Generally, mosquito salivary extracts when administrated with the virus potentiate infectivity and pathogenicity, resulting in murine morbidity and mortality to otherwise innocuous arboviruses (Schneider and Higgs, 2008). The underlying reasons identified so far are accelerated immune cell recruitment, upregulation of viral recognition receptors, and modulation of interferon type I production.

Aedes Saliva

It is the most studied mosquito saliva (Briant et al., 2014). Different molecules have been identified in *Aedes* saliva. A vasodilatory sialokinin (Champagne and Ribeiro, 1994), a D7 protein (Calvo et al., 2007), an apyrase that inhibits adenosine diphosphate (ADP)-dependent platelet aggregation and inhibits platelet-aggregating factors (Champagne and Ribeiro, 1994), and a 48-kDa factor Xa inhibitor belonging to the serpin family of serine protease inhibitors (Stark and James, 1998) are the best characterized. In a study on dengue virus, *Aedes aegypti* mosquito salivary gland extracts increased the recruitment of monocytes and neutrophils, however, only in the presence of antiviral antibody mimicking antibody-dependent enhancement of dengue virus infection (Schmid et al., 2016). Semliki Forest Virus intradermal injection at bite sites triggered increased recruitment of neutrophils and inflammatory monocytes that are permissive to productive infection (Pingen et al., 2016). However, another study showed that the presence of *Ae. aegypti* salivary extracts during West Nile virus inoculation reduced the numbers of neutrophils and T cells in the skin but increased the production of IL-10 (Schneider et al., 2010). A strong IL-4 expression was observed in mouse ears infected with chikungunya virus via mosquito bites compared to needle injection alone (Thangamani et al., 2010). As expected for IL-4 production, the dermal infiltrate was characterized by numerous eosinophils. Since a Th2 inflammatory milieu positively affects C-type lectin expression of resident and recruited monocytes, it is possible that IL-4 has a direct positive impact on arbovirus infection via C-type lectin upregulation (Schaeffer et al., 2015). *Ae. aegypti* mosquito salivary gland extracts reduced the production of mast cell TNFα (Bissonnette et al., 1993). Mast cells are likely activated by immunoglobulins recognizing salivary components bound to mast cell immunoglobulin receptors. An observation indicating that dengue virus could bind to *Aedes* salivary gland components raises the idea that antibodies directed against such components could indirectly mediate cell infection through immunoglobulin receptors (Cao-Lormeau, 2009). In a study of vesicular stomatitis virus–infected fibroblasts it was suggested that salivary gland extracts from *Aedes triseriatus* mosquito reduces the production of type I interferons (Limesand et al., 2003). However, the presence of *Ae. aegypti* salivary gland extracts enhanced cutaneous interferon beta transcript induction by Semliki Forest Virus (Pingen et al., 2016).

Anopheles Saliva

It also caused the degranulation of mast cells, which led to the influx of leukocytes to the bite site (Demeure et al., 2005). Its role has been particularly studied in malaria, a disease caused by *Plasmodium* spp. (Schneider et al., 2011) (see Chapter 4).

Culex Saliva

When infection of West Nile virus was tested in mice presensitized to salivary gland proteins of *Culex tarsalis* mosquito, there was an increase in viremia after virus administration (Styer et al., 2011). Similarly, in another animal model, higher West Nile virus (WNV) titers were observed in the serum of chickens infected by *Culex pipiens* mosquito feeding, as compared to needle-inoculated animals (Styer et al., 2006). This increase in virus infectivity can be explained at the cellular level by the presence of saliva that is associated with enhanced early viral replication, especially in the skin (Schneider et al., 2010). The results of the latter study corroborate the capacity of saliva to increase IL-10 production, to dysregulate antiviral signaling by antigen-presenting cells and to elevate influx of WNV-susceptible cell types to the inoculation site, providing further insight into the role of mosquito cofactors in the acute pathogenesis of the infection. Therefore, and perhaps not surprisingly, attempts to vaccinate mice against the *Aedes* salivary D7 protein had inverse effects, namely increased West Nile virus infectivity by virus-infected mosquitoes (Reagan et al., 2012).

CONCLUDING REMARKS

The unprecedented large-scale outbreak in the Americas of the Zika virus underlines the continuous threat of arboviruses. Despite its emergence as an important public health problem and increasing knowledge about its biology and physiopathology, neither vaccine nor drug is available to treat the infection. Yet, the efficacy of the yellow fever vaccine (Collins and Barrett, 2017) and Japanese encephalitis vaccine (Hegde and Gore, 2017) and recent success with a dengue-based vaccines (Torresi et al., 2017) shows that it is possible to generate effective arboviral vaccines against mosquito-borne viruses. The study of infection events of arbovirus in the skin has generated a considerable collection of data. Mouse studies have played an important part in this, but experimentation with human skin is gaining attention. Among the issues that have so far been little studied are those that question the skin as infectious reservoir, such as viral particles that remain dormant in certain metabolically inactive cells. Also the question of whether arthropods become (re)infected through the skin rather than through blood is of interest. Studies of arbovirus infection in the skin underline the critical role played by the immune cells in mediating infection and in determining the level of viremia. However, infection and the resulting cell response are influenced by a plethora of factors, which include the diversity of immune cells depending on skin type, site, age, prior exposure to virus, the complexity of mosquito salivary proteins, the physical injury created by the arthropod on blood searching, and feeding. At first view, this would make a prediction difficult of whether or not an infection will be asymptomatic or whether it will cause life-threating complications. However, just because the cutaneous immune reaction reflects all these parameters, it may on the contrary serve a predictive value for subsequent pathological consequences. Another outcome could be that topically applied drugs provide efficacy not only to counteract productive infection but also to avoid those immunological reactions that contribute to the harmful pathological events of the ongoing or the future viral infections. Further research should be fostered into these directions.

REFERENCES

Albornoz, A., Hoffmann, A.B., Lozach, P.Y., Tischler, N.D., 2016. Early Bunyavirus-host cell interactions. Viruses 8 (5), E143.

Arnold, S.J., Osvath, S.R., Hall, R.A., King, N.J., Sedger, L.M., 2004. Regulation of antigen processing and presentation molecules in West Nile virus-infected human skin fibroblasts. Virology 324, 286–296.

Bente, D.A., Forrester, N.L., Watts, D.M., McAuley, A.J., Whitehouse, C.A., Bray, M., 2013. Crimean-Congo hemorrhagic fever: history, epidemiology, pathogenesis, clinical syndrome and genetic diversity. Antiviral Research 100, 159–189.

Bernard, E., Hamel, R., Neyret, A., Ekchariyawat, P., Moles, J.P., Simmons, G., Chazal, N., Despres, P., Misse, D., Briant, L., 2015. Human keratinocytes restrict chikungunya virus replication at a post-fusion step. Virology 476, 1–10.

Bhatt, S., Gething, P.W., Brady, O.J., Messina, J.P., Farlow, A.W., Moyes, C.L., Drake, J.M., Brownstein, J.S., Hoen, A.G., Sankoh, O., Myers, M.F., George, D.B., Jaenisch, T., Wint, G.R., Simmons, C.P., Scott, T.W., Farrar, J.J., Hay, S.I., 2013. The global distribution and burden of dengue. Nature 496, 504–507.

Bissonnette, E.Y., Rossignol, P.A., Befus, A.D., 1993. Extracts of mosquito salivary gland inhibit tumour necrosis factor alpha release from mast cells. Parasite Immunology 15, 27–33.

Braverman, I., 1993. The dermal microvascular unit: relationship to immunological processes and dermal dendrocytes. In: Nickoloff, B.J. (Ed.), Dermal Immune System. CRC Press, Boca Raton, FL, pp. 91–110.

Briant, L., Després, P., Choumet, V., Missé, D., 2014. Role of skin immune cells on the host susceptibility to mosquito-borne viruses. Virology 464–465, 26–32.

Bustos-Arriaga, J., Garcia-Machorro, J., Leon-Juarez, M., Garcia-Cordero, J., Santos-Argumedo, L., Flores-Romo, L., Mendez-Cruz, A.R., Juarez-Delgado, F.J., Cedillo-Barron, L., 2011. Activation of the innate immune response against DENV in normal non-transformed human fibroblasts. Plos Neglected Tropical Diseases 5, e1420.

Calvo, E., Dao, A., Pham, V., Ribeiro, J., February 2007. An insight into the sialome of *Anopheles funestus* reveals an emerging pattern in anopheline salivary protein families. Insect Biochemistry and Molecular Biology 37 (2), 164–175.

Cao-Lormeau, V.M., 2009. Dengue viruses binding proteins from *Aedes aegypti* and *Aedes polynesiensis* salivary glands. Journal of Virology 6.

Cao-Lormeau, V.M., Roche, C., Teissier, A., Robin, E., Berry, A.L., Mallet, H.P., Sall, A.A., Musso, D., 2014. Zika virus, French Polynesia, South pacific, 2013. Emerging Infectious Diseases 20, 1085–1086.

Carod-Artal, F.J., Wichmann, O., Farrar, J., Gascon, J., 2013. Neurological complications of dengue virus infection. Lancet Neurology 12, 906–919.

Cerny, D., Haniffa, M., Shin, A., Bigliardi, P., Tan, B.K., Lee, B., Poidinger, M., Tan, E.Y., Ginhoux, F., Fink, K., 2014. Selective susceptibility of human skin antigen presenting cells to productive dengue virus infection. PLoS Pathogens 10, e1004548.

Chaaithanya, I.K., Muruganandam, N., Surya, P., Anwesh, M., Alagarasu, K., Vijayachari, P., 2016. Association of oligoadenylate synthetase gene cluster and DC-SIGN (CD209) gene polymorphisms with clinical symptoms in chikungunya virus infection. DNA and Cell Biology 35, 44–50.

Champagne, D., Ribeiro, J., 1994. Sialokinin I and II: vasodilatory tachykinins from the yellow fever mosquito *Aedes aegypti*. Proceedings of the National Academy of Sciences of the United States of America 91, 138–142.

Cheng, Y., King, N.J., Kesson, A.M., 2004. The role of tumor necrosis factor in modulating responses of murine embryo fibroblasts by flavivirus, West Nile. Virology 329, 361–370.

Clark, K.B., Onlamoon, N., Hsiao, H.M., Perng, G.C., Villinger, F., 2013. Can non-human primates serve as models for investigating dengue disease pathogenesis? Frontiers in Microbiology 4, 305.

Collins, N.D., Barrett, A.D., 2017. Live attenuated yellow fever 17D vaccine: a legacy vaccine still controlling outbreaks in modern day. Current Infectious Disease Reports 19, 14.

Couderc, T., Chretien, F., Schilte, C., Disson, O., Brigitte, M., Guivel-Benhassine, F., Touret, Y., Barau, G., Cayet, N., Schuffenecker, I., Despres, P., Arenzana-Seisdedos, F., Michault, A., Albert, M.L., Lecuit, M., 2008. A mouse model for Chikungunya: young age and inefficient type-I interferon signaling are risk factors for severe disease. PLoS Pathogens 4, e29.

Cox, J., Mota, J., Sukupolvi-Petty, S., Diamond, M.S., Rico-Hesse, R., 2012. Mosquito bite delivery of dengue virus enhances immunogenicity and pathogenesis in humanized mice. Journal of Virology 86, 7637–7649.

Danthi, P., Guglielmi, K.M., Kirchner, E., Mainou, B., Stehle, T., Dermody, T.S., 2010. From touchdown to transcription: the reovirus cell entry pathway. Current Topics in Microbiology and Immunology 343, 91–119.

Davis, C.W., Nguyen, H.Y., Hanna, S.L., Sanchez, M.D., Doms, R.W., Pierson, T.C., 2006. West Nile virus discriminates between DC-SIGN and DC-SIGNR for cellular attachment and infection. Journal of Virology 80, 1290–1301.

Demeure, C.E., Brahimi, K., Hacini, F., Marchand, F., Peronet, R., Huerre, M., St-Mezard, P., Nicolas, J.F., Brey, P., Delespesse, G., Mecheri, S., 2005. Anopheles mosquito bites activate cutaneous mast cells leading to a local inflammatory response and lymph node hyperplasia. The Journal of Immunology 174, 3932–3940.

do Valle, T.Z., Billecocq, A., Guillemot, L., Alberts, R., Gommet, C., Geffers, R., Calabrese, K., Schughart, K., Bouloy, M., Montagutelli, X., Panthier, J.J., 2010. A new mouse model reveals a critical role for host innate immunity in resistance to Rift Valley fever. The Journal of Immunology 185, 6146–6156.

Duffy, M.R., Chen, T.H., Hancock, W.T., Powers, A.M., Kool, J.L., Lanciotti, R.S., Pretrick, M., Marfel, M., Holzbauer, S., Dubray, C., Guillaumot, L., Griggs, A., Bel, M., Lambert, A.J., Laven, J., Kosoy, O., Panella, A., Biggerstaff, B.J., Fischer, M., Hayes, E.B., 2009. Zika virus outbreak on Yap island, Federated States of Micronesia. The New England Journal of Medicine 360, 2536–2543.

Gaunt, M.W., Sall, A.A., de Lamballerie, X., Falconar, A.K., Dzhivanian, T.I., Gould, E.A., 2001. Phylogenetic relationships of flaviviruses correlate with their epidemiology, disease association and biogeography. The Journal of General Virology 82, 1867–1876.

Goncalves, A.R., Moraz, M.L., Pasquato, A., Helenius, A., Lozach, P.Y., Kunz, S., 2013. Role of DC-SIGN in Lassa virus entry into human dendritic cells. Journal of Virology 87, 11504–11521.

Gubler, D.J., 2001. Human arbovirus infections worldwide. Annals of the New York Academy of Sciences 951, 13–24.

Hamel, R., Dejarnac, O., Wichit, S., Ekchariyawat, P., Neyret, A., Luplertlop, N., Perera-Lecoin, M., Surasombatpattana, P., Talignani, L., Thomas, F., Cao-Lormeau, V.M., Choumet, V., Briant, L., Despres, P., Amara, A., Yssel, H., Misse, D., 2015. Biology of Zika virus infection in human skin cells. Journal of Virology 89, 8880–8896.

Hegde, N., Gore, M., 2017. Japanese encephalitis vaccines: immunogenicity, protective efficacy, effectiveness, and impact on the burden of disease. Human Vaccines and Immunotherapeutics 13, 1–18.

Hoover, L.I., Fredericksen, B.L., 2014. IFN-dependent and -independent reduction in West Nile virus infectivity in human dermal fibroblasts. Viruses 6, 1424–1441, doi: 10.3390/v6031424.

Jessie, K., Fong, M.Y., Devi, S., Lam, S.K., Wong, K.T., 2004. Localization of dengue virus in naturally infected human tissues, by immunohistochemistry and in situ hybridization. The Journal of Infectious Diseases 189, 1411–1418.

Kwan, W.H., Navarro-Sanchez, E., Dumortier, H., Decossas, M., Vachon, H., Barreto dos Sanchez, F., Fridman, W.H., Rey, F.A., Harris, E., Despres, P., Mueller, C.G., 2008. Dermal-type macrophages expressing CD209/DC-SIGN show inherent resistance to dengue virus growth. Plos Neglected Tropical Diseases 2, e311.

Lambrechts, L., Scott, T.W., Gubler, D.J., 2010. Consequences of the expanding global distribution of *Aedes albopictus* for dengue virus transmission. Plos Neglected Tropical Diseases 4, e646.

Lazear, H.M., Pinto, A.K., Vogt, M.R., Gale Jr., M., Diamond, M.S., 2011. Beta interferon controls West Nile virus infection and pathogenesis in mice. Journal of Virology 85, 7186–7194.

Lequime, S., Paul, R.E., Lambrechts, L., 2016. Determinants of arbovirus vertical transmission in mosquitoes. PLoS Pathogens 12, e1005548.

Lessler, J., Chaisson, L.H., Kucirka, L.M., Bi, Q., Grantz, K., Salje, H., Carcelen, A.C., Ott, C.T., Sheffield, J.S., Ferguson, N.M., Cummings, D.A., Metcalf, C.J., Rodriguez-Barraquer, I., 2016. Assessing the global threat from Zika virus. Science 353, aaf816014.

Lim, P.Y., Behr, M.J., Chadwick, C.M., Shi, P.Y., Bernard, K.A., 2011. Keratinocytes are cell targets of West Nile virus in vivo. Journal of Virology 85, 5197–5201.

Limesand, K.H., Higgs, S., Pearson, L.D., Beaty, B.J., 2003. Effect of mosquito salivary gland treatment on vesicular stomatitis New Jersey virus replication and interferon alpha/beta expression in vitro. Journal of Medical Entomology 40, 199–205.

Limon-Flores, A.Y., Perez-Tapia, M., Estrada-Garcia, I., Vaughan, G., Escobar-Gutierrez, A., Calderon-Amador, J., Herrera-Rodriguez, S.E., Brizuela-Garcia, A., Heras-Chavarria, M., Flores-Langarica, A., Cedillo-Barron, L., Flores-Romo, L., 2005. Dengue virus inoculation to human skin explants: an effective approach to assess in situ the early infection and the effects on cutaneous dendritic cells. International Journal of Experimental Pathology 86, 323–334.

Londono-Renteria, B., Marinez-Angarita, J., Troupin, A., Colpitts, T., 2017. Role of mast cells in dengue virus pathogenesis. DNA and Cell Biology 36, 423–427.

Lumley, S., Horton, D.L., Hernandez-Triana, L.L.M., Johnson, N., Fooks, A.R., Hewson, R., 2017. Rift Valley fever virus: strategies for maintenance, survival and vertical transmission in mosquitoes. The Journal of General Virology 98, 875–887. https://doi.org/10.1099/jgv.1090.000765. Epub 002017 May 000730.

Marsh, M., Bolzau, E., Helenius, A., 1983. Penetration of Semliki Forest virus from acidic prelysosomal vacuoles. Cell 32, 931–940.

Mathiot, C.C., Grimaud, G., Garry, P., Bouquety, J.C., Mada, A., Daguisy, A.M., Georges, A.J., 1990. An outbreak of human Semliki Forest virus infections in Central African Republic. The American Journal of Tropical Medicine and Hygiene 42, 386–393.

Miller, J.L., Dewet, B.J., Martinez-Pomares, L., Radcliffe, C.M., Dwek, R.A., Rudd, P.M., Gordon, S., 2008. The Mannose receptor mediates dengue virus infection of macrophages. PLoS Pathogens 4, e17.

Nuttall, P.A., Jones, L.D., Labuda, M., Kaufman, W.R., 1994. Adaptations of arboviruses to ticks. Journal of Medical Entomology 31, 1–9.

Pialoux, G., Gauzere, B.A., Jaureguiberry, S., Strobel, M., 2007. Chikungunya, an epidemic arbovirosis. The Lancet Infectious Diseases 7, 319–327.

Pingen, M., Bryden, S.R., Pondeville, E., Schnettler, E., Kohl, A., Merits, A., Fazakerley, J.K., Graham, G.J., McKimmie, C.S., 2016. Host inflammatory response to mosquito bites enhances the severity of arbovirus infection. Immunity 44, 1455–1469.

Reagan, K.L., Machain-Williams, C., Wang, T., Blair, C.D., 2012. Immunization of mice with recombinant mosquito salivary protein D7 enhances mortality from subsequent West Nile virus infection via mosquito bite. PLoS Neglected Tropical Diseases 6, e1935.

Sakuntabhai, A., Turbpaiboon, C., Casademont, I., Chuansumrit, A., Lowhnoo, T., Kajaste-Rudnitski, A., Kalayanarooj, S.M., Tangnararatchakit, K., Tangthawornchaikul, N., Vasanawathana, S., Chaiyaratana, W., Yenchitsomanus, P.T., Suriyaphol, P., Avirutnan, P., Chokephaibulkit, K., Matsuda, F., Yoksan, S., Jacob, Y., Lathrop, G.M., Malasit, P., Despres, P., Julier, C., 2005. A variant in the CD209 promoter is associated with severity of dengue disease. Nature Genetics 37, 507–513.

Saleem, K., Shaikh, I., 2008. Skin lesions in hospitalized cases of dengue Fever. Journal of the College of Physicians and Surgeons pakistan 18, 608–611.

San Martin, J.L., Brathwaite, O., Zambrano, B., Solorzano, J.O., Bouckenooghe, A., Dayan, G.H., Guzman, M.G., 2010. The epidemiology of dengue in the americas over the last three decades: a worrisome reality. The American Journal of Tropical Medicine and Hygiene 82, 128–135.

Schaeffer, E., Flacher, V., Papageorgiou, V., Decossas, M., Fauny, J., Krämer, M., Mueller, C.G., 2015. Dermal CD14+ dendritic cell and macrophage infection by dengue virus is stimulated by Interleukin-4. The Journal of Investigative Dermatology 135, 1743–1751.

Scherbik, S.V., Stockman, B.M., Brinton, M.A., 2007. Differential expression of interferon (IFN) regulatory factors and IFN-stimulated genes at early times after West Nile virus infection of mouse embryo fibroblasts. Journal of Virology 81, 12005–12018.

Schmid, M.A., Glasner, D.R., Shah, S., Michlmayr, D., Kramer, L.D., Harris, E., 2016. Mosquito saliva increases endothelial permeability in the skin, immune cell migration, and dengue pathogenesis during antibody-dependent enhancement. PLoS Pathogens 12, e1005676.

Schmid, M.A., Harris, E., 2014. Monocyte recruitment to the dermis and differentiation to dendritic cells increases the targets for dengue virus replication. PLoS Pathogens 10, e1004541.

Schneider, B.S., Higgs, S., 2008. The enhancement of arbovirus transmission and disease by mosquito saliva is associated with modulation of the host immune response. Transactions of the Royal Society of Tropical Medicine and Hygiene 102, 400–408.

Schneider, B.S., Soong, L., Coffey, L.L., Stevenson, H.L., McGee, C.E., Higgs, S., 2010. *Aedes aegypti* saliva alters leukocyte recruitment and cytokine signaling by antigen-presenting cells during West Nile virus infection. PLoS One 5, e11704.

Schneider, B., Mathieu, C., Peronet, R., Mécheri, S., 2011. *Anopheles stephensi* saliva enhances progression of cerebral malaria in a murine model. Vector Borne and Zoonotic Diseases 11, 423–432.

Shimojima, M., Takenouchi, A., Shimoda, H., Kimura, N., Maeda, K., 2014. Distinct usage of three C-type lectins by Japanese encephalitis virus: DC-SIGN, DC-SIGNR, and LSECtin. Archives of Virology 159, 2023–2031.

Shirato, K., Miyoshi, H., Goto, A., Ako, Y., Ueki, T., Kariwa, H., Takashima, I., 2004. Viral envelope protein glycosylation is a molecular determinant of the neuroinvasiveness of the New York strain of West Nile virus. The Journal of General Virology 85, 3637–3645.

Shirato, K., Miyoshi, H., Kariwa, H., Takashima, I., 2006. The kinetics of proinflammatory cytokines in murine peritoneal macrophages infected with envelope protein-glycosylated or nonglycosylated West Nile virus. Virus Research 121, 11–16.

Sourisseau, M., Schilte, C., Casartelli, N., Trouillet, C., Guivel-Benhassine, F., Rudnicka, D., Sol-Foulon, N., Le Roux, K., Prevost, M.C., Fsihi, H., Frenkiel, M.P., Blanchet, F., Afonso, P.V., Ceccaldi, P.E., Ozden, S., Gessain, A., Schuffenecker, I., Verhasselt, B., Zamborlini, A., Saib, A., Rey, F.A., Arenzana-Seisdedos, F., Despres, P., Michault, A., Albert, M.L., Schwartz, O., 2007. Characterization of reemerging chikungunya virus. PLoS Pathogens 3, e89.

Srikiatkhachorn, A., Kelley, J.F., 2014. Endothelial cells in dengue hemorrhagic fever. Antiviral Research 109, 160–170.

St John, A.L., Rathore, A.P., Yap, H., Ng, M.L., Metcalfe, D.D., Vasudevan, S.G., Abraham, S.N., 2011. Immune surveillance by mast cells during dengue infection promotes natural killer (NK) and NKT-cell recruitment and viral clearance. Proceedings of the National Academy of Sciences of the United States of America 108, 9190–9195.

Stark, K., James, A., 1998. Isolation and characterization of the gene encoding a novel factor Xa-directed anticoagulant from the yellow fever mosquito, *Aedes aegypti*. Journal of Biological Chemistry 273, 20802–20809.

Stiasny, K., Heinz, F.X., 2006. Flavivirus membrane fusion. The Journal of General Virology 87, 2755–2766.

Styer, L., Bernard, K., Kramer, L., 2006. Enhanced early West Nile virus infection in young chickens infected by mosquito bite: effect of viral dose. The American Journal of Tropical Medicine and Hygiene 75, 337–345.

Styer, L.M., Lim, P.Y., Louie, K.L., Albright, R.G., Kramer, L.D., Bernard, K.A., 2011. Mosquito saliva causes enhancement of West Nile virus infection in mice. Journal of Virology 85, 1517–1527.

Surasombatpattana, P., Hamel, R., Patramool, S., Luplertlop, N., Thomas, F., Despres, P., Briant, L., Yssel, H., Misse, D., 2011. Dengue virus replication in infected human keratinocytes leads to activation of antiviral innate immune responses. Infection Genetics and Evolution 11, 1664–1673.

Tassaneetrithep, B., Burgess, T.H., Granelli-Piperno, A., Trumpfheller, C., Finke, J., Sun, W., Eller, M.A., Pattanapanyasat, K., Sarasombath, S., Birx, D.L., Steinman, R.M., Schlesinger, S., Marovich, M.A., 2003. DC-SIGN (CD209) mediates dengue virus infection of human dendritic cells. The Journal of Experimental Medicine 197, 823–829.

Tay, S.S., Roediger, B., Tong, P.L., Tikoo, S., Weninger, W., 2013. The skin-resident immune network. Current Dermatology Reports 3, 13–22.

Thangamani, S., Higgs, S., Ziegler, S., Vanlandingham, D., Tesh, R., Wikel, S., 2010. Host immune response to mosquito-transmitted chikungunya virus differs from that elicited by needle inoculated virus. PLoS One 5, e12137.

Torresi, J., Ebert, G., Pellegrini, M., 2017. Vaccines licensed and in clinical trials for the prevention of dengue. Human Vaccines and Immunotherapeutics 14, 1–14.

Troupin, A., Shirley, D., Londono-Renteria, B., Watson, A.M., McHale, C., Hall, A., Hartstone-Rose, A., Klimstra, W.B., Gomez, G., Colpitts, T.M., 2016. A role for human skin mast cells in dengue virus infection and systemic spread. The Journal of Immunology 197, 4382–4391.

Tsetsarkin, K.A., Chen, R., Weaver, S.C., 2016. Interspecies transmission and chikungunya virus emergence. Current Opinion in Virology 16, 143–150.

Weaver, S.C., Barrett, A.D., 2004. Transmission cycles, host range, evolution and emergence of arboviral disease. Nature Reviews Microbiology 2, 789–801.

Welte, T., Reagan, K., Fang, H., Machain-Williams, C., Zheng, X., Mendell, N., Chang, G.J., Wu, P., Blair, C.D., Wang, T., 2009. Toll-like receptor 7-induced immune response to cutaneous West Nile virus infection. The Journal of General Virology 90, 2660–2668.

Wu, S.J., Grouard-Vogel, G., Sun, W., Mascola, J.R., Brachtel, E., Putvatana, R., Louder, M.K., Filgueira, L., Marovich, M.A., Wong, H.K., Blauvelt, A., Murphy, G.S., Robb, M.L., Innes, B.L., Birx, D.L., Hayes, C.G., Frankel, S.S., 2000. Human skin Langerhans cells are targets of dengue virus infection. Nature Medicine 6, 816–820.

Xia, H., Beck, A.S., Gargili, A., Forrester, N., Barrett, A.D., Bente, D.A., 2016. Transstadial transmission and long-term association of Crimean-Congo hemorrhagic fever virus in ticks shapes genome plasticity. Scientific Reports 6, 35819.

Chapter 9

Tick-Borne Bacteria and Host Skin Interface

Quentin Bernard[1], Ema Helezen[2], Nathalie Boulanger[2]
[1]University of Maryland, College Park, MD, United States;
[2]Université de Strasbourg, Strasbourg, France

INTRODUCTION

Ticks are the most important vectors of pathogens affecting animals world-wide and second after mosquitoes where humans are concerned (Dantas-Torres et al., 2012; Jongejan and Uilenberg, 2004). They can transmit a large variety of pathogenic organisms, including viruses, bacteria, and parasites (protozoa and helminths). The risk of tick-borne diseases (TBDs) has markedly increased due to the modifications of the ecosystem by humans, the intensification of human and animal mobility, and socioeconomic changes (Kilpatrick and Randolph, 2012; Léger et al., 2013; Lindgren et al., 2012; Madder et al., 2011; Rizzoli et al., 2014). Among pathogens transmitted by ticks, bacteria are responsible of the most prevalent diseases, such as Lyme borreliosis in the northern hemi-sphere (Stanek et al., 2012; Steere et al., 2016) or the worldwide present rickett-sioses (Balraj et al., 2009; Blanton, 2013).

The skin constitutes a key interface in arthropod-borne diseases (ABD) (Bernard et al., 2014; Frischknecht, 2007). This organ normally acts as a physi-cal barrier and, even more importantly, as a very efficient immune barrier (see Chapter 1). Indeed, the recent discovery of innate immunity and the Toll-like receptors (TLRs) in mammals has evidenced the major role of the skin in the control of infections. In ABDs, the skin of the vertebrate host represents the first contact where the arthropod co-inoculates the pathogen with its saliva. The skin is even more important in TBDs since hard ticks stay anchored to the skin for a period of 3–10 days (Brossard and Wikel, 2004; Wikel, 1999, 2013). Tick saliva modulates the pharmacology and the immunity of the vertebrate host (de la Fuente et al., 2017a,b; Kazimírová and Štibrániová, 2013). Many of these TBDs start with a skin inflammation, e.g., erythema migrans (EM) in Lyme borreliosis (Mullegger, 2004) or eschar in most of rickettsiosis (Balraj et al., 2009), pointing out the importance of the skin in the physiopathology of

Skin and Arthropod Vectors. https://doi.org/10.1016/B978-0-12-811436-0.00009-5

these diseases. Recent studies also demonstrated the major role of the skin as an amplification site (Kern et al., 2011, 2015) and a latency site for TBDs (Grillon et al., 2017). This chapter will focus on the transmission of the main bacteria transmitted by ticks, i.e., *Borrelia*, *Anaplasma*, and *Rickettsia*. The transmission of pathogens generally requires the maturation of bacteria in the tick midgut and the subsequent migration to the salivary glands. This explains the delay of pathogen transmission on the hard tick bite. In soft ticks taking a quick blood meal, the pathogens are already present in the salivary glands and therefore the inoculation of pathogens takes only a few minutes (Table 9.1).

Borrelia, Ticks and Host Skin

Borreliae are spirochetes, spiral bacteria belonging to the order of Spirochaetales and to the family of Spirochaetaceae. They are slowly growing extracellular

TABLE 9.1 Main Bacteria Transmitted to Human and Animals With a Known Transmission Time and Skin Inflammation

Pathogen	Disease	Main Tick Vector	Pathogen Transmission Time	Skin Manifestations
Borrelia burgdorferi sensu lato	Lyme borreliosis	*Ixodes* spp.	16–72 h	Erythema migrans, lymphocytoma, acrodermatitis
Borrelia miyamotoi	Relapsing fever (RF) (hard ticks)	*Ixodes* spp.	ND	None
Borrelia turicatae, *Borrelia hermsii*	RF (soft ticks)	*Ornithodoros* spp.	15–90 min	None
Anaplasma phagocytophilum		*Ixodes* spp.	24–48 h	None
Anaplasma marginale		*Rhipicephalus* spp.		
		Dermacentor spp.		
Rickettsia rickettsii	Rocky Mountain spotted fever		10 h	Eschar

ND, no data.
Adapted from de la Fuente, J., Contreras, M., Estrada-Peña, A., Cabezas-Cruz, A., 2017b. Targeting a global health problem: vaccine design and challenges for the control of tick-borne diseases. Vaccine 17, S0264–410X(17)31023–X.

bacteria, pathologically subdivided into two groups: One is responsible for Lyme borreliosis, transmitted only by hard ticks of the genus *Ixodes* (Radolf et al., 2012), and the other group includes those bacteria that are the causative agents for relapsing fevers (RFs). Soft ticks (*Ornithodoros* and *Argas*) and hard ticks can transmit these RFs (Barbour, 2005; Ogden et al., 2014), and one of these infections is even known to be transmitted by a louse.

Lyme Borreliosis

This is the most important vector-borne disease of the northern hemisphere (Stanek et al., 2012). The infectious agent is a spirochete bacterium named *Borrelia burgdorferi* sensu lato, whereof 21 genospecies are currently identified (Table 9.2). Human Lyme borreliosis is the most common tick-borne infection affecting human in Europe and in the United States. The estimated incidence is 300,000 cases per annum in the United States (Mead, 2015) and 65,500 cases in Europe (Rizzoli et al., 2011). *Borrelia afzelii*, *B. garinii*, *B. burgdorferi* sensu stricto (ss), *B. bavariensis*, and *B. spielmanii* are the major species involved in human pathogenesis. The first local clinical manifestation is the so-called erythema migrans (EM), which corresponds to a skin inflammation expanding from the site of the tick bite. In the absence of an efficient immune response, the bacteria can disseminate via the blood and the skin to distant organs: heart, joints, central nervous system, and distant skin (Stanek et al., 2012; Steere et al., 2016). Lyme borreliosis is transmitted by hard ticks belonging to the *Ixodes* genus, including mostly *Ixodes scapularis/pacificus* in North America, *Ixodes ricinus* in Europe, and *Ixodes persulcatus* in Asia. They transmit the bacteria to a vertebrate host during their blood meal via the skin interface, a well-defined immune organ that *Borrelia* has to bypass. A well-established mouse model is widely used to analyze *Borrelia* transmission to the skin and the dissemination to target organs (distant skin, heart, joint, and bladder). During the tick blood meal, *Borrelia* can be detected in the skin from 16 to 42 h depending on the tick and the *Borrelia* species (Cook, 2015; Crippa et al., 2002; Ohnishi et al., 2001). For a successful transmission, bacteria have to express different surface lipoproteins, including OspC, to assure their survival within the host skin (Tilly et al., 2006). Without OspC, *Borrelia* is completely cleared from the skin within 48 h (Tilly et al., 2007), pointing out the importance of this protein in the early bacteria transmission (Xu et al., 2008). OspC also promotes the dissemination notably through the binding of plasminogen (Önder et al., 2012), making *Borrelia* an active organism able to degrade tissue barriers such as extracellular matrices (Coleman et al., 1997; Gebbia et al., 2001; Klempner et al., 1993). Moreover, OspC might be associated with an antiphagocytic activity against macrophages (Yang et al., 2013). Following *Borrelia* inoculation into mouse skin, monocyte chemoattractant protein-1 (MCP-1), keratinocyte-derived chemokine (CXCL1), and vascular endothelial growth factor (VEGF) expression are increased. The VEGF increase is highly dependent on the presence of OspC.

TABLE 9.2 Species of the Complex *Borrelia burgdorferi* sensu lato (Duneau et al., 2008; Rudenko et al., 2011; Franke et al., 2013)

Borrelia Species	Main Vectors	Hosts/Main Reservoirs	Pathogenicity	Geographic Distribution
Borrelia afzelii	*Ixodes ricinus*	Rodents	Lyme borreliosis	Europe
	Ixodes persulcatus			Asia
Borrelia americana	*Ixodes minor*	Birds	Nonpathogenic	North America
Borrelia andersoni	*Ixodes dentatus*	Rabbit	Nonpathogenic	North America
Borrelia bavariensis	*I. ricinus*	Rodents	Lyme borreliosis (neuroborreliosis)	Europe
Borrelia bissettii	*I. ricinus*	Rodents	Potentially pathogenic	Europe
	Ixodes scapularis			North America
	Ixodes pacificus			
Borrelia burgdorferi sensu stricto	*I. ricinus*	Rodents, lizards, big mammals	Lyme borreliosis (mainly arthritis)	Europe
	Ixodes uriae (Sweden, Norway)			
	I. scapularis			North America: East coast
	I. pacificus			North America: West coast
Borrelia californiensis	*Ixodes jellisonii*	Cervids	Nonpathogenic	USA
	I. pacificus			
	Ixodes spinipalpis			

Species	Tick vector	Host	Pathogenicity	Region
Borrelia carolinensis	*I. minor*	Rodents, birds	Nonpathogenic	USA
Borrelia finlandensis	*I. ricinus*	Hares	Nonpathogenic	Europe
Borrelia garinii	*I. ricinus*	Birds	Lyme borreliois (neuroborreliosis)	Europe
	I. persulcatus			Asia, North America and artic and sub-arctic islands (sea birds)
	I. uriae			
Borrelia japonica	*Ixodes ovatus*	Rodents	Nonpathogenic	Japan
Borrelia kurtenbachii	*I. scapularis*	Rodents	Potentially pathogenic	USA
Borrelia lusitaniae	*I. ricinus*	Lizards	Potentially pathogenic	Europe
	I. uriae	Rodents, sea birds		North Africa
Borrelia mayonii	*I. scapularis*	Rodents	Lyme borreliosis	North America
Borrelia sinica	*I. ovatus*	Rodents	Nonpathogenic	China
Borrelia spielmanii	*I. ricinus*	Rodents	Lyme borreliosis (cutaneous)	Europe
Borrelia tanukii	*Ixodes tanuki*	Unknown	Nonpathogenic	Japan
Borrelia turdi	*Ixodes turdus*	Birds	Nonpathogenic	Japan
Borrelia valaisiana	*I. ricinus*	Birds, lizards	Potentially pathogenic	Europe
	Ixodes columnae			Asia
	Ixodes nipponensis			
Borrelia yangzte	*Ixodes granulatus*	Rodents	Nonpathogenic	Asia
	I. nipponensis			

VEGF could help *Borrelia* to disseminate by altering vascular permeability into the skin (Antonara et al., 2010). In the past few years, the number of identified *Borrelia* genes potentially involved in skin persistence or dissemination has largely increased. DpbA and DpbB, both adhesins expressed by *Borrelia* during transmission, are able to bind decorin from the skin and are essential for an effective load of the bacteria into the skin. The ability of the bacteria to interact with skin extracellular matrix is then required for a successful infection (Xu et al., 2008). Other lipoproteins such as bba57, bb0744, lmp1, or bb0347 have also been recently characterized for their role in the bacterial persistence in the skin (Gaultney et al., 2013; Kern et al., 2015; Wager et al., 2015; Yang et al., 2009, 2013).

The immune response of the vertebrate host to *Borrelia* has been particularly well investigated not only in mouse models but also *in vitro*, by coincubating *Borrelia* not only with different specific immune cells but also with resident skin cells, as described in the following chapter.

In Vitro Studies

Borrelia as an extracellular bacterium interacts with different immune cells and resident skin cells (keratinocytes and fibroblasts).

Fibroblasts. They are essential cells responsible of extracellular matrix and collagen synthesis in many host tissues including the skin dermis. These cells can respond to the bacteria by secreting chemokines (interleukin (IL-8) and CXCL1), proinflammatory cytokines (tumor necrosis factor-alpha (TNF-α)), or matrix metalloproteinases (MMP-1,-3,-9 and -12), probably through a nuclear factor-kappa B (NF-κB)–dependent mechanism (Ebnet et al., 1997; Schramm et al., 2012; Zhao et al., 2007). MMPs could help to the pathogen dissemination and persistence in the skin. However, it has been shown that tick saliva has a lytic effect on these immune cells. The saliva could thus create a feeding pit allowing the bacteria to adapt and multiply before disseminating (Schramm et al., 2012). Bacteria themselves could also participate to clear the fibroblast inflammatory response by inducing their apoptosis (Rozwadowska et al., 2017). Finally, some recent studies suggest that the bacteria might be able to invade fibroblasts to protect themselves from the immune system, when saliva is not present. *Borrelia* could invade fibroblasts through a β(1) integrin and Src kinase-dependent mechanism (Wu et al., 2011). However, such a fibroblast invasion by *Borrelia* has not been shown *in vivo* yet.

Keratinocytes. This cell type also plays an important role in the recognition of pathogens. They constitute 90% of the epidermis, which is the top layer of the skin (Nestle et al., 2009), forming an effective physical and chemical barrier against bacteria. During *Borrelia* transmission, keratinocytes can be disrupted by the tick hypostome and the two chelicerae (Sonenshine and Anderson, 2014). On the other hand, unaffected healthy keratinocytes might be able to participate to the inflammatory process against the bacteria and the tick bite injury (Bernard et al., 2016). These cells are able to secrete many inflammatory cytokines and antimicrobial peptides in response to *Borrelia*, including IL-8, MCP-1, hBD2, hBD3, and MMP-9 (Gebbia et al., 2001; Marchal et al., 2011). However, this response mediated in part by the *Borrelia* OspC protein is further inhibited by tick saliva, which could help the bacteria to disseminate by altering immune cell attraction to the infection site. It has been shown in mice that tick saliva also modulates inflammatory molecules secretion of epidermal cells, favoring the Th2 subtype and, thus, the survival of bacteria (Pechová et al., 2002). The ability to express TLR molecules, including TLR2 (Bernard et al., 2016), and to present antigens (Gaspari and Katz, 1988) might enable keratinocytes to amplify the immune response against *Borrelia* during early infection.

Neutrophils. Similar to many other ABDs, neutrophils have been well investigated in Lyme borreliosis. Neutrophils are usually the first cells attracted to an infection site. They have been shown to be present at the tick bite site (Glatz et al., 2017). However, infiltrates of EM skin lesions have been reported to contain only few neutrophils (Salazar et al., 2003). After *Borrelia* inoculation into the dermis, neutrophils are attracted together with basophils and eosinophils to the site within 6 h. They are then rapidly substituted by eosinophils, while at 16 h after inoculation, the response becomes mostly driven by macrophages. The inoculation of *Borrelia* into the skin induces the expression of the chemokine CXCL1, a neutrophil chemoattractant that has been shown to efficiently improve the control of the initial infection by the host (Xu et al., 2007). Neutrophils, when present, are able to entrap the bacteria within their neutrophil extracellular traps (NETs). The NET formation is not affected by tick saliva even if the saliva decreases the reactive oxygen species production during the process. NETs are able to kill *Borrelia*, although some bacteria may survive in these traps by forming round bodies (Menten-Dedoyart et al., 2012). Some neutrophil functions are affected by two of the tick saliva proteins: ISL 929 and ISL 1373 (Guo et al., 2009). The uptake of *Borrelia* by neutrophils is also affected by tick saliva in downregulating beta2 integrins (Montgomery et al., 2004). *Borrelia* itself is also able to inhibit neutrophils functions. Thanks to its surface protein OspB, it decreases phagocytosis, oxidative burst, and serum sensitivity. Moreover, bacteria can inhibit the neutrophil chemotaxis toward fMLP in an Osp-independent mechanism (Hartiala et al., 2008). With all these neutrophil inhibitory mechanisms, *Borrelia* is able to avoid its clearance despite its susceptibility to H_2O_2, NO (nitric oxidase), elastase, and LL-37 (human cathelicidin) (Lusitani et al., 2002).

Macrophages. These cells are a critical component of the innate immune system present in several host tissues. They seem to play a major role in *Borrelia* clearance, as they phagocyte *Borrelia* in part through a complement receptor 3–CD14 mechanism (Hawley et al., 2013). The phagocytosis of bacteria also depends on formins Daam1, mDia1, and FMNL1, which control formation of the filopodia, which are used to capture *Borrelia* (Hoffmann et al., 2014; Naj et al., 2013). After phagocytosis, macrophages can secrete proinflammatory cytokines (IL-1β, TNF-α, IL-6) (Strle et al., 2009), type I interferons (IFNs), and chemokines (CCL3, CCL4, CCL5, CXCL9, CXCL10) (Gautam et al., 2012). On the other hand, *Borrelia* upregulates the IL-10 expression, which suppresses the activation of murine macrophages (Lazarus et al., 2006, 2008). The *Borrelia*-elicited IL-10 is also able to inhibit its phagocytosis, inflammatory mediator expressions, and NO production by macrophages (Chung et al., 2013). This mechanism acts in an autocrine mode since macrophages are, at least in part, the source of the secreted IL-10. *Borrelia* also uses the lipoprotein OspC to protect itself from phagocytosis during early infection (Carrasco et al., 2015).

Dendritic cells (DCs). They constitute an essential link between innate and adaptive immunity. Langerhans cells (LCs) and dermal DCs are among the most abundant cells moving to the infection site during *Borrelia* transmission (Hulínská et al., 1994; Salazar et al., 2003). They are attracted to the bite site through a TLR2-dependent mechanism (Mason et al., 2016). DCs are able to recognize and to phagocyte *Borrelia* to induce T cell activation (Mason et al., 2013; Tel et al., 2010). However, *Borrelia* has developed strategies to control this activation. Disseminating strains of *Borrelia* are able to activate tolerogenic indoleamine 2,3-dioxygenase (IDO) producing DCs through a type I and II IFN-dependent response. The ability of these cells to produce IDO has been associated with reduced immune response against some pathogens (Love et al., 2015). Moreover, the bacteria recognition by TLR2 does not induce CD38 expression by DCs as Gram-negative bacteria usually do. The weak expression of CD38, an ectoenzyme involved in DC migration to draining lymph nodes, could help *Borrelia* to effectively control the expansion of the adaptive response (Hartiala et al., 2010). Moreover, *Borrelia* decreases the major histocompatibility complex (MHC) II expression by LCs, increasing its survivability in early and late skin manifestations (e.g., acrodermatitis chronica atrophicans (ACA)) (Silberer et al., 2000). The MHC-II downregulation could be linked to the IL-10 secreted during the infection. *Borrelia* has also been shown to induce CD1b and CD1c expression by myeloid cells. This activation occurs through a TLR2/IL-1β mechanism (Yakimchuk et al., 2011). CD1 molecules are involved in self- and foreign lipid antigen presentation. The overexpression of this molecule could lead to autoimmune processes explaining in part the overinflammation during skin manifestation (e.g., EM and ACA). If *Borrelia* is able to modulate DC activation itself, it is helped by tick saliva. Indeed, the tick saliva inhibits DC maturation, phagocytosis, and cytokines production (TNF-α, IL-6, and IL-10) (Lieskovska and Kopecky, 2012a; Slámová et al., 2011). The TNF-α secretion might be inhibited by the tick saliva suppression of Erk1/2, Akt, and NF-κB expression (Lieskovska and Kopecky, 2012a). Moreover, the TNF-α, IL-6, and IL-12 suppression seems to be mediated by the Salp15 protein present in tick saliva (Hovius et al., 2008a,b). Also, tick saliva interferes with the pathway that mediates IFN-β notably through the action of sialostatin L2 (Lieskovská et al., 2015; Lieskovská and Kopecký, 2012b). LCs have been shown to drive T cell response toward Th2 subtype in the tick bite context. This mechanism is mediated by tick saliva in part through prostaglandin E2. However, it seems to happen only when *Borrelia* is absent, at least in male mice (Vesely et al., 2009).

Monocytes. They are rapidly recruited following an infection. Arrived at the infection site, they can differentiate into macrophages or DCs or directly participate to inflammatory processes (Swirski et al., 2009). *Borrelia* can be efficiently phagocytosed by monocytes inducing TNF-α, IL-6, IL-10, IL-18, and IL-1β expression owing in part to a CD14-dependent mechanism not necessarily involving TLR2 (Giambartolomei et al., 1999; Salazar et al., 2009). Moreover, IFNs are induced on activation of the TLR2–TLR8–MyD88 pathway in monocytes following internalization of *Borrelia* into phagosomal vacuoles (Cervantes et al., 2011, 2013). The TLR2 activation by *Borrelia* decreases the expression of TLR5-recognizing flagellin on monocytes. This mechanism could help the bacteria and the host to control the inflammation (Cabral et al., 2006). *Borrelia* also induces the expression of IL-22 but not IL-17 by peripheral blood mononuclear cells including monocytes through an IL-1β and caspase-1–dependent mechanism (Bachmann et al., 2010). IL-22 could help the bacteria to disseminate since it is known to increase vascular permeability and MMP expression (Andoh et al., 2005; Geboes et al., 2009). However, IL-22, especially in the absence of IL-17 (Sonnenberg et al., 2010), is also known to exhibit protective and antibacterial properties and could also affect in a negative way the bacterial growth and dissemination from the skin. The lack of IL-17 could be responsible for the insufficient neutrophil recruitment and could promote the bacterial dissemination. *Borreliae* are able to induce monocyte apoptosis following their uptake. This could be beneficial not only for the host, avoiding excess damages associated to the pathogen inflammation, but also for *Borrelia*, which could thus escape from these cells (Cruz et al., 2008).

Mast cells. They are well represented in the dermis. Their functions have been almost solely associated with allergic manifestations. However, their possible role toward *Borrelia* infections has been explored in a few studies. Mouse mast cells can degranulate and secrete TNF-α in response to living *Borrelia* (Talkington and Nickell, 1999). This mechanism is independent of OspA protein, a major surface lipoprotein expressed in the tick midgut (Ohnishi et al., 2001), but might be mediated in part through a direct binding of the bacteria to Fc$_\gamma$ receptors (Talkington and Nickell, 2001). It has recently been shown that *Borrelia* is able to alter the murine mast cells degranulation response to tick saliva *in vitro*. However, an *in vivo* knockout model for mast cells did not show a significant role of these cells in controlling the bacteria dissemination in mice at least during primary tick infestation (Bernard et al., 2017).

In Vivo Studies

After the discovery of spirochetes in young patients in the mid-1970s, different animal models (rabbits, hamsters, rat, dogs, monkeys) have been developed to get insight into the pathogenesis of Lyme disease. The genetic susceptibility and environmental factors explain the absence of a really appropriate model to mimic human Lyme borreliosis (Borchers et al., 2015). The rhesus macaque presents skin manifestations such as EM, but due to its cost and ethical issues, this model is rarely used (Philipp and Johnson, 1994). Rabbits also develop EM but they eventually clear the infection (Foley et al., 1995). Therefore, mice represent up to now the most suitable model because of their well-known immunology and genetics. Among the different genetic backgrounds tested, the three-weeks-old C3H/HeN mouse appeared as the most satisfying model. After intraperitoneal inoculation of *in vitro*-cultured *B. burgdorferi* ss, these mice developed carditis and severe arthritis (Barthold et al., 1990). To mimic natural infection through tick bite, intradermal inoculation has been performed and revealed tissue persistence for at least 360 days particularly in the skin (Barthold et al., 1991, 1993). *Borrelia* was observed to disseminate either hematogenously persisting transiently in the blood (Barthold et al., 1993; Wormser et al., 2005) or contiguously through the dermis (Bockenstedt et al., 2014; Straubinger et al., 1997). After needle inoculation, bacteria first multiply locally in the skin and then disseminate to target organs (Kern et al., 2011, 2015), including the whole skin where they persist for months (Barthold et al., 1993; Grillon et al., 2017). The mouse model also allows to investigate the role of tick saliva on the transmission since mice can be infected by nymph ticks. The tick saliva has a clear immunosuppressive effect on skin innate immunity (Christe et al., 2000; Kern et al., 2011). The mouse model also showed the key role of the skin to the persistence of *Borrelia*; using *Borrelia* labeled by the green fluorescent protein (GFP), it was shown that bacteria remained viable and motile for months in the skin in absence of antibiotic treatment (M. Wooten, University of Toledo, USA—personal communication). This model also demonstrated the virulence variability between the different species of *Borrelia* and individual strains (Wang et al., 2002). It also illustrated the role of the skin as a filter for these different strains, bringing out the biodiversity of *Borrelia* found in ticks, in the skin and distant organs (Brisson et al., 2011; Rego et al., 2014). However, this model is not perfect since the mouse does not present the typical skin manifestations of human Lyme borreliosis: EM, borrelial lymphocytoma, and ACA.

Understanding the mechanism of pathogen transmission in ABDs is essential to develop efficient tools to control these diseases. The process of transmission from infected ticks to vertebrate host has been particularly well investigated. But it is also interesting to explore the acquisition process of *Borrelia* from the infected animals to uninfected ticks to better understand the maintenance of the enzootic cycle. Recent intravital imaging studies explored this area (Chong et al., 2013). Owing to two-photon microscopy, it has been shown that *Borrelia* in the host skin can move toward the tick hypostome within 6 h after the tick attachment. The study also suggests as others that the true source of *Borrelia* might not be the blood but the skin (Bockenstedt et al., 2014). However, it was demonstrated that in natural transmission, *Borrelia* does not reach the tick midgut before 12 h of attachment (Schwan, 1996). The skin, as the environment where *Borrelia* evolves before acquisition by ticks, might be able to modulate the bacteria gene expression. For example, BBD18, an OspC repressor, appears to be only expressed during the process of host to tick transmission (Hayes et al., 2014) and thus in the skin. However, the skin properties involved in the regulation of bacterial gene homeostasis need to be investigated further.

Relapsing Fevers

RFs are zoonoses that are common in humans and in numerous animal species. The vectors for these diseases are mainly soft ticks of the genus *Ornithodoros*, which live in the burrows of rodents, cracks of the ground or walls, predominantly in warm regions of the world (Schwan and Piesman, 2002). The risk of being bitten by soft ticks increases during sleeping time in housing environments infested with rodents. Several species of *Borrelia* are associated with specific soft ticks (Barbour, 2005). Every region where RFs are endemic can have its own "tick/*borrelia*" association. Some regions of the world seem to be more affected by these diseases such as West Africa with *Borrelia crocidurae* or East Africa with *Borrelia duttonii*. In North America, *Borrelia parkeri* and *Borrelia hermsii* are responsible of most of the cases (Table 9.3). It is important to note that *Borrelia miyamotoi* and *Borrelia lonestari* look like exceptions among these *Borreliae*, as they are transmitted by hard ticks (Krause et al., 2015; Platonov et al., 2011). The main symptom, which has given the disease its name, is triggered by the ability of these bacteria to continuously change their surface protein expression, thus perpetually escaping the immune response of the host in a cyclic manner (Barbour, 2005).

RF spirochetes are mainly transmitted by *Ornithodoros*, a fast-feeding argasid tick. Bacteria colonize the midgut of unfed ticks, but they persist also in other tissues including the salivary glands (Schwan and Piesman, 2002). From there, spirochetes are efficiently and rapidly transmitted to the vertebrate host within minutes of attachment. While numerous groups have studied the role of the hard tick *Ixodes* saliva in assisting the transmission of Lyme disease spirochetes (Hovius et al., 2008a,b; Kazimírová and Štibrániová, 2013; Liu and Bonnet, 2014; Šimo et al., 2017), only few have investigated the potential role of the soft tick saliva in RF *Borrelia* transmission. The first study on *Ornithodoros* saliva demonstrated the presence of antihemostatic activity and the presence of apyrase (Ribeiro et al., 1991). Later on, with the progress in proteomics and transcriptomics techniques, soft tick saliva has been more thoroughly investigated (Francischetti et al., 2008, 2009; Mans et al., 2002; Oleaga et al., 2007). Two recent studies looked at the process of *Borrelia turicatae* transmission (Boyle et al., 2014; Krishnavajhala et al., 2017). The transmission occurs within a minute and dissemination into the blood is also very quick. During acquisition, *B. turicatae* enters the tick midgut and invades the salivary glands during the following weeks (Boyle et al., 2014). Using GFP, the systemic and persistent infection of ticks in the midgut and salivary glands for at least 18 months was confirmed (Krishnavajhala et al., 2017). After blood feeding on the vertebrate host, the lumen of the salivary glands still

TABLE 9.3 Main *Borrelia* Species and Their Associated Vectors Involved in Relapsing Fever Diseases

Borrelia Species	Vector	Hosts	Pathogenicity	Geographic Distribution
Borrelia recurrentis	*Pediculus humanus* (louse)	Human		Cosmopolite
Old World				
Borrelia caucasica	*Ornithodoros asperus* *Ornithodoros verrucosus*	Rodents, humans	Flu-like syndrome	Caucasia, Armenia, Azerbaijan, Georgia
Borrelia crocidurae	*Ornithodoros sonrai*	Rodents, fox	Mild symptomatology, neurological symptoms	Western and Northern Africa
Borrelia duttonii	*Ornithodoros moubata* sensu lato complex	Humans	Neurological signs, ocular complications, neonatal infections,	East, Central, and Southern Africa, Madagascar
Borrelia hispanica	*Ornithodoros erraticus,* *Ornithodoros marocanus*	Rodents, insectivores, weasels, foxes, bats, jackals, dogs, human	Ocular complications, neurological signs (rare)	Maghreb, Spain, Portugal, Greece, Cyprus, Turkey
Borrelia latyschewi	*Ornithodoros tartakovskyi*	Rodents, human, hedgehog	Flu-like syndrome	Central Asia, Middle East
Borrelia persica	*Ornithodoros tholozani*	Rodents, humans	Neurological signs (rare), respiratory distress syndrome (rare)	Middle East, Egypt, Central Asia, India

New World				
Borrelia hermsii	*Ornithodoros hermsi*	Rodents, deer, dog, human	Flu-like syndrome	British Columbia (Canada), Western USA
Borrelia mazottii	*Ornithodoros talaje*	Bats, rodents, birds	Likely pathogenic	Mexico and Guatemala
Borrelia parkeri	*Ornithodoros parkeri*	Rodents, horses, humans	Flu-like syndrome	Western USA
Borrelia turicatae	*Ornithodoros turicata*	Rodents, dogs, humans	Ocular complications, neurological signs	British Columbia (Canada), USA, Mexico
Bothrops venezuelensis	*Ornithodoros rudis*	Rodents	Relapsing fever	Panama, Columbia, Venezuela, Ecuador
Relapsing-Like Fever Transmitted by Hard Ticks				
Borrelia miyamotoi	*Ixodes persulcatus*	Rodents, birds, humans	Flu-like syndrome, tick-borne relapsing fever, neurological signs	Asia, Europe, USA
	Ixodes ricinus			
	Ixodes scapularis			
Borrelia lonestari	*Amblyomma americanum*	Birds, deers, swine	Not pathogenic	USA

Adapted from Rebaudet, S., Parola, P., October 2006. Epidemiology of relapsing fever borreliosis in Europe. FEMS Immunology and Medical Microbiology 48 (1), 11–15.

contains a high number of bacteria. Therefore, it is concluded that only few spirochetes are transmitted. Overall, the role of the tick saliva in the process of bacteria transmission and the role of skin innate immunity require further investigation. Animal models already exist and should help to obtain a better understanding of the processes involved in RF *Borrelia* transmission.

Anaplasma, Ticks and Skin Interface

The family of Anaplasmataceae (order of Rickettsiales) includes Gram-negative bacteria of different genera: *Anaplasma*, *Ehrlichia*, *Neorickettsia*, and *Wolbachia*. They are strictly intracellular organisms and multiply within vacuoles in the cytoplasm of the eukaryotes (Dumler et al., 2001). Among these bacteria, it is the transmission process of *Anaplasma* by hard ticks, which has been most investigated. These bacteria can affect mammalian hosts and cause severe and sometimes deadly diseases in humans, domestic livestock, companion animals, and wildlife (Battilani et al., 2017; Gaff et al., 2014) (Table 9.4).

Anaplasma phagocytophilum is the most important species. The first documentation as an animal disease dates from 1932, when it was described as "tick fever" in Scottish sheep, before it was discovered worldwide in different animal species (Battilani et al., 2017; Woldehiwet, 2010). In Europe, the prevalence is 85% in deer populations (Petrovec et al., 2002) and between 0.5% and 19.2% in rodents (Rar and Golovljova, 2011; Stuen et al., 2013). *Anaplasma* life cycle starts in wild mammals such as white-footed mice and white-tailed deer that serve as primary reservoirs. Domestic animals (dogs) are secondary reservoirs for human infections (Dugat et al., 2015).

Human granulocytic anaplasmosis (HGA) caused by *A. phagocytophilum*, which is transmitted by *Ixodes* ticks, represents the most important disease in terms of animal and veterinary incidence in temperate regions (Dugat et al., 2015). HGA symptoms are variable and nonspecific. Patients present with influenza-like symptoms (fever, headache, myalgia, malaise) and sometimes also with neurological symptoms. These symptoms can be associated with biological modifications of the blood (decreased blood cell counts, thrombocytopenia, leukopenia, and anemia). HGA can be fatal but 60% of patients present moderate manifestations (Dugat et al., 2015). In the United States, it has been reported that 3% of patients develop life-threatening complications and nearly 1% dies from the infection every year (Bakken and Dumler, 2015). However, a high seroprevalence estimated between 11% and 15% indicates that human anaplasmosis might be underestimated in the United States (Aguero-Rosenfeld et al., 2002). In Asia, the disease can be more severe with multiple organ dysfunction as shown in Chinese patients (Zhang et al., 2013).

A. phagocytophilum is not the most important animal anaplasmosis worldwide. Granulocytic anaplasmosis in domestic ruminants has only been described in Europe and in other domestic animals in Europe and the United States (Dugat et al., 2015). Other anaplasmoses of veterinary importance and/or zoonotic (Rar

TABLE 9.4 Main Anaplasmoses of Medical and Veterinary Importance (Battilani et al., 2017)

Pathogen	Vectors	Disease	Main Hosts	Geographic Distribution
Anaplasma bovis	*Amblyomma* spp., *Rhipicephalus* spp., *Hyalomma* spp.	Bovine anaplasmosis	Cattle, buffaloes	Africa, USA, Europe, South America, Asia
Anaplasma centrale	*Rhipicephalus simus*	Mild anaplasmosis (accine strain)	Cattle	Worldwide in tropical and subtropical regions
Anaplasma marginale	*Dermacentor* spp. *Rhipicephalus* spp.	Bovine anaplasmosis	Cattle, wild ruminants	Worldwide in tropical and subtropical regions
Anaplasma ovis	Mainly *Dermacentor* spp. et *Rhipicephalus* spp.	Ovine anaplasmosis	Sheep, goats, wild ruminants	Europe, USA, Africa, Asia
Anaplasma phagocytophilum	*Ixodes* spp.	Human granulocytic anaplasmosis, equine anaplasmosis, dog and cat anaplasmosis, tick-borne fever of ruminants	Humans, horses, ruminants, rodents, carnivores, insectivores	Worldwide
Anaplasma platys	*Rhipicephalus sanguineus*	Cyclic thrombocytopenia in dogs	Dogs	Worldwide

The two most studied are in gray.

and Golovljova, 2011) are presented in Table 9.4. *Anaplasma marginale* has a high impact in tropical and subtropical regions where it can affect domestic and wild ruminants. The target cells for the bacteria are erythrocytes (Battilani et al., 2017). A prepatent stage (7–60 days) precedes the invasion into erythrocytes. Infected erythrocytes are phagocytized and destroyed by reticuloendothelial cells, which results in anemia. The spectrum of clinical symptoms includes fever, decreased milk production, weight loss, gastrointestinal signs, and possibly death. The severity of the disease is age-dependent, with younger animals normally having only mild symptoms, but becoming permanently infected, they constitute a reservoir (Battilani et al., 2017).

Transmission to the Vertebrate Host

Members of the *Anaplasma* genus present a remarkable diversity in cell tropism, geographical distribution, host range, vectors, and pathogenicity. Ticks of the *Ixodes* species (*I. scapularis*, *I. pacificus*, *Ixodes persulcatus*, and *Ixodes ovatus*) transmit *A. phagocytophilum*, whereas *Rhipicephalus* and *Dermacentor* hard ticks transmit *A. marginale* (Battilani et al., 2017). The Gram-negative, intracellular, and polymorphic *Anaplasma* cocci are inoculated during tick bite. Initially small-dense cored bacteria enter white blood cells (granulocytes in the case of *A. phagocytophilum*) via host cellular P-selectin glycoprotein ligand 1 receptors interacting with unknown bacterial ligands (Rikihisa, 2010). Once internalized, bacteria form large reticulate cells. These cells mature into dense-cored cells and aggregate into clumps, dense intracellular microcolonies called "morulae" (Rikihisa, 2011).

Anaplasma lacks lipopolysaccharide and pili, therefore other envelope proteins are a critical interface between the bacteria and their host cells. Among surface proteins, the MSP (major surface protein) 2 superfamily is unique to *Anaplasmatacae* (*Anaplasma* and *Ehrlichia*). More interestingly, this family of proteins contains one highly polymorphic gene *p44* (Battilani et al., 2017). *p44* gene codes for a MSP composed of β-barrel with porin activity leading to passive diffusion of hydrophilic small molecular nutriments across the outer membrane. Thus, antigenic variations lead to immune evasion (Brown, 2012). *P44* is a good candidate for explaining *Anaplasma* persistence.

Anaplasma evades host defenses and promotes chemotactic mechanisms that assist the attraction of neutrophils to bite site. Subsequent degranulation of neutrophils leads to increased permeability of blood vessels, which in turn increases also cell infiltration (Granquist et al., 2010). *A. phagocytophilum* has been shown to compromise different intracellular mechanisms to facilitate its survival within neutrophils. It inhibits host–cell apoptosis, subverts autophagy, and dissociates eukaryote cholesterol from low-density lipoprotein (Rikihisa, 2010).

Anaplasma phagocytophilum

One reported study focused on transmission of *A. phagocytophilum* from *I. ricinus* to lambs (Granquist et al., 2010). Skin biopsies showed that not only neutrophils and macrophages were infected but also bacteria were found in the lumen of blood vessels and in adventical tunic. The endothelium is therefore a transition site where *A. phagocytophilum* is transferred to neutrophils that are loosely bound to the endothelium and then released into the blood stream. The long-term survival of *A. phagocytophilum* is assured in the skin. The skin constitutes a good reservoir during infection and a source for transmission to other feeding ticks. Interestingly, all lambs were positive for MSP2, indicating its possible importance for *Anaplasma* persistence. Another interesting protein is MSP4, a highly conserved surface protein that is involved in cell binding (Battilani et al., 2017; de la Fuente et al., 2016).

Anaplasma marginale

A. marginale is a pathogen of ruminants and bovine that targets erythrocytes. Little is known about *A. marginale* persistence. *A. marginale* contains also MSP2 family but this family of genes is different from MSP2 of *A. phagocytophilum* (Battilani et al., 2017). However, the mechanism of antigenic variation is the same as for *A. phagocytophilum* (Brown, 2012). Other interesting proteins are MSP1 complex that is unique to *A. marginale*. This complex is a heterodimer composed of MSP1a and MSP1b. MSP1a shows a high variability among strains and serves as an adhesin for bovine and tick cells like MSP1b (Battilani et al., 2017). MSP1a is considered as an interesting candidate for a vaccine.

Role of the Tick in Bacteria Transmission

Anaplasma is transmitted by hard ticks. While *Ixodes* tick saliva has been thoroughly investigated in the transmission of *B. burgdorferi* sl, responsible of Lyme borreliosis, the precise role of this saliva in *A. phagocytophilum* transmission is poorly understood so far. An early study showed that *Anaplasma* migration from the gut to the salivary glands takes approximately 24 h (Hodzic et al., 1998). The migration of *A. phagocytophilum* inside the tick was also investigated. A tick protein Salp16 found in the salivary gland was shown to be essential for *Anaplasma* transmission to the vertebrate host (Sukumaran et al., 2006). It was also described that the bacteria need another tick protein, called P11, to be internalized into tick hemocytes. The silencing of this tick protein reduces the invasion of salivary glands (Liu et al., 2011). Another protein of *I. scapularis* tick, sialostatin L2, was shown to inhibit the activation of the inflammatory cascade by targeting caspase-1. It then decreases the secretion of IL-1β and IL-18 by macrophages. The inhibition of this cascade seems to facilitate the invasion of macrophages by *A. phagocytophilum* in the skin and decreases also the local inflammation (Chen et al., 2014). An *in vitro* study on macrophages isolated from the bone marrow and stimulated by *A. phagocytophilum* confirmed the antiinflammatory activity of *I. scapularis* saliva. The saliva inhibited cytokine secretion by macrophages (Chen et al., 2012).

Concerning *A. marginale*, only very little work has been done so far to evaluate the role of the tick on bacterial transmission, although the disease is very important in cattle in tropical countries. One study used an *I. scapularis* cell line to explore the interaction of *A. marginale* with ticks. Various bacterial surface proteins were identified as important for the adhesion to tick cells, and the system was further tested to measure the effect of tick saliva on the interaction (Blouin et al., 2002). More recently, five genes including glutathione S-transferase, cytochrome c oxidase sub III (COXIII), dynein, synaptobrevin, and phosphatidylinositol-3,4,5-triphosphate 3-phosphatase were found upregulated on the incubation of tick cells with the bacteria using tick embryonic cell line (BME26) and suppression subtractive hybridization libraries (Bifano et al., 2014). RNA interference was then tested in *Rhipicephalus microplus*, an important biological vector of *A. marginale*, to evaluate the relevance of these genes in bacterial transmission. Only ticks inoculated with RNAi corresponding to COXIII failed to transmit *A. marginale* to naïve calves.

Rickettsiae, Ticks and Skin Interface

The genus *Rickettsia* consists of a set of small Gram-negative, obligate intracellular bacteria, belonging to the alpha subdivision of the proteobacteria, not cultivable and responsible for rickettsioses. They are small rods or coccobacilli, surrounded by a translucent zone, the slime layer, consisting of polysaccharides (Macaluso and Paddock, 2014).

The life cycle of these bacteria includes necessarily an arthropod vector and a vertebrate host. Until today, 27 species involved in arthropod-borne rickettsial diseases and distributed in 3 groups have been characterized (Balraj et al., 2009): (1) the typhus group containing the flea-borne murine and louse-borne epidemic typhus, (2) the group containing *Rickettsia tsutsugamushi*, transmitted by trombiculid mites and responsible for scrub typhus, and (3) the spotted fever group (SFG) transmitted by hard ticks, most of them being both vector and main reservoir (Table 9.5). The first two groups will not be discussed further here, as ticks are not involved in their life cycle.

Rickettsiae are prevalent on all five continents, and their geographical distribution is closely related to the presence of their vectors (Parola et al., 2013). Certain species of *Rickettsiae* are associated with a single vector tick while others can be transmitted by different tick species. For example, *Rickettsia conorii conorii* is transmitted only by *Rhipicephalus sanguineus* in the Mediterranean region, whereas in sub-Saharan Africa, *Haemaphysalis leachi* and *Rhipicephalus simus* are the two vectors. *R. conorii* causes Mediterranean spotted fever that is clinically similar to Rocky Mountain spotted fever (RMSF) and is characterized by "eschar," a lesion with necrotic center (Merhej et al., 2014).

Rickettsia rickettsii, as another example, is transmitted by *Dermacentor andersoni*, *Dermacentor variabilis*, and *R. sanguineus* sl (Merhej et al., 2014; Parola et al., 2013). *R. rickettsii* is responsible of RMSF, the most severe tick-borne rickettsioses causing vasculitic disease and the death of 2%–31% of patients in North, Central, and South America (Macaluso and Paddock, 2014).

The following mechanisms of *Rickettsia* transmission have been described so far. Inside the tick, transmission occurs via the transovarial and transstadial route.

TABLE 9.5 Main Rickettsioses of the Spotted Fever Group

Rickettsia Species	Main Vectors	Disease	Geographic Distribution
Rickettsia africae	Amblyomma spp. Rhipicephalus spp.	African tick bite fever or African rickettsiosis	Sub-Saharan Africa, Ethiopia, Antilles islands, Reunion island, New Caledonia
Rickettsia australis	Ixodes holocyclus, Ixodes spp.	Queensland tick typhus, Australian spotted fever	Australia
Complexe de Rickettsia conorii	Rhipicephalus sanguineus Haemaphysalis leachi	Mediterranean spotted fever	Mediterranean region, Africa, Middle-east
Rickettsia helvetica	Ixodes spp.	Aneruptive rickettsiosis	Europe, Japan, Thailand
Rickettsia massiliae	R. sanguineus, Rhipicephalus spp.	Rickettsiosis	Mediterranean region, Mali, Canaries islands
Rickettsia rickettsii	Dermacentor spp. R. sanguineus Amblyomma spp.	Rocky Mountain spotted fever	North and South America
Rickettsia sibirica mongolitimonae	Hyalomma spp. Rhizomucor pusillus	Eruptive fever Lar (lymphangitis-associated rickettsioses)	Mongolia, Europe
Rickettsia slovaca	Dermacentor marginatus, Dermacentor reticulatus	Tibola (tick-borne lymphadenopathy), senlat	Slovakia, France, Hungary, Switzerland, Eurasia
Rickettsia raoultii	Rhabdomys pumilio		
Rickettsia sibirica	Dermacentor spp. Hyalomma spp. Haemaphysalis concinna	Siberian typhus	Siberia, Mongolia, Pakistan

Adapted from Parola, P., Raoult, D., May–June 2006. Tropical rickettsioses. Clinics in Dermatology 24 (3), 191–200; Parola, P., Paddock, C.D., Socolovschi, C., Labruna, M., Mediannikov, O., Kernif, T., Abdad, M., Stenos, J., Bitam, I., Fournier, P., Raoult, D., 2013. Update on tick-borne rickettsioses around the world: a geographic approach. Clinical Microbiology Reviews 26, 657–702.

At the vertebrate level, various processes have been identified, the most common being the feeding during the tick bite, but also more surprising ones such as eye contamination following crushing of a tick, or contamination of abraded skin by tick-borne *Rickettsiae* present in tick feces, and finally aerosols have been found (Balraj et al., 2009; Karbowiak et al., 2016). Thus, it should be emphasized that ticks are not only vectors but also reservoirs (Karbowiak et al., 2016).

To enter the human endothelial cells, *Rickettsiae* are phagocytized by the host cells (Balraj et al., 2009). As opposed to *Anaplasma* and *Ehrlichia* that replicate in membrane-bound vesicles inside host cells, *Rickettsia* escapes from phagosomes and replicates directly in the cytoplasm (Balraj et al., 2009; Rikihisa, 2011).

Transmission to the Vertebrate Host

The general process of *Rickettsia* transmission to the vertebrate host is still under investigation. *Rickettsia* transmission usually occurs 10 h after the beginning of the infective blood meal as shown, for example, in *R. rickettsia* (de la Fuente et al., 2017b). Interestingly, among the different arthropod-transmitted *Rickettsia* species, 16 of them cause long-lasting skin eschars, whereas 5 others induce an inflammatory lesion when inoculated to guinea pigs. This indicates that the skin plays an important role in the transmission process of the cutaneous manifestations and the persistence of bacteria (La Scola et al., 2009).

Rickettsiae are surfaced by a layer (slime layer) composed not only of polysaccharides that elicit protective immune responses but also of so-called adhesins responsible for host endothelial cell invasion. Early studies in *R. rickettsii* described changes in this layer during the tick feeding process (Hayes and Burgdorfer, 1982). It contains proteins such as Omp (outer membrane protein) A, OmpB, Sca (surface cell antigen) 1, and Sca2, which act as adhesins promoting attachment to and invasion into host cells (Balraj et al., 2009). They belong to a family also called autotransporters, which is typically found in Gram-negative bacteria. Antibodies to particular epitopes of OmpA and OmpB may protect against reinfection, but they appear not to play a key role in immunity against primary infection.

Rickettsiae exploit the host cell actin cytoskeleton to penetrate and to migrate inside and between the cells. The surface protein RickA is a virulence factor, which hijacks the actin of the host cells to induce actin polymerization and rearrangements enabling the invasion of the host cell and then the formation of actin filopodia. This newly acquired motility enables the pathogen to migrate inside and between the host cells. The filopodia push bacteria to the surface of the host cell, deforming the membrane outward and invaginating into an adjacent cell. The deformed membrane then disrupts, permitting *Rickettsiae* to enter adjacent cell without an exposure to extracellular environment (Uchiyama, 2012). RickA could be a major protein involved in persistence (Balraj et al., 2009). More details on the physiopathology at the cellular and molecular level are presented in the review of Mansueto et al (Mansueto et al., 2012).

Except the above description of an eschar at the site of inoculation, very few studies have been performed to understand the process of *Rickettsia* transmission at the skin interface. Recently, the transmission of *Rickettsia parkeri* has been analyzed in different animal models. The tick saliva has been shown to increase the level of *Rickettsia* in the skin (Banajee et al., 2015; Grasperge et al., 2014).

Tick–Rickettsia *Interactions*

Ticks can become infected by different routes: transovarial, venereal, not only during blood feeding on an infected vertebrate host but also by cofeeding close to a *Rickettsia*-infected tick on a nonrickettsemic vertebrate host. After infection, the bacteria invade rapidly the cells of the midgut epithelium where they multiply (Macaluso and Paddock, 2014). Baldridge et al. (2007) used the SFG endosymbiont, *Rickettsia monacensis*, genetically modified to express GFP, to show the invasion of salivary glands and of the respiratory system in *I. scapularis*. The precise mechanism of tissue invasion has not been elucidated yet. However, the generalized infection by *Rickettsia* has clearly been demonstrated in different tick–*Rickettsia* models (Macaluso and Paddock, 2014). *In vitro* studies on tick cells showed that *Rickettsia felis* interacts with histone H2B through its OmpB protein. The invasion of host cells by the bacteria could be associated with an actin tail that propels the bacteria from cell to cell, as shown in the vertebrate host (Macaluso and Paddock, 2014). Recent studies also demonstrated the role of tick saliva in *Rickettsia* transmission. *R. parkeri*, belonging to the pathogenic SFG and transmitted by *Amblyomma maculatum*, induced a cellular infiltrate composed of neutrophils and macrophages at the site of inoculation in C3H/HeN mice. However, the tick saliva proteins involved in this process are not yet characterized (Banajee et al., 2016). Similarly, the infection of *R. conorii* was studied in a mouse model in the presence and absence of *Rhipicephalus* saliva. The cytokine expression was measured following two different infection protocols: intradermal inoculation of *Rickettsia* and infestation with infected nymphs. The mRNA levels of IL-1β, IL-10, and NF-κB were analyzed in C3H/HeJ lung samples. IL-1β and NF-κB expression were higher in mice infected in absence of tick saliva (Milhano et al., 2015). As shown here and also in the previous model of *B. burgdorferi* sl in *I. ricinus* complex, the tick saliva obviously has an immunosuppressive effect on transmission and dissemination (Barthold et al., 1991; Kern et al., 2011).

CONCLUSIONS

Ticks act as particularly efficient vectors in the transmission of pathogens. Thanks to their long-lasting blood meal, the effect of the hard tick saliva on the transmission of bacteria such as *Borrelia* involved in Lyme borreliosis could be particularly well investigated. By its pharmaco- and immunomodulatory properties, tick saliva helps the bacteria to multiply locally and to persist in absence of an inflammatory response (Kern et al., 2011, 2015). The role of the skin in the transmission and persistence of *Borrelia* has also been very well studied over the last years. However, the pathogenicity of *Borrelia* transmitted by soft ticks in the context of RF is still poorly investigated. The skin constitutes an immunotolerant organ where the pathogen is coinoculated with the tick saliva. Further studies are necessary to decipher the mechanisms of transmission and skin persistence of other bacteria such as *Rickettsia* and *Anaplasma*. Even more neglected is the transmission of parasites such as *Babesia* and *Theileria*, which affect not only domestic animals and impair agricultural economy in tropical countries (Perez de Leon et al., 2014) but also human health (Vannier and Krause, 2012). All these data should improve our understanding of the pathogen transmission and help to develop new strategies to fight these diseases. The identification of pathogen proteins specifically expressed in the skin during the early transmission or during pathogen persistence should lead to new diagnosis tools (Grillon et al., 2017; Schnell et al., 2015) and to the establishment of transmission blocking vaccines (de la Fuente et al., 2008, 2017a,b; de la Fuente and Merino, 2013).

REFERENCES

Aguero-Rosenfeld, M.E., Donnarumma, L., Zentmaier, L., Jacob, J., Frey, M., Noto, R., Carbonaro, C.A., Wormser, G.P., July 2002. Seroprevalence of antibodies that react with Anaplasma phagocytophila, the agent of human granulocytic ehrlichiosis, in different populations in Westchester County, New York. Journal of Clinical Microbiology 40 (7), 2612–2615.

Andoh, A., Zhang, Z., Inatomi, O., Fujino, S., Deguchi, Y., Araki, Y., Tsujikawa, T., Kitoh, K., Kim-Mitsuyama, S., Takayanagi, A., Shimizu, N., Fujiyama, Y., September 2005. Interleukin-22, a member of the IL-10 subfamily, induces inflammatory responses in colonic subepithelial myofibroblasts. Gastroenterology 129 (3), 969–984.

Antonara, S., Ristow, L., Mccarthy, J., Coburn, J., 2010. Effect of *Borrelia burgdorferi* OspC at the site of inoculation in mouse skin. Infection and Immunity 78, 4723–4733. https://doi.org/10.1128/IAI.00464-10.

Bachmann, M., Horn, K., Rudloff, I., Goren, I., Holdener, M., Christen, U., Darsow, N., Hunfeld, K.P., Koehl, U., Kind, P., Pfeilschifter, J., Kraiczy, P., Mühl, H., October 14, 2010. Early production of IL-22 but not IL-17 by peripheral blood mononuclear cells exposed to live Borrelia burgdorferi: the role of monocytes and interleukin-1. PLoS Pathogens 6 (10), e1001144.

Bakken, J., Dumler, J., 2015. Human granulocytic anaplasmosis. Infectious Disease Clinics of North America 29, 341–355.

Baldridge, G., Kurtti, T., Burkhardt, N., Baldridge, A., Nelson, C., Oliva, A., Munderloh, U., 2007. Infection of *Ixodes scapularis* ticks with *Rickettsia monacensis* expressing green fluorescent protein: a model system. Journal of Invertebrate Pathology 94, 163–174.

Balraj, P., Renesto, P., Raoult, D., 2009. Advances in Rickettsia pathogenicity. In: Rickettsiology and Rickettsial Diseases-Fifth International Conference. Ann. NY. Acad SCi, pp. 94–105.

Banajee, K., Embers, M., Langohr, I., Doyle, L., Hasenkampf, N.R., Macaluso, K., 2015. *Amblyomma maculatum* feeding augments Rickettsia parkeri infection in a rhesus macaque model: a pilot study. PLoS One 10, e0135175.

Banajee, K., Verhoeve, V., Harris, E., Macaluso, K., 2016. Effect of *Amblyomma maculatum* (Acari: Ixodidae) saliva on the acute cutaneous immune response to Rickettsia parkeri infection in a murine model. Journal of Medical Entomology 53, 1252–1260.

Barbour, A.G., 2005. Relapsing fever. In: In Tick Borne Diseases of Humans, pp. 268–291.

Barthold, S., Beck, D., Hansen, G., Terwilliger, G., Moody, K., 1990. Lyme borreliosis in selected strains and ages of laboratory mice. The Journal of Infectious Diseases 162, 133–138.

Barthold, S.W., Persing, D.H., Armstrong, A.L., Peeples, R.A., 1991. Kinetics of *Borrelia burgdorferi* dissemination and evolution of disease after intradermal inoculation of mice. The American Journal of Pathology 139, 263–273.

Barthold, S., de Souza, M., Janotka, J., Smith, A., Persing, D., 1993. Chronic Lyme borreliosis in the laboratory mouse. The American Journal of Pathology 143, 959–971.

Battilani, M., De Arcangeli, S., Balboni, A., Dondi, F., 2017. Genetic diversity and molecular epidemiology of Anaplasma. Infection, Genetics and Evolution 49, 195–211.

Bernard, Q., Jaulhac, B., Boulanger, N., 2014. Smuggling across the border: how arthropod-borne pathogens evade and exploit the host defense system of the skin. The Journal of Investigative Dermatology 1–9. https://doi.org/10.1038/jid.2014.36.

Bernard, Q., Gallo, R., Jaulhac, B., Nakatsuji, T., Luft, B., Yang, X., Boulanger, N., 2016. Ixodes tick saliva suppresses the keratinocyte cytokine response to TLR2/TLR3 ligands during early exposure to Lyme borreliosis. Experimental Dermatology 25, 26–31.

Bernard, Q., Wang, Z., Di Nardo, A., Boulanger, N., June 27, 2017. Interaction of primary mast cells with Borrelia burgdorferi (sensu stricto): role in transmission and dissemination in C57BL/6 mice. Parasites and Vectors 10 (1), 313.

Bifano, T., Ueti, M., Esteves, E., Reif, K., Braz, G., Scoles, G., Bastos, R., White, S., Daffre, S., 2014. Knockdown of the *Rhipicephalus microplus* cytochrome c oxidase subunit III gene is associated with a failure of Anaplasma marginale transmission. PLoS One 9, e98614.

Blanton, L., 2013. Rickettsial infections in the tropics and in the traveler. Current Opinion in Infectious Diseases 26, 435–440.

Blouin, E., de la Fuente, J., Garcia-Garcia, J., Sauer, J., Saliki, J., Kocan, K., 2002. Applications of a cell culture system for studying the interaction of Anaplasma marginale with tick cells. Animal Health Research Reviews 3, 57–68.

Bockenstedt, L., Gonzalez, D., Mao, J., Li, M., Belperron, A., Haberman, A., 2014. What ticks do under your skin: two-photon intravital imaging of *Ixodes scapularis* feeding in the presence of the Lyme disease spirochete. The Yale Journal of Biology and Medicine 87, 3–13.

Borchers, A., Keen, C., Huntley, A., Gershwin, M., 2015. Lyme disease: a rigorous review of diagnostic criteria and treatment. Journal of Autoimmunity 57, 82–115.

Boyle, W., Wilder, H., Lawrence, A., Lopez, J., 2014. Transmission dynamics of Borrelia turicatae from the arthropod vector. PLoS Neglected Tropical Diseases 8, e2767.

Brisson, D., Baxamusa, N., Schwartz, I., Wormser, G.P., 2011. Biodiversity of *Borrelia burgdorferi* strains in tissues of Lyme disease patients. PLoS One 6, e22926. https://doi.org/10.1371/journal.pone.0022926.

Brossard, M., Wikel, S.K., 2004. Tick immunobiology. Parasitology (129 Suppl.), S161–S176. https://doi.org/10.1017/S0031182004004834.

Brown, W.C., May 2012. Adaptive immunity to Anaplasma pathogens and immune dysregulation: implications for bacterial persistence. Comparative Immunology, Microbiology and Infectious Diseases 35 (3), 241–252. https://doi.org/10.1016/j.cimid.2011.12.002.

Cabral, E.S., Gelderblom, H., Hornung, R.L., Munson, P.J., Martin, R., Marques, A.R., 2006. *Borrelia burgdorferi* lipoprotein-mediated TLR2 stimulation causes the down-regulation of TLR5 in human monocytes. The Journal of Infectious Diseases 193, 849–859. https://doi.org/10.1086/500467.

Carrasco, S.E., Troxell, B., Yang, Y., Brandt, S.L., Li, H., Sandusky, G.E., Condon, K.W., Serezani, C.H., Yang, X.F., December 2015. Outer surface protein OspC is an antiphagocytic factor that protects Borrelia burgdorferi from phagocytosis by macrophages. Infection and Immunity 83 (12), 4848–4860.

Cervantes, J.L., Dunham-Ems, S.M., La Vake, C.J., Petzke, M.M., Sahay, B., Sellati, T.J., Radolf, J.D., Salazar, J.C., 2011. Phagosomal signaling by *Borrelia burgdorferi* in human monocytes involves Toll-like receptor (TLR) 2 and TLR8 cooperativity and TLR8-mediated induction of IFN-beta. Proceedings of the National Academy of Sciences of the United States of America 108, 3683–3688. https://doi.org/10.1073/pnas.1013776108.

Cervantes, J.L., La Vake, C.J., Weinerman, B., Luu, S., O'Connell, C., Verardi, P.H., Salazar, J.C., 2013. Human TLR8 is activated upon recognition of *Borrelia burgdorferi* RNA in the phagosome of human monocytes. Journal of Leukocyte Biology 94, 1231–1241. https://doi.org/10.1189/jlb.0413206.

Chen, G., Severo, M., Sohail, M., Sakhon, O., Wikel, S., Kotsyfakis, M., Pedra, J., 2012. *Ixodes scapularis* saliva mitigates inflammatory cytokine secretion during *Anaplasma phagocytophilum* stimulation of immune cells. Parasites and Vectors 5, 229.

Chen, G., Wang, X., Severo, M., Sakhon, O., Sohail, M., Brown, L., Sircar, M., Snyder, G., Sundberg, E., Ulland, T., Olivier, A., Andersen, J., Zhou, Y., Shi, G., Sutterwala, F., Kotsyfakis, M., Pedra, J., 2014. The tick salivary protein sialostatin L2 inhibits caspase-1-mediated inflammation during *Anaplasma phagocytophilum* infection. Infection and Immunity 82, 2553–2564.

Chong, S., Evrard, M., Ng, L., 2013. Lights, camera, and action: vertebrate skin sets the stage for immune cell interaction with arthropod-vectored pathogens. Frontiers in Immunology 4, 286.

Christe, M., Rutti, B., Brossard, M., 2000. Cytokines (IL-4 and IFN-gamma) and antibodies (IgE and IgG2a) produced in mice infected with *Borrelia burgdorferi* sensu stricto via nymphs of *Ixodes ricinus* ticks or syringe inoculations. Parasitology Research 86, 491–496.

Chung, Y., Zhang, N., Wooten, R.M., 2013. *Borrelia burgdorferi* elicited-IL-10 suppresses the production of inflammatory mediators, phagocytosis, and expression of co-stimulatory receptors by murine macrophages and/or dendritic cells. PLoS One 8, e84980. https://doi.org/10.1371/journal.pone.0084980.

Coleman, J., Gebbia, J., Piesman, J., Degen, J., Bugge, T., Benach, J., 1997. Plasminogen is required for efficient dissemination of *B. burgdorferi* in ticks and for enhancement of spirochetemia in mice. Cell 89, 1111–1119.

Cook, M.J., 2015. Lyme borreliosis: a review of data on transmission time after tick attachment. International Journal of General Medicine 8, 1–8. https://doi.org/10.2147/IJGM.S73791.

Crippa, M., Rais, O., Gern, L., 2002. Investigations on the mode and dynamics of transmission and infectivity of *Borrelia burgdorferi* sensu stricto and *Borrelia afzelii* in *Ixodes ricinus* ticks. Vector Borne and Zoonotic Diseases 2, 3–9.

Cruz, A.R., Moore, M.W., La Vake, C.J., Eggers, C.H., Salazar, J.C., Radolf, J.D., 2008. Phagocytosis of *Borrelia burgdorferi*, the Lyme disease spirochete, potentiates innate immune activation and induces apoptosis in human monocytes. Infection and Immunity 76, 56–70. https://doi.org/10.1128/IAI.01039-07.

Dantas-Torres, F., Chomel, B.B., Otranto, D., 2012. Ticks and tick-borne diseases: a one health perspective. Trends in Parasitology 28, 437–446. https://doi.org/10.1016/j.pt.2012.07.003.

de la Fuente, J., Estrada-Peña, A., Cabezas-Cruz, A., Kocan, K.M., March 2016. Anaplasma phagocytophilum Uses Common Strategies for Infection of Ticks and Vertebrate Hosts. Trends in Microbiology 24 (3), 173–180. https://doi.org/10.1016/j.tim.2015.12.001.

de la Fuente, J., Kocan, K., Almazan, C., Blouin, E., 2008. Targeting the tick-pathogen interface for novel control strategies. Frontiers in Bioscience 13, 6947–6956.

de la Fuente, J., Antunes, S., Bonnet, S., Cabezas-Cruz, A., Domingos, A., Estrada-Peña, A., Johnson, N., Kocan, K., Mansfield, K., Nijhof, A., Papa, A., Rudenko, N., Villar, M., Alberdi, P., Torina, A., Ayllón, N., Vancova, M., Golovchenko, M., Grubhoffer, L., Caracappa, S., Fooks, A., Gortazar, C., Rego, R., 2017a. Tick-pathogen interactions and vector competence: identification of molecular drivers for tick-borne diseases. Frontiers in Cellular and Infection Microbiology 7, 114.

de la Fuente, J., Contreras, M., Estrada-Peña, A., Cabezas-Cruz, A., 2017b. Targeting a global health problem: vaccine design and challenges for the control of tick-borne diseases. Vaccine 17 S0264–410X(17)31023–X.

de la Fuente, J., Merino, O., December 5, 2013. Vaccinomics, the new road to tick vaccines. Vaccine 31 (50), 5923–5929.

Dugat, T., Lagrée, A., Maillard, R., Boulouis, H., Haddad, N., 2015. Opening the black box of *Anaplasma phagocytophilum* diversity: current situation and future perspectives. Frontiers in Cellular and Infection Microbiology 5, 61.

Duneau, D., Boulinier, T., Gómez-Díaz, E., Petersen, A., Tveraa, T., Barrett, R.T., McCoy, K.D., May 2008. Prevalence and diversity of Lyme borreliosis bacteria in marine birds. Infection, Genetics and Evolution 8 (3), 352–359. https://doi.org/10.1016/j.meegid.2008.02.006.

Dumler, J., Barbet, A., Bekker, C., Dasch, G., Palmer, G., Ray, S., Rikihisa, Y., Rurangirwa, F., 2001. Reorganization of genera in the families Rickettsiaceae and Anaplasmataceae in the order Rickettsiales: unification of some species of Ehrlichia with Anaplasma, Cowdria with Ehrlichia and Ehrlichia with Neorickettsia, descriptions of six new species combi. International Journal of Systematic and Evolutionary Microbiology 51, 2145–2165.

Ebnet, K., Brown, K.D., Siebenlist, U.K., Simon, M.M., Shaw, S., 1997. *Borrelia burgdorferi* activates nuclear factor-kappa B and is a potent inducer of chemokine and adhesion molecule gene expression in endothelial cells and fibroblasts. Journal of Immunology 158, 3285–3292.

Foley, D., Gayek, R., Skare, J., Wagar, E., Champion, C., Blanco, D., Lovett, M., Miller, J., 1995. Rabbit model of Lyme borreliosis: erythema migrans, infection-derived immunity, and identification of *Borrelia burgdorferi* proteins associated with virulence and protective immunity. Journal of Clinical Investigation 96, 965–975.

Francischetti, I.M., Mans, B.J., Meng, Z., Gudderra, N., Veenstra, T.D., Pham, V.M., Ribeiro, J.M., 2008. An insight into the sialome of the soft tick, Ornithodorus parkeri. Insect Biochemistry and Molecular Biology 38, 1–21. https://doi.org/10.1016/j.ibmb.2007.09.009.

Francischetti, I.M.B., Sá-Nunes, A., Mans, B.J., Santos, I.M., Ribeiro, J.M.C., 2009. The role of saliva in tick feeding. Frontiers in Bioscience 14, 2051–2088.

Franke, J., Hildebrandt, A., Dorn, W., February 2013. Exploring gaps in our knowledge on Lyme borreliosis spirochaetes–updates on complex heterogeneity, ecology, and pathogenicity. Ticks and Tick-borne Diseases 4 (1–2), 11–25. https://doi.org/10.1016/j.ttbdis.2012.06.007.

Frischknecht, F., 2007. The skin as interface in the transmission of arthropod-borne pathogens. Cellular Microbiology 9, 1630–1640.

Gaff, H., Kocan, K., Sonenshine, D., 2014. Tick-borne rickettsioses II (Anaplasmataceae). In: Biol. Ticks.

Gaspari, A., Katz, S., 1988. Induction and functional characterization of class II MHC (Ia) antigens on murine keratinocytes. Journal of Immunology 140, 2956–2963.

Gaultney, R., Gonzalez, T., Floden, A., Brissette, C., 2013. BB0347, from the Lyme disease spirochete *Borrelia burgdorferi*, is surface exposed and interacts with the CS1 heparin-binding domain of human fibronectin. PLoS One 8, e75643.

Gautam, A., Dixit, S., Embers, M., Gautam, R., Philipp, M.T., Singh, S.R., Morici, L., Dennis, V.A., 2012. Different patterns of expression and of IL-10 modulation of inflammatory mediators from macrophages of Lyme disease-resistant and -susceptible mice. PLoS One 7, e43860. https://doi.org/10.1371/journal.pone.0043860.

Gebbia, J.A., Coleman, J.L., Benach, J.L., 2001. Borrelia spirochetes upregulate release and activation of matrix metalloproteinase gelatinase B (MMP-9) and collagenase 1 (MMP-1) in human cells. Infection and Immunity 69, 456–462.

Geboes, L., Dumoutier, L., Kelchtermans, H., Schurgers, E., Mitera, T., Renauld, J.C., Matthys, P., February 2009. Proinflammatory role of the Th17 cytokine interleukin-22 in collagen-induced arthritis in C57BL/6 mice. Arthritis and Rheumatism 60 (2), 390–395.

Giambartolomei, G.H., Dennis, V.A., Lasater, B.L., Philipp, M.T., 1999. Induction of pro- and anti-inflammatory cytokines by *Borrelia burgdorferi* lipoproteins in monocytes is mediated by CD14. Infecttion and Immunity 67, 140–147.

Glatz, M., Means, T., Haas, J., Steere, A.C., Mullegger, R.R., 2017. Characterization of the early local immune response to *Ixodes ricinus* tick bites in human skin. Experimental Dermatology 26, 263–269. https://doi.org/10.1111/exd.13207.

Granquist, E., Aleksandersen, M., Bergström, K., Dumler, S., Torsteinbø, W., Stuen, S., 2010. A morphological and molecular study of *Anaplasma phagocytophilum* transmission events at the time of *Ixodes ricinus* tick bite. Acta Veterinaria Scandinavica 17, 43.

Grasperge, B.J., Morgan, T., Paddock, C., Peterson, K., Macaluso, K., 2014. Feeding by *Amblyomma maculatum* (Acari: Ixodidae) enhances Rickettsia parkeri (Rickettsiales: Rickettsiaceae) infection in the skin. Journal of Medical Entomology 51, 855–863.

Grillon, A., Westermann, B., Cantero, P., Jaulhac, B., Voordouw, M., Kapps, D., Collin, E., Barthel, C., Ehret-Sabatier, L., Boulanger, N., 2017. Identification of Borrelia protein candidates in mouse skin for potential diagnosis of disseminated Lyme borreliosis. Scientific Reports 7, 16719.

Guo, X., Booth, C.J., Paley, M.A., Wang, X., DePonte, K., Fikrig, E., Narasimhan, S., Montgomery, R.R., 2009. Inhibition of neutrophil function by two tick salivary proteins. Infection and Immunity 77, 2320–2329. https://doi.org/10.1128/IAI.01507-08.

Hartiala, P., Hytönen, J., Suhonen, J., Leppäranta, O., Tuominen-Gustafsson, H., Viljanen, M.K., 2008. *Borrelia burgdorferi* inhibits human neutrophil functions. Microbes and Infection 10, 60–68. https://doi.org/10.1016/j.micinf.2007.10.004.

Hartiala, P., Hytönen, J., Yrjänäinen, H., Honkinen, M., Terho, P., Söderström, M., Penttinen, M.A., Viljanen, M.K., 2010. TLR2 utilization of Borrelia does not induce p38- and IFN-beta autocrine loop-dependent expression of CD38, resulting in poor migration and weak IL-12 secretion of dendritic cells. Journal of Immunology 184, 5732–5742. https://doi.org/10.4049/jimmunol.0803944.

Hawley, K.L., Martín-Ruiz, I., Iglesias-Pedraz, J.M., Berwin, B., Anguita, J., 2013. CD14 targets complement receptor 3 to lipid rafts during phagocytosis of *Borrelia burgdorferi*. International Journal of Biological Sciences 9, 803–810. https://doi.org/10.7150/ijbs.7136.

Hayes, B., Dulebohn, D., Sarkar, A., Tilly, K., Bestor, A., Ambroggio, X., Rosa, P., April 1, 2014. Regulatory protein BBD18 of the Lyme disease spirochete: essential role during tick acquisition? MBio 5 (2) e01017–14.

Hayes, S., Burgdorfer, W., 1982. Reactivation of *Rickettsia rickettsii* in *Dermacentor andersoni* ticks: an ultrastructural analysis. Infection and Immunity 37, 779–785.

Hodzic, E., Fish, D., Maretzki, C., De Silva, A., Feng, S., Barthold, S., 1998. Acquisition and transmission of the agent of human granulocytic ehrlichiosis by *Ixodes scapularis* ticks. Journal of Clinical Microbiology 36, 3574–3578.

Hoffmann, A.-K., Naj, X., Linder, S., 2014. Daam1 is a regulator of filopodia formation and phagocytic uptake of *Borrelia burgdorferi* by primary human macrophages. The FASEB Journal 28, 3075–3089. https://doi.org/10.1096/fj.13-247049.

Hovius, J.R., De Jong, M.A.W.P., Dunnen, J., Litjens, M., Fikrig, E., Van Der Poll, T., Gringhuis, S.I., Geijtenbeek, T.B.H., 2008a. Salp15 binding to DC-SIGN inhibits cytokine expression by impairing both nucleosome remodeling and mRNA stabilization. PLoS Pathogens 4.

Hovius, J.W.R., Levi, M., Fikrig, E., 2008b. Salivating for knowledge: potential pharmacological agents in tick saliva. PLoS Medicine 5, 0202–0208.

Hulínská, D., Barták, P., Hercogová, J., Hancil, J., Basta, J., Schramlová, J., 1994. Electron microscopy of Langerhans cells and *Borrelia burgdorferi* in Lyme disease patients. Zentralblatt für Bakteriologie-International Journal of Medical Education 280, 348–359.

Jongejan, F., Uilenberg, G., 2004. The global importance of ticks. Parasitology 129, 14.

Karbowiak, G., Biernat, B., Stańczak, J., Szewczyk, T., Werszko, J., 2016. The role of particular tick developmental stages in the circulation of tick-borne pathogens affecting humans in Central Europe. 3. Rickettsiae. Annals of Parasitology 62 (2), 89–100. https://doi.org/10.17420/ap6202.38.

Kazimírová, M., Štibrániová, I., 2013. Tick salivary compounds: their role in modulation of host defences and pathogen transmission. Frontiers in Cellular and Infection Microbiology 3, 43. https://doi.org/10.3389/fcimb.2013.00043.

Kern, A., Collin, E., Barthel, C., Michel, C., Jaulhac, B., Boulanger, N., 2011. Tick saliva represses innate immunity and cutaneous inflammation in a murine model of Lyme disease. Vector Borne and Zoonotic Diseases 11, 1343–1350. https://doi.org/10.1089/vbz.2010.0197. Larchmont, NY.

Kern, A., Schnell, G., Bernard, Q., Bœuf, A., Jaulhac, B., Collin, E., Barthel, C., Ehret-Sabatier, L., Boulanger, N., 2015. Heterogeneity of *Borrelia burgdorferi* sensu stricto population and its involvement in Borrelia pathogenicity: study on murine model with specific emphasis on the skin interface. PLoS One 10, e0133195.

Kilpatrick, A., Randolph, S., 2012. Drivers, dynamics, and control of emerging vector-borne zoonotic diseases. Lancet 380, 1946–1955.

Klempner, M.S., Noring, R., Rogers, R.A., 1993. Invasion of human skin fibroblasts by the Lyme disease spirochete, *Borrelia burgdorferi*. The Journal of Infectious Diseases 167, 1074–1081.

Krause, P.J., Fish, D., Narasimhan, S., Barbour, A.G., 2015. Borrelia miyamotoi infection in nature and in humans. Clinical Microbiology and Infection 21, 631–639. https://doi.org/10.1016/j.cmi.2015.02.006.

Krishnavajhala, A., Wilder, H.K., Boyle, W.K., Damania, A., Thornton, J.A., Pérez de Leon, A.A., Teel, P.D., Lopez, J.E., 2017. Imaging of Borrelia turicatae producing the green fluorescent protein reveals persistent colonization of the *Ornithodoros turicata* midgut and salivary glands from nymphal acquisition through transmission. Applied and Environmental Microbiology 83, e02503–e02516.

La Scola, B., Bechah, Y., Lepidi, H., Raoult, D., 2009. Prediction of rickettsial skin eschars in humans using an experimental Guinea pig model. Microbial Pathogenesis 47, 128–133.

Lazarus, J.J., Meadows, M.J., Lintner, R.E., Wooten, R.M., 2006. IL-10 deficiency promotes increased *Borrelia burgdorferi* clearance predominantly through enhanced innate immune responses. Journal of Immunology 177, 7076–7085.

Lazarus, J.J., Kay, M.A., McCarter, A.L., Wooten, R.M., 2008. Viable *Borrelia burgdorferi* enhances interleukin-10 production and suppresses activation of murine macrophages. Infection and Immunity 76, 1153–1162. https://doi.org/10.1128/IAI.01404-07.

Léger, E., Vourc'h, G., Vial, L., Chevillon, C., McCoy, K., 2013. Changing distributions of ticks: causes and consequences. Experimental and Applied Acarology 59, 219–244.

Lieskovska, J., Kopecky, J., 2012a. Effect of tick saliva on signalling pathways activated by TLR-2 ligand and *Borrelia afzelii* in dendritic cells. Parasite Immunology 34, 421–429. https://doi.org/10.1111/j.1365-3024.2012.01375.x.

Lieskovská, J., Kopecký, J., 2012b. Tick saliva suppresses IFN signalling in dendritic cells upon *Borrelia afzelii* infection. Parasite Immunology 34, 32–39. https://doi.org/10.1111/j.1365-3024.2011.01345.x.

Lieskovská, J., Pálenková, J., Širmarová, J., Elsterová, J., Kotsyfakis, M., Campos Chagas, A., Calvo, E., Růžek, D., Kopecký, J., 2015. Tick salivary cystatin sialostatin L2 suppresses IFN responses in mouse dendritic cells. Parasite Immunology 37, 70–78. https://doi.org/10.1111/pim.12162.

Lindgren, E., Andersson, Y., Suk, J.E., Sudre, B., Semenza, J.C., 2012. Public health. Monitoring EU emerging infectious disease risk due to climate change. Science 336, 418–419. https://doi.org/10.1126/science.1215735.

Liu, L., Narasimhan, S., Dai, J., Zhang, L., Cheng, G., Fikrig, E., 2011. *Ixodes scapularis* salivary gland protein P11 facilitates migration of *Anaplasma phagocytophilum* from the tick gut to salivary glands. EMBO Reports 12, 1196–1203.

Liu, X.Y., Bonnet, S.I., 2014. Hard tick factors implicated in pathogen transmission. PLoS Neglected Tropical Diseases 8, e2566. https://doi.org/10.1371/journal.pntd.0002566.

Love, A.C., Schwartz, I., Petzke, M.M., 2015. Induction of indoleamine 2,3-dioxygenase by *Borrelia burgdorferi* in human immune cells correlates with pathogenic potential. Journal of Leukocyte Biology 97, 379–390. https://doi.org/10.1189/jlb.4A0714-339R.

Lusitani, D., Malawista, S.E., Montgomery, R.R., 2002. *Borrelia burgdorferi* are susceptible to killing by a variety of human polymorphonuclear leukocyte components. The Journal of Infectious Diseases 185, 797–804. https://doi.org/10.1086/339341.

Macaluso, K., Paddock, C., 2014. Tick-borne spotted fever group rickettsioses and Rickettsia species. In: Biol. Ticks, pp. 211–250.

Madder, M., Thys, E., Achi, L., Toure, A., De Deken, R., 2011. Rhipicephalus (Boophilus) microplus: a most successful invasive tick species in West-Africa. Experimental and Applied Acarology 53, 139–145. https://doi.org/10.1007/s10493-010-9390-8.

Mans, B.J., Louw, A.I., Neitz, A.W., 2002. Savignygrin, a platelet aggregation inhibitor from the soft tick Ornithodoros savignyi, presents the RGD integrin recognition motif on the Kunitz-BPTI fold. Journal of Biological Chemistry 277, 21371–21378. https://doi.org/10.1074/jbc.M112060200.

Mansueto, P., Vitale, G., Cascio, A., Seidita, A., Pepe, I., Carroccio, A., di Rosa, S., Rini, G., Cillari, E., Walker, D., 2012. New insight into immunity and immunopathology of Rickettsial diseases. Clinical and Developmental Immunology 2012, 967852.

Marchal, C., Schramm, F., Kern, A., Luft, B.J., Yang, X., Schuijt, T.J., Hovius, J., Jaulhac, B., Boulanger, N., 2011. Antialarmin effect of tick saliva during the transmission of Lyme disease. Infection and Immunity 79, 774–785. https://doi.org/10.1128/IAI.00482-10.

Mason, L.M.K., Veerman, C.C., Geijtenbeek, T.B.H., Hovius, J.W.R., 2013. Ménage à trois: Borrelia, dendritic cells, and tick saliva interactions. Trends in Parasitology 1–9. https://doi.org/10.1016/j.pt.2013.12.003.

Mason, L.M., Wagemakers, A., van 't Veer, C., Oei, A., van der Pot, W.J., Ahmed, K., van der Poll, T., Geijtenbeek, T.B., Hovius, J.W., October 3, 2016. Borrelia burgdorferi Induces TLR2-Mediated Migration of Activated Dendritic Cells in an Ex Vivo Human Skin Model. PLoS One 11 (10), e0164040.

Mead, P., 2015. Epidemiology of Lyme disease. Infectious Disease Clinics of North America 29, 187–210.

Menten-Dedoyart, C., Faccinetto, C., Golovchenko, M., Dupiereux, I., Van Lerberghe, P., Dubois, S., Desmet, C., Elmoualij, B., Baron, F., Rudenko, N., Oury, C., Heinen, E., Couvreur, B., 2012. Neutrophil extracellular traps entrap and kill *Borrelia burgdorferi* sensu stricto spirochetes and are not affected by *Ixodes ricinus* tick saliva. Journal of Immunology 189, 5393–5401.

Merhej, V., Angelakis, E., Socolovschi, C., Raoult, D., 2014. Genotyping, evolution and epidemiological findings of Rickettsia species. Infection, Genetics and Evolution 25, 122–137.

Milhano, N., Saito, T., Bechelli, J., Fang, R., Vilhena, M., De Sousa, R., Walker, D., 2015. The role of Rhipicephalus sanguineus sensu lato saliva in the dissemination of Rickettsia conorii in C3H/HeJ mice. Medical and Veterinary Entomology 29, 225–229.

Montgomery, R.R., Lusitani, D., De Boisfleury Chevance, A., Malawista, S.E., 2004. Tick saliva reduces adherence and area of human neutrophils. Infection and Immunity 72, 2989–2994.

Mullegger, R., 2004. Dermatological manifestations of Lyme borreliosis. European Journal of Dermatology 14, 296–309.

Naj, X., Hoffmann, A.-K., Himmel, M., Linder, S., 2013. The formins FMNL1 and mDia1 regulate coiling phagocytosis of Borrelia burgdorferi by primary human macrophages. Infection and Immunity 81, 1683–1695. https://doi.org/10.1128/IAI.01411-12.

Nestle, F.O., Di Meglio, P., Qin, J.-Z., Nickoloff, B.J., 2009. Skin immune sentinels in health and disease. Nature Reviews Immunology 9, 679–691. https://doi.org/10.1038/nri2622.

Ogden, N., Artsob, H., Margos, G., Tsao, J., 2014. Non-ricketsial tick-borne bacteria and the diseases they cause. In: Sonenshine, D., Roe, M.R. (Eds.), Biology of Ticks. Oxford University Press, pp. 278–312.

Ohnishi, J., Piesman, J., de Silva, A.M., 2001. Antigenic and genetic heterogeneity of Borrelia burgdorferi populations transmitted by ticks. Proceedings of the National Academy of Sciences of the United States of America 98, 670–675.

Oleaga, A., Escudero-Poblacion, A., Camafeita, E., Perez-Sanchez, R., 2007. A proteomic approach to the identification of salivary proteins from the argasid ticks Ornithodoros moubata and Ornithodoros erraticus. Insect Biochemistry and Molecular Biology 37, 1149–1159. https://doi.org/10.1016/j.ibmb.2007.07.003.

Önder, Ö., Humphrey, P.T., McOmber, B., Korobova, F., Francella, N., Greenbaum, D.C., Brisson, D., 2012. OspC is potent plasminogen receptor on surface of Borrelia burgdorferi. Journal of Biological Chemistry 287, 16860–16868.

Parola, P., Paddock, C.D., Socolovschi, C., Labruna, M., Mediannikov, O., Kernif, T., Abdad, M., Stenos, J., Bitam, I., Fournier, P., Raoult, D., 2013. Update on tick-borne rickettsioses around the world: a geographic approach. Clinical Microbiology Reviews 26, 657–702.

Parola, P., Raoult, D., May–June, 2006. Tropical rickettsioses. Clinics in Dermatology 24 (3), 191–200.

Pechová, J., Stepanova, G., Kovar, L., Kopecky, J., Kovár, L., Kopecký, J., 2002. Tick salivary gland extract-activated transmission of Borrelia afzelii spirochaetes. Folia Parasitologica 49 (49), 153–159.

Perez de Leon, A., Vannier, E., Almazan, C., Krause, P., 2014. Tick-borne protozoa. In: Sonenshine, D., Roe, M. (Eds.), Biology of Ticks, vol. 2. Oxford University Press.

Petrovec, M., Bidovec, A., Sumner, J.W., Nicholson, W.L., Childs, J.E., Avsic-Zupanc, T., July 31, 2002. Infection with Anaplasma phagocytophila in cervids from Slovenia: evidence of two genotypic lineages. Wiener Klinische Wochenschrift 114 (13–14), 641–647.

Philipp, M., Johnson, B., 1994. Animal models of Lyme disease: pathogenesis and immunoprophylaxis. Trends in Microbiology 2, 431–437.

Platonov, A., Karan, L., Kolyasnikova, N., Makhneva, N., Toporkova, M., Maleev, V., Fish, D., Krause, P.J., 2011. Humans infected with relapsing fever spirochete Borrelia miyamotoi, Russia. Emerging Infectious Diseases 17, 1816–1823.

Radolf, J.D., Caimano, M.J., Stevenson, B., Hu, L.T., 2012. Of ticks, mice and men: understanding the dual-host lifestyle of Lyme disease spirochaetes. Nature Reviews Microbiology 10, 87–99. https://doi.org/10.1038/nrmicro2714.

Rar, V., Golovljova, I., December 2011. Anaplasma, Ehrlichia, and "Candidatus Neoehrlichia" bacteria: pathogenicity, biodiversity, and molecular genetic characteristics, a review. Infection, Genetics and Evolution 11 (8), 1842–1861. https://doi.org/10.1016/j.meegid.2011.09.019.

Rebaudet, S., Parola, P., October 2006. Epidemiology of relapsing fever borreliosis in Europe. FEMS Immunology and Medical Microbiology 48 (1), 11–15.

Rego, R.O.M., Bestor, A., Štefka, J., Rosa, P.A., 2014. Population bottlenecks during the infectious cycle of the Lyme disease spirochete Borrelia burgdorferi. PLoS One 9. https://doi.org/10.1371/journal.pone.0101009.

Ribeiro, J.M., Endris, T.M., Endris, R., 1991. Saliva of the soft tick, *Ornithodoros moubata*, contains anti-platelet and apyrase activities. Comparative Biochemistry and Physiology Part A: Physiology 100 (1), 109–112.

Rikihisa, Y., 2010. *Anaplasma phagocytophilum* and *Ehrlichia chaffeensis*: subversive manipulators of host cells. Nature Reviews Microbiology 8, 328–339.

Rikihisa, Y., 2011. Mechanisms of obligatory intracellular infection with *Anaplasma phagocytophilum*. Clinical Microbiology Reviews 24, 469–489.

Rizzoli, A., Hauffe, H., Carpi, G., Vourc'h, G., Neteler, M., Rosà, R., July 7, 2011. Lyme borreliosis in Europe. Euro Surveillance 16 (27), 16.

Rizzoli, A., Silaghi, C., Obiegala, A., Rudolf, I., Hubálek, Z., Földvári, G., Plantard, O., Vayssier-Taussat, M., Bonnet, S., Spitalská, E., Kazimírová, M., 2014. *Ixodes ricinus* and its transmitted pathogens in urban and peri-urban areas in Europe: new hazards and relevance for public health. Frontiers in Public Health 2, 251.

Rozwadowska, B., Albertyńska, M., Okła, H., Jasik, K.P., Swinarew, A.S., Mazurek, U., Dudek, S., Urbańska-Jasik, D., Poprawa, I., April 2017. Induction of Apoptosis in Normal Human Dermal Fibroblasts Infected with Borrelia burgdorferi Sensu Lato. Vector Borne and Zoonotic Diseases 17 (4), 237–242.

Rudenko, N., Golovchenko, M., Grubhoffer, L., Oliver Jr., J.H., September 2011. Updates on Borrelia burgdorferi sensu lato complex with respect to public health. Ticks and Tick-borne Diseases 2 (3), 123–128. https://doi.org/10.1016/j.ttbdis.2011.04.002.

Salazar, J.C., Pope, C.D., Sellati, T.J., Feder, H.M., Kiely, T.G., Dardick, K.R., Buckman, R.L., Moore, M.W., Caimano, M.J., Pope, J.G., Krause, P.J., Radolf, J.D., 2003. Coevolution of markers of innate and adaptive immunity in skin and peripheral blood of patients with erythema migrans. Journal of Immunology 171, 2660–2670.

Salazar, J.C., Duhnam-Ems, S., La Vake, C., Cruz, A.R., Moore, M.W., Caimano, M.J., Velez-Climent, L., Shupe, J., Krueger, W., Radolf, J.D., 2009. Activation of human monocytes by live *Borrelia burgdorferi* generates TLR2-dependent and -independent responses which include induction of IFN-beta. PLoS Pathogens 5, e1000444. https://doi.org/10.1371/journal.ppat.1000444.

Schnell, G., Boeuf, A., Westermann, B., Jaulhac, B., Carapito, C., Boulanger, N., Ehret-Sabatier, L., 2015. Discovery and targeted proteomics on cutaneous biopsies: a promising work toward an early diagnosis of Lyme disease. Molecular and Cellular Proteomics 14, 1254–1264.

Schramm, F., Kern, A., Barthel, C., Nadaud, S., Meyer, N., Jaulhac, B., Boulanger, N., 2012. Microarray analyses of inflammation response of human dermal fibroblasts to different strains of *Borrelia burgdorferi* sensu stricto. PLoS One 7, e40046. https://doi.org/10.1371/journal.pone.0040046.

Schwan, T., 1996. Ticks and Borrelia: model systems for investigating pathogen-arthropod interactions. Infectious Agents and Disease 5, 167–181.

Schwan, T., Piesman, J., 2002. Vector interactions and molecular adaptations of Lyme disease and relapsing fever spirochetes associated with transmission by ticks. Emerging Infectious Diseases 8, 115–121. https://doi.org/10.3201/eid0802.010198.

Silberer, M., Koszik, F., Stingl, G., Aberer, E., 2000. Downregulation of class II molecules on epidermal Langerhans cells in Lyme borreliosis. British Journal of Dermatology 143, 786–794.

Šimo, L., Kazimirova, M., Richardson, J., Bonnet, S., 2017. The essential role of tick salivary glands and saliva in tick feeding and pathogen transmission. Frontiers in Cellular and Infection Microbiology 7, 281.

Slámová, M., Skallová, A., Páleníková, J., Kopecký, J., 2011. Effect of tick saliva on immune interactions between *Borrelia afzelii* and murine dendritic cells. Parasite Immunology 33, 654–660. https://doi.org/10.1111/j.1365-3024.2011.01332.x.

Sonenshine, D.E., Anderson, J.M., 2014. Mouthparts and digestive system. In: Sonenshine, D.E., Roe, R.M. (Eds.), Biology of Ticks. Oxford University Press, New York 1, pp. 122–162.

Sonnenberg, G.F., Fouser, L.A., Artis, D., 2010. Functional biology of the IL-22-IL-22R pathway in regulating immunity and inflammation at barrier surfaces. Advances in Immunology 107, 1–29.

Stanek, G., Wormser, G., Gray, J., Strle, F., 2012. Lyme borreliosis. Lancet 379, 461–473. https://doi.org/10.1016/S0140-6736(11)60103-7.

Steere, A., Strle, F., Wormser, G., Hu, L., Branda, J., Hovius, J., Li, X., Mead, P., 2016. Lyme borreliosis. Nature Reviews Disease Primers 2, 1–13.

Straubinger, R., Summers, B., Chang, Y., Appel, M., 1997. Persistence of *Borrelia burgdorferi* in experimentally infected dogs after antibiotic treatment. The Journal of Microbiology 35, 111–116.

Stuen, S., Pettersen, K.S., Granquist, E.G., Bergström, K., Bown, K.J., Birtles, R.J., April 2013. Anaplasma phagocytophilum variants in sympatric red deer (*Cervus elaphus*) and sheep in southern Norway. Ticks and Tick-borne Diseases 4 (3), 197–201. https://doi.org/10.1016/j.ttbdis.2012.11.014.

Strle, K., Drouin, E.E., Shen, S., El Khoury, J., McHugh, G., Ruzic-Sabljic, E., Strle, F., Steere, A.C., 2009. *Borrelia burgdorferi* stimulates macrophages to secrete higher levels of cytokines and chemokines than *Borrelia afzelii* or *Borrelia garinii*. The Journal of Infectious Diseases 200, 1936–1943. https://doi.org/10.1086/648091.

Sukumaran, B., Narasimhan, S., Anderson, J., DePonte, K., Marcantonio, N., Krishnan, M., Fish, D., Telford, S., Kantor, F., Fikrig, E., 2006. An *Ixodes scapularis* protein required for survival of *Anaplasma phagocytophilum* in tick salivary glands. The Journal of Experimental Medicine 203, 1507–1517.

Swirski, F.K., Nahrendorf, M., Etzrodt, M., Wildgruber, M., Cortez-Retamozo, V., Panizzi, P., Figueiredo, J.-L., Kohler, R.H., Chudnovskiy, A., Waterman, P., Aikawa, E., Mempel, T.R., Libby, P., Weissleder, R., Pittet, M.J., 2009. Identification of splenic reservoir monocytes and their deployment to inflammatory sites. Science 325, 612–616. https://doi.org/10.1126/science.1175202.

Talkington, J., Nickell, S.P., 1999. *Borrelia burgdorferi* spirochetes induce mast cell activation and cytokine release. Infection and Immunity 67, 1107–1115.

Talkington, J., Nickell, S.P., 2001. Role of Fc gamma receptors in triggering host cell activation and cytokine release by *Borrelia burgdorferi*. Infection and Immunity 69, 413–419. https://doi.org/10.1128/IAI.69.1.413-419.2001.

Tel, J., Lambeck, A.J.A., Cruz, L.J., Tacken, P.J., de Vries, I.J.M., Figdor, C.G., 2010. Human plasmacytoid dendritic cells phagocytose, process, and present exogenous particulate antigen. Journal of Immunology 184, 4276–4283. https://doi.org/10.4049/jimmunol.0903286.

Tilly, K., Krum, J.G., Bestor, A., Jewett, M.W., Grimm, D., Bueschel, D., Byram, R., Dorward, D., Vanraden, M.J., Stewart, P., Rosa, P., 2006. *Borrelia burgdorferi* OspC protein required exclusively in a crucial early stage of mammalian infection. Infection and Immunity 74, 3554–3564. https://doi.org/10.1128/IAI.01950-05.

Tilly, K., Bestor, A., Jewett, M.W., Rosa, P., 2007. Rapid clearance of Lyme disease spirochetes lacking OspC from skin. Infection and Immunity 75, 1517–1519. https://doi.org/10.1128/IAI.01725-06.

Uchiyama, T., 2012. Tropism and pathogenicity of rickettsiae. Frontiers in Microbiology 3, 230.

Vannier, E., Krause, P., 2012. Human babesiosis. The New England Journal of Medicine 366, 2397–2407.

Vesely, D.L., Fish, D., Shlomchik, M.J., Kaplan, D.H., Bockenstedt, L.K., 2009. Langerhans cell deficiency impairs *Ixodes scapularis* suppression of Th1 responses in mice. Infection and Immunity 77, 1881–1887. https://doi.org/10.1128/IAI.00030-09.

Wager, B., Shaw, D.K., Groshong, A.M., Blevins, J.S., Skare, J.T., 2015. BB0744 affects tissue tropism and spatial distribution of *Borrelia burgdorferi*. Infection and Immunity 83, 3693–3703. https://doi.org/10.1128/IAI.00828-15.

Wang, G., Ojaimi, C., Wu, H., Saksenberg, V., Iyer, R., Liveris, D., McClain, S., Wormser, G.P., Schwartz, I., 2002. Disease severity in a murine model of Lyme borreliosis is associated with the genotype of the infecting *Borrelia burgdorferi* sensu stricto strain. The Journal of Infectious Diseases 186, 782–791.

Wikel, S.K., 1999. Tick modulation of host immunity: an important factor in pathogen transmission. International Journal of Parasitology 29, 851–859.

Wikel, S.K., 2013. Ticks and tick-borne pathogens at the cutaneous interface: host defenses, tick countermeasures, and a suitable environment for pathogen establishment. Frontiers and Microbiology 4, 337. https://doi.org/10.3389/fmicb.2013.00337.

Woldehiwet, Z., 2010. The natural history of *Anaplasma phagocytophilum*. Veterinary Parasitology 167, 108–122. https://doi.org/10.1016/j.vetpar.2009.09.013.

Wormser, G.P., Masters, E., Nowakowski, J., McKenna, D., Holmgren, D., Ma, K., Ihde, L., Cavaliere, L.F., Nadelman, R.B., 2005. Prospective clinical evaluation of patients from Missouri and New York with erythema migrans-like skin lesions. Clinical Infectious Diseases: An Official Publication of the Infectious Diseases Society of America 41, 958–965. https://doi.org/10.1086/432935.

Wu, J., Weening, E.H., Faske, J.B., Höök, M., Skare, J.T., 2011. Invasion of eukaryotic cells by *Borrelia burgdorferi* requires β1 integrins and Src kinase activity. Infection and Immunity 79, 1338–1348. https://doi.org/10.1128/IAI.01188-10.

Xu, Q., Seemanapalli, S.V., Reif, K.E., Brown, C.R., Liang, F.T., 2007. Increasing the recruitment of neutrophils to the site of infection dramatically attenuates *Borrelia burgdorferi* infectivity. Journal of Immunology 178, 5109–5115.

Xu, Q., McShan, K., Liang, F.T., 2008. Essential protective role attributed to the surface lipoproteins of *Borrelia burgdorferi* against innate defences. Molecular Microbiology 69, 15–29. https://doi.org/10.1111/j.1365-2958.2008.06264.x.

Yakimchuk, K., Roura-Mir, C., Magalhaes, K.G., de Jong, A., Kasmar, A.G., Granter, S.R., Budd, R., Steere, A., Pena-Cruz, V., Kirschning, C., Cheng, T.-Y., Moody, D.B., 2011. *Borrelia burgdorferi* infection regulates CD1 expression in human cells and tissues via IL1-β. European Journal of Immunology 41, 694–705. https://doi.org/10.1002/eji.201040808.

Yang, X., Coleman, A.S., Anguita, J., Pal, U., 2009. A chromosomally encoded virulence factor protects the Lyme disease pathogen against host-adaptive immunity. PLoS Pathogens 5, e1000326. https://doi.org/10.1371/journal.ppat.1000326.

Yang, X., Qin, J., Promnares, K., Kariu, T., Anderson, J.F., Pal, U., 2013. Novel microbial virulence factor triggers murine Lyme arthritis. The Journal of Infectious Diseases 207, 907–918. https://doi.org/10.1093/infdis/jis930.

Zhang, L., Wang, G., Liu, Q., Chen, C., Li, J., Long, B., Yu, H., Zhang, Z., He, J., Qu, Z., Yu, J., Liu, Y., Dong, T., Yao, N., Wang, Y., Cheng, X., Xu, J., 2013. Molecular analysis of *Anaplasma phagocytophilum* isolated from patients with febrile diseases of unknown etiology in China. PLoS One 8, e57155.

Zhao, Z., McCloud, B., Fleming, R., Klempner, M.S., 2007. *Borrelia burgdorferi*-induced monocyte chemoattractant protein-1 production in vivo and in vitro. Biochemical and Biophysical Research Communications 358, 528–533. https://doi.org/10.1016/j.bbrc.2007.04.150.

Chapter 10

Tick-Borne Viruses and Host Skin Interface

Mária Kazimírová[1], Pavlína Bartíková[2], Iveta Štibrániová[2]
[1]Institute of Zoology, Slovak Academy of Sciences, Bratislava, Slovakia; [2]Biomedical Research Centre, Institute of Virology, Slovak Academy of Sciences, Bratislava, Slovakia

INTRODUCTION

Ticks transmit a broad spectrum of arboviruses (ar=arthropod, bo=borne viruses). Some of the viruses have significant medical and veterinary impact by causing serious diseases in humans and animals (Nuttall, 2014; Brackney and Armstrong, 2016). Both soft ticks (Argasidae) and hard ticks (Ixodidae) are able to transmit viruses, but hard ticks are vectors of the majority of viruses of medical and veterinary importance (Labuda and Nuttall, 2004, 2008; Nuttall, 2014). Vector-borne diseases are among the most complex of all infectious diseases to prevent and control because of outbreaks in new regions and countries due to increasing global travel, urbanization, climate change, and a few available vaccines. Their transmission is also unique through arthropod bite.

Tick-borne viruses (TBVs) are maintained in nature through circulation between vector ticks and vertebrate hosts. Thanks to their feeding behavior and a complex life cycle, ticks can also act as long-term virus reservoirs, retaining the viruses for their whole life span, i.e., after molting to next instars (transstadial transmission) and/or transmitting them through eggs (transovarially) to their progeny (Nuttall et al., 1994; Nuttall and Labuda, 2003; Turell, 2015). Due to their unique biology, ticks probably shaped the evolution of TBVs (Nuttall and Labuda, 2003; Kuno and Chang, 2005).

TBVs have adapted to infect and replicate in both vertebrate and tick cells so as to be able to survive and circulate in natural foci. Neither the molecular mechanisms that allow TBVs to switch between ticks and vertebrate hosts nor the mechanism of viral persistence in different environments is fully understood, but it is suggested that viral, tick, and vertebrate host factors are involved in these complex processes (Nuttall et al., 1994; Labuda and Nuttall, 2004; Robertson et al., 2009; Mlera et al., 2014; Nuttall, 2014). Tick cell lines became an important complementary tool to *in vivo* research of the molecular

Skin and Arthropod Vectors. https://doi.org/10.1016/B978-0-12-811436-0.00010-1
325

basis of the tick–host–arbovirus relationships (Bell-Sakyi et al., 2012). Tick cell lines can be used, e.g., to investigate the suitability of potential vector ticks to acquire and amplify particular viruses (Lawrie et al., 2004; Růžek et al., 2008) to study the differences in virus properties while infecting vertebrate and tick cells (Šenigl et al., 2004, 2006), to analyze pathways in ticks that are associated with virus infection (Grabowski et al., 2016), and to investigate tick antiviral responses (Weisheit et al., 2015).

Hard ticks require several hours to days to complete their blood meal and need to counteract multiple host defense responses, which would normally disrupt tick feeding and cause rejection of the ticks. Modulation of host defenses is not restricted to hard ticks and was also documented for the short-feeding soft ticks (Mans and Neitz, 2004; Mans et al., 2008). Ticks generally succeed in completing their blood meal thanks to the presence of biologically active molecules in their saliva that modulate host hemostasis and immune responses (Kazimírová and Štibrániová, 2013; Wikel, 2014; Šimo et al., 2017).

Host skin provides the first line of defense against injury caused by penetration of tick mouthparts and against infection by pathogenic agents, including arboviruses (Frischknecht, 2007; Wikel, 2014). Thus, the early interactions between host immunity and initial tick-mediated immunomodulation occurring at the tick–vertebrate skin interface seem to be essential in successful transmission of arboviruses and their establishment and dissemination in vertebrate hosts (Wikel, 2013, 2014). Indeed, transmission of viruses via a tick bite is not a mechanistic process but is promoted by tick saliva. The "saliva-assisted transmission" (SAT) phenomenon was demonstrated first for Thogotovirus (THOV) (Jones et al., 1989) and subsequently for tick-borne encephalitis virus (TBEV) (Alekseev and Chunikhin, 1991; Labuda et al., 1993c). Moreover, transmission of viruses from virus-infected to noninfected ticks cofeeding in close proximity in an immunomodulated host skin was observed and was found to occur even in absence of host viremia. This phenomenon (nonviremic transmission—NVT) was suggested to be an indirect evidence of SAT (Nuttall and Labuda, 2008). The attachment sites of ticks in the host skin, which are modulated by tick saliva, are reported to be important foci of virus replication early after virus transmission (Labuda et al., 1996; Hermance and Thangamani, 2014; Hermance et al., 2016). Cellular infiltration and cell migration at the tick attachment site are likely to facilitate NVT (Labuda et al., 1996). Mediators of SAT probably vary according to tick species and may define their vector competence for certain TBVs (Nuttall and Labuda, 2008). The underlying molecular mechanisms of SAT and NVT are, however, still poorly understood, and future research should be aimed at the identification of the key tick salivary molecules triggering tick-mediated skin immunomodulation and enhancement of TBV transmission. In addition, the knowledge on tick immunomodulators exploited by TBVs could be utilized in development of anti-tick and anti-TBV vaccines.

TAXONOMY OF TICK-BORNE VIRUSES

TBVs or "tiboviruses" (Hubálek and Rudolf, 2012) represent a large group of viruses characterized by their specific biological transmission among competent hematophagous hard or soft ticks and vertebrate hosts and by their ability to infect and replicate in both vertebrate and in arthropod cells. First described TBVs, Nairobi sheep disease virus (NSDV) (1910) and Louping ill virus (LIV) (1929) and the mosquito-borne yellow fever virus (1928) triggered the avalanche of discoveries of more than 500 arboviruses during the next 30 years or more (Bichaud et al., 2014). At least 160 named viruses are tick-borne, of which about 50 are recognized or probable "virus species" (Nuttall, 2014). Taxonomically, it is a heterogenous group of vertebrate viruses found in several viral families: *Asfarviridae, Reoviridae, Flaviviridae, Orthomyxoviridae, Rhabdoviridae*, the newly formed *Nyamiviridae* (order *Mononegavirales*) (King et al., 2011), and the families *Nairoviridae, Phenuiviridae*, and *Peribunyaviridae* included in the new order *Bunyavirales* (Adams et al., 2017). With only one exception (African swine fever virus [ASFV]), all TBVs are RNA viruses. Taxonomy of viruses is created by the International Committee on Taxonomy of Viruses (ICTV) established in 1966. According to the latest ICTV report (Adams et al., 2017), virus taxonomy comprises 8 orders, 122 families, 35 subfamilies, 735 genera, and 4404 species of viruses and viroids. Besides ICTV, the International Catalogue of Arboviruses Including Certain Other Viruses of Vertebrates (https://wwwn. cdc.gov/Arbocat/Default.aspx) was established to collate data on the isolation and biological characterization of new arboviruses. During its two decades of existence (the heyday of arbovirology), most of the current major arboviruses were discovered, characterized, studied, and included in this catalog. However, since the newly discovered potential arboviruses are recorded as genomic sequences in other databases, no registration has longer occurred. Each virus family is characterized by a unique genome organization and replication strategy. This implies that TVB lineages have evolved independently at least seven times (Nuttall, 2014). Almost 25% of TBVs are associated with diseases. Several TBVs cause very serious human or animal diseases, while others play definitely a role in human or animal pathology though the disease is usually either less serious or infrequently reported. Some TBVs had not proven medical or veterinary significance. However, certain viral diseases of wildlife and of domestic and companion animals and humans may often pass unnoticed or misdiagnosed and eventually may appear as emerging diseases (Dörrbecker et al., 2010; Hubálek and Rudolf, 2012).

Arboviruses, including TBVs, represent the largest biological group of vertebrate viruses that replicate successfully in two phylogenetically distinct systems (vertebrates and invertebrates) irrespective of whether they have an RNA genome double-stranded or single-stranded, segmented or nonsegmented, or of positive or negative polarity (Nuttall, 2009). Thus, it is reasonable to assume that their lifestyle is successful.

Animal viruses are generally transmitted between hosts directly, mechanically, and/or vertically. Arboviruses undergo biological transmission rather than mechanical or direct (Kuno and Chang, 2005; Nuttall, 2014).

In recent decades, a number of recognized TBVs have reemerged and/or spread and pose an increasing threat to human and animal health. Meanwhile, new TBVs are being discovered and unclassified viruses are being allocated to genera or families thanks to improvements in molecular technologies. At present, more than 16 specific tick-borne diseases (TBD) of humans and 19 TBD of veterinary importance have been described (Nicholson et al., 2009; Sonenshine and Roe, 2014). The latest emerging TBD, caused by Bourbon virus, was reported in Kansas in 2014 (Kosoy et al., 2015). This trend of emerging TBD will likely continue.

DNA Viruses

African Swine Fever Virus

As mentioned above, there is only one known DNA virus transmitted by ticks, ASFV. It belongs to the *Asfarviridae* family with the single genus *Asfivirus* and is transmitted by soft ticks (*Ornithodoros* spp.) or directly by the oral route (Vinuela, 1985; Dixon and Wilkinson, 1988; Dixon et al., 2005, 2011). ASFV, a large enveloped double-stranded DNA virus, is the causative agent of African swine fever, a highly contagious hemorrhagic disease of swine. Early studies of ASFV experimental infection of ticks showed primary localization in the midgut (Greig, 1972) and hemocytes (Endris et al., 1987). ASFV circulates among warthogs and bushpigs (sylvatic cycle) without any apparent effects on their health; however, in domestic pigs (domestic cycle) it causes severe hemorrhagic disease with high mortality (Anderson et al., 1998; Costard et al., 2013). ASFV infection of the tick can result in a persistent infection, depending on the ASFV isolate and the tick combination (Burrage, 2013).

Recent studies suggest that other DNA viruses may also be transmitted by ticks—Lumpy skin disease virus (LSDV) and Murid herpesvirus 4 (MuHV 4).

Lumpy Skin Disease Virus

This virus belongs to the genus *Capripoxvirus*, member of the subfamily *Chordopoxvirinae* and family *Poxviridae*, which involve large enveloped viruses with linear double-stranded DNA (Skinner et al., 2011). It causes lumpy skin disease of cattle in Africa and the Middle East. Currently, it is widely accepted that LSDV is associated with blood-feeding insects such as mosquitoes and stable flies (Carn and Kitching, 1995; Chihota et al., 2001, 2003). However, recent transmission studies have demonstrated the first evidence for a role of the hard ticks *Amblyomma hebraeum* and *Rhipicephalus appendiculatus* in mechanical and transstadial transmission and of *Rhipicephalus decoloratus* in transovarial transmission of LSDV (Lubinga et al., 2014b, 2015; Tuppurainen et al., 2011, 2013a,b). Detection of the virus in different organs (salivary glands (SG), hemocytes, synganglion, midgut, ovaries, testes) of ticks fed on experimentally infected cattle and in naturally infected ticks collected from the field indicates the potential for biological transmission of LSDV by ticks (Lubinga et al., 2013, 2014a; Tuppurainen et al., 2011, 2013a,b).

Murid Herpesvirus 4

Another DNA virus potentially transmitted by ticks is MuHV 4 strain 68 (murine herpesvirus 68 (MHV-68), genus *Rhadinovirus,* subfamily *Gammaherpesvirinae,* family *Herpesviridae,* order *Herpesvirales*) (Hajnická et al., 2017), a natural pathogen of rodents of the family Muridae that are hosts for immature tick stages. Transmission of the virus in rodent populations is direct and occurs mainly via the intranasal route. During acute respiratory infection, MHV-68 infects macrophages, B-lymphocytes, lung alveolar, and endothelial cells. Similarly, to other gammaherpesviruses, MHV-68 causes a lifelong latent infection in host B-lymphocytes that may lead to lymphoproliferative disorders and tumor development (Rajčáni et al., 1985; Rajčáni and Kúdelová, 2007). Thus, MHV-68 is an animal model for the study of human lymphotropic diseases, such as Burkitt lymphoma. The first detection of MHV-68 DNA in *Ixodes ricinus* ticks collected from green lizards was described by Ficová et al. (2011). Presence of the virus DNA in *Dermacentor reticulatus* ticks and of live virus in the tick organs was reported by Kúdelová et al. (2015). Vrbová et al. (2016) detected MHV-68 DNA in *Haemaphysalis concinna* ticks.

RNA Viruses

Flaviviridae, Genus Flavivirus (ssRNA+)

The family *Flaviviridae* represents a diverse group of small enveloped viruses with positive-strand RNA genomes classified into four genera: *Flavivirus*, *Pestivirus*, *Hepacivirus*, and *Pegivirus* (Simmonds et al., 2011). Only one genus, *Flavivirus,* contains arboviruses, of which about 28% are tick-borne. By phylogenetic analysis, tick-borne flaviviruses formed three distinct groups, i.e., group associated with seabirds and mammals, respectively, and the Kadam virus that forms a third evolutionary lineage (Gaunt et al., 2001; Grard et al., 2007) (Table 10.1). TBEV causes tick-borne encephalitis (TBE) that is one of the most dangerous human neural infections. At least 11,000 human cases of TBE have been reported annually in Russia and about 3000 cases in the rest of Europe (Mansfield et al., 2009; Dörrbecker et al., 2010). TBEV has been subdivided into three subtypes, European, Siberian, and Far Eastern. The clinical outcome of TBE in humans depends on the TBEV subtype. Related viruses within the same group, LIV, Langat virus (LGTV), and Powassan virus (POWV), also cause encephalitis in humans, whereas three other viruses, Omsk hemorrhagic fever virus, Kyasanur Forest disease virus (KFDV), and Alkhurma

TABLE 10.1 *Flaviviridae* (ssRNA+)

Genus/Species	Vector	Geographic Localization	References
Flavivirus, tick-borne virus (TBV) group			
Mammalian TBV group			
Tick-borne encephalitis virus	*Ixodes* spp., many ixodid ticks	Europe, Asia	a
Louping ill virus	*I. ricinus*	Ireland, Spain, Scotland, Greece	b
Langat virus	*I. granulatus*	Malaysia, Russia, Thailand	c
Powassan virus	*I. cookei, I. marxi*	Canada, Russia, USA	d
Kyasanur Forest disease virus	*H. spinigera*, many other ixodid ticks	India	e
Omsk hemorrhagic fever virus	*D. reticulatus*	Russia, USA, Canada, Australia	f
Gadgets Gully virus	*I. uriae*	Australia	g
Royal Farm virus	*Ar. hermanni*	Afghanistan	h
Related, unclassified			
Karshi virus	*O. tholozani (papillipes)*	Uzbekistan, Kazakhstan	i
Seabird TBV group			
Tyuleniy virus	*I. uriae*	Russia, USA, Norway	j
Meaban virus	*C. maritimus*	France	k
Saumarez Reef virus	*C. capensis, I. eudyptidis*	Australia	l
Kadam TBV groups			
Kadam virus	*R. parvus*	Uganda	m

Ar., Argas; C., Carios; D., Dermacentor; H., Haemaphysalis; I., Ixodes; O., Ornithodoros; R., Rhipicephalus.

[a]*https://wwwn.cdc.gov/arbocat/VirusDetails.aspx?ID=404&SID=13.*
[b]*https://wwwn.cdc.gov/arbocat/VirusDetails.aspx?ID=271&SID=13.*
[c]*https://wwwn.cdc.gov/arbocat/VirusDetails.aspx?ID=259&SID=13.*
[d]*https://wwwn.cdc.gov/arbocat/VirusDetails.aspx?ID=381&SID=13.*
[e]*https://wwwn.cdc.gov/arbocat/VirusDetails.aspx?ID=252&SID=13.*
[f]*https://wwwn.cdc.gov/arbocat/VirusDetails.aspx?ID=347&SID=13.*
[g]*https://wwwn.cdc.gov/arbocat/VirusDetails.aspx?ID=155&SID=13.*
[h]*https://wwwn.cdc.gov/arbocat/VirusDetails.aspx?ID=403&SID=13.*
[i]*https://wwwn.cdc.gov/arbocat/VirusDetails.aspx?ID=228&SID=13.*
[j]*https://wwwn.cdc.gov/arbocat/VirusDetails.aspx?ID=496&SID=13.*
[k]*https://wwwn.cdc.gov/arbocat/VirusDetails.aspx?ID=295&SID=13.*
[l]*https://wwwn.cdc.gov/arbocat/VirusDetails.aspx?ID=425&SID=13.*
[m]*https://wwwn.cdc.gov/arbocat/VirusDetails.aspx?ID=218&SID=13.*

virus (ALKV) cause fatal hemorrhagic fevers (Gritsun et al., 2003; Lani et al., 2014). Antigenic and genetic similarity of LIV to TBEV suggested that LIV is another subtype of TBEV (Grard et al., 2007; Hubálek and Rudolf, 2012). Viruses within the TBEV serocomplex are considered emerging and reemerging pathogens due to the recent rise in the incidence of human infection (POWV in the United States; Hermance and Thangamani, 2017), the recognition of TBEV in new geographic areas, and the emergence of new viruses such as ALKV, a subtype of KFDV (Charrel et al., 2001), and Deer tick virus, a subtype of POWV (Pugliese et al., 2007; Robertson et al., 2009). In 2011, severe disease and mortality was reported in a herd of goats in Spain. According to genome sequencing and phylogenetic analysis, the virus was significantly divergent from LIV and Spanish sheep encephalitis virus and was named as Spanish goat encephalitis virus (Mansfield et al., 2015).

Reoviridae *(dsRNA, segmented)*

Reoviridae is a large family containing nonenveloped viruses with very diverse biological properties infecting vertebrates, plants, and insects (Attoui et al., 2011). It contains 15 genera of viruses with genomes composed of 9–12 segments of linear double-stranded RNA. TBVs belong to two genera—*Coltivirus* and *Orbivirus* (Table 10.2). *Coltivirus* comprises two species, Colorado tick fever virus (CTFV) that causes acute febrile illness in humans and Eyach virus (EYAV).

CTFV, found primarily in the Rocky Mountain region of the United States and southwestern Canada, is transmitted mainly by adult and nymphal *Dermacentor andersoni* ticks. With 200–400 case reports per year, it is the second most important arboviral infection in the United States after West Nile (Meagher and Decker, 2012), and the prevalence of the disease is directly dependent on the seasonal activity and geographical distribution of the vector tick. Although humans most frequently become infected by biting of infected ticks, other ways of transmission, such as contact with infected animal blood or tissues and person-to-person transmission via blood transfusion, have been reported (Emmons, 1988; Cimolai et al., 1988).

EYAV was first isolated from *I. ricinus* in Germany in 1972 (Rehse-Küpper et al., 1976) and later in France (in 1981) from *Ixodes ventalloi* and *I. ricinus* (Chastel et al., 1984). Antibodies to EYAV have been found in patients with meningoencephalitis and polyneuritis in the former Czechoslovakia, but a causal relationship to the virus has not been established (Málková et al., 1980). Vertebrate hosts are rodents and the European rabbit (*Oryctolagus cuniculus*) (Attoui et al., 2002; Hubálek and Rudolf, 2012). No permissive mammalian cell lines have been found for replication of EYAV, thus the virus can be isolated only from brain of suckling mice after intracranial injection (Charrel et al., 2004; Moutallier et al., 2016).

Orbivirus currently includes 22 distinct virus species and 10 probably members. Approximately 60 tick-borne orbiviruses have been identified and divided to 5 TBV species (Table 10.2). Chenuda virus includes seven different serotypes

TABLE 10.2 *Reoviridae* (segmented dsRNA)

Subfamily/ Genus/Species	Vector	Geographic Localization	References
Spinareovirinae			
Coltivirus			
Colorado tick fever virus	*Dermacentor* spp., *H. leporispalustris*	USA	a
Eyach virus	*I. ricinus, I. ventalloi*	France, Germany	Rehse-Küpper et al. (1976)[b]
Sedoreovirinae			
Orbivirus			
Chenuda virus	*Ar. reflexus, Hy. asiaticum*	Egypt, South Africa, Russia	Taylor et al. (1966)[c]
Chobar Gorge virus	*Ornithodoros* spp.	Canada	d
Great Island virus	*I. uriae*	Egypt, Sudan, India, Jamaica, Russia, Pakistan	e
Wad Medani virus	*R. sanguineus, Hyalomma* spp., *Amblyomma* spp.	Malaysia, Singapore	Taylor et al. (1966)[f]
St. Croix River virus	*I. scapularis*	USA	Munderloh et al. (1994)[g]
Unassigned			
Lake Clarendon virus	*Ar. robertsi*	Australia	g
Matucare virus	*O. boliviensis*	Bolivia	h

Ar., Argas; H., Haemaphysalis; Hy., Hyalomma; I., Ixodes; O., Ornithodoros; R., Rhipicephalus.
[a]*https://wwwn.cdc.gov/arbocat/VirusDetails.aspx?ID=115&SID=13.*
[b]*https://wwwn.cdc.gov/arbocat/VirusDetails.aspx?ID=146&SID=13.*
[c]*https://wwwn.cdc.gov/arbocat/VirusDetails.aspx?ID=107&SID=13.*
[d]*https://wwwn.cdc.gov/arbocat/VirusDetails.aspx?ID=111&SID=13.*
[e]*https://wwwn.cdc.gov/arbocat/VirusDetails.aspx?ID=167&SID=13.*
[f]*https://wwwn.cdc.gov/arbocat/VirusDetails.aspx?ID=514&SID=13.*
[g]*https://wwwn.cdc.gov/arbocat/VirusDetails.aspx?ID=257&SID=13.*
[h]*https://wwwn.cdc.gov/arbocat/VirusDetails.aspx?ID=292&SID=13.*

from soft ticks of the genera *Argas, Carios*, and *Ornithodoros* parasitizing birds. Chenuda virus was originally isolated from *Argas reflexus hermanni*, collected from a pigeon house in Egypt (Labuda and Nuttall, 2004, 2008). The most diverse species, Great Island virus, transmitted by hard ticks parasitizing on seabirds (Labuda and Nuttall, 2004, 2008), comprises 36 serotypes.

Interestingly, St. Croix River virus (SCRV) was isolated only from tick cell lines IDE2 derived from *Ixodes scapularis* ticks (Munderloh et al., 1994). The vertebrate host is unknown and SCRV can therefore be considered as a possible "tick-only virus" (Nuttall, 2009).

Bunyavirales *(ssRNA-, segmented)*

Bunyaviridae, comprising around 530 viruses infecting vertebrates, insects, and plants and formerly divided into five genera, has been revised by the ICTV Bunyaviridae study group. The family has been elevated to the order *Bunyavirales* with 9 families (8 new families and 1 renamed) and 13 genera (Briese et al., 2016; Walker et al., 2016b; Junglen, 2016). TBVs together with other arboviruses are included in three families—*Nairoviridae, Phenuiviridae,* and *Peribunyaviridae.* Nairoviruses are primarily transmitted by ticks, whereas orthobunyaviruses (*Peribunyaviridae*) are generally mosquito-borne and phleboviruses (Phenuiviridae) are vectored mainly by sand flies (Labuda, 1991). These viruses cause a broad spectrum of clinical syndromes (febrile illness, encephalitis, hemorrhagic fever) in infected humans and animals and differ greatly in their geographic distribution (Horne and Vanlandingham, 2014). The genome of these viruses consists of single-stranded negative sense three RNA segments—large, medium, and small. The *Nairoviridae* comprises the genus *Orthonairovirus* with 12 species and several putative nairoviruses (Table 10.3) (Kuhn et al., 2016a,b; Walker et al., 2015b, 2016b). Both, hard and soft ticks host these viruses. Genetic comparison revealed the high diversity of the nairoviruses, which correlates with serologic data and tick host associations (Honig et al., 2004). The most medically important member of the genus and the best studied representative is Crimean–Congo hemorrhagic fever virus (CCHFV) associated with a series of outbreaks across Europe, Middle East, Asia, and Africa (Hoogstraal, 1979; Papa et al., 2010; Tekin et al., 2010; Bente et al., 2013). The main vectors of CCHFV are hard ticks of the *Hyalomma* genus, which have a very wide geographic distribution. The virus demonstrates very low vector specificity and has been isolated from 31 ixodid tick species and two argasid tick species (Bente et al., 2013). CCHFV is nonpathogenic in its natural hosts but highly pathogenic in humans; transmission to humans occurs through tick bite, crushing of engorged ticks, or contact with infected animal blood (Whitehouse, 2004). The most veterinary important nairovirus is NSDV, causing lethal hemorrhagic gastroenteritis in small ruminants in Africa and India.

In addition, several newly discovered viruses belong to this family, e.g., Huángpí tick virus 1, Tǎchéng tick virus 1, and Wēnzhōu tick virus (Li et al., 2015) of the new *Burana orthonairovirus* species, and Soft tick bunyavirus isolated from *Argas vespertilionis* ticks (Oba et al., 2016) of the *Keterah orthonairovirus* species, later identified as an isolate of Keterah virus (Kuhn et al., 2016a).

TABLE 10.3 Bunyavirales, (segmented ssRNA-)

Family/Genus/Species	Vector	Geographic Localization	References
Nairoviridae			
Orthonairovirus			
Burana orthonairovirus	H. doenitzi, H. hystricis, D. marginatus	China	Li et al. (2015)
Crimean–Congo hemorrhagic fever orthonairovirus	Hy. marginatum, many ixodid spp.	Europe, Asia, Africa, Middle East	a,b
Dera Ghazi Khan orthonairovirus	Argas spp.	Egypt, Iran, Pakistan, Taiwan, Australia, Java, South Africa, Thailand	c
Dugbe orthonairovirus	A. variegatum, Hy. intermedia	Africa, India	d
Hazara orthonairovirus	I. redikorzevi	West Pakistan	e
Hughes orthonairovirus	Carios spp., I. uriae, Argas spp.	USA, Mexico, Peru, Ireland, Wales, Trinidad	f
Keterah orthonairovirus	C. pusillus	Malaysia	g
Nairobi sheep disease orthonairovirus	R. appendiculatus, H. intermedia	Ethiopia, Kenya, Rwanda, Tanzania	h
Qalyub orthonairovirus	Ornithodoros spp., I. uriae	Senegal, Uzbekistan, Ethiopia, Egypt	i
Sakhalin orthonairovirus	I. uriae, I. signatus	Russia, USA, Canada, Australia	j
Putative nairoviruses			
Artashat virus	O. alactagalis, O. verrucosus	Armenia, Azerbaijan	Al'khovskii et al. (2014)
Burana virus	H. punctata, H. concinna	Kyrgyzstan	L'vov et al. (2014a)
Chim virus	R. turanicus, Ornithodoros spp.,	Uzbekistan	k

Ellidaey virus ELL 80-3b	I. uriae	Iceland	Nuttall (2014)
Foula virus F 80-1	I. uriae	Scotland	Nuttall (2014)
Geran virus	O. verrucosus	Azerbaijan	L'vov et al. (2014b)
Grimsey virus GRIMS82-1b	I. uriae	Iceland	Nuttall (2014)
Inner Farne Island virus IF 80-3, IF 80-4	I. uriae	England	Nuttall (2014)
Isle of May virus IM81	I. uriae	Scotland	Nuttall (2014)
Kachemak Bay virus	I. signatus	USA	Nuttall (2014)
Kao Shuan virus	Ar. robertsi	Taiwan, Australia, Java	l
Mykines virus M82-2	I. uriae	Denmark	m
Pathum Thani virus	Ar. robertsi	Thailand	n
Pretoria virus	Ar. africolumbae	South Africa	o
Puffin Island virus	C. maritimus	Wales	
Nàyún tick virus	Rhipicephalus spp.	China	Xia et al. (2015)
South Bay virus	I. scapularis	USA	Tokarz et al. (2014a)
Tamdy virus	Hy. asiaticum, Hy. plumbeum	Turkmenistan, Uzbekistan	p
Peribunyaviridae			
Orthobunyavirus			
Bakau orthobunyavirus	Ar. abdulsalami	Pakistan	q
Estero Real orthobunyavirus	O. tadaridae	Czech rep	r
Tete orthobunyavirus	Hyalomma spp.	Egypt, Cyprus, Italy	s

Continued

TABLE 10.3 *Bunyavirales, (segmented ssRNA-)*—cont'd

Family/Genus/Species	Vector	Geographic Localization	References
Phenuiviridae			
Phlebovirus			
Uukuniemi phlebovirus	*Argas* spp., *Ixodes* spp., *Rhipicephalus* spp.	USA, Africa, Europa, Russia, Pakistan,	t
SFTS phlebovirus	*H. longicornis, R. microplus*	China, USA	Zhang et al. (2011) and McMullan et al. (2012)
Unsigned/grouped			
Bhanja virus	*Haemaphysalis* spp., *A. variegatum, Rhipicephalus* spp., *Hyalomma* spp.	India, Italy, Nigeria, Senegal, Russia, Bulgaria, Central Asia,	u
Forecariah virus	*Boophilus geigy*	Guinea	v
Kismayo virus	*R. pulchellus*	Somalia	w
Kaisodi virus	*H. spingera, H. turturis*	India	x
Lanjan virus	*D. auratus, Haemaphysalis* spp., *I. granulatus*	Malaysia	y
Silverwater virus	*H. leporispalustris*	Canada, USA	z
Ungrouped			
Lone Star virus	*A. americanum*	USA	aa
Razdan virus	*D. marginatus*	Armenia	bb
Sunday Canyon virus	*Ar. cooleyi*	USA	cc
Wanowrie virus	*Hyalomma* spp.	India, Sri Lanka, Egypt, Iran	dd

A., Amblyomma; Ar., Argas; C., Carios; D., Dermacentor; H., Haemaphysalis; Hy., Hyalomma; I., Ixodes; O., Ornithodoros; R., Rhipicephalus.

[a] https://wwwn.cdc.gov/arbocat/VirusDetails.aspx?ID=122&SID=13.
[b] https://wwwn.cdc.gov/arbocat/VirusDetails.aspx?ID=116&SID=13.
[c] https://wwwn.cdc.gov/arbocat/VirusDetails.aspx?ID=132&SID=13.
[d] https://wwwn.cdc.gov/arbocat/VirusDetails.aspx?ID=135&SID=13.
[e] https://wwwn.cdc.gov/arbocat/VirusDetails.aspx?ID=178&SID=13.
[f] https://wwwn.cdc.gov/arbocat/VirusDetails.aspx?ID=181&SID=13.
[g] https://wwwn.cdc.gov/arbocat/VirusDetails.aspx?ID=234&SID=13.
[h] https://wwwn.cdc.gov/arbocat/VirusDetails.aspx?ID=320&SID=13.
[i] https://wwwn.cdc.gov/arbocat/VirusDetails.aspx?ID=391&SID=13.
[j] https://wwwn.cdc.gov/arbocat/VirusDetails.aspx?ID=409&SID=13.
[k] https://wwwn.cdc.gov/arbocat/VirusDetails.aspx?ID=110&SID=13.
[l] https://wwwn.cdc.gov/arbocat/VirusDetails.aspx?ID=226&SID=13.
[m] https://wwwn.cdc.gov/arbocat/VirusDetails.aspx?ID=319&SID=13.
[n] https://wwwn.cdc.gov/arbocat/VirusDetails.aspx?ID=368&SID=13.
[o] https://wwwn.cdc.gov/arbocat/VirusDetails.aspx?ID=383&SID=13.
[p] https://wwwn.cdc.gov/arbocat/VirusDetails.aspx?ID=462&SID=13.
[q] https://wwwn.cdc.gov/arbocat/VirusDetails.aspx?ID=43&SID=13.
[r] https://wwwn.cdc.gov/arbocat/VirusDetails.aspx?ID=143&SID=13.
[s] https://wwwn.cdc.gov/arbocat/VirusDetails.aspx?ID=473&SID=13.
[t] https://wwwn.cdc.gov/arbocat/VirusDetails.aspx?ID=505&SID=13.
[u] https://wwwn.cdc.gov/arbocat/VirusDetails.aspx?ID=64&SID=13.
[v] https://wwwn.cdc.gov/arbocat/VirusDetails.aspx?ID=150&SID=13.
[w] https://wwwn.cdc.gov/arbocat/VirusDetails.aspx?ID=240&SID=13.
[x] https://wwwn.cdc.gov/arbocat/VirusDetails.aspx?ID=222&SID=13.
[y] https://wwwn.cdc.gov/arbocat/VirusDetails.aspx?ID=260&SID=13.
[z] https://wwwn.cdc.gov/arbocat/VirusDetails.aspx?ID=440&SID=13.
[aa] https://wwwn.cdc.gov/arbocat/VirusDetails.aspx?ID=270&SID=13.
[bb] https://wwwn.cdc.gov/arbocat/VirusDetails.aspx?ID=394&SID=13.
[cc] https://wwwn.cdc.gov/arbocat/VirusDetails.aspx?ID=456&SID=13.
[dd] https://wwwn.cdc.gov/arbocat/VirusDetails.aspx?ID=516&SID=13.

The *Orthobunyavirus* genus (*Peribunyaviridae*) comprises 48 species, 3 of them are TBVs—Bakau, Estero Real, and Tete. The third genus, *Phlebovirus* (*Phenuiviridae*), contains two tick-borne species, Uukuniemi phlebovirus and the newly described severe fever with thrombocytopenia syndrome virus (SFTSV, also named Huaiyangshan virus or Henan virus) that was reported first in China (Yu et al., 2011; Zhang et al., 2011; Xu et al., 2011). The viral genomic RNA was detected mainly in *Haemaphysalis longicornis* and *Rhipicephalus (Boophilus) microplus* ticks. However, there is evidence of direct transmission through infected blood. Recently, a new virus closely related to SFTSV named Heartland was isolated from two severely febrile patients (McMullan et al., 2012) and from field collected nymphs of *Amblyomma americanum* (Savage et al., 2013) in the United States.

At present, 40 viruses have not been assigned to genera or approved as species (Plyusnin et al., 2011), among them are Bhanja virus (BHAV), Forecariah virus, and Kismayo virus. Based on serologic tests, they are antigenically related to each another and form a Bhanja serogroup together with Palma virus (PALV). Moreover, by phylogenetic and serological analyses, BHAV has been found to be closely related to both SFTSV and Heartland virus (Dilcher et al., 2012; Matsuno et al., 2013). Virome studies of some tick species lead to discovery of several novel bunyaviruses in the genera *Orthonairovirus*, i.e., South Bay virus, and *Phlebovirus*, i.e., Blacklegged tick phlebovirus (BTPV) and the *Dermacentor variabilis*-associated virus the American dog tick phlebovirus (ADTPV) (Tokarz et al., 2014b). Phleboviruses have traditionally been classified into two groups consisting of sand fly/mosquito-borne viruses and TBVs (Nichol et al., 2005). At least three phylogenetic clusters of tick-borne phleboviruses have been identified, each comprised of several potential species: the Uukuniemi group, the Bhanja group, and the SFTSV group (Dilcher et al., 2012; McMullan et al., 2012; Palacios et al., 2013). Phylogenetic analysis revealed that BTPV does not cluster with any of these groups and forms a separate monophyletic clade outside all tick-borne and sand fly/mosquito-borne phleboviruses, similar to Gouleako and Cumuto mosquito-borne viruses. ADTPV is more similar to viruses within the Uukuniemi group but forms a distinct monophyletic clade outside this group (Tokarz et al., 2014b).

Growth characteristics and genome sequencing analysis of Lone Star virus (LSV), an unclassified bunyavirus originally isolated from the lone star tick *A. americanum*, definitively identified LSV as a phlebovirus and by phylogenetic analysis to the Bhanja group viruses (Swei et al., 2013).

Orthomyxoviridae *(ssRNA-, segmented)*

The family *Orthomyxoviridae*, in addition to three genera of influenza viruses, includes the genus *Thogotovirus* with two species (Table 10.4)—THOV and Dhori virus (DHOV)—which are arboviruses transmitted biologically by

TABLE 10.4 *Orthomyxoviridae* (segmented sRNA-)

Genus/Species	Vector	Geographic Localization	Reference
Thogotovirus			
Thogotovirus	*A. variegatum, Hyalomma* spp., *Rhipicephalus* spp.	Cameroon, Egypt, Ethiopia, Iran, Kenya, Sicily, Uganda	a
Dhori virus	*D. marginatus, Hyalomma* spp.	Armenia, Azerbaijan, Egypt, India, Portugal, Russia, Saudi Arabia	b
Quaranjavirus			
Quaranfil virus	*Argas* spp.	Egypt, Nigeria, South Africa, Afghanistan, Iran	Taylor et al. (1966)[c]
Johnston Atoll virus	*C. capensis, C. denmarki*	Central Pacific Islands, Australia, New Zealand, southwest Africa, France, Germany	d

A., Amblyomma; C., Carios; D., Dermacentor.
[a]https://wwwn.cdc.gov/arbocat/VirusDetails.aspx?ID=477&SID=13.
[b]https://wwwn.cdc.gov/arbocat/VirusDetails.aspx?ID=133&SID=13.
[c]https://wwwn.cdc.gov/arbocat/VirusDetails.aspx?ID=392&SID=13.
[d]https://wwwn.cdc.gov/arbocat/VirusDetails.aspx?ID=211&SID=13.

Rhipicephalus spp. and *Hyalomma* spp. ticks, respectively (Davies et al., 1986; Jones et al., 1989). Virions are relatively large, enveloped, and depending on the genus, they contain different numbers of segments of linear, negative-sense single-stranded RNA (DHOV seven, THOV, and Quaranjavirus six segments). THOV has been found to occur in the Central African Republic, Cameroon, Uganda, Ethiopia, and in southern Europe and has been isolated from *Rhipicephalus* sp. in Kenya and Sicily, from *Amblyomma variegatum* in Nigeria and from *Hyalomma* sp. in Nigeria and Egypt. THOV is known to infect humans and animals (including cattle, sheep, donkeys, camels, buffaloes, and rats). DHOV has different but overlapping geographic distribution that includes India, eastern Russia, Egypt, and southern Portugal. DHOV has been isolated from *Hyalomma* sp. THOV contains six single-stranded RNA segments; one of them encodes a surface glycoprotein, a remarkable homolog to the glycoprotein (gp64) of baculoviruses (Morse et al., 1992), probably associated with the ability of thogotoviruses to infect ticks (Nuttall, 2009). THOV has been used in experimental studies of SAT (see below).

In 2012, according to a proposal (Presti et al., 2009; McCauley et al., 2012), the new genus *Quaranjavirus* was created in this virus family. This genus comprises two species, Quaranfil virus and Johnston Atoll virus. Quaranfil virus was originally isolated from two children with febrile illness from the villages of Quaranfil and Sindbis in Egypt in 1953 (Taylor et al., 1966). Subsequently several strains of the virus have been isolated from ticks and seabirds in a number of countries throughout Africa and the Middle East. Johnston Atoll virus is serologically related to Quaranfil virus. It was originally isolated from *Ornithodoros capensis* ticks collected in Noddy Tern bird nests (Sand Island, Johnston Atoll, Central Pacific). Multiple strains have subsequently been isolated from ticks from eastern Australia, New Zealand, and Hawaii (rev. in Presti et al., 2009).

In 2014, a new *Thogotovirus*, Bourbon virus, associated with febrile illness and death was described in United States (Kosoy et al., 2015).

Mononegavirales: Rhabdoviridae *and* Nyamiviridae *(ssRNA-)*

The viral order *Mononegavirales* was established in 1991 by ICTV to accommodate related viruses (assigned in three families, *Filoviridae, Paramyxoviridae,* and *Rhabdoviridae*) with nonsegmented, linear, single-stranded negative-sense RNA genomes. Another two families *Bornaviridae* and *Nyamiviridae* joined the other three mononegavirales families in 1996 and 2014, respectively. In 2016, the order *Mononegavirales* was edited by including of two new families, *Mymonaviridae* and *Sunviridae*, and upgrading of subfamily *Pneumovirinae* (*Paramyxoviridae*) to family status *Pneumoviridae* (Afonso et al., 2016).

Members of the family *Rhabdoviridae* (Table 10.5) infect a wide range of vertebrates, invertebrates, and plants. Their transmission can occur via various arthropod vectors. Many members of the genus *Vesiculovirus* are typical arboviruses, such as vesicular stomatitis virus (VSV). Isfahan vesiculovirus has been isolated from sandflies and also from *Hyalomma asiaticum* ticks in Turkmenia (Karabatsos, 1985). None of the recognized tick-borne rhabdoviruses is known to cause disease. Within this family, a new genus *Ledantevirus* was created comprising 14 new species, four of which are TBVs—*Barur ledantevirus, Kern Canyon ledantevirus, Kolente ledantevirus,* and *Yongjia ledantevirus* (Blasdell et al., 2015; Walker et al., 2015a, 2016a). The formerly unassigned TBVs, Barur and Kern Canyon, were assigned to this new genus as *Barur ledantevirus* and *Kern Canyon ledantevirus*. In recent years, a number of novel rhabdoviruses have been identified from various animal species, but so far only few tick-borne rhabdoviruses have been described. Ghedin et al. (2013) isolated Kolente virus (*K. ledantevirus*) from *A. variegatum* ticks and bats collected in Guinea, West Africa. However, little is known about its ecology, mode of transmission, host range, or epidemiology. Yongjia tick virus 2 was detected by next generation sequencing in a pool of *Haemaphysalis hystricis* ticks collected from wild or domestic animals in Zhejiang Province, China, between 2011 and 2013 (Li et al., 2015). Next new probably TBV, Long Island tick rhabdovirus, was detected in *A. americanum* ticks (Tokarz et al., 2014a). Dilcher et al. (2015) isolated a novel rhabdovirus named Zahedan virus, from *Hyalomma anatolicum anatolicum* ticks collected in Iran, which is closely related to Moussa virus isolated from *Culex* mosquitoes

TABLE 10.5 *Mononegavirales* (ssRNA-), *Rhabdoviridae*, and *Nyamiviridae*

Order/Family/ Genus/Species	Vector	Geographic Localization	References
Mononegavirales			
Rhabdoviridae			
Vesiculovirus			
Isfahan vesiculovirus	*Hy. asiaticum*	Russia	a
Ledantevirus			
Barur ledantevirus	*H. intermedia,* *R. pulchellus*	India, Kenya, Somalia	b
Kern Canyon *ledantevirus*	unknown		Labuda and Nuttall (2004)
Kolente ledantevirus	*A. variegatum*	Guinea	Ghedin et al. (2013)
Yongjia ledantevirus	*H. hystricis*	China	Li et al. (2015)
Unassigned			
Connecticut virus	*I. dentatus*	USA	c
Kwatta	*H. spinigera,* *H. turturis*	Surinam	d
New Minto virus	*H. leporispalustris*	USA	e
Sawgras virus	*D. variabilis,* *H. leporispalustris*	USA	f
Nyamiviridae			
Nyavirus			
Nyamanini virus	*Argas* spp.	South Africa, Egypt, Thailand, Nigeria, Nepal, Sri Lanka	Taylor et al. (1966)g
Midway nyavirus	*Ornithodoros* spp.	Hawaii, USA, Japan	Rehse-Küpper et al. (1976)
Sierra Nevada nyavirus	*O. coriaceus*	USA	Rogers et al. (2014)

D., Dermacentor; H., Haemaphysalis; Hy., Hyalomma; I., Ixodes; O., Ornithodoros;
R., Rhipicephalus.
[a]*https://wwwn.cdc.gov/arbocat/VirusDetails.aspx?ID=196&SID=13.*
[b]*https://wwwn.cdc.gov/arbocat/VirusDetails.aspx?ID=51&SID=13.*
[c]*https://wwwn.cdc.gov/arbocat/VirusDetails.aspx?ID=117&SID=13.*
[d]*https://wwwn.cdc.gov/arbocat/VirusDetails.aspx?ID=251&SID=13.*
[e]*https://wwwn.cdc.gov/arbocat/VirusDetails.aspx?ID=329&SID=13.*
[f]*https://wwwn.cdc.gov/arbocat/VirusDetails.aspx?ID=426&SID=13.*
[g]*https://wwwn.cdc.gov/arbocat/VirusDetails.aspx?ID=340&SID=13.*

from West Africa (Quan et al., 2010) and Long Island tick rhabdovirus. However, further studies are needed to confirm if these viruses are tick-borne.

In 2013, by the proposals of Kuhn et al. (2013) and Mihindukulasuriya et al. (2009), the new family *Nyamiviridae* (ssRNA-), created in the order *Mononegavirales* (Table 10.5), comprises the genus *Nyavirus* including two TBVs, Nyamanini nyavirus (NYMV) and Midway nyavirus (MIDWV). NYMV was discovered in 1957 and repeatedly isolated from land birds and *Argas* spp. ticks. It is endemic in South Africa, Egypt, Thailand, Nigeria, Nepal, and Sri Lanka. MIDWV was discovered in 1966 and repeatedly isolated from seabirds and *Ornithodoros* spp. ticks. It is endemic in Hawaii, United States, and Japan. NYMV and MIDWV are serologically related but clearly distinct from each other and not related serologically to any other virus tested (rev. in Kuhn et al., 2013). In 2014, Tesh et al. proposed a new virus species in this family, Sierra Nevada nyavirus (SNVV). Based on its genomic structure and phylogeny, Sierra Nevada virus is closely related to NYMV and MIDWV, indicating that it is a third member of the *Nyavirus* genus (Rogers et al., 2014). SNVV was originally isolated at the University of California, from Vero cell cultures inoculated with a homogenate of *Ornithodoros coriaceus* ticks collected in Northern California. The virus caused a viral cytopathic effect in both Vero and Baby Hamster Kidney cells within 48 h after inoculation, and intracranial inoculation of newborn mice with SNVV leads to disease and death within 2–3 days (Tesh et al., 2014).

By virome analysis of *I. scapularis*, Tokarz et al. (2014b) identified a new mononegavirales-like virus with the greatest similarity to the Nyamanini and Midway viruses (17% amino acid identity).

TICKS AS VECTORS OF ARBOVIRUSES

Persistence of Arboviruses in the Tick Population

Ticks possess several unique features (remarkable longevity, long feeding period, hematophagy in all postembryonic life stages, blood digestion within midgut cells) (Sonenshine et al., 2002; Nuttall and Labuda, 2003). The complex life history and feeding biology of ticks predetermine their success as vectors of viruses. Once infected, many vectors remain infected for the rest of life. Because of the exceptional longevity of ticks, they can carry viruses over prolonged periods (e.g., during winter or other unfavorable weather conditions) and transmit them transstadially prior to active virus transmission. Thus, besides their vector role, ticks also act as excellent long-term reservoirs of TBVs. A specific mode of TBV persistence in the tick population is via transovarial transmission in which the virus from infected females is transmitted through eggs to their progeny. Although there is evidence of transovarial transmission from experimental studies for a number of TBVs, the levels of this mode of transmission in nature seem to be low (Nuttall et al., 1994; Kuno and Chang, 2005). NVT and cofeeding transmission (see description below) have also important implications for the survival of TBVs in nature.

Ticks and Vector Competence

Isolation of an arbovirus from a tick is not an evidence of vector competence; the virus may simply be present in the blood meal (Nuttall, 2009). To determine vector competence, the following parameters have to be demonstrated: (1) acquisition of the virus during blood feeding on an infected host and (2) transmission of the virus to a host by the tick after its molting to the next developmental stage. During the "extrinsic incubation period" (i.e., between acquisition of the virus through infected blood and its transmission to a new host) the tick is unable to transmit the viruses (Nuttall, 2009).

The association between the tick and the transmitted virus is very intimate and highly specific. Comparatively few tick species have been proved to be vectors of TBVs. Indeed, fewer than 10% of the known tick species are suggested to be competent virus vectors. They are mainly members of the large tick genera. For soft ticks these are *Ornithodoros*, *Carios*, and *Argas* and, among hard ticks, virus vectors have been found mostly in the genera *Ixodes*, *Haemaphysalis*, *Hyalomma*, *Amblyomma*, *Dermacentor*, and *Rhipicephalus* (Labuda and Nuttall, 2004, 2008; Nuttall, 2014). Majority of TBVs are transmitted either by hard ticks or by soft ticks but rarely by both (Labuda and Nuttall, 2004). Moreover, some tick species are known to be vectors of a few TBV species (e.g., *I. ricinus, A. variegatum*) while others can transmit many different TBV species (e.g., *Ixodes uriae* is the vector of at least seven TBVs) (Labuda and Nuttall, 2008).

Transmission of Tick-Borne Viruses

The transmission cycle of TBVs can be represented by a triangle involving a tick, a virus and a vertebrate host, and their interfaces. During coevolution, molecular interactions have developed between ticks, TBVs, and vertebrates, which mediate the transmission of TBVs and survival of all three members of the triangle. The tick–host interface is represented by the tick attachment site in the host skin. The virus is taken up by the tick through an infected blood meal while feeding on an infected vertebrate host. It enters the midgut, disseminates in the body of the tick, and reaches the SG so as to be transmitted during feeding via saliva to a next host. On its route, the virus must overcome several barriers, such as the midgut infection barrier, midgut release barrier, midgut escape barrier, SG infection barrier, and SG release barrier (Nuttall, 2014). In addition, the virus must evade tick innate immune responses to survive, persist, and be transmitted (Hynes, 2014).

During the coevolution with their vectors, tick-borne microorganisms have developed different strategies so as to cope with the immune responses of ticks. High-throughput techniques have provided insight into both the immune responses of ticks evoked by bacteria and bacterial evasion strategies (Smith and Pal, 2014). In contrast, information on molecular mechanisms involved in interactions of TBVs with ticks is scarce (Hynes, 2014; Gulia-Nuss et al., 2016).

RNA interference (RNAi) with an expansion of Argonaute genes appears to be the main antiviral mechanism in ticks against arboviruses (Schnettler et al., 2014; Gulia-Nuss et al., 2016).

The most studied and the best understood TBV transmission cycle is probably that of the TBEV and its principal vectors, *Ixodes persulcatus* and *I. ricinus*. Several experimental studies have been carried out to explain the host–vector–TBEV interactions. For example, tick infestation of viremic laboratory animals indicated that majority of the tested hard tick species (*Ixodes* spp., *Haemaphysalis* spp., *Dermacentor* spp., *R. appendiculatus*) were able to acquire the virus through the infected blood meal and transmit it transstadially (Rajčáni et al., 1976; Nosek et al., 1984; Kožuch and Nosek, 1985; Alekseev et al., 1988, 1991; Alekseev and Chunikhin, 1990a; Labuda et al., 1993a). TBEV was found to infect tick SG prior to attachment and can be transmitted to a vertebrate host by saliva soon after commencement of tick feeding (Řeháček, 1965). Similarly, successful transmission of POWV occurs as soon as within 15 min of tick attachment (Ebel and Kramer, 2004), and transmission of THOV by *R. appendiculatus* is likely to occur within 24 h of attachment (Kaufman and Nuttall, 2003). Onset of feeding was found to enhance amplification of TBEV in SG of *I. persulcatus* and *I. ricinus* (Alekseev and Chunikhin, 1990b; Khasnatinov et al., 2009; Slovák et al., 2014a) and of THOV in SG of *R. appendiculatus* (Kaufmann and Nuttall, 2003). However, the mechanisms by which TBVs disseminate in various tick tissues to reach the SG where their replication is upregulated by feeding are unknown and need to be further investigated (Nuttall, 2014; Slovák et al., 2014a).

Once transmitted to a vertebrate host, TBVs face host immune responses, but the interactions of TBVs with vertebrate hosts and pathogenesis of the viral infections are beyond the scope of this chapter.

Tools to Study Tick–Virus Interactions

Experimental Infection of Ticks With Arboviruses

In addition to infecting ticks by feeding on viremic laboratory hosts previously inoculated with a virus suspension, direct injection of TBV suspensions through the tick coxa into the hemocoel (Khasnatinov et al., 2009; Belova et al., 2012; Slovák et al., 2014a), or through the anus into the rectum (Gonzalez et al., 1991; Belova et al., 2012), and immersion of ticks in virus-containing medium (Mitzel et al., 2007; Belova et al., 2012) have been adopted. These alternative methods of infection serve for the purposes of vector competence and virus transmission studies in the laboratory. However, in contrast to advantages, there are several drawbacks of the listed procedures. Parenteral inoculation was found to be the most efficient method to infect *I. ricinus* adults with TBEV (Khasnatinov et al., 2009; Belova et al., 2012; Slovák et al., 2014a), but it is very artificial as some of the natural barriers are bypassed and usually large doses of virus are used. Moreover, inoculation enables to infect exotic species that are not natural vectors of particular viruses. On the other hand, inoculation into the hemocoel allows the use of less pathogenic variants of viruses in laboratory models. Virus inoculation has been usually used to infect adult ticks due to difficulties in handling of subadult stages, although nymphs are more abundant in nature than adults and are considered as the main stage involved in pathogen transmission. Moreover, there may be differences in vector competence between nymphs and adults. Immersion of ticks in media containing virus has been proved as suitable to infect tick larvae and nymphs and was successfully applied, e.g., in infecting *I. scapularis* ticks with LGTV. High infection and transstadial transmission rates were achieved, suggesting that the immersion technique is suitable to study virus replication, traffic, and survival in ticks (Mitzel et al., 2007).

Cell Cultures

Tick cell cultures have a valuable and irreplaceable role in many aspects of tick and tick-borne pathogen research. Since primary tick cell or tissue explant cultures have been developed, propagation of both arboviruses and non–arthropod-transmitted viruses has been attempted (Řeháček and Kožuch, 1964; Řeháček, 1965; Yunker and Cory, 1967; Cory and Yunker, 1971). In total, 38 TBVs, 16 mosquito-borne viruses, and 1 non–vector-borne virus have been propagated in tick cell lines (Bell-Sakyi et al., 2012). Current advances in molecular virology and arthropod genomics and proteomics, together with methods for genetic manipulation, started to elucidate the complex interactions between arboviruses and their vectors at the cellular and molecular level (Nuttall, 2009). Tick cells became useful tools to identify and characterize tick genes associated with infection with TBVs, including those which mediate antiviral activity, and the pathways exploited by TBVs. By using LGTV-infected *I. scapularis*-derived cell line, antiviral RNAi responses in tick cells were characterized (Schnettler et al., 2014). The study revealed the production of virus-derived small interfering RNAs, which are key molecules of the antiviral RNAi response. In a proteomics study of an *I. scapularis* cell line infected with LGTV, 264 differentially expressed tick proteins were identified, out of which majority were downregulated (Grabowski et al., 2016). The proteins with upregulated expression in infected cells were associated with cellular metabolic pathways and glutaminolysis. Transcriptomes and proteomes of cell lines derived from *I. ricinus* and *I. scapularis* infected with TBEV were analyzed and compared by Weisheit et al. (2015). Several molecules that may be involved in the tick cell innate immune response against flaviviruses and cell stress responses were identified. Moreover, it has been shown that, in addition to RNAi, ticks probably possess other antiviral responses. These may involve the heat-shock proteins HSP90, HSP70, and gp96, the complement-associated protein factor H, and trypsin.

SALIVA-ASSISTED TRANSMISSION

The established route of TBV transmission by infected ticks is via tick saliva secreted into the feeding lesion in the host skin that is modified by the pharmacological properties of tick saliva. Accordingly, SAT was defined as "the indirect promotion of arthropod-borne pathogen transmission via the actions of arthropod saliva molecules on the vertebrate host" (Nuttall and Labuda, 2008).

Development of viremia in a vertebrate host had previously been considered as an important requirement for biological transmission of arboviruses (World Health Organization Scientific Group, 1985). However, by mimicking natural conditions of transmission using THOV-infected *R. appendiculatus* adults or nymphs that cofed with uninfected nymphs or larvae on the same naïve guinea pig, a high percentage of the uninfected ticks became infected, although no viremia was detected in the host animals (Jones et al., 1987). Moreover, nonviremic guinea pigs supported virus transmission between cofeeding ticks to a higher degree than viremic hamsters. Based on this pioneer work demonstrating that a vertebrate host refractory to infection can play a role in the epidemiology of an arbovirus, a novel mode of arbovirus transmission, NVT, was proposed (Jones et al., 1987) and has been considered as an indirect evidence of SAT (Nuttall and Labuda, 2008). In this mode of tick-to-tick transmission, TBVs are transmitted between ticks cofeeding on a vertebrate, whereby ticks are both reservoirs and vectors, and the vertebrate serves as a transient bridge (Randolph, 2011).

Majority of the SAT events have been found to be related to hard ticks (Table 10.6). This may indicate that there are differences in the capacity of hard- and soft ticks to mediate NVT. Due to their feeding biology (long-lasting blood meal, attachment to hosts in aggregates and in close proximity), hard ticks appear to be the most suitable vectors for NVT (Nuttall and Labuda, 2003). Since the first reports on NVT and SAT for THOV, the direct and indirect effects of SAT have been experimentally demonstrated for a few other TBVs, including TBEV, LIV, CCHFV, BHAV, PALV transmitted by hard ticks, and West Nile virus and ASFV vectored by soft ticks (Table 10.6).

Although the knowledge on the significance or frequency of NVT in natural conditions is still limited, the phenomenon appears to play a significant role in the survival of TBVs due to creating less pathological effects on vertebrate hosts (Randolph et al., 1996; Labuda and Randolph, 1999; Nuttall and Labuda, 2003; Randolph, 2011).

Thogotovirus

Direct evidence of the SAT (originally referred to as "saliva-activated transmission"; Jones et al., 1990b) was first demonstrated in an experiment where increased acquisition of THOV by noninfected *R. appendiculatus* nymphs was observed when the ticks fed on naïve guinea pigs inoculated with a mixture of the virus and salivary gland extract (SGE) of partially fed *R. appendiculatus* or *A. variegatum* females in comparison to ticks feeding on hosts inoculated with the virus alone (Jones et al., 1989). However, this effect was obvious only when the virus was inoculated with SGE into the same skin site where the ticks fed. Viremia was not detected in any of the tested guinea pigs, indicating that THOV transmission was enhanced by tick salivary factors that may

TABLE 10.6 Evidence of Saliva-Assisted and Nonviremic Transmission of Tick-Borne Viruses

Virus	Tick Species	Described Effect	References
	Argasidae		
Asfarviridae			
African Swine Fever Virus	*O. porcinus*	SGE, Modulation of the immune response in pigs, increased macrophage recruitment in the dermis, promotion of viral infection	Bernard et al. (2016)
Flaviridae			
West Nile virus	*O. moubata*	NVT	Lawrie et al. (2004)
	Ixodidae		
Orthomyxoviridae			
Togoto virus	*R. appendiculatus*	SGE, enhanced transmission and infectivity	Jones et al. (1987)
		NVT	Jones et al. (1989)
	A. variegatum	NVT	Jones et al. (1990a)
	R. evertsi	NVT	Jones et al. (1992a)
	A. hebraeum	NVT	Jones et al. (1992a)
	A. cajannense	NVT	Jones et al. (1992a)
	H. dromedarii	NVT	Jones et al. (1992a)
	H. marginatum rufipes	NVT	Jones et al. (1992a)
	A. variegatum	SAT	Jones et al. (1992c)
	R. (B.) microplus	SAT	Jones et al. (1992c)
Flaviridae			
Tick-borne encephalitis virus	*I. persulcatus*	NVT	Alekseev and Chunikhin (1990b)
	I. ricinus	SGE, enhanced transmission and infectivity	Alekseev et al. (1991)
		SGE, enhanced transmission and infectivity	Labuda et al. (1993b)

TABLE 10.6 Evidence of Saliva-Assisted and Nonviremic Transmission of Tick-Borne Viruses—cont'd

Virus	Tick Species	Described Effect	References
		NVT	Labuda et al. (1993a,c)
		Saliva, *in vitro* modulation of infection rate of DCs and production of cytokines	Fialová et al. (2010)
	I. scapularis	Sialostatin L2, interference with IFN action, enhanced replication of TBEV in DCs	Lieskovská et al. (2015)
	D. reticulatus	SAT	Labuda et al. (1993b)
	D. marginatus	SAT	Labuda et al. (1993a,c)
	R. appendiculatus	SAT	Labuda et al. (1993a,c)
Louping ill virus	*I. ricinus*	NVT	Jones et al. (1997)
Powassan virus	*I. scapularis*	SGE, virus dose-dependent SAT	Hermance and Thangamani (2015)
Bunyavirales			
Crimean–Congo haemorrhagic fever virus	*H. marginatum*	NVT	Gordon et al. (1993)
Palma virus	*R. appendiculatus*	NVT	Labuda et al. (1997a)
	D. marginatus	NVT	Labuda et al. (1997a)
Bhanja virus	*R. appendiculatus*	NVT	Labuda et al. (1997a)
	D. marginatus	NVT	Labuda et al. (1997a)
Heartland Virus	*A. americanum*	NVT	Godsey et al. (2016)

A., Amblyomma; B., Boophilus; D., Dermacentor; DCs, dendritic cells; H, Hyalomma; I., Ixodes; IFN, interferon; NVT, nonviremic transmission; O., Ornithodoros; R., Rhipicephalus; SAT, saliva-assisted transmission; SGE, salivary gland extract.

also mediate NVT. NVT of THOV was found to be reduced when cofeeding of ticks took place on virus-immune guinea pigs (Jones and Nuttall, 1989b; Jones et al., 1992b). The effect of the SAT factor(s) on virus transmission appears to persist in the host skin for several days. This was demonstrated by an increasing proportion of infected *R. appendiculatus* ticks feeding at the skin site where THOV was inoculated 2–3 days after inoculation of SGE (Jones and Nuttall, 1989a). Furthermore, it was suggested that SAT factors that probably involve one or more proteins or peptides enhance virus transmission through immunomodulation of the host and not by a direct effect of salivary molecules on the virus (Jones et al., 1990b). In addition, *R. appendiculatus* was found to be more efficient in mediating NVT than *A. variegatum*, indicating species-specific differences between the vector ticks (Jones et al., 1990a).

Tick-Borne Encephalitis Virus

Transmission of flaviviruses is also likely to be mediated by factor(s) associated with the saliva of feeding ticks (Alekseev et al., 1991; Labuda et al., 1993c). SAT was demonstrated in experiments where naïve guinea pigs were inoculated with a mixture of TBEV and SGE derived from partially fed uninfected *I. ricinus*, *D. reticulatus*, or *R. appendiculatus* females or with virus alone and were infested with uninfected *R. appendiculatus* nymphs. As an outcome, increased acquisition of the virus was observed in ticks feeding on animals inoculated with the mixture of SGE and virus (Labuda et al., 1993c).

Studies similar to those carried out with THOV demonstrated NVT for TBEV by using *I. persulcatus*, *I. ricinus*, *Dermacentor marginatus*, *D. reticulatus*, and *R. appendiculatus* (Alekseev and Chunikhin, 1990b; Labuda et al., 1993b). To reproduce natural conditions of TBEV transmission, infected and uninfected *I. ricinus* ticks were allowed to cofeed on naïve wild rodent species that are considered as the main natural hosts of the tick's immature stages. Acquisition of the virus was found to be high in ticks feeding on susceptible *Apodemus* mice (*Apodemus flavicollis* and *Apodemus agrarius*) that had undetectable or very low levels of viremia. In contrast, NVT was four times lower to ticks feeding on tick-resistant bank voles (*Clethrionomys glareolus*) that displayed significantly higher viremia and virus infection in lymph nodes and spleen than *Apodemus* mice (Labuda et al., 1993d).

In addition to rodents, other wild-living vertebrate species, which are natural hosts of *I. ricinus*, were examined for their ability to support NVT. The tested animals were previously not exposed to TBEV. During the experiments, the animals were infested with infected *I. ricinus* females and uninfected nymphs that were allowed to cofeed and were subsequently tested for the presence of TBEV. Rodents, mainly *Apodemus* mice, were found to be the most efficient amplifying hosts, although they had a very low or no detectable viremia. In contrast, hedgehogs and pheasants either did not support NVT or they supported it to a low level. Certain degree of NVT was also registered for bank voles, but their viremia was higher compared with the mice (Labuda et al., 1993d). The aim of another study was to find out if virus-immune natural rodents can participate in transmission of TBEV. For this purpose, yellow-necked mice and bank voles were immunized against TBEV and were infested with infected and uninfected *I. ricinus* as previously described. In spite of the presence of virus-specific neutralizing antibodies, these animals supported NVT, suggesting that also hosts immune to TBEV can participate in the TBEV transmission cycles in nature. However, most of the NVT on the immune animals was localized to ticks feeding in close proximity (Labuda et al., 1997b). Additionally, species-specific differences in the dissemination of TBEV in skin of mice and voles after attachment of infected *I. ricinus* ticks and their consequences for NVT were demonstrated by examination of rodent skin samples. Delayed dissemination of the virus from the attachment site of infected ticks to sites where uninfected ticks had fed was confirmed for bank voles but not for mice (Labuda et al., 1996). Moreover, TBEV was found to be recruited preferentially to the tick attachment sites, suggesting that the local skin site of tick attachment is an important area for viral replication after initial TBEV transmission by ticks (Labuda et al., 1996).

Powassan Virus

Enhancement of POWV transmission by SGE derived from *I. scapularis* has been documented recently by using laboratory mice, but the efficiency of SAT depended on the inoculated virus dose (Hermance and Thangamani, 2015). Mice inoculated with a mixture of a high virus dose and SGE and with virus alone displayed severe neurological signs of the disease. In contrast, severe clinical signs of the disease were observed in mice inoculated with a low dose of POWV plus SGE, whereas mice inoculated only with a low dose of the virus showed no signs of the disease and displayed low level of viremia.

Louping III Virus

NVT may play a role also in the persistence of LIV in red grouse (*Lagopus lagopus scoticus*) and mountain hare (*Lepus timidus*) populations (Hudson et al., 1995). While grouse are not able to support virus transmission, the virus can persist through NVT on mountain hares. Mountain hares that did not develop detectable viremia were shown to support NVT of LIV between cofeeding infected and noninfected *I. ricinus* ticks, but the efficiency of NVT was significantly reduced when ticks cofeed on virus-immune hares (Jones et al., 1997).

Bunyaviruses

Evidence of SAT has also been documented for bunyaviruses. Transmission of CCHFV from apparently nonviremic ground-feeding birds to *Hyalomma marginatum rufipes* (Zeller et al., 1994) and between adult *Hyalomma truncatum* cofeeding on naïve rabbits (Gonzalez et al., 1992) was shown. Furthermore,

Gordon et al. (1993) were able to demonstrate transmission of the virus from infected adults of *Hyalomma* spp. to larvae and nymphs while cofeeding on nonviremic guinea pigs, although the proportion of recipient ticks that acquired the virus was relatively low. Still, NVT appears to be an important amplification mechanism of CCHFV in nature (Bente et al., 2013).

NVT of Palma and Bhanja viruses on laboratory mice was shown by using various donor and recipient tick species (*D. marginatus, D. reticulatus, Rhipicephalus sanguineus, R. appendiculatus,* and *I. ricinus*) (Labuda et al., 1997a). Transmission of the newly described Heartland virus from experimentally infected *A. americanum* nymphs to cofeeding larvae has been documented recently (Godsey et al., 2016).

Saliva-Assisted Transmission and Soft Ticks

In soft ticks, evidence of SAT has been rare. The first case was reported for the mosquito-borne West Nile virus and transmission between cofeeding infected and uninfected *Ornithodoros moubata* ticks (Lawrie et al., 2004). In a recent study on ASFV, modulation of the systemic immune response of domestic pigs and of skin inflammation and cellular responses at the tick bite site by SGE or feeding of *Ornithodoros porcinus* ticks was observed (Bernard et al., 2016). The immunomodulatory effect of tick SGE was shown in a greater hyperthermia in pigs inoculated with the virus and SGE than in pigs inoculated with the virus alone. The density and recruitment of Langerhans cells after infection with ASFV were inhibited by SGE at the tick bite and at the SGE inoculation site. Both SGE and virus were found to induce macrophage recruitment.

MOLECULAR ASPECTS OF SALIVA-ASSISTED TRANSMISSION

Laboratory studies suggest that SAT correlates with the vector competence of certain tick species for particular viruses. For example, SAT for THOV was demonstrated for *R. appendiculatus* and *A. variegatum* (natural vectors) but not for *I. ricinus* or soft ticks (noncompetent vectors) (Jones et al., 1992a). In contrast, SAT for TBEV was observed not only with the natural vectors, *I. persulcatus* and *I. ricinus* (Prostriata), but also with species belonging to Metastriata (i.e., *D. reticulatus, R. appendiculatus*) (Alekseev et al., 1991; Labuda et al., 1993c), although in distant NVT *I. persulcatus* and *I. ricinus* appeared to be more efficient donors and recipients of TBEV than Amblyomminae species (Alekseev and Chunikhin, 1992).

In contradiction to these findings, *D. reticulatus* SGE was found to promote the replication of the insect-borne VSV *in vitro* (Hajnická et al., 1998) and the production of the nucleocapsid viral protein (Kocáková et al., 1999; Sláviková et al., 2002) by a yet not explained mechanism.

The reported cases of SAT associated with TBVs provide no explanation of the molecular mechanisms of the processes. During the last two decades, modern molecular genetic and high-throughput techniques have been applied in the systemic characterization of tick salivary composition and to shed light on the underlying molecular mechanisms of exploitation of tick salivary molecules by tick-borne pathogens. A wide range of bioactive proteins have been discovered in tick saliva, and different expression profiles for a number of genes, depending on the presence or absence of a microorganism, have been described in various tick tissues, including SG (Chmelař et al., 2016). However, research in this field is more advanced for the tick–tick-borne bacteria interactions than for TBVs (see Chapters 5 and 9). One of the reasons might be the high pathogenicity of TBVs of medical and veterinary importance that requires strict conditions for their handling and usage in animal experimentation (e.g., laboratories and animal facilities of biosafety levels 3 and 4). Ethical issues connected with the usage of animals in experiments may be another reason for limited studies in TBVs in the laboratory. Considering these constraints, usage of less pathogenic models instead of highly pathogenic TBVs and research on cell lines became alternative tools to explore processes at the tick–TBV interface.

Tick Saliva Molecules

The SG transcript expression profile in response to feeding and infection with a TBV has been inferred from whole body transcriptome analysis of *I. scapularis* nymphs infected with LGTV (McNally et al., 2012). The study revealed that in nymphs feeding for 3 days on naïve mice the number of transcripts related to metabolism sharply increased in comparison to unfed ticks. A total of 578 transcripts were upregulated and 151 transcripts were downregulated in response to feeding. Differences in expression profiles were revealed also between LGTV-infected and -uninfected ticks during the 3 days of feeding. The differently regulated transcripts included putative secreted salivary proteins, lipocalins, Kunitz domain-containing proteins, antimicrobial peptides, and transcripts of unknown function. A transcript upregulated in LGTV-infected nymphs that belonged to the 5.3 kDa family was detected. The same transcript was found to be upregulated in *Borrelia burgdorferi*-infected *I. scapularis* nymphs, suggesting a role of this protein in tick immunity or host defense (Ribeiro et al., 2006). However, because whole *I. scapularis* nymphs were used, it is not possible to say with absolute certainty that the observed differentially regulated transcription is SG specific. Thus, the specific tick molecule(s) assisting in virus replication and transmission needs to be identified.

Virus Proteins Involved in Saliva-Assisted Transmission

The mechanisms of adaptation of TBVs to their vectors and hosts are important aspects that need to be studied so as to understand the processes of TBV transmission. For example, specific mutations in the viral envelope protein of TBEV have been found to affect NVT between cofeeding *I. ricinus* for Siberian and European TBEV strains (Khasnatinov et al., 2010). Furthermore, it has recently been demonstrated that the structural genes of the European TBEV strain Hypr may determine high NVT rates of the virus between cofeeding *I. ricinus* ticks, whereas the region of the TBEV genome encoding nonstructural proteins determines cytotoxicity in cultured mammalian cells (Khasnatinov et al., 2016).

Application to Vaccine Development

Knowledge of the tick salivary compounds that are important for feeding and facilitate the transmission of pathogens may be applied in the development of novel transmission-blocking vaccines to control tick populations and prevent from infections with TBVs. The protective effect of antitick immunity against infection with a TBV has been demonstrated with a cement protein (64p) derived from *R. appendiculatus* ticks. Immunization with the recombinant protein (TRP64) was shown to impair tick feeding (Trimnell et al., 2002, 2005) and protected laboratory mice against lethal infection with TBEV (Labuda et al., 2006). Immunization with TRP64 appeared to impair both transmission of TBEV from infected ticks to mice and NVT. The interference with TBEV transmission has probably been mediated by disruption of the immunomodulated initial skin infection site of tick attachment, leading to a local cutaneous inflammatory response.

SKIN INTERFACE: IMMUNOMODULATION OF THE TICK ATTACHMENT SITE

The tick–host–pathogen interface represents a very complex environment. Host immune responses generate all host wound healing phases (hemostasis, inflammation, reactions of innate and acquired immunity, proliferation, contraction, and migration phases) (Shaw and Martin, 2009), which are evoked by tick mouthpart penetration. All arms of the host defense can be neutralized by tick immunomodulators. Thus, the composition of the tick saliva appears to provide the reaction to the challenges posed by redundant defense mechanisms in the host skin (Ribeiro and Francischetti, 2003; Francischetti et al., 2009; Kotál et al., 2015). Ticks are able to modulate cutaneous and systemic immune defenses of their hosts that involve keratinocytes, natural killer (NK) cells, dendritic cells (DCs), T cell subpopulations (T helper type 1 [Th1], Th2, Th17, T regulatory cells), B cells, neutrophils, mast cells, basophils, endothelial cells, cytokines, chemokines, complement, and extracellular matrix (Kazimírová and Štibrániová, 2013; Štibrániová et al., 2013; Wikel, 2013; Heinze et al., 2014; Šimo et al., 2017). Although defenses may share common features across host species, differences likely result in the evolutionary adaptation of tick saliva molecules to specific tick–host relationships. Hundreds of different tick salivary proteins have been found to be differentially expressed during blood feeding, whereby differences in expression of SG genes and in saliva composition exist across and within tick genera (Alarcon-Chaidez et al., 2007; Mans et al., 2008;

Peterková et al., 2008; Vančová et al., 2007, 2010b; Francischetti et al., 2009; Kazimírová and Štibrániová, 2013; Wikel, 2013; Šimo et al., 2017).

The biology of ticks and the pharmacological properties of their saliva indicate that ticks are not passive partners in the vector–host–pathogen relationship. It is suggested that immunomodulatory activities of tick saliva can be exploited by tick-borne pathogens, including TBVs, to facilitate their transmission and replication in the host. Indeed, the dynamic balance between acquired resistance of the host and tick modulation of host immunity has been found to affect both tick feeding and pathogen transmission (Bowman et al., 1997; Ramamoorthi et al., 2005; Brossard and Wikel, 2008; Nuttall and Labuda, 2008; Wikel, 1999, 2013; Šimo et al., 2017). Complex interactions between the host immune responses and early tick-mediated immunomodulation, all of which initially occur at the skin interface, underlie the successful transmission of tick-borne pathogens (Fig. 10.1). The direct evidence of the exploitation of this unique environment created by tick saliva (SAT) at the feeding site by TBVs and NVT and the indirect evidence of SAT have been demonstrated and were described above.

Many of the different vertebrate skin cell types play crucial roles in the defense against invading pathogenic microorganisms (Frischknecht, 2007; Bernard et al., 2014, 2015). Penetration through the skin brings tick mouthparts into contact with keratinocytes, which possess receptors of innate immune responses, antimicrobial peptides, and proinflammatory cytokines (Merad et al., 2008; Martinon et al., 2009; Nestle et al., 2009). TBVs delivered into the skin

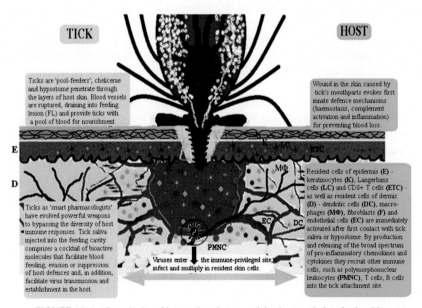

FIGURE 10.1 Complexity of interactions between ticks, host, and virus in the skin.

also encounter a number of different cell types, including rich DCs networks and neutrophils, which play roles in pathogen elimination during the early stages of infection (Labuda et al., 1996; Wu et al., 2000; Robertson et al., 2009).

Dendritic Cells and Keratinocytes

Studies on transmission of the tick-borne bacterium *B. burgdorferi* have already indicated that saliva of *I. ricinus* inhibited cutaneous innate immunity and migration of immune cells to the infectious tick bite site by an antialarmin effect on human primary keratinocytes (Marchal et al., 2011, 2009). TBEV, the most notable member of the TBE serocomplex, is naturally delivered into the host skin during blood feeding of an infected tick. The virus was found to replicate at the tick bite site within keratinocytes, dermal macrophages, Langerhans DCs, and neutrophils (Wikel et al., 1994; Labuda et al., 1996; Wu et al., 2000; Ho et al., 2001; Libraty et al., 2001; Marovich et al., 2001).

In general, recognition of a virus by immature DCs in the periphery via pattern recognition receptor systems, such as Toll-like receptors (TLRs) located at the cell surface and within endosomes or the retinoic acid-inducible gene I (*Rig-1*)-like helicases detecting nucleic acids within the cytosol (Kochs et al., 2010), is known to result in taking up viral antigens and result in DCs activation and their migration to local lymphoid tissues. As DCs are the key players in the induction of protective immunity to viral infection, tick salivary molecules that modulate DCs functions might also be exploited by viruses to circumvent host immune responses.

During the early phase of infection, a virus replicates within the dermis and subsequently in the skin draining lymph nodes. Activation of DCs confers their ability to activate naive T cells into Th1, Th2, and cytotoxic T lymphocyte effector cells. This interaction activates signaling pathways that lead to high expression of major histocompatibility complex (MHC) class II (required for antigen presentation), T cell costimulatory molecules (i.e., CD80 and CD86), and proinflammatory cytokines, such as type I interferon (IFN), interleukin (IL) 6, and IL12, which drive antiviral Th1 responses (Johnston et al., 2000; Masson et al., 2008). Together with IL1, tumor necrosis factor (TNF)-alpha, a powerful cytokine secreted by several cell types after viral infection, is known to promote DCs migration from the skin into regional lymph nodes. Significantly higher production of TNF-alpha and IL6 but undetectable levels of IL10 and IL12p70 were observed as results of infection of DCs by the virulent Hypr TBEV strain compared to infection with the less virulent Neudoerfl strain (Fialová et al., 2010). Simultaneously, a significant impact of *I. ricinus* saliva on the levels of TNF-alpha and IL6 was also demonstrated, whereby bystander DCs kept immature phenotype as assessed by low expression of B7-2 (CD86) and MHC class II molecules. However, no significant effect of the tick saliva on virus-induced upregulation of MHC class II and B7-2 molecules has been observed (Fialová et al., 2010). The release of proinflammatory cytokines, TNF-alpha and IL6, has been reported previously for DCs treated with uninfected *I. scapularis* saliva (Sá-Nunes et al., 2007; Hovius et al., 2008). The findings might suggest that in the presence of tick saliva, DCs keep a less mature phenotype and thus remain permissive for the virus.

Cavassani et al. (2005) showed that saliva of *R. sanguineus* inhibited differentiation, maturation, and function of murine bone marrow-derived DCs. Coincubation of DCs with saliva of *R. sanguineus* directed attenuation of antigen-specific T cells cytokine production stimulated by DCs (Oliveira et al., 2008). Moreover, *R. sanguineus* saliva was found to impair the maturation of DCs stimulated with lipopolysaccharide (LPS), a TLR4 ligand, by inhibition of the activation of the extracellular signal–regulated kinases (ERK) 1/2 and p38 mitogen-activated protein (MAP) kinases, leading to increased production of IL10 and reduced synthesis of IL12p70 and TNF-alpha (Oliveira et al., 2010). Prostaglandin E2 from *I. scapularis* saliva was identified as a major inhibitor of DCs maturation and function (Sá-Nunes et al., 2007). In addition, Skallová et al. (2008) suggested that *I. ricinus* saliva impairs maturation of murine DCs through affecting TLR3, TLR7, TLR9, or CD40 ligation and found that tick saliva reduced TBEV-mediated DCs apoptosis.

Tick saliva was shown to inhibit the chemotactic function of chemokines and selectively impair chemotaxis of immature DCs by downregulating cell surface receptors. Saliva of *R. sanguineus* was found to inhibit immature DCs migration in response to CCL3 (macrophage inflammatory protein 1-alpha (MIP-1-alpha)) (migration via receptors CCR1 or CCR5), to CCL4 (macrophage inflammatory protein-1 beta [MIP-1 beta]) (via CCR1), and to CCL5 (RANTES) (migration via CCR1, CCR3, CCR5) (Oliveira et al., 2008). Evasin-1 (derived from *R. sanguineus*) is also able to bind to human CCL3 and mouse CCL3 (Dias et al., 2009). Two salivary cystatins (cysteine protease inhibitors) derived from *I. scapularis* have been functionally characterized and have been shown to inhibit cathepsins L and S, impair inflammation, and suppress DCs maturation (Kotsyfakis et al., 2006, 2008; Sá-Nunes et al., 2009). Thus, tick saliva can impair early migration of DCs from inflamed skin.

The reported findings suggest that in presence of tick saliva, TBEV-infected DCs might stay in the skin for a prolonged time and thus enhance chances of NVT among cofeeding ticks (Fialová et al., 2010). The observations also support previous findings of Labuda et al. (1996) who suggested that the tick bite site in the host skin is crucial for TBEV replication and subsequent NVT transmission to cofeeding *I. ricinus* ticks. Indeed, analysis of murine skin explants taken from the bite sites of infected and noninfected ticks showed viral replication in monocytes/macrophages and transport of the viral antigen to the bite site of noninfected ticks via Langerhans cells and neutrophils. Thus, cellular infiltration observed at the tick feeding sites and cell migration from these sites probably facilitate NVT of the virus (Labuda et al., 1996).

It is obvious that the interactions between DCs, IFN responses, and the virus are likely to substantially influence the outcome of the infection. Early IFN and

DCs responses are modulated not only by the virus (Best et al., 2005) but also by the tick salivary immunomodulatory compounds inoculated with the virus into the host skin. The decrease in the number of DCs observed around the attachment sites of *D. andersoni* ticks suggests that Langerhans cells migration to lymph nodes occurs after contact with tick saliva components and T cells responses. Indeed, *in vitro* treatment of DCs obtained from the lymph nodes of tick-bite sensitized tick-resistant guinea pigs with *D. andersoni* saliva induced T cell proliferation (Nithiuthai and Allen, 1985), and coincubation of DCs with tick saliva directed attenuation of antigen-specific T cells cytokine production stimulated by DCs (Oliveira et al., 2008).

A novel mechanism of immunomodulation facilitating pathogen transmission has been proposed recently by Preston et al. (2013). Japanin, a lipocalin from *R. appendiculatus* ticks, was found to specifically reprogram DCs responses to a wide variety of *in vitro* stimuli. Japanin altered the expression of costimulatory and coinhibitory transmembrane molecules, modulated secretion of proinflammatory, antiinflammatory. and T cell polarizing cytokines, and inhibited the differentiation of DCs from monocytes. Based on these findings it has been suggested that the failure of DCs to mature in response to viral or to tick immunomodulators has important implications for induction of effective antiviral T cell-mediated immunity; it may lead to an aberrant antiviral immune response and ineffective virus clearance.

Interferon Pathways

DCs represent an early target of TBEV infection and are major producers of IFN. Generally, following virus infection, the host cell deploys the rapid response to limit virus replication in both the infected cells and in neighboring cells.

Type I Interferon

Although the innate immune response is multifaceted, type I IFN (including multiple IFN-alpha molecules and IFN-beta) has a central role (Akira et al., 2006; Kawai and Akira, 2006). Although type I IFN signaling is generally recognized as an important component of antiviral innate immunity, previous studies indicate that its role during vector-borne flavivirus infections is complex and varies from one type of infection to another. The IFN-dependent innate immune response is essential for protection against flavivirus infections. Type I IFNs are the first line of defense against many viral infections and contribute to initial survival. Type I and II IFNs inhibit flavivirus infection in cell culture and in animals. Type I IFNs (alpha or beta) block flavivirus infection by preventing translation and replication of infectious viral RNA, which occurs at least partially through an RNAse L, Mx1, and protein kinase R-independent mechanism. For example, Mx1 or MxA proteins have been determined as the innate resistance factors in mammalian cells against tick-borne myxoviruses (THOV, Batken virus) (Halle et al., 1995; Frese et al., 1997) and bunyaviruses (CCHFV,

Dogbe virus) (Andersson et al., 2004; Bridgen et al., 2004). Possible manipulation of IFN signaling by tick SGE was indicated by Dessens and Nuttall (1998) who have shown THOV transmission to uninfected ticks feeding on Mx1 A2G mice (a strain resistant to infection) following needle- or virus challenge, probably due to *Mx1* gene manipulation after injection of virus mixed with tick SGE.

LGTV, a member of the TBE complex of viruses, was found to be able to block STAT-1 phosphorylation in response to either IFN-alpha or IFN-gamma by interaction of viral nonstructural protein NS5 with both the IFN-alpha/beta receptor subunit, IFNAR2 (IFNAR2-2 or IFNAR2c), and the IFN-gamma receptor subunit, IFNGR1 (Best et al., 2005). Other TBVs, e.g., CCHFV, NSDV/ Ganjam virus, cause delayed type I IFN response not only by inactivation of the interferon regulatory factor (IRF)3 pathway (Andersson et al., 2004) but also by inhibition of the RNase L pathway (Habjan et al., 2008) and the ISG15 pathway (Frias-Staheli et al., 2007; Holzer et al., 2011). Thus, the IFN pathways manipulated by TBVs might also be targets for tick salivary molecules.

Arboviruses, however, are generally not recognized as strong inducers of IFN-alpha/beta, with one exception, VSV, an insect-borne rhabdovirus. Using this virus, Hajnická et al. (1998, 2000) were the first who provided evidence that SGE of partially fed adult *R. appendiculatus* or *D. reticulatus* ticks increased viral yields by 100- to 1000-fold in murine cell cultures. The effect appeared to result from inhibition of the antiviral effect of IFN by SGE compounds, possibly acting through the IFN-alpha/beta receptor rather than directly affecting IFN. Recently, Lieskovská et al. (2015) observed and enhanced replication of TBEV in bone marrow DCs in the presence of tick sialostatin L2, possibly as a consequence of impaired IFN-beta signaling. They found decreasing of STAT-1 and STAT-2 phosphorylation and induction of IFN-stimulated genes, *Irf-7* and *Ip-10* in LPS-stimulated DCs, by both sialostatin L and sialostatin L2. The inhibitory effect of tick cystatins on IFN responses in host DCs appears to be a novel mechanism that might elucidate the role of tick saliva in the transmission of TBVs.

Type II Interferon

Immune IFN, known as type II IFN or IFN-gamma, is secreted mostly by activated NK cells and macrophages during the early stages of infection (Darwich et al., 2009; Malmgaard, 2004). Suppression of NK cells activity by SGE from *D. reticulatus, R. appendiculatus*, and *I. ricinus* was demonstrated by Kubeš et al. (1994). During later stages of infection, IFN-gamma is produced by activated T lymphocytes (Boehm et al., 1997) in answer to receptor-mediated stimulation (through T cell receptors or NK cell receptors) or in response to early produced cytokines, such as IL12, IL18, and IFN-alpha/beta (Darwich et al., 2009). IFN-gamma inhibits flavivirus replication via the generation of proinflammatory and antiviral molecules including nitric oxide (NO). Although antiviral activity is not the primary biological function of IFN-gamma, it stimulates cell-mediated immune responses that are critical for the development of immunity against pathogenic intracellular microorganisms, inducing the activation of macrophages for microbicidal activity and increasing the expression of MHC for more effective antigen presentation. It was shown that SGE of *I. ricinus* reduced polyinosinic–polycytidylic acid-induced production of IFN-alpha, IFN-beta, and IFN-gamma (Kopecký and Kuthejlová, 1998), and SGE from female *D. reticulatus* inhibited antiviral effects of IFN-alpha and IFN-beta produced by mouse fibroblasts (Hajnická et al., 2000). SGE from 5-days fed *D. reticulatus* and *I. ricinus* females was shown to inhibit concanavalin (Con)A-stimulated IFN-gamma production by mouse splenocytes (Vančová, unpublished).

Type III Interferon

IFN-lambda genes are expressed in response to many classes of viruses and to a variety of TLR agonists; in fact, the same stimuli are responsible for expression of type I IFN genes induced by transcriptional mechanisms involving IRFs and nuclear factor kappa B (NF-κB) (Onoguchi et al., 2007; Osterlund et al., 2007; Kotenko, 2011). IFN-lambda was found to reduce CCHFV replication in human lung carcinoma (A549) and human hepatocyte derived cellular carcinoma (HuH7) cells in a dose-dependent manner but with lower antiviral effect than IFN-alpha alone, which support the antagonism between the IFN types against lethal CCHFV infection *in vitro* (Bordi et al., 2015).

Among skin cell populations, keratinocytes and melanocytes but not fibroblasts, endothelial cells, or subcutaneous adipocytes are targets of IFN-lambda (Witte et al., 2009). According to results provided by Lim et al. (2011), Limon-Flores et al. (2005), and Surasombatpattana et al. (2011), keratinocytes were proposed as key players of early arboviral infection capable to produce high levels of infectious viruses in the skin favoring viral dissemination to the entire body. Keratinocytes are cells that both produce and respond to type III IFN (Odendall et al., 2014). IFN-lambda probably acts primarily as a protection of mucosal entities, such as the lung, skin, or digestive tract (Hermant and Michiels, 2014). Manipulation of type III IFN functions by tick saliva has not been documented.

Macrophages, Fibroblasts, and Neutrophils

TBV-infected cells were shown to migrate to draining lymph nodes, where they encounter early immune responses (Johnston et al., 2000). Immune cells were found to carry the TBEV antigen from one tick bite site to the other without dissemination. This route of virus transmission was still maintained, although at a reduced level, even if the animals were immunized and produced an antibody response against TBEV (Labuda et al., 1997b). POWV, one of the less studied human pathogenic flaviviruses, is transmitted to the host very early during tick feeding. The histopathology of *I. scapularis* feeding sites demonstrated that neutrophil and mononuclear cell infiltrates were recruited to the feeding sites of POWV-infected ticks earlier than to the attachment sites of uninfected ticks (Hermance et al., 2016). Macrophages and fibroblasts also contained POWV antigens, which may suggest that they are early cellular targets of infection at the tick feeding site.

Macrophages

Macrophages are among potential targets for TBVs. It has been demonstrated that mononuclear/macrophage lineages are important sources of local TBEV replication before viremia occurs (Dörrbecker et al., 2010). Infection of macrophages with TBEV was shown to enhance NO production, which inhibits virus replication (Plekhova et al., 2008), but the exact role of NO production by macrophages in the context of TBEV infection remains to be elucidated. Nevertheless, TBVs might also exploit immunomodulatory properties of tick saliva to subvert macrophage's defensive activities and facilitate TBV transmission.

Saliva of *I. ricinus* was found to decrease the oxidative activity of mouse macrophages (Kuthejlová et al., 2001). The tick macrophage migration inhibitor factor identified in SG of *A. americanum* (Jaworski et al., 2001) was found to block migration of macrophages to the tick attachment site and might also impair macrophage functions during virus infection. Macrophages recruit in increased numbers to the site of injury in response to inflammatory and immune stimulation and produce cytokines and chemokines that attract inflammatory cells to the tick bite site. Resident macrophages in the skin act as antigen-presenting cells that were found to elicit a potent proliferative response during secondary tick infestation (Wikel et al., 1978; Francischetti et al., 2009). Saliva of *I. ricinus* was found to induce the production of CCL2 (monocytes chemoattractant protein 1 [MCP-1]) and CXCL2 (macrophage inflammatory protein 2-alpha [MIP-2-alpha]) attracting neutrophils (Langhansova et al., 2015), which could be involved in dissemination of TBEV (Dörrbecker et al., 2010). Moreover, SGE of *R. microplus*, depending on host species, differently affected macrophage CD86 molecules (Brake et al., 2010; Brake and Pérez de León, 2012), which could be attributed to specificity of the host immune response toward a less proinflammatory Th2 profile.

Growth Factors

Activated macrophages release platelet-derived growth factor (PDGF) and transforming growth factor (TGF)-beta1, which attract fibroblasts and smooth muscle cells to the wound site. TGF-beta1 that controls signals of fibroblast functions is produced by activated platelets, macrophages, and T lymphocytes and affects extracellular matrix deposition, increases collagen, proteoglycans, and fibronectin gene transcription. Furthermore, TGF-beta1 stimulates the tissue metalloprotease inhibitor and other cytokines (ILs, fibroblast growth factor (FGF), TNF-beta3). TGF-beta1 binding activity and other growth factor binding activities (PDGF, hepatocyte growth factor [HGF], FGF2) have been detected in SGE of *D. reticulatus, R. appendiculatus, I. ricinus, I. scapularis,*

A. variegatum, and *Hyalomma excavatum* ticks (Hajnická et al., 2011; Slovák et al., 2014b). Kramer et al. (2011) identified enhancing effect of *D. variabilis* saliva on basal-and PDGF-stimulated migration of macrophage-derived IC-21 cell line. Saliva factors that regulate cell signaling, phagocytosis, and gene expression skewed immune response toward a Th2 response, which is characterized by production of antiinflammatory cytokines IL4 and IL10 (Kramer et al., 2011). Many of the abovementioned growth factors and cytokines are involved in epithelial wound healing and vasculature repair, together with cytoskeletal elements (Werner and Grose, 2003). During early and late primary tick infestation, some of genes (*Cyr61, SMAD5, TNFrsf 12, Junb, Epgnc*) involved in the wound healing processes were found to be upregulated at the tick bite site. These molecules may be related to TNF-alpha, activator protein 1 (AP-1), and growth factors responses. Other genes encoding cytoskeletal elements (*collagen type 1 gene, laminin beta2*), signaling molecules, growth factor receptor (*Pdgfrb*), or growth factor (*Tgfb3*) were downregulated (Heinze et al., 2012, 2014). The findings suggest that tick salivary molecules impairing wound healing processes might be exploited by TBVs by a yet unknown mechanism(s).

Neutrophils

Beside DCs and macrophages, neutrophils are recruited to the site of TBEV infection (Dörrbecker et al., 2010; Labuda et al., 1996). Neutrophils probably play a role in complementing the cytokine and chemokine responses soon after TBEV infection. They may also be involved in the peripheral spread of TBEV. We can only speculate about the exploitation of tick saliva neutrophil inhibitors by TBVs. Neutrophils at the tick attachment site are activated by thrombin from the blood coagulation cascade, by platelet-activating factor, by releasing of proteases modulating platelet function, such as cathepsin G, and or enzymes that act on the tissue matrix, such as elastase. Neutrophils are the most abundant cells in the acute inflammatory infiltrate induced by the primary tick infestation but not during subsequent tick infestations, at least not by all ticks species and not in all tick–host associations (Brown, 1982; Brown et al., 1983, 1984; Gills and Walker, 1985). Ticks are also able to generate a neutrophil chemotactic factor in their saliva by cleavage of C5-convertase (Berenberg et al., 1972). Neutrophil infiltration and activation is orchestrated by chemokines such as CCL3 and CXCL8/KC (IL-8). It has been demonstrated that SGE of different hard tick species can effectively bind and block in action a broad spectrum of proinflammatory cytokines and chemokines, whereby all of the tested tick species were shown to possess anti-CXCL8 activity mediated by one or more molecules (Hajnická et al., 2001, 2005; Vančová et al., 2010a). Earlier studies confirmed an inhibition of CXCL8-coordinated neutrophil migration by *D. reticulatus* SGE due to inhibition of CXCL8-binding to the cell receptors (Kocáková et al., 2003). Evasin-1 and Evasin-3 were identified as potent inhibitors CCL3 and/or CXCL8-induced recruitment of human and murine neutrophils (Déruaz et al., 2008).

Early Cutaneous Responses and Virus Transmission

Due to the fact that flaviviruses can be transmitted within 15 min of tick attachment (Ebel and Kramer, 2004), attention has to be focused on early stages of tick feeding.

Microarray analysis of mouse immune responses to early tick feeding that were investigated 1, 3, 6, and 12 h after initial attachment of *I. scapularis* nymphs showed upregulation of genes for keratinocytes migration 6 h postattachment (Heinze et al., 2012). Furthermore, genes related to signaling pathways such as NF-κB and cation homeostasis were upregulated 1 and 3 h after attachment of *I. scapularis*, suggesting activation of proinflammatory pathways. Six hours after tick attachment, upregulation of cytoskeletal elements, cell signaling, transcription, and antimicrobial immune response were prominent, while 12 h postattachment inflammation and chemotaxis were significant (Heinze et al., 2012).

During the early phases of primary infestation by *D. andersoni* adults, keratin intermediate filaments but no other cytoskeletal molecules were found to be activated in the murine skin (Heinze et al., 2014). Furthermore, significant upregulation of the genes for chemokines (*Ccr1, Ccl2, Ccl6, Ccl7, Ccl12, Cxcl1, Cxcl2,* and *Cxcl4/Pfx4*), cytokines (*Il1b*), and antimicrobial molecules, consistent with migration of monocytes and neutrophils into the *D. andersoni* bite site, has been confirmed (Heinze et al., 2014). Moreover, genes related to DNA repair, transcription, chromatin remodeling, transcription factor binding, RNA splicing, and mRNA metabolism were downregulated, suggesting a tick-induced modulation of early host responses.

NF-κB and nuclear factor of activated T-cells (NFAT) were previously identified as two of the most important factors coordinating mechanisms of viral evasion by regulation of proinflammatory molecules and cytokines, which evoke inflammatory response and recruitment of immune cells (Kopp and Ghosh, 1995). Only few of the genes (*Nfkbia* and *Tsc22d3*) that are involved in inhibition of NF-κB and AP-1 proinflammatory pathways were found to be upregulated by feeding of *D. andersoni* (Heinze et al., 2014). Attachment of *I. scapularis* nymphs to mouse skin lead to upregulation of AP-1 expression via the MAP kinase pathway and, in addition, to upregulation of members of the NF-κB family such as *Nfkbia, Ikbz,* and *Nfkbiz* and many proinflammatory cytokines and chemokines (IL1 beta, IL6, and CCL2) (Heinze et al., 2012).

In addition to cutaneous responses to uninfected tick feeding, Hermance and Thangamani (2014) investigated the murine cutaneous immune responses to infestation by POWV-infected *I. scapularis* nymphs during 6 h postattachment to characterize tick-induced changes in cutaneous gene expression at the early stages of feeding of infected ticks. Majority of the upregulated genes 3 h postattachment were found to be associated with inflammatory responses, such as *Il1b,* *Il6,* and *Il36a.* These molecules are known to influence the quantity of phagocytes and neutrophils during inflammation (Fielding et al., 2008; Rider et al., 2011).

TLR4 molecules were also upregulated in the skin of mice soon after attachment of *I. scapularis* (Hermance and Thangamani, 2014). The TLR4 associated signaling pathway is also known to be involved in the innate immune response to viral infection (Okumura et al., 2010). In addition, TLR4 with upregulated CCR3 contributes to chemotaxis of lymphocytes and eosinophils. These in turn help to establish a proinflammatory environment and subsequent inflammation at the attachment site of POWV-infected ticks feeding for 3 h. At later time points of tick attachment, majority of upregulated proinflammatory genes were downregulated, including *Il1b*, *Il18*, *Ifn-gamma*, and *Tnf*. The cytokines IL1B, IFN-gamma, and TNF were strongly influenced by molecules that regulate the inflammatory response and cell-to-cell signaling 6 h after attachment of POWV-infected ticks. The slight upregulation of CCL2 suggests that CCL2 induced recruitment of monocytes to the tick bite site. Both IFNGR2 and TNF were significantly upregulated. This suggests that these cytokines are linked to the host's induction of nitric oxide synthase (NOS)2 after 3 h of feeding of POWV-infected ticks. At this time point the NOS2 pattern was consistent with the overall upregulation of other molecules associated with cell death. In general, feeding of POWV-infected ticks was found to recruit immune cells much earlier than the feeding of uninfected ticks, probably due to virus infection or changes in tick saliva secretion or an effect of both (Hermance and Thangamani, 2014). The findings were confirmed by histopathology of the *I. scapularis* attachment site showing that neutrophil and mononuclear cell infiltrates were recruited to the feeding site of POWV-infected ticks earlier than to the attachment site of uninfected ticks. The results suggest that macrophages and fibroblasts may contain viral antigens and are the early targets of infection at the tick feeding site (Hermance et al., 2016).

In contrast to the previous findings, during secondary infestation of *D. andersoni* nymphs lasting for 120 h, the cutaneous responses were characterized by increased activity of host complement and of the factors involved in coagulation pathways, activation of TLR, enhanced acute phase response, and antigen presentation and activation of lymphocytes (Heinze et al., 2014).

CONCLUSIONS

Ticks succeeded in their role as blood feeders and vectors of TBVs thanks to their complex life history and feeding biology. Studies on the sialotranscriptome of *I. scapularis* (Ribeiro et al., 2006; Valenzuela et al., 2002) and the *I. scapularis* genome project (Gulia-Nuss et al., 2016) demonstrated the complexity and the redundancy in tick saliva protein functions within gene families.

Components in tick saliva play a crucial role in tick feeding and mediating transmission of TBVs. An increasing body of knowledge exists to date on (1) manipulation of host defenses by ticks to enhance feeding and promote pathogen transmission and (2) strategies used by tick-borne pathogens to evade host immunity and assure survival in different biological systems.

The immunomodulatory factors that promote SAT of tick-borne pathogens vary for different pathogens and different tick vector species (Nuttall and Labuda, 2004). However, in comparison with tick-borne bacteria, information on interactions of TBVs with ticks is still limited. The mechanisms of exploitation of tick molecules that manipulate the host cytokine/chemokine/growth factors networks for transmission and replication of TBVs are not clear. No tick SAT factors enhancing virus transmission have been identified.

The general pattern of tick infestation- or tick saliva-induced immunomodulation consists of downregulation of Th1 cytokines and upregulation of Th2 cytokines leading to suppression of host antibody responses. Tick salivary gene polymorphism is one of the possible mechanisms of avoiding host rejection (Bergman et al., 2000; Daix et al., 2007; Alarcon-Chaidez et al., 2007; Couvreur et al., 2008). The rapid variation of the polymorphic proteins likely contributes to tick survival through immune evasion. These proteins belong to multigene families, and gene conversion is the proposed mechanism of sequence polymorphism.

The systems biology approach employing transcriptomics and proteomics will potentially reveal molecular mechanisms of the survival strategy of TBVs in their vectors and vertebrate hosts. Identification of SAT factors enhancing TBV transmission and understanding of the relationships between ticks, TBVs, and hosts may lead to the development of novel strategies controlling ticks and viral TBD.

ACKNOWLEDGMENTS

The work was supported by the Slovak Research and Development Agency (contract no. APVV-0737-12), VEGA 2/0089/13, and VEGA 2/0199/15.

REFERENCES

Adams, M.J., Lefkowitz, E.J., King, A.M.Q., Harrach, B., Harrison, R.L., Knowles, N.J., Kropinski, A.M., Krupovic, M., Kuhn, J.H., Mushegian, A.R., Nibert, M., Sabanadzovic, S., Sanfaçon, H., Siddell, S.G., Simmonds, P., Varsani, A., Zerbini, F.M., Gorbalenya, A.E., Davison, A.J., 2017. Changes to taxonomy and the International Code of Virus Classification and Nomenclature ratified by the International Committee on Taxonomy of Viruses. Archives of Virology 162, 2505–2538.
Afonso, C.L., Amarasinghe, G.K., Bányai, K., et al., 2016. Taxonomy of the order Mononegavirales: update 2016. Archives of Virology 161, 2351–2360.
Akira, S., Uemat, S., Takeuchi, O., 2006. Pathogen recognition and innate immunity. Cell 124, 783–801.
Al'khovskii, S.V., L'vov, D.K., Shchelkanov, M.Iu., Shchetinin, A.M., Deriabin, P.G., Gitel'man, A.K., Botikov, A.G., Samokhvalov, E.I., Zakarian, V.A., 2014. Taxonomic status of the Artashat virus (ARTSV) (Bunyaviridae, Nairovirus) isolated from the ticks *Omithodoros alactagalis* Issaakjan, 1936 and *O. verrucosus* Olenev, Sassuchin et Fenuk, 1934 (Argasidae Koch, 1844) collected in Transcaucasia. Voprosy Virusologii 59 (3), 24–28 (In Russian).

Alarcon-Chaidez, F.J., Sun, J., Wikel, S.K., 2007. Transcriptome analysis of the salivary glands of *Dermacentor andersoni* Stiles (Acari: Ixodidae). Insect Biochemistry and Molecular Biology 37, 48–71.

Alekseev, A.N., Chunikhin, S.P., 1990a. Transmission of the tick-borne encephalitis virus by ixodid ticks in the experiment (mechanisms, terms, species and sexual distinctions). Parazitologia 24 (3), 177–185 (In Russian).

Alekseev, A.N., Chunikhin, S.P., 1990b. Exchange of the tick-borne encephalitis virus between Ixodidae simultaneously feeding on the animals with subthreshold levels of viremia. Meditsinskaia Parazitologiia i Parazitarnye Bolezni (Moscow) 2, 48–50 (In Russian).

Alekseev, A.N., Chunikhin, S.P., 1991. Virus exchange between feeding ticks in the absence of viremia in a vertebrate host (distant transmission). Meditsinskaia Parazitologiia (Moscow) 2, 50–54 (In Russian).

Alekseev, A.N., Chunikhin, S.P., 1992. Difference in distant transmission ability of tick-borne encephalitis virus by ixodid ticks belonging to different subfamilies. Parazitologia 26, 506–515 (In Russian).

Alekseev, A.N., Razumova, I.V., Chunikhin, S.P., Reshetnikov, I.A., 1988. Behavior of the tick-borne encephalitis virus in *Dermacentor marginatus* Sulz. (Ixodidae) ticks of different physiological ages. Meditsinskaia Parazitologiia (Moscow) 3, 17–21 (In Russian).

Alekseev, A.N., Chunikhin, S.P., Rukhkian, M., Stefutkina, L.F., 1991. The possible role of the salivary gland substrate in ixodid ticks as an adjuvant enhancing arbovirus transmission. Meditsinskaia Parazitologiia (Moscow) 1, 28–31 (In Russian).

Anderson, E.C., Hutchings, G.H., Mukarati, N., Wilkinson, P.J., 1998. African swine fever virus infection of the bushpig (*Potamochoerus porcus*) and its significance in the epidemiology of the disease. Veterinary Microbiology 62, 1–15.

Andersson, I., Bladh, L., Mousavi-Jazi, M., Magnusson, K.-E., Lundkvist, A., Haller, O., Mirazimi, A., 2004. Human MxA protein inhibits the replication of Crimean-Congo hemorrhagic fever virus. Journal of Virology 78, 4323–4329.

Attoui, H., Mohd Jaafar, F., Biagini, P., Cantaloube, J.F., De Micco, P., Murphy, F.A., De Lamballerie, X., 2002. Genus *Coltivirus* (family *Reoviridae*): genomic and morphologic characterization of old world and new world viruses. Archives of Virology 147, 533–561.

Attoui, H., Mertens, P.P.C., Becnel, J., Belaganahalli, S., Bergoin, M., Brussaard, C.P., Chappell, J.D., Ciarlet, M., del Vas, M., Dermody, T.S., Dormitzer, P.R., Duncan, R., Fang, Q., Graham, R., Guglielmi, K.M., Harding, R.M., Hillman, B., Makkay, A., Marzachì, C., Matthijnssens, J., Milne, R.G., Mohd Jaafar, F., Mori, H., Noordeloos, A.A., Omura, T., Patton, J.T., Rao, S., Maan, M., Stoltz, D., Suzuki, N., Upadhyaya, N.M., Wei, C., Zhou, H., 2011. Reoviridae. In: King, A.M.Q., Adams, M.J., Carstens, E.B., Lefkowitz, E.J. (Eds.), Virus Taxonomy: Classification and Nomenclature of Viruses: Ninth Report of the International Committee on Taxonomy of Viruses. Elsevier Inc., pp. 541–637.

Bell-Sakyi, L., Kohl, A., Bente, D.A., Fazakerley, J.K., 2012. Tick cell lines for study of Crimean-Congo hemorrhagic fever virus and other arboviruses. Vector Borne and Zoonotic Diseases 12, 769–781.

Belova, O.A., Burenkova, L.A., Karganova, G.G., 2012. Different tick-borne encephalitis virus (TBEV) prevalences in unfed versus partially engorged ixodid ticks – evidence of virus replication and changes in tick behavior. Ticks and Tick-Borne Diseases 3, 240–246.

Bente, D.A., Forrester, N.L., Watts, D.M., McAuley, A.J., Whitehouse, C.A., Bray, M., 2013. Crimean-Congo hemorrhagic fever: history, epidemiology, pathogenesis, clinical syndrome and genetic diversity. Antiviral Research 100, 159–189.

Berenberg, J.L., Ward, P.A., Sonenshine, D.E., 1972. Tick-bite injury: mediation by a complement-derived chemotactic factor. The Journal of Immunology 109, 451–456.

Bergman, D.K., Palmer, M.J., Caimano, M.J., Radolf, J.D., Wikel, S.K., 2000. Isolation and molecular cloning of a secreted immunosuppressant protein from *Dermacentor andersoni* salivary gland. The Journal of Parasitology 86, 516–525.

Bernard, Q., Jaulhac, B., Boulanger, N., 2014. Smuggling across the border: how arthropod-borne pathogens evade and exploit the host defense system of the skin. Journal of Investigative Dermatology 134, 1211–1219.

Bernard, Q., Jaulhac, B., Boulanger, N., 2015. Skin and arthropods: an effective interaction used by pathogens in vector-borne diseases. European Journal of Dermatology (Suppl. 1), 18–22.

Bernard, J., Hutet, E., Paboeuf, F., Randriamparany, T., Holzmuller, P., Lancelot, R., Rodrigues, V., Vial, L., Le Potier, M.F., 2016. Effect of *O. porcinus* tick salivary gland extract on the African swine fever virus infection in domestic pig. PLoS One 11, 1–19.

Best, S.M., Morris, K.L., Shannon, J.G., Robertson, S.J., Mitzel, D.N., Park, G.S., Boer, E., Wolfinbarger, J.B., Bloom, M.E., 2005. Inhibition of interferon-stimulated JAK-STAT signaling by a tick-borne flavivirus and identification of NS5 as an interferon antagonist. Journal of Virology 79, 12828–12839.

Bichaud, L., de Lamballerie, X., Alkan, C., Izri, A., Gould, E.A., Charrel, R.N., 2014. Arthropods as a source of new RNA viruses. Microbial Pathogenesis 77, 136–141.

Blasdell, K.R., Guzman, H., Widen, S.G., Firth, C., Wood, T.G., Holmes, E.C., Tesh, R.B., Vasilakis, N., Walker, P.J., 2015. Ledantevirus: a proposed new genus in the Rhabdoviridae has a strong ecological association with bats. The American Journal of Tropical Medicine and Hygiene 92, 405–410.

Boehm, U., Klamp, T., Groot, M., Howard, J.C., 1997. Cellular responses to interferon-γ. Annual Review of Immunology 15, 749–795.

Bordi, L., Lalle, E., Caglioti, C., Travaglini, D., Lapa, D., Marsella, P., Quartu, S., Kis, Z., Arien, K.K., Huemer, H.P., Meschi, S., Ippolito, G., Di Caro, Capobianchi, M.R., Castilletti, C., 2015. Antagonistic antiviral activity between IFN-lambda and IFN-alpha against lethal Crimean-Congo hemorrhagic fever virus in vitro. PLoS One 10, e0116816.

Bowman, A.S., Coons, L.B., Needham, G.R., Sauer, J.R., 1997. Tick saliva: recent advances and implications for vector competence. Medical and Veterinary Entomology 11, 277–285.

Brackney, D.E., Armstrong, P.M., 2016. Transmission and evolution of tick-borne viruses. Current Opinion in Virology 21, 67–74.

Brake, D.K., Pérez de León, A.A., 2012. Immunoregulation of bovine macrophages by factors in the salivary glands of *Rhipicephalus microplus*. Parasites and Vectors 5, 38.

Brake, D.K., Wikel, S.K., Tidwell, J.P., Perez de Leon, A.A., 2010. *Rhipicephalus microplus* salivary gland molecules induce differential CD86 expression in murine macrophages. Parasites and Vectors 3, 103.

Bridgen, A., Dalrymple, D.A., Weber, F., Elliott, R.M., 2004. Inhibition of *Dugbe nairovirus* replication by human MxA protein. Virus Research 99, 47–50.

Briese, T., Alkhovskii, S.V., Beer, M., Calisher, C.H., Charrel, R., Ebi-hara, H., Jain, R., Kuhn, J.H., Lambert, A., Maes, P., Nunes, M., Plyusnin, A., Schmaljohn, C., Tesh, R.B., Yeh, S.-D., Elbeaino, T., Digiaro, M., Martelli, G.P., Muehlbach, H.-P., Mielke-Ehret, N., Sasaya, T., Choi, I.R., Haenni, A.-L., Jonson, G., Shirako, Y., Wei, T., Zhou, X., Junglen, S., 2016. Create the Order Bunyavirales, Including Eight New Families, and One Renamed Family ICTV Taxonomic Proposal 2016.030a-vM.A.v6. http://www.ictv.global/proposals-16/2016.030a-vM.A.v6.Bunyavirales.pdf.

Brossard, M., Wikel, S., 2008. Tick immunobiology. In: Bowman, A.S., Nuttall, P.A. (Eds.), Ticks: Biology, DiseaseandControl. Cambridge UniversityPress, Cambridge, pp. 186–204.

Brown, S.J., 1982. Antibody and cell-mediated immune resistance by guinea pigs to adult *Amblyomma americanum* ticks. The American Journal of Tropical Medicine and Hygiene 31, 1285–1290.

Brown, S.J., Worms, M.J., Askenase, W.P., 1983. *Rhipicephalus appendiculatus*: larval feeding sites in guinea pigs actively sensitized and receiving immune serum. Experimental Parasitology 55, 111–120.

Brown, S.J., Bagnall, B.G., Askenase, P.W., 1984. *Ixodes holocyclus*: kinetics of cutaneous basophil responses in naive and actively and passively sensitized guinea-pigs. Experimental Parasitology 57, 40–47.

Burrage, T.G., 2013. African swine fever virus infection in Ornithodoros ticks. Virus Research 173, 131–139.

Carn, V.M., Kitching, R.P., 1995. An investigation of possible routes of transmission of lumpy skin disease virus (neethling). Epidemiology and Infection 114, 219–226.

Cavassani, K.A., Aliberti, J.C., Dias, A.R.V., Silva, J.S., Ferreira, B.R., 2005. Tick saliva inhibits differentiation, maturation and function of murine bone-marrow-derived dendritic cells. Immunology 114, 235–245.

Charrel, R.N., Zaki, A.M., Attoui, H., Fakeeh, M., Billoir, F., Yousef, A.I., de Chesse, R., De Micco, P., Gould, E.A., de Lamballerie, X., 2001. Complete coding sequence of the Alkhurma virus, a tick-borne Flavivirus causing severe hemorrhagic fever in humans in Saudi Arabia. Biochemical and Biophysical Research Communication 287, 455–461.

Charrel, R.N., Attoui, H., Butenko, A.M., Clegg, J.C., Deubel, V., Frolova, T.V., Gould, E.A., Gritsun, T.S., Heinz, F.X., Labuda, M., Lashkevich, V.A., Loktev, V., Lundkvist, A., Lvov, D.V., Mandl, C.W., Niedrig, M., Papa, A., Petrov, V.S., Plyusnin, A., Randolph, S., Süss, J., Zlobin, V.I., de Lamballerie, X., 2004. Tick-borne virus diseases of human interest in Europe. Clinical Microbiology and Infection 10, 1040–1055.

Chastel, C., Main, A.J., Couatarmanac'h, A., Le Lay, G., Knudson, D.L., Quillien, M.C., Beaucournu, J.C., 1984. Isolation of Eyach virus (Reoviridae, Colorado tick fever group) from *Ixodes ricinus* and *I. ventalloi* ticks in France. Archives of Virology 82, 161–171.

Chihota, C.M., Rennie, L.F., Kitching, R.P., Mellor, P.S., 2001. Mechanical transmission of lumpy skin disease virus by *Aedes aegypti* (Diptera: Culicidae). Epidemiology and Infection 126, 317–321.

Chihota, C.M., Rennie, L.F., Kitching, R.P., Mellor, P.S., 2003. Attempted mechanical transmission of lumpy skin disease virus by biting insects. Medical and Veterinary Entomology 17, 294–300.

Chmelař, J., Kotál, J., Karim, S., Kopacek, P., Francischetti, I.M., Pedra, J.H., Kotsyfakis, M., 2016. Sialomes and mialomes: a systems-biology view of tick tissues and tick-host interactions. Trends in Parasitology 32, 242–254.

Cimolai, N., Anand, C.M., Gish, G.J., Calisher, C.H., Fishbein, D.B., 1988. Human colorado tick fever in southern Alberta. Canadian Medical Association Journal 139, 45–46.

Cory, J., Yunker, C.E., 1971. Primary cultures of tick hemocytes as systems for arbovirus growth. Annals of the Entomological Society of America 64, 1249–1254.

Costard, S., Mur, L., Lubroth, J., Sanchez-Vizcaino, J.M., Pfeiffer, D.U., 2013. Epidemiology of African swine fever virus. Virus Research 173, 191–197.

Couvreur, B., Beaufays, J., Charon, C., Lahaye, K., Gensale, F., Denis, V., Charloteaux, B., Decrem, Y., Prévôt, P.P., Brossard, M., Vanhamme, L., Godfroid, E., 2008. Variability and action mechanism of a family of anticomplement proteins in *Ixodes ricinus*. PLoS One 3, e1400.

Daix, V., Schroeder, H., Praet, N., Georgin, J.P., Chiappino, I., Gillet, L., De Fays, K., Decrem, Y., Leboulle, G., Godfroid, E., Bollen, A., Pastoret, P.P., Gern, L., Sharp, P.M., Vanderplasschen, A., 2007. Ixodes ticks belonging to the *Ixodes ricinus* complex encode a family of anticomplement proteins. Insect Molecular Biology 16, 155–166.

Darwich, L., Com, G., Peña, R., Bellido, R., Blanco, E.J., Este, J.A., Borras, F.E., Clotet, B., Ruiz, L., Rosell, A., Andreo, F., Parkhouse, R.M., Bofill, M., 2009. Secretion of interferon-gamma by human macrophages demonstrated at the single-cell level after costimulation with interleukin (IL)-12 plus IL-18. Immunology 126, 386–393.

Davies, C.R., Jones, L.D., Nuttall, P.A., 1986. Experimental studies on the transmission cycle of Thogoto virus, a candidate orthomyxovirus, in *Rhipicephalus appendiculatus* ticks. The American Journal of Tropical Medicine and Hygiene 35, 1256–1262.

Déruaz, M., Frauenschuh, A., Alessandri, A.L., Dias, J.M., Coelho, F.M., Russo, R.C., Ferreira, B.R., Graham, G.J., Shaw, J.P., Wells, T.N.C., Teixeira, M.M., Power, C., Proudfoot, A.E.I., 2008. Ticks produce highly selective chemokine binding proteins with antiinflammatory activity. Journal of Experimental Medicine 205, 2019–2031.

Dessens, J.T., Nuttall, P.A., 1998. Mx1-based resistance to Thogoto virus in A2G mice is bypassed in tick-mediated virus delivery. Journal of Virolology 72, 8362–8364.

Dias, J.M., Losberger, C., Déruaz, M., Power, C.A., Proudfoot, A.E.I., Shaw, J.P., 2009. Structural basis of chemokine sequestration by a tick chemokine binding protein: the crystal structure of the complex between Evasin-1 and CCL3. PLoS One 4, e8514.

Dilcher, M., Alves, M.J., Finkeisen, D., Hufert, F., Weidmann, M., 2012. Genetic characterization of Bhanja virus and Palma virus, two tick-borne phleboviruses. Virus Genes 45, 311–315.

Dilcher, M., Faye, O., Faye, O., Weber, F., Koch, A., Sadegh, C., Weidmann, M., Sall, A.A., 2015. Zahedan rhabdovirus, a novel virus detected in ticks from Iran. Virology Journal 12, 183.

Dixon, L.K., Wilkinson, P.J., 1988. Genetic diversity of African swine fever virus isolates from soft ticks (*Ornithodoros moubata*) inhabiting warthog burrows in Zambia. Journal of General Virology 69, 2981–2993.

Dixon, L.K., Escribano, J.M., Martins, C., Rock, D.L., Salas, M.L., Wilkinson, P.J., 2005. Asfarviridae. In: Fauquet, M., Mayo, M.A., Maniloff, J., Desselberg, U., Ball, L.A. (Eds.), Virus Taxonomy. Eighth Report of the International Committee on Taxonomy of Viruses. Elsevier/Academic Press, London.

Dixon, L.K., Alonso, C., Escribano, J.M., Martins, C., Revilla, Y., Salas, M.L., Takamatsu, H., 2011. The Asfarviridae. In: King, A.M.Q., Adams, M.J., Carstens, E.B., Lefkowitz, E.J. (Eds.), Virus Taxonomy: Classification and Nomenclature of Viruses: Ninth Report of the International Committee on Taxonomy of Viruses, pp. 153–162.

Dörrbecker, B., Dobler, G., Spiegel, M., Hufert, F.T., 2010. Tick-borne encephalitis virus and the immune response of the mammalian host. Travel Medicine and Infectious Diseases 8, 213–322.

Ebel, G., Kramer, L., 2004. Short report: duration of tick attachment required for transmission of Powassan virus by deer ticks. The American Journal of Tropical Medicine and Hygiene 71, 268–271.

Emmons, R.W., 1988. Ecology of Colorado tick fever. Annual Review of Microbiology 42, 49–64.

Endris, R.G., Haslett, T.M., Geering, G., Hess, W.R., Monahan, M.J., 1987. A hemolymph test for the detection of African swine fever virus in *Ornithodoros coriaceus*. Journal of Medical Entomology 24, 192–197.

Fialová, A., Cimburek, Z., Iezzi, G., Kopecký, J., 2010. *Ixodes ricinus* tick saliva modulates tick-borne encephalitis virus infection of dendritic cells. Microbes and Infection 12, 580–585.

Ficová, M., Betáková, T., Pančík, P., Václav, R., Prokop, P., Halásová, Z., Kúdelová, M., 2011. Molecular detection of Murine Herpesvirus 68 in ticks feeding on free-living reptiles. Microbial Ecology 62, 862–867.

Fielding, C., McLoughlin, R., McLeod, L., Colmont, C.S., Najdovska, M., Grail, D., Ernst, M., Jones, S.A., Topley, N., Jenkins, B.J., 2008. IL-6 regulates neutrophil trafficking during acute inflammation via STAT3. The Journal of Immunology 181, 2189–2195.

Francischetti, I.M.B., Sa-Nunes, A., Mans, B.J., Santos, I.M., Ribeiro, J.M.C., 2009. The role of saliva in tick feeding. Frontiers in Bioscience 14, 2051–2088.

Frese, M., Weeber, M., Weber, F., Speth, V., Haller, O., 1997. Mx1 sensitivity: Batken virus is an orthomyxovirus closely related to Dhori virus. Journal of General Virology 78, 2453–2458.

Frias-Staheli, N., Giannakopoulos, N.V., Kikkert, M., Taylor, S.L., Bridgen, A., Paragas, J., Richt, J.A., Rowland, R.R., Schmaljohn, C.S., Lenschow, D.J., Snijder, E.J., García-Sastre, A., Virgin 4th, H.W., 2007. Ovarian tumor domain-containing viral proteases evade ubiquitin-and ISG15-dependent innate immune responses. Cell Host and Microbe 2, 404–416.

Frischknecht, F., 2007. The skin as interface in the transmission of arthropod-borne pathogens. Cellular Microbiology 9, 1630–1640.

Gaunt, M., Sall, A.A., De Lamballerie, X., De, Falconar, A.K.I., Dzhiranian, T.I., Gould, E.A., 2001. Phylogenetic relationships of aviviruses correlate with their epidemiology, disease association and biogeography. Journal of General Virology 82, 1867–1876.

Ghedin, E., Rogers, M.B., Widen, S.G., Guzman, H., da Rosa, A.P.A.T., Wood, T.G., Fitch, A., Popov, V., Holmes, E.C., Walker, P.J., Vasilakis, N., Tesh, R.B., 2013. Kolente virus, a rhabdovirus species isolated from ticks and bats in the Republic of guinea. Journal of General Virology 94, 2609–2615.

Gills, H.S., Walker, A.R., 1985. Differential cellular responses at *Hyalomma anatolicum anatolicum* feeding sites on susceptible and tick-resistant rabbits. Parasitology 91, 591–607.

Godsey, M.S., Savage, H.M., Burkhalter, K.L., Bosco-Lauth, A.M., Delorey, M.J., 2016. Transmission of Heartland virus (Bunyaviridae: phlebovirus) by experimentally infected *Amblyomma americanum* (Acari: Ixodidae). Journal of Medical Entomology 53, 1226–1233.

Gonzalez, J.P., Cornet, J.P., Wilson, M.L., Camicas, J.L., 1991. Crimean-Congo haemorrhagic fever virus replication in adult *Hyalomma truncatum* and *Amblyomma variegatum* ticks. Research in Virology 142, 483–488.

Gonzalez, J.P., Camicas, J.L., Cornet, J.P., Faye, O., Wilson, M.L., 1992. Sexual and transovarian transmission of Crimean-Congo haemorrhagic fever virus in *Hyalomma truncatum* ticks. Research in Virology 143, 23e28.

Gordon, S.W., Linthicum, K.J., Moulton, J.R., 1993. Transmission of Crimean-Congo hemorrhagic fever virus in two species of *Hyalomma ticks* from infected adults to cofeeding immature forms. The American Journal of Tropical Medicine and Hygiene 48, 576–580.

Grabowski, J.M., Perera, R., Roumani, A.M., Hedrick, V.E., Inerowicz, H.D., Hill, C.A., Kuhn, R.J., 2016. Changes in the proteome of Langat-infected *Ixodes scapularis* ISE6 cells: metabolic pathways associated with flavivirus infection. PLoS Neglected Tropical Diseases 10, e0004180.

Grard, G., Moureau, G., Charrel, R.N., Lemasson, J.J., Gonzalez, J.P., Gallian, P., Gritsun, T.S., Holmes, E.C., Gould, E.A., de Lamballerie, X., 2007. Genetic characterization of tick-borne flaviviruses: new insights into evolution, pathogenetic determinants and taxonomy. Virology 361, 80–92.

Grieg, A., 1972. The localization of African swine fever virus in the tick *Ornithodoros moubata porcinus*. Archiv für die gesamte Virusforschung 39, 240–247.

Gritsun, T.S., Lashkevich, V.A., Gould, E.A., 2003. Tick-borne encephalitis. Antiviral Research 57, 129–146.

Gulia-Nuss, M., Nuss, A.B., Meyer, J.M., Sonenshine, D.E., Roe, R.M., Waterhouse, R.M., Sattelle, D.B., de la Fuente, J., Ribeiro, J.M., Megy, K., Thimmapuram, J., Miller, J.R., Walenz, B.P., Koren, S., Hostetler, J.B., Thiagarajan, M., Joardar, V.S., Hannick, L.I., Bidwell, S., Hammond, M.P., Young, S., Zeng, Q., Abrudan, J.L., Almeida, F.C., Ayllón, N., Bhide, K., Bissinger, B.W., Bonzon-Kulichenko, E., Buckingham, S.D., Caffrey, D.R., Caimano, M.J., Croset, V., Driscoll, T., Gilbert, D., Gillespie, J.J., Giraldo-Calderón, G.I., Grabowski, J.M., Jiang, D., Khalil, S.M.S., Kim, D., Kocan, K.M., Koči, J., Kuhn, R.J., Kurtti, T.J., Lees, K., Lang, E.G., Kennedy, R.C., Kwon, H., Perera, R., Qi, Y., Radolf, J.D., Sakamoto, J.M., Sánchez-Gracia, A., Severo,

M.S., Silverman, N., Šimo, L., Tojo, M., Tornador, C., Van Zee, J.P., Vázquez, J., Vieira, F.G., Villar, M., Wespiser, A.R., Yang, Y., Zhu, J., Arensburger, P., Pietrantonio, P.V., Barker, S.C., Shao, R., Zdobnov, E.M., Hauser, F., Grimmelikhuijzen, C.J.P., Park, Y., Rozas, J., Benton, R., Pedra, J.H.F., Nelson, D.R., Unger, M.F., Tubio, J.M.C., Tu, Z., Robertson, H.M., Shumway, M., Sutton, G., Wortman, J.R., Lawson, D., Wikel, S.K., Nene, V.M., Fraser, C.M., Collins, F.H., Birren, B., Nelson, K.E., Caler, E., Hill, C.A., 2016. Genomic insights into the *Ixodes scapularis* tick vector of Lyme disease. Nature Communications 7, 10507.

Habjan, M., Anderson, I., Klingsrom, J., Schumann, M., Martin, A., Zimmermann, P., Wagner, V., Pichlmair, A., Schneider, U., Mühlberger, E., Mirazimi, A., Weber, F., 2008. Processing of genome 5′termini as a strategy of negative-strand RNA viruses to avoid RIG-1-dependent interferon induction. PLoS One 3, e 2032.

Hajnická, V., Fuchsberger, N., Slovák, M., Kocáková, P., Labuda, M., Nuttall, P.A., 1998. Tick salivary gland extracts promote virus growth in vitro. Parasitology 116, 533–538.

Hajnická, V., Kocáková, P., Slovák, M., Labuda, M., Fuchsberger, N., Nuttall, P.A., 2000. Inhibition of the antiviral action of interferon by tick salivary gland extract. Parasite Immunology 22, 201–206.

Hajnická, V., Kocáková, P., Sláviková, M., Slovák, M., Gašperík, J., Fuchsberger, N., Nuttall, P.A., 2001. Anti-interleukin-8 activity of tick salivary gland extracts. Parasite Immunology 23, 483–489.

Hajnická, V., Kúdelová, M., Štibrániová, I., Slovák, M., Bartíková, P., Halásová, Z., Pančík, P., Belvončíková, P., Vrbová, M., Holikova, V., Hails, R., Nuttall, P.A., 2017. Tick-borne transmission of murine gammaherpesvirus 68. Frontiers in Cellular and Infection Microbiology. https://doi.org/10.3389/fcimb.2017.00458.

Hajnická, V., Vančová, I., Kocáková, P., Slovák, M., Gašperík, J., Sláviková, M., Hails, R.S., Labuda, M., Nuttall, P.A., 2005. Manipulation of host cytokine network by ticks: a potential gateway for pathogen transmission. Parasitology 130, 333–342.

Hajnická, V., Vančová-Štibrániová, I., Slovák, M., Kocáková, P., Nuttall, P.A., 2011. Ixodid tick salivary gland products target host wound healing growth factors. International Journal for Parasitology 41, 213–223.

Halle, O., Frese, M., Rost, D., Nuttall, P.A., Kochs, G., 1995. Tick-borne Thogoto virus infection in mice is inhibited by the orthomyxovirus resistance gene product Mx1. Journal of Virology 69, 2596–2601.

Heinze, D.M., Carmical, J.R., Aronson, J.F., Thangamani, S., 2012. Early immunologic events at the tick-host interface. PLoS One 7, e47301.

Heinze, D.M., Carmical, J.R., Aronson, J.F., Alarcon-Chaidez, F., Wikel, S., Thangamani, S., 2014. Murine cutaneous responses to the rocky mountain spotted fever vector, *Dermacentor andersoni*, feeding. Frontiers in Microbiology 5, 198.

Hermance, M.E., Thangamani, S., 2014. Proinflammatory cytokines and chemokines at the skin interface during Powassan virus transmission. Journal of Investigative Dermatology 134, 2280–2283.

Hermance, M.E., Thangamani, S., 2015. Tick saliva enhances Powassan virus transmission to the host, influencing its dissemination and the course of disease. Journal of Virology 89, 7852–7860.

Hermance, M.E., Thangamani, S., 2017. Powassan Virus: an emerging arbovirus of public health concern in North America. Vector Borne and Zoonotic Diseases. https://doi.org/10.1089/vbz.2017.2110.

Hermance, M.E., Santos, R.I., Kelly, B.C., Valbuena, G., Thangamani, S., 2016. Immune cell targets of infection at the tick-skin interface during Powassan virus transmission. PLoS One 11, e0155889.

Hermant, P., Michiels, T., 2014. Interferon-λ in the context of viral infections: production, response and therapeutic implications. Journal of Innate Immunity 6, 563–574.

Ho, L.J., Wang, J.J., Shaio, M.F., Kao, C.L., Chang, D.M., Han, S.W., Lai, J.H., 2001. Infection of human dendritic cells by dengue virus causes cell maturation and cytokine production. The Journal of Immunology 166, 1499–1506.

Holzer, B., Bakshi, S., Bridgen, A., Baron, M.D., 2011. Inhibition of Interferon induction and action by the *Nairovirus* Nairobi sheep disease virus/Ganjam virus. PLoS One 6, e28594.

Honig, J.E., Osborne, J.C., Nichol, S.T., 2004. The high genetic variation of viruses of the genus *Nairovirus* reflects the diversity of their predominant tick hosts. Virology 318, 10–16.

Hoogstraal, H., 1979. The epidemiology of tick-borne Crimean-Congo hemorrhagic fever in Asia, Europe, and Africa. Journal of Medical Entomology 15, 307–417.

Horne, K.M.E., Vanlandingham, D.L., 2014. Bunyavirus-vector interactions. Viruses 6, 4373–4397.

Hovius, J.W., de Jong, M.A., den Dunnen, J., Litjens, M., Fikrig, E., van der Poll, T., Gringhuis, S.I., Geijtenbeek, T.B., 2008. Salp15 binding to DC-SIGN inhibits cytokine expression by impairing both nucleosome remodeling and mRNA stabilization. PLoS Pathogens 4, e31.

Hubálek, Z., Rudolf, I., 2012. Tick-borne viruses in Europe. Parasitology Research 111, 9–36.

Hudson, P.J., Norman, R., Laurenson, M.K., Newborn, D., Gaunt, M., Jones, L., Reid, H., Gould, E., Bowers, R., Dobson, A., 1995. Persistence and transmission of tick-borne viruses: *Ixodes ricinus* and louping-ill virus in red grouse populations. Parasitology 111, 549–558.

Hynes, W.L., 2014. How ticks control microbes: innate immune responses. Chapter 5. In: Sonnenshine, D., Roe, R.M. (Eds.), Biology of Ticks, vol. 2. Oxford University Press, Oxford, New York, pp. 129–146.

Jaworski, D.C., Jasinskas, A., Metz, C.N., Bucala, R., Barbour, A.G., 2001. Identification and characterization of a homologue of the proinflammatory cytokine, macrophage migration inhibitory factor in the tick, *Amblyomma americanum*. Insect Molecular Biology 10, 323–331.

Johnston, L.J., Halliday, G.M., King, N.J., 2000. Langerhans cells migrate to local lymph nodes following cutaneous infection with an arbovirus. The Journal of Investigative Dermatology 114, 560–568.

Jones, L.D., Nuttall, P.A., 1989a. Non-viraemic transmission of Thogoto virus: influence of time and distance. Transactions of the Royal Society of Tropical Medicine and Hygiene 83, 712–714.

Jones, L.D., Nuttall, P.A., 1989b. The effect of virus-immune hosts on Thogoto virus infection of the tick, *Rhipicephalus appendiculatus*. Virus Research 14, 129–139.

Jones, L.D., Davies, C.R., Steele, G.M., Nuttall, P.A., 1987. A novel mode of arbovirus transmission involving a nonviremic host. Science 237, 775–777.

Jones, L.D., Hodgson, E., Nuttall, P.A., 1989. Enhancement of virus transmission by tick salivary glands. Journal of General Virology 70, 1895–1898.

Jones, L.D., Davies, C.R., Williams, T., Cory, J., Nuttall, P.A., 1990a. Non-viraemic transmission of Thogoto virus: vector efficiency of *Rhipicephalus appendiculatus* and *Amblyomma variegatum*. Transactions of the Royal Society of Tropical Medicine and Hygiene 84, 846–848.

Jones, L.D., Hodgson, E., Nuttall, P.A., 1990b. Characterization of tick salivary gland factor(s) that enhance Thogoto virus transmission. Archives of Virology (Suppl. 1), 227–234.

Jones, L.D., Hodgson, E., Williams, T., Higgs, S., Nuttall, P.A., 1992a. Saliva activated transmission (SAT) of Thogoto virus: relationship with vector potential of different haematophagous arthropods. Medical and Veterinary Entomology 6, 261–265.

Jones, L.D., Kaufman, W.R., Nuttall, P.A., 1992b. Modification of the skin feeding site by tick saliva mediates virus transmission. Experientia 48, 779–782.

Jones, L.D., Matthewson, M., Nuttall, P.A., 1992c. Saliva-activated transmission (SAT) of Thogoto virus: dynamics of SAT factor activity in the salivary glands of *Rhipicephalus appendiculatus, Amblyomma variegatum*, and *Boophilus microplus* ticks. Experimental and Applied Acarology 13, 241–248.

Jones, L.D., Gaunt, M., Hails, R.S., Laurenson, K., Hudson, P.J., Reid, H., Henbest, P., Gould, E.A., 1997. Transmission of louping ill virus between infected and uninfected ticks co-feeding on mountain hares. Medical and Veterinary Entomology 11, 172–176.

Junglen, S., 2016. Evolutionary origin of pathogenic arthropod-borne viruses – a case study in the family Bunyaviridae. Archives of Virology 154, 1719–1727.

Karabatsos, N. (Ed.), 1985. International Catalogue of Arboviruses, in- Cluding Certain Other Viruses of Vertebrates, third ed. American Society for Tropical Medicine and Hygiene, San Antonio. With the 1986–1995 Supplements to the International catalogue. CDC Div. Vector-Borne Infect Dis., Ft Collins.

Kaufman, W., Nuttall, P.A., 2003. *Rhipicephalus appendiculatus* (Acari: Ixodidae): dynamics of Thogoto virus infection in female ticks during feeding on guinea pigs. Experimental Parasitology 104, 20–25.

Kawai, T., Akira, S., 2006. Innate immune recognition of viral infection. Nature Immunology 7, 131–137.

Kazimírová, M., Štibrániová, I., 2013. Tick salivary compounds: their role in modulation of host defences and pathogen transmission. Frontiers of Cellular and Infectious Microbiology 3, 43.

Khasnatinov, M.A., Ustanikova, K., Frolova, T.V., Pogodina, V.V., Bochkova, N.G., Levina, L.S., Slovak, M., Kazimirova, M., Labuda, M., Klempa, B., Eleckova, E., Gould, E.A., Gritsun, T.S., 2009. Non-hemagglutinating flaviviruses: molecular mechanisms for the emergence of new strains via adaptation to European ticks. PLoS One 4, e7295.

Khasnatinov, M.A., Ustanikova, K., Frolova, T.V., Pogodina, V.V., Bochkova, N.G., Levina, L.S., Slovak, M., Kazimirova, M., Labuda, M., Klempa, B., Eleckova, E., Gould, E.A., Gritsun, T.S., 2010. Specific point mutations in the envelope protein of Tick-borne encephalitis virus enhance non-viraemic transmission efficiency in a tick vector. In: 14th International Congress on Infectious Diseases (ICID) Abstracts.

Khasnatinov, M.A., Tuplin, A., Gritsun, S.J., Slovak, M., Kazimirova, M., Lickova, M., Havlikova, S., Klempa, B., Labuda, M., Gould, E.A., Gritsun, T.S., 2016. Tick-borne encephalitis virus structural proteins are the primary viral determinants of non-viraemic transmission between ticks whereas non-structural proteins affect cytotoxicity. PLoS One 11, e0158105.

King, A.M.Q., Adams, M.J., Carstens, E.B., Lefkowitz, E.J. (Eds.), 2011. Virus Taxonomy: Classification and Nomenclature of Viruses: Ninth Report of the International Committee on Taxonomy of Viruses. Elsevier Inc., Amsterdam.

Kocáková, P., Hajnická, V., Slovák, M., Nuttall, P.A., Fuchsberger, N., 1999. Promotion of vesicular stomatitis virus nucleocapsid protein production by arthopod saliva. Acta Virologica 43, 251–254.

Kocáková, P., Sláviková, M., Hajnická, V., Slovák, M., Gašperík, J., Vančová, I., Fuchsberger, N., Nuttall, P.A., 2003. Effect of fast protein liquid chromatography fractionated salivary gland extracts from different ixodid tick species on interleukin-8 binding to its cell receptors. Folia Parasitologica (Praha) 50, 79–84.

Kochs, G., Bauer, S., Vogt, C., Frenz, T., Tschopp, J., Kalinke, U., Waibler, Z., 2010. Thogoto virus infection induces sustained type I Interferon responses that depend on RIG-I-like helicase signaling of conventional dendritic cells. Journal of Virology 84, 12344–12350.

Kopecký, J., Kuthejlová, M., 1998. Suppressive effect of *Ixodes ricinus* salivary gland extract on mechanisms of natural immunity in vitro. Parasite Immunology 20, 169–174.

Kopp, E.B., Ghosh, S., 1995. NF-KB and Rel proteins in innate immunity. In: Dixon, F.J., Alt, F., Austen, K.F., Kishimoto, T., Melchers, F., Uhr, J.W. (Eds.), Advances in Immunology, vol. 58. Academic Press, Inc., San Diego.

Kosoy, O.I., Lambert, A.J., Hawkinson, D.J., Pastula, D.M., Goldsmith, C.S., Hunt, D.C., Staples, J.E., 2015. Novel Thogotovirus associated with febrile illness and death, United States, 2014. Emerging Infectious Diseases 21, 760–764.

Kotál, J., Langhansová, H., Lieskovská, J., Andersen, J.F., Francischetti, I.M., Chavakis, T., Kopecký, J., Pedra, J.H., Kotsyfakis, M., Chmelař, J., 2015. Modulation of host immunity by tick saliva. Journal of Proteomics 128, 58–68.

Kotenko, S.V., 2011. IFNs-lambda. Current Opinion in Immunology 23, 583–590.

Kotsyfakis, M., Sá-Nunes, A., Francischetti, I.M.B., Mather, T.N., Andersen, J.F., Ribeiro, J.M.C., 2006. Antiinflammatory and immunosuppressive activity of sialostatin L, a salivary cystatin from the tick *Ixodes scapularis*. Journal of Biological Chemistry 281, 26298–26307.

Kotsyfakis, M., Anderson, J.M., Andersen, J.F., Calvo, E., Francischetti, I.M.B., Mather, T.N., Valenzuela, J.G., Ribeiro, J.M.C., 2008. Cutting edge: immunity against a 'silent' salivary antigen of the Lyme vector *Ixodes scapularis* impairs its ability to feed. The Journal of Immunology 181, 5209–5212.

Kožuch, O., Nosek, J., 1985. Replication of tick-borne encephalitis (TBE) virus in *Ixodes ricinus* ticks. Folia Parasitologica (Praha) 32, 373–375.

Kramer, C.D., Poole, N.M., Coons, L.B., Cole, J.A., 2011. Tick saliva regulates migration, phagocytosis, and gene expression in the macrophage-like cell line, IC-21. Experimental Parasitology 127, 665–671.

Kubeš, M., Fuchsberger, N., Labuda, M., Žuffová, E., Nuttall, P.A., 1994. Salivary gland extracts of partially fed *Dermacentor reticulatus* ticks decrease natural killer cell activity in vitro. Immunology 82, 113–116.

Kúdelová, M., Belvončíková, P., Vrbová, M., Kovaľová, A., Štibrániová, I., Kocáková, P., Slovák, M., Špitalská, E., Lapuníková, B., Matúšková, R., Šupolíková, M., 2015. Detection of murine herpesvirus 68 (MHV-68) in *Dermacentor reticulatus* ticks. Microbial Ecology 70, 785–794.

Kuhn, J.H., Bekal, S., Caı, Y., Clawson, A.N., Domier, L.L., Herrel, M., Jahrling, P.B., Kondo, H., Lambert, K.N., Mihindukulasuriya, K.A., Nowotny, N., Radoshitzky, S.R., Schneider, U., Staeheli, P., Suzuki, N., Tesh, R.B., Wang, D., Wang, L.F., Dietzgen, R.G., 2013. Nyamiviridae: proposal for a new family in the order Mononegavirales. Archives of Virology 158, 2209–2226.

Kuhn, J.H., Alkhovskii, S.V., Bao, Y., Palacios, G., Tesh, R.B., Vasilakis, N., Walker, P.J., 2016a. Create 5 Species in the Genus Nairovirus; Change the Genus Name to Orthonairovirus and Rename Its Constituent Species Similarly. ICTV Taxonomic Proposal 2016.026a, bM.A.v3. Nairovirus_5sp. http://www.ictv.global/proposals-16/2016.026a,bM.A.v3.Nairovirus_5sp.pdf.

Kuhn, J.H., Wiley, M.R., Rodriguez, S.E., Bao, Y., Prieto, K., Travassos da Rosa, A.P., et al., 2016b. Genomic characterization of the genus *Nairovirus* (family Bunyaviridae). Viruses 8, 164.

Kuno, G., Chang, G.J.J., 2005. Biological transmission of arboviruses: reexamination of and new insights into components, mechanisms, and unique traits as well as their evolutionary trends. Clinical Microbiology Reviews 18, 608–637.

Kuthejlová, M., Kopecký, J., Štepánová, G., Macela, A., 2001. Tick salivary gland extract inhibits killing of *Borrelia afzelii* spirochetes by mouse macrophages. Infection and Immunity 69, 575–578.

L'vov, D.K., Al'khovskiĭ, S.V., Shchelkanov, M.Iu., Shchetinin, A.M., Deriabin, P.G., Gitel'man, A.K., Aristova, V.A., Botikov, A.G., 2014a. Taxonomic status of the Burana virus (BURV) (Bunyaviridae, *Nairovirus*, Tamdy group) isolated from the ticks Haemaphysalis punctata Canestrini et Fanzago, 1877 and Haem. concinna Koch, 1844 (Ixodidae, Haemaphysalinae) in Kyrgyzstan. Voprosy Virusologii 59 (4), 10–15(In Russian)

L'vov, D.K., Al'khovskiĭ, S.V., Shchelkanov, M.Iu., Deriabin, P.G., Shchetinin, A.M., Samokhvalov, E.I., Aristova, V.A., Gitel'man, A.K., Botikov, A.G., 2014b. Genetic characterization of the Geran virus (GERV, Bunyaviridae, nairovirus, Qalyub group) isolated from the ticks *Ornithodoros verrucosus* Olenev, Zasukhin and Fenyuk, 1934 (Argasidae) collected in the burrow of Meriones erythrourus Grey, 1842 in Azerbaijan. Voprosy Virusologii 59 (5), 13–18 (In Russian).

Labuda, M., 1991. Arthropod vectors in the evolution of bunyaviruses. Acta Virologica 35, 98–105.

Labuda, M., Nuttall, P.A., 2004. Tick-borne viruses. Parasitology (129 Suppl.), S221–S245.

Labuda, M., Nuttall, P.A., 2008. Viruses transmitted by ticks. In: Bowman, A.S., Nuttall, P.A. (Eds.), Ticks: Biology, Disease and Control. Cambridge University Press, Cambridge, UK, pp. 253–280.

Labuda, M., Randolph, S.E., 1999. Survival strategy of tick-borne encephalitis virus: cellular basis and environmental determinants. Zentralblatt für Bakteriologie 289, 513–524.

Labuda, M., Danielova, V., Jones, L.D., Nuttall, P.A., 1993a. Amplification of tick-borne encephalitis virus infection during co-feeding of ticks. Medical and Veterinary Entomology 7, 339–342.

Labuda, M., Jones, L.D., Williams, T., Danielova, V., Nuttall, P.A., 1993b. Efficient transmission of tick-borne encephalitis virus between cofeeding ticks. Journal of Medical Entomology 30, 295–299.

Labuda, M., Jones, L.D., Williams, T., Nuttall, P.A., 1993c. Enhancement of tick-borne encephalitis virus transmission by tick salivary gland extracts. Medical and Veterinary Entomology 7, 193–196.

Labuda, M., Nuttall, P.A., Kožuch, O., Elečková, E., Williams, T., Žuffová, E., Sabó, A., 1993d. Non-viraemic transmission of tick-borne encephalitis virus: a mechanism for arbovirus survival in nature. Experientia 49, 802–805.

Labuda, M., Austyn, J.M., Zuffova, E., Kozuch, O., Fuchsberger, N., Lysy, J., Nuttall, P.A., 1996. Importance of localized skin infection in tick-borne encephalitis virus transmission. Virology 219, 357–366.

Labuda, M., Alves, M.J., Elečková, E., Kožuch, O., Filipe, A.R., 1997a. Transmission of tick-borne bunyaviruses by cofeeding ixodid ticks. Acta Virologica 41, 325–328.

Labuda, M., Kozuch, O., Zuffová, E., Eleckóva, E., Hails, R.S., Nuttall, P.A., 1997b. Tick-borne encephalitis virus transmission between ticks cofeeding on specific immune natural rodent hosts. Virology 235, 138–143.

Labuda, M., Trimnell, A.R., Ličková, M., Kazimírová, M., Davies, G.M., Lissina, O., Hails, R.S., Nuttal, P.A., 2006. An antivector vaccine protects against a lethal vector-borne pathogen. PLoS Pathogens 2, e27.

Langhansová, H., Bopp, T., Schmitt, E., Kopecký, J., 2015. Tick saliva increases production of three chemokines including monocyte chemoattractant protein-1, a histamine-releasing cytokine. Parasite Immunology 37, 92–96.

Lani, R., Moghaddam, E., Haghani, A., Chang, L.Y., AbuBakar, S., Zandi, K., 2014. Tick-borne viruses: a review from the perspective of therapeutic approaches. Ticks and Tick-Borne Diseases 5, 457–465.

Lawrie, C.H., Uzcátegui, N.Y., Armesto, M., Bell-Sakyi, L., Gould, E.A., 2004. Susceptibility of mosquito and tick cell lines to infection with various flaviviruses. Medical and Veterinary Entomology 18, 268–274.

Li, C.X., Shi, M., Tian, J.H., Lin, X.D., Kang, Y.J., Chen, L.J., Qin, X.-C., Xu, J., Holmes, E.C., Zhang, Y.-Z., 2015. Unprecedented genomic diversity of RNA viruses in arthropods reveals the ancestry of negative-sense RNA viruses. eLife 4, e05378.

Libraty, D.H., Pichyangkul, S., Ajariyakhajorn, C., Endy, T.P., Ennis, F.A., 2001. Human dendritic cells are activated by dengue virus infection: enhancement by gamma interferon and implications for disease pathogenesis. Journal of Virology 75, 3501–3508.

Lieskovská, J., Páleniková, J., Langhansová, H., Campos Chagas, A., Calvo, E., Kotsyfakis, M., Kopecký, J., 2015. Tick sialostatins L and L2 differentially influence dendritic cell responses to *Borrelia* spirochetes. Parasites and Vectors 8, 275.

Lim, P.Y., Behr, M.J., Chadwick, C.M., Shi, P.Y., Bernard, K.A., 2011. Keratinocytes are -cell targets of West Nile virus in vivo. Journal of Virology 85, 5197–5201.

Limon-Flores, A.Y., Perez-Tapia, M., Estrada-Garcia, I., Vaughan, G., Escobar-Gutierrez, A., Calderon-Amador, J., Herrera-Rodriguez, S.E., Brizuela-Garcia, A., Heras-Chavarria, M., Flores-Langarica, A., Cedillo-Barron, L., Flores-Romo, L., 2005. Dengue virus inoculation to human skin explants: an effective approach to assess in situ the early infection and the effects on cutaneous dendritic cells. International Journal of Experimental Pathology 86, 323–334.

Lubinga, J.C., Tuppurainen, E.S.M., Stoltsz, W.H., Ebersohn, K., Coetzer, J.A.W., Venter, E.H., 2013. Detection of lumpy skin disease virus in saliva of ticks fed on lumpy skin disease virus-infected cattle. Experimental and Applied Acarology 61, 129–138.

Lubinga, J.C., Clift, S.J., Tuppurainen, E.S.M., Stoltsz, W.H., Babiuk, S., Coetzer, J.A.W., Venter, E.H., 2014a. Demonstration of lumpy skin disease virus infection in *Amblyomma hebraeum* and *Rhipicephalus appendiculatus* ticks using immunohistochemistry. Ticks and Tick Borne Diseases 5, 113–120.

Lubinga, J.C., Tuppurainen, E.S.M., Coetzer, J.A.W., Stoltsz, W.H., Venter, E.H., 2014b. Transovarial passage and transmission of LSDV by *Amblyomma hebraeum, Rhipicephalus appendiculatus* and *Rhipicephalus decoloratus*. Experimental and Applied Acarology 62, 67–75.

Lubinga, J.C., Tuppurainen, E.S.M., Mahlare, R., Coetzer, J.A.W., Stoltsz, W.H., Venter, E.H., 2015. Evidence of transstadial and mechanical transmission of lumpy skin disease virus by *Amblyomma hebraeum* ticks. Transboundary and Emerging Diseases 62, 174–182.

Málková, D., Holubová, J., Kolman, J.M., Marhoul, Z., Hanzal, F., Kulková, H., Markvart, K., Simková, L., 1980. Antibodies against some arboviruses in persons with various neuropathies. Acta Virologica 24, 298.

Malmgaard, L., 2004. Induction and regulation of IFNs during viral infections. Journal of Interferon and Cytokine Research 24, 439–454.

Mans, B.J., Neitz, A.W.H., 2004. Adaptation of ticks to a blood-feeding environment: evolution from a functional perspective. Insect Biochemistry and Molecular Biology 34, 1–17.

Mans, B.J., Andersen, J.F., Francischetti, I.M.B., Valenzuela, J.G., Schwan, T.G., Pham, V.M., Garfield, M.K., Hammer, C.H., Ribeiro, J.M.C., 2008. Comparative sialomics between hard and soft ticks: implications for the evolution of blood-feeding behavior. Insect Biochemistry and Molecular Biology 38, 42–58.

Mansfield, K.L., Johnson, N., Phipps, L.P., Stephenson, J.R., Fooks, A.R., Solomon, T., 2009. Tick-borne encephalitis virus - a review of an emerging zoonosis. Journal of General Virology 90, 1781–1794.

Mansfield, K.L., Morales, A.B., Johnson, N., Ayllón, N., Höfle, U., Alberdi, P., Fernández de Mera, I.G., Marín, J.F.G., Gortázar, C., de la Fuente, J., Fooks, A.R., 2015. Identification and characterization of a novel tick-borne flavivirus subtype in goats (*Capra hircus*) in Spain. Journal of General Virology 96, 1676–1681.

Marchal, C.M., Luft, B.J., Yang, X., Sibilia, J., Jaulhac, B., Boulanger, N.M., 2009. Defensin is suppressed by tick salivary gland extract during the in vitro interaction of resident skin cells with *Borrelia burgdorferi*. Journal of Investigative Dermatology 129, 2515–25177.

Marchal, C., Schramm, F., Kern, A., Luft, B.J., Yang, X., Schuijt, T.J., Hovius, J.W., Jaulhac, B., Boulanger, N., 2011. Antialarmin effect of tick saliva during the transmission of Lyme disease. Infections and Immunity 79, 774–785.

Marovich, M., Grouard-Vogel, G., Louder, M., Eller, M., Sun, W., Wu, S.J., Putvatana, R., Murphy, G., Tassaneetrithep, B., Burgess, T., Birx, D., Hayes, C., Schlesinger-Frankel, S., Mascola, J., 2001. Human dendritic cells as targets of dengue virus infection. The Journal of Investigative Dermatology Symposium Proceedings 6, 219–224.

Martinon, F., Mayor, A., Tschopp, J., 2009. The inflammasomes: guardians of the body. Annual Review of Immunology 27, 229–265.

Masson, F., Mount, A.M., Wilson, N.S., Belz, G.T., 2008. Dendritic cells: driving the differentiation programme of T-cells in viral infections. Immunology and Cell Biology 86, 333–342.

Matsuno, K., Weisend, C., Travassos da Rosa, A.P., Anzick, S.L., Dahlstrom, E., Porcella, S.F., Dorward, D.W., Yu, X.J., Tesh, R.B., Ebihara, H., 2013. Characterization of the Bhanja serogroup viruses (Bunyaviridae), a novel species of the genus Phlebovirus and its relationship with other emerging tick-borne phleboviruses. Journal of Virology 87, 3719–3728.

McCauley, J.W., Hongo, S., Kaverin, N.V., Kochs, G., Lamb, R.A., Matrosovich, M., Palese, P., Perez, D., Presti, R., Rimstad, E., Smith, G., 2012. Create 2 New Species in the Proposed New Genus Quaranjavirus. ICTV Taxonomic Proposal 2011.012a-dV.A.v3.Quaranjavirus. https://talk.ictvonline.org/ICTV/proposals/2011.012a-dV.A.v3.Quaranjavirus.pdf.

McMullan, L.K., Folk, S.M., Kelly, A.J., MacNeil, A., Goldsmith, C.S., Metcalfe, M.G., Batten, B.C., Albariño, C.G., Zaki, S.R., Rollin, P.E., Nicholson, W.L., Nichol, S.T., 2012. A new Phlebovirus associated with severe febrile illness in Missouri. The New England Journal of Medicine 367, 834–841.

McNally, K.L., Mitzel, D.N., Anderson, J.M., Ribeiro, J.M.C., Valenzuela, J.G., Myers, T.G., Godinez, A., Wolfinbarger, J.B., Best, S.M., Bloom, M.E., 2012. Differential salivary gland transcript expression profile in *Ixodes scapularis* nymphs upon feeding or flavivirus infection. Ticks and Tick Borne Diseases 3, 18–26.

Meagher, K.E., Decker, C.F., 2012. Other tick-borne illnesses, tularemia, Colorado tick fever, tick paralysis. Disease-a-Month 58, 370–376.

Merad, M., Ginhoux, F., Collin, M., 2008. Origin homeostasis and function of Langerhans cells and other langerin-expressing dendritic cells. Nature Reviews Immunology 8, 935–947.

Mihindukulasuriya, K.A., Nguyen, N.L., Wu, G., Huang, H.V., Travassos, A.P.A., Popov, V.L., Tesh, R.B., Wang, D., 2009. Nyamanini and midway viruses define a novel taxon of RNA viruses in the order Mononegavirales. Journal of Virology 83, 5109–5116.

Mitzel, D.N., Wolfinbarger, J.B., Long, R.D., Masnick, M., Best, S.M., Bloom, M.E., 2007. Tick-borne flavivirus infection in *Ixodes scapularis* larvae: development of a novel method for synchronous viral infection of ticks. Virology 365, 410–418.

Mlera, L., Melik, W., Bloom, M.E., 2014. The role of viral persistence in flavivirus biology. Pathogens and Disease 71, 135–161.

Morse, M.A., Marriott, A.C., Nuttall, P.A., 1992. The glycoprotein of Thogoto virus (a tick-borne orthomyxo-like virus) is related to the baculovirus glycoprotein gp64. Virology 186, 640–646.

Moutailler, S., Popovici, I., Devillers, E., Vayssier-Taussat, M., Eloit, M., 2016. Diversity of viruses in *Ixodes ricinus*, and characterization of a neurotropic strain of Eyach virus. New Microbes and New Infections 11, 71–81.

Munderloh, U.G., Liu, Y.-J., Wang, M., Chen, C.T., Kurtti, J., 1994. Establishment, maintenance and description of cell lines from the tick *Ixodes scapularis*. Journal of Parasitology 80, 533–543.

Nestle, F.O., DiMeglio, P., Qin, J.Z., Nickoloff, B.J., 2009. Skin immune sentinels in health and disease. Nature Reviews Immunology 9, 679–691.

Nichol, S.T., Beaty, B.J., Elliot, R.M., Goldbach, R., Plyusin, A., Schmaljohn, C.S., Tesh, R., 2005. Family Bunyaviridae. Elsevier Academic Press, London, United Kingdom.

Nicholson, W.L., Sonenshine, D.E., Lane, R.S., Uilenberg, G., 2009. Ticks (Ixodida). In: Mullen, G.R., Durden, L.A. (Eds.), Medical and Veterinary Entomology. Academic Press, Burlington, pp. 493–542.

Nithiuthai, S., Allen, J.R., 1985. Langerhans cells present tick antigens to lymph node cells from tick-sensitized Guinea-pigs. Immunology 55, 157–163.

Nosek, J., Korolev, M.B., Chunikhin, S.P., Kožuch, O., Čiampor, F., 1984. The replication and eclipse-phase of the tick-borne encephalitis virus in *Dermacentor reticulatus*. Folia Parasitologica (Praha) 31, 187–189.

Nuttall, P.A., 2009. Molecular characterization of tick-virus interactions. Frontiers in Bioscience 14, 2466–2483.

Nuttall, P.A., 2014. Tick-borne viruses. In: Sonenshine, D.E., Roe, R.M. (Eds.), Biology of Ticks, vol. 2. Oxford University Press, New York, USA, pp. 180–210.

Nuttall, P.A., Labuda, M., 2003. Dynamics of infection in tick vectors and at the tick–host interface. Advances in Virus Research 60, 233–272.

Nuttall, P.A., Labuda, M., 2004. Tick-host interactions: saliva-activated transmission. Parasitology (129 Suppl.), S177–S189.

Nuttall, P.A., Labuda, M., 2008. Saliva-assisted transmission of tick-borne pathogens. In: Bowman, A.S., Nuttall, P.A. (Eds.), Ticks: Biology, Disease and Control. Cambridge University Press, Cambridge, UK, pp. 205–219.

Nuttall, P.A., Jones, L.D., Labuda, M., Kaufman, W.R., 1994. Adaptations of arboviruses to ticks. Journal of Medical Entomology 31, 1–9.

Oba, M., Omatsu, T., Takano, A., Fujita, H., Sato, K., Nakamoto, A., Takahashi, M., Takada, N., Kawabata, H., Ando, S., Mizutani, T., 2016. A novel bunyavirus from the soft tick, *Argas vespertilionis*, in Japan. The Journal of Veterinary Medical Science 78, 443–445.

Odendall, C., Dixit, E., Stavru, F., Bierne, H., Franz, K.M., Durbin, A.F., Boulant, S., Gehrke, L., Cossart, P., Kagan, J.C., 2014. Diverse intracellular pathogens activate type III interferon expression from peroxisomes. Nature Immunology 15, 717–726.

Okumura, A., Pitha, P., Yoshimura, A., Harty, R.N., 2010. Interaction between Ebola virus glycoprotein and host toll-like receptor 4 leads to induction of proinflammatory cytokines and SOCS1. The Journal of Virology 84, 27–33.

Oliveira, C.J., Cavassani, K.A., More, D.D., Garlet, G.P., Aliberti, J.C., Silva, J.S., Ferreira, B.R., 2008. Tick saliva inhibits the chemotactic function of MIP-1 alpha and selectively impairs chemotaxis of immature dendritic cells by down-regulating cell-surface CCR5. International Journal of Parasitology 38, 705–716.

Oliveira, C.J., Carvalho, W., Garcia, G., Ferreira, B., 2010. Tick saliva induces regulatory dendritic cells: MAP-kinases and Toll-like receptor-2 expression as potential targets. Veterinary Parasitology 167, 288–297.

Onoguchi, K., Yoneyama, M., Takemura, A., Akira, S., Taniguchi, T., Namiki, H., Fujita, T., 2007. Viral infections activate types I and III interferon genes through a common mechanism. Journal of Biological Chemistry 282, 7576–7581.

Osterlund, P.I., Pietilä, T.E., Veckman, V., Kotenko, S.V., Julkunen, I., 2007. IFN regulatory factor family members differentially regulate the expression of type III IFN (IFN-lambda) genes. The Journal of Immunology 179, 3434–3442.

Palacios, G., Savji, N., Travassos da Rosa, A., Guzman, H., Yu, X., Desai, A., Rosen, G.E., Hutchison, S., Lipkin, W.I., Tesh, R., 2013. Characterization of the Uukuniemi virus group (Phlebovirus: Bunyaviridae): evidence for seven distinct species. Journal of Virology 87, 3187–3195.

Papa, A., Dalla, V., Papadimitriou, E., Kartalis, G.N., Antoniadis, A., 2010. Emergence of Crimean-Congo hemorrhagic fever in Greece. Clinical Microbiology and Infection 16, 843–847.

Peterková, K., Vančová, I., Hajnická, V., Slovák, M., Šimo, L., Nuttall, P.A., 2008. Immunomodulatory arsenal of nymphal ticks. Medical and Veterinary Entomology 22, 167–171.

Plekhova, N.G., Somova, L.M., Zavorueva, D.V., Krylova, N.V., Leonova, G.N., 2008. NO-producing activity of macrophages infected with tick-borne encephalitis virus. Bulletine Experimental Biology of Medicine 145, 344–347.

Plyusnin, A., Beaty, B.J., Elliott, R.M., Goldbach, R., Kormelink, R., Lundkvist, Å., Schmaljohn, C.S., Tesh, R.B., 2011. Bunyaviridae. In: King, A.M.Q., Adams, M.J., Carstens, E.B., Lefkowitz, E.J. (Eds.), Virus Taxonomy: Classification and Nomenclature of Viruses: Ninth Report of the International Committee on Taxonomy of Viruses. Elsevier Inc., pp. 725–741.

Presti, R.M., Zhao, G., Beatty, W.L., Mihindukulasuriya, K.A., Travassos da Rosa, A.P.A., Popov, V.L., Tesh, R.B., Virgin, H.W., Wang, D., 2009. Quaranfil, Johnston Atoll, and lake Chad viruses are novel members of the family Orthomyxoviridae. Journal of Virology 83, 11599–11606.

Preston, S.G., Majtán, J., Kouremenou, C., Rysnik, O., Burger, L.F., Cabezas Cruz, A., Guzman, M.C., Nunn, M.A., Paesen, G.C., Nuttall, P.A., Austyn, J.M., 2013. Novel immunomodulators from hard ticks selectively reprogramme human dendritic cell responses. PLoS Pathogens 9, e 1003450.

Pugliese, A., Beltramo, T., Torre, D., 2007. Emerging and re-emerging viral infections in Europe. Cell Biochemistry and Function 25, 1–13.

Quan, P.-L., Junglen, S., Tashmukhamedova, A., Conlan, S., Hutchison, S.K., Kurth, A., Ellerbrok, H., Egholm, M., Briese, T., Leendertz, F.H., Lipkin, W.I., 2010. Moussa virus: a new member of the Rhabdoviridae family isolated from *Culex decens* mosquitoes in Côte d'Ivoire. Virus Research 147, 17–24.

Rajčáni, J., Kúdelová, M., 2007. Murid herpesvirus 4 (MHV-4): an animal model for human gammaherpesvirus research. In: Minarovits, J., Gonczol, E., Valyi-Nagy, T. (Eds.), Latency Strategies of Herpesviruses. Springer, Berlin, pp. 102–136. Chapter V.

Rajčáni, J., Nosek, J., Kožuch, O., Waltinger, H., 1976. Reaction of the host to the tick-bite. II. Distribution of tick borne encephalitis virus in sucking ticks. Zentralblatt für Bakteriologie A 236, 1–9.

Rajčáni, J., Blaškovič, D., Svobodová, J., Čiampor, F., Hučková, D., Staneková, D., 1985. Pathogenesis of acute and persistent murine herpesvirus infection in mice. Acta Virologica 29, 51–60.

Ramamoorthi, N., Narasimhan, S., Pal, U., Bao, F., Yang, X.F., Fish, D., Anguita, J., Norgard, M.V., Kantor, F.S., Anderson, J.F., Koski, R.A., Fikrig, E., 2005. The Lyme disease agent exploits a tick protein to infect the mammalian host. Nature 436, 573–577.

Randolph, S.E., 2011. Transmission of tick-borne pathogens between co-feeding ticks: Milan Labuda's enduring paradigm. Ticks and Tick Borne Diseases 2, 179–182.

Randolph, S.E., Gern, L., Nuttall, P.A., 1996. Co-feeding ticks: epidemiological significance for tick-borne pathogen transmission. Parasitology Today 12, 472–479.

Rehse-Küpper, B., Casals, J., Rehse, E., Ackermann, R., 1976. Eyach, an arthropod-borne virus related to Colorado tick fever virus in the Federal Republic of Germany. Acta Virologica 20, 339–342.

Ribeiro, J.M.C., Francischetti, I.M.B., 2003. Role of arthropod saliva in blood feeding: sialome and post-sialome perspectives. Annual Review of Entomology 48, 73–88.

Ribeiro, J.M.C., Alarcon-Chaidez, F., Francischetti, I.M., Mans, B.J., Mather, T.N., Valenzuela, J.G., Wikel, S.K., 2006. An annotated catalog of salivary gland transcripts from *Ixodes scapularis* ticks. Insect Biochemistry and Molecular Biology 36, 111–129.

Rider, P., Carmi, Y., Guttman, O., Braiman, A., Cohen, I., Voronov, E., White, M.R., Dinarello, C.A., Apte, R.N., 2011. IL-1a and IL-1b recruit different myeloid cells and promote different stages of sterile inflammation. The Journal of Immunology 187, 4835–4843.

Robertson, S.J., Mitzel, D.N., Taylor, R.T., Best, S.M., Bloom, M.E., 2009. Tick-borne flaviviruses: dissecting host immune responses and virus countermeasures. Immunology Research 43, 172–186.

Rogers, M.B., Cui, L., Fitch, A., Popov, V., Travassos da Rosa, A.P.A., Vasilakis, N., Tesh, R.B., Ghedin, E., 2014. Short report: whole genome analysis of Sierra Nevada virus, a novel Mononegavirus in the family Nyamiviridae. The American Journal of Tropical Medicine and Hygiene 91, 159–164.

Růžek, D., Bell-Sakyi, L., Kopecký, J., Grubhoffer, L., 2008. Growth of tick-borne encephalitis virus (European subtype) in cell lines from vector and non-vector ticks. Virus Research 137, 142–146.

Sá-Nunes, A., Bafica, A., Lucas, D.A., Conrads, T.P., Veenstra, T.D., Andersen, J.F., Mather, T.N., Ribeiro, J.M., Francischetti, I.M., 2007. Prostaglandin E2 is a major inhibitor of dendritic cell maturation and function in *Ixodes scapularis* saliva. The Journal of Immunology 179, 1497–1505.

Sá-Nunes, A., Bafica, A., Antonelli, L.R., Choi, E.Y., Francischetti, I.M.B., Andersen, J.F., Shi, G.P., Chavakis, T., Ribeiro, J.M., Kotsyfakis, M., 2009. The immunomodulatory action of Sialostatin L on dendritic cells reveals its potential to interfere with autoimmunity. The Journal of Immunology 182, 7422–7429.

Savage, H.M., Godsey, M.S., Amy, L., Panella, N.A., Burkhalter, K.L., Harmon, J.R., Lash, R.R., Ashley, D.C., Nicholson, W.L., 2013. First detection of Heartland virus (Bunyaviridae: Phlebovirus) from field collected arthropods. The American Journal of Tropical Medicine and Hygiene 89, 445–452.

Schnettler, E., Tykalová, H., Watson, M., Sharma, M., Sterken, M.G., Obbard, D.J., Lewis, S.H., McFarlane, M., Bell-Sakyi, L., Barry, G., Weisheit, S., Best, S.M., Kuhn, R.J., Pijlman, G.P., Chase-Topping, M.E., Gould, E.A., Grubhoffer, L., Fazakerley, J.K., Kohl, A., 2014. Induction and suppression of tick cell antiviral RNAi responses by tick-borne flaviviruses. Nucleic Acids Research 42, 9436–9446.

Šenigl, F., Kopecký, J., Grubhoffer, L., 2004. Distribution of E and NS1 proteins of TBE virus in mammalian and tick cells. Folia Microbiologica (Praha) 49, 213–216.

Šenigl, F., Grubhoffer, L., Kopecký, J., 2006. Differences in maturation of tick-borne encephalitis virus in mammalian and tick cell line. Intervirology 49, 239–248.

Shaw, T.J., Martin, P., 2009. Wound repair at a glance. Journal of Cell Science 122, 3209–3213.

Simmonds, P., Becher, P., Collett, M.S., Gould, E.A., Heinz, F.X., Meyers, G., Monath, T., Pletnev, A., Rice, C.M., Stiasny, K., Thiel, H.-J., Weiner, A., Bukh, J., 2011. Flaviviridae. In: King, A.M.Q., Adams, M.J., Carstens, E.B., Lefkowitz, E.J. (Eds.), Virus Taxonomy: Classification and Nomenclature of Viruses: Ninth Report of the International Committee on Taxonomy of Viruses. Elsevier Inc., pp. 1003–1020.

Šimo, L., Kazimirova, M., Richardson, J., Bonnet, S.I., 2017. The essential role of tick salivary glands and saliva in tick feeding and pathogen transmission. Frontiers in Cellular and Infection Microbiology 7, 281.

Skallová, A., Iezzi, G., Ampenberger, F., Kopf, M., Kopecký, J., 2008. Tick saliva inhibits dendritic cell migration, maturation and function while promoting development of Th2 responses. The Journal of Immunology 180, 6186–6192.

Skinner, M.A., Buller, R.M., Damon, I.K., Lefkowitz, E.J., McFadden, G., McInnes, C.J., Mercer, A.A., Moyer, R.W., Upton, C., 2011. The Poxviridae. In: King, A.M.Q., Adams, M.J., Carstens, E.B., Lefkowitz, E.J. (Eds.), Virus Taxonomy: Classification and Nomenclature of Viruses: Ninth Report of the International Committee on Taxonomy of Viruses. Elsevier Inc., Amsterdam, pp. 291–307.

Sláviková, M., Kocáková, P., Slovák, M., Vančová, I., Hajnická, V., Gasperík, J., Fuchsberger, N., Nuttall, P.A., 2002. Vesicular stomatitis virus nucleocapsid protein production in cells treated with selected fast protein liquid chromatography fractions of tick salivary gland extracts. Acta Virologica 46, 117–120.

Slovák, M., Kazimírová, M., Siebenstichová, M., Ustaníková, K., Klempa, B., Gritsun, T., Gould, E.A., Nuttall, P.A., 2014a. Survival dynamics of tick-borne encephalitis virus in Ixodes ricinus ticks. Ticks and Tick Borne Diseases 5, 962–969.

Slovák, M., Štibrániová, I., Hajnická, V., Nuttall, P.A., 2014b. Antiplatelet-derived growth factor (PDGF) activity in the saliva of ixodid ticks is linked with their long mouthparts. Parasite Immunology 36, 32–42.

Smith, A.A., Pal, U., 2014. Immunity-related genes in *Ixodes scapularis* - perspectives from genome information. Frontiers in Cellular and Infection Microbiology 4, 116.

Sonenshine, D.R., Roe, R.M., 2014. Overview. Ticks, people, and animals. In: Sonenshine, D.R., Roe, R.M. (Eds.), Biology of Ticks, vol. 1. Oxford University Press, New York, USA, pp. 3–17.

Sonenshine, D.E., Lane, R.S., Nicholson, W.L., 2002. Ticks (Ixodida). In: Mullen, G., Durden, L. (Eds.), Medical and Veterinary Entomology. Academic Press, San Diego, pp. 517–558.

Štibrániová, I., Lahová, M., Bartíková, P., 2013. Immunomodulators in tick saliva and their benefits. Acta Virologica 57, 200–216.

Surasombatpattana, P., Hamel, R., Patramool, S., Luplertlop, N., Thomas, F., Despres, P., Briant, L., Yssel, H., Misse, D., 2011. Dengue virus replication in infected human keratinocytes leads to activation of antiviral innate immune responses. Infection. Genetics and Evolution 11, 1664–1673.

Swei, A., Russell, B.J., Naccache, S.N., Kabre, B., Veeraraghavan, N., Pilgard, M.A., Johnson, B.J.B., Chiu, C.Y., 2013. The genome sequence of lone star virus, a highly divergent bunyavirus found in the *Amblyomma americanum* tick. PLoS One 8, e62083.

Taylor, R.M., Hurlbut, H.S., Work, T.H., Kingston, J.R., Hoogstraal, H., 1966. Arboviruses isolated from *Argas* ticks in Egypt: Quaranfil, Chenuda, and Nyamanini. The American Journal of Tropical Medicine and Hygiene 15, 76–86.

Tekin, S., Barut, S., Bursali, A., Gul Aydogan, G., Yuce, O., Demir, F., Yildirim, B., 2010. Seroprevalence of Crimean-Congo hemorrhagic fever (CCHF) in risk groups in Tokat Province of Turkey. African Journal of Microbiology Research 4, 214–217.

Tesh, R.B., Ghedin, E., Vasilakis, N., Rogers, M.B., Cui, L., Fitch, A., Popov, V., Pravassos da Rosa, A.P.A., 2014. ICTV Taxonomic Proposal 2014.007aV.A.v2. Nyavirus_sp. Create 1 New Species in the Genus *Nyavirus*, Family Nyamiviridae, Order Mononegavirales. http://www.ictvonline.org/proposals-14/2014.007aV.A.v2.Nyavirus_sp.pdf.

Tokarz, R., Sameroff, S., Leon, M.S., Jain, K., Lipkin, W.I., 2014a. Genome characterization of Long Island tick rhabdovirus, a new virus identified in *Amblyomma americanum* ticks. Virology Journal 11, 1–5.

Tokarz, R., Williams, S.H., Sameroff, S., Sanchez Leon, M., Jain, K., Lipkin, W.I., 2014b. Virome analysis of *Amblyomma americanum, Dermacentor variabilis*, and *Ixodes scapularis* ticks reveals novel highly divergent vertebrate and invertebrate viruses. Journal of Virology 88, 11480–14492.

Trimnell, A.R., Hails, R.S., Nuttall, P.A., 2002. Dual action ectoparasite vaccine targeting 'exposed' and 'concealed' antigens. Vaccine 20, 3560–3568.

Trimnell, A.R., Davies, G.M., Lissina, O., Hails, R.S., Nuttall, P.A., 2005. A cross-reactive tick cement antigen is a candidate broad-spectrum tick vaccine. Vaccine 23, 4329–4341.

Tuppurainen, E.S.M., Stoltsz, W.H., Troskie, M., Wallace, D.B., Oura, C.A.L., Mellor, P.S., Coetzer, J.A.W., Venter, E.H., 2011. A potential role for ixodid (hard) tick vectors in the transmission of Lumpy Skin Disease Virus in cattle. Transboundary and Emerging Diseases 58, 93–104.

Tuppurainen, E.S.M., Lubinga, J.C., Stoltsz, W.H., Troskie, M., Carpenter, S.T., Coetzer, J.A.W., Venter, E.H., Oura, C.A.L., 2013a. Mechanical transmission of lumpy skin disease virus by *Rhipicephalus appendiculatus* male ticks. Epidemiology and Infection 141, 425–430.

Tuppurainen, E.S.M., Lubinga, J.C., Stoltsz, W.H., Troskie, M., Carpenter, S.T., Coetzer, J.A.W., Venter, E.H., Oura, C.A.L., 2013b. Evidence of vertical transmission of lumpy skin disease virus in *Rhipicephalus decoloratus* ticks. Ticks and Tick Borne Diseases 4, 329–333.

Turell, M.J., 2015. Experimental transmission of Karshi (mammalian tick-borne flavivirus group) virus by *Ornithodoros* ticks >2,900 days after initial virus exposure supports the role of soft ticks as a long-term maintenance mechanism for certain flaviviruses. PLoS Neglected Tropical Diseases 9, 1–8.

Valenzuela, J.G., Francischetti, I.M.B., Pham, V.M., Garfield, M.K., Mather, T.N., Ribeiro, J.M.C., 2002. Exploring the sialome of the tick *Ixodes scapularis*. Journal of Experimental Biology 205, 2843–2864.

Vančová, I., Slovák, M., Hajnická, V., Labuda, M., Šimo, L., Peterková, K., Hails, R.S., Nuttall, P.A., 2007. Differential anti-chemokine activity of *Amblyomma variegatum* adult ticks during blood-feeding. Parasite Immunology 29, 169–177.

Vančová, I., Hajnická, V., Slovák, M., Kocáková, P., Paesen, G.C., Nuttall, P.A., 2010a. Evasin-3-like anti-chemokine activity in salivary gland extracts of ixodid ticks during blood-feeding: a new target for tick control. Veterinary Parasitology 167, 274–278.

Vančová, I., Hajnická, V., Slovák, M., Nuttall, P.A., 2010b. Anti-chemokine activities of ixodid ticks depend on tick species, developmental stage, and duration of feeding. Veterinary Parasitology 167, 274–278.

Vinuela, E., 1985. African swine fever. Current Topics in Microbiology and Immunology 116, 456–461.

Vrbová, M., Belvončíková, P., Kovaľová, A., Matúšková, R., Slovák, M., Kúdelová, M., 2016. Molecular detection of murine gammaherpesvirus 68 (MHV-68) in *Haemaphysalis concinna* ticks collected in Slovakia. Acta Virologica 60, 426–428.

Walker, P.J., Firth, C., Widen, S.G., Blasdell, K.R., Guzman, H., Wood, T.G., Paradkar, P.N., Holmes, E.C., Tesh, R.B., Vasilakis, N., 2015a. Evolution of genome size and complexity in the *Rhabdoviridae*. PLoS Pathogens 11, e1004664.

Walker, P.J., Widen, S.G., Firth, C., Blasdell, K.R., Wood, T.G., Travassos da Rosa, A.P., Guzman, H., Tesh, R.B., Vasilakis, N., 2015b. Genomic characterization of Yogue, Kasokero, Issyk-Kul, Keterah, Gossas, and Thiafora viruses: nairoviruses naturally infecting bats, shrews, and ticks. The American Journal of Tropical Medicine and Hygiene 93, 1041–1051.

Walker, P.J., Blasdell, K.R., Vasilakis, N., Tesh, R.B., Calisher, C.H., et al., 2016a. One New Genus (Ledantevirus) Including 14 New Species in the Family Rhabdoviridae. ICTV 2016.006a-dM .A.v2.*Ledantevirus*.

Walker, P.J., Widen, S.G., Wood, T.G., Guzman, H., Tesh, R.B., Vasilakis, N., 2016b. A global genomic characterization of nairoviruses identifies nine discrete genogroups with distinctive structural characteristics and host-vector associations. The American Journal of Tropical Medicine and Hygiene 94, 1107–1122.

Weisheit, S., Villar, M., Tykalova, H., Popara, M., Loecherbach, J., Watson, M., Ruzek, D., Grubhoffer, L., de la Fuente, J.J., Fazakerley, J.K., Bell-Sakyi, L., Tykalová, H., Popara, M., Loecherbach, J., Watson, M., Růžek, D., Grubhoffer, L., de la Fuente, J.J., Fazakerley, J.K., Bell-Sakyi, L., 2015. *Ixodes scapularis* and *Ixodes ricinus* tick cell lines respond to infection with tick-borne encephalitis virus: transcriptomic and proteomic analysis. Parasites and Vectors 8, 599.

Werner, S., Grose, R., 2003. Regulation of wound healing by growth factors and cytokines. Physiological Reviews 83, 835–870.

Whitehouse, C.A., 2004. Crimean-Congo hemorrhagic fever. Antiviral Research 64, 145–160.

Wikel, S.K., 1999. Tick modulation of host immunity: an important factor in pathogen transmission. International Journal for Parasitology 29, 851–859.

Wikel, S., 2013. Ticks and tick-borne pathogens at the cutaneous interface: host defenses, tick countermeasures, and a suitable environment for pathogen establishment. Frontiers in Microbiology 4, 337.

Wikel, S.K., 2014. Tick-host interactions. In: Sonenshine, D.E., Roe, R.M. (Eds.), Biology of Ticks, vol. 2. Oxford University Press, New York, USA, pp. 88–128.

Wikel, S.K., Graham, J.E., Allen, J.R., 1978. Acquired resistance to ticks. IV. Skin reactivity and in vitro lymphocyte responsiveness to salivary gland antigen. Immunology 34, 257–263.

Wikel, S.K., Ramachandra, R.N., Bergman, D.K., 1994. Tick-induced modulation of the host immune response. International Journal for Parasitology 24, 59–66.

Witte, K., Gruetz, G., Volk, H.-D., Looman, A.C., Asadullah, K., Sterry, W., Sabat, R., Wolk, K., 2009. Despite IFN-lambda receptor expression, blood immune cells, but not keratinocytes or melanocytes, have an impaired response to type III interferons: implications for therapeutic applications of these cytokines. Genes Immunology 10, 702/714.

World Health Organization Scientific Group, 1985. Arthropod-Borne and Rodent-Borne Viral Diseases. World Health Organization Tech. Rep. Ser. No. 719. World Health Organization, Geneva, Switzerland.

Wu, S.J., Grouard-Vogel, G., Sun, W., Mascola, J.R., Brachtel, E., Putvatana, R., Louder, M.K., Filgueira, L., Marovich, M.A., Wong, H.K., Blauvelt, A., Murphy, G.S., Robb, M.L., Innes, B.L., Birx, D.L., Hayes, C.G., Frankel, S.S., 2000. Human skin Langerhans cells are targets of dengue virus infection. Nature Medicine 6, 816–820.

Xia, H., Hu, C., Zhang, D., Tang, S., Zhang, Z., Kou, Z., Fan, Z., Bente, D., Zeng, C., Li, T., 2015. Metagenomic profile of the viral communities in *Rhipicephalus* spp. Ticks from Yunnan, China. PLoS One 10, e0121609.

Xu, B., Liu, L., Huang, X., Ma, H., Zhang, Y., Du, Y., Wang, P., Tang, X., Wang, H., Kang, K., Zhang, S., Zhao, G., Wu, W., Yang, Y., Chen, H., Mu, F., Chen, W., 2011. Metagenomic analysis of fever, thrombocytopenia and leukopenia syndrome (FTLS) in Henan Province, China: discovery of a new bunyavirus. PLoS Pathogens 7, e1002369.

Yu, X.J., Liang, M.F., Zhang, S.Y., Liu, Y., Li, J.D., Sun, Y.L., Zhang, L., Zhang, Q.F., Popov, V.L., Li, C., Qu, J., Li, Q., Zhang, Y.P., Wang, X.J., Kan, B., Wang, S.W., Wan, K.L., et al., 2011. Fever with thrombocytopenia associated with a novel bunyavirus in China. The New England Journal of Medicine 364, 1523–1532.

Yunker, C.E., Cory, J., 1967. Growth of Colorado tick fever virus in primary tissue cultures of its vector, *Dermacentor andersoni* Stiles (Acarina: Ixodidae), with notes on tick tissue culture. Experimental Parasitology 20, 267–277.

Zeller, H.G., Cornet, J.P., Camicas, J.L., 1994. Experimental transmission of Crimean-Congo hemorrhagic fever virus by West African wild ground-feeding birds to *Hyalomma marginatum rufipes* ticks. The American Journal of Tropical Medicine and Hygiene 50, 676–681.

Zhang, Y.Z., Zhou, D.J., Xiong, Y., Chen, X.P., He, Y.W., Sun, Q., Yu, B., Li, J., Dai, Y.A., Tian, J.H., Qin, X.C., Jin, D., Cui, Z., Luo, X.L., Li, W., Lu, S., Wang, W., Peng, J.S., Guo, W.P., Li, M.H., Li, Z.J., Zhang, S., Chen, C., Wang, Y., de Jong, M.D., Xu, J., 2011. Hemorrhagic fever caused by a novel tick-borne Bunyavirus in Huaiyangshan, China. Zhonghua Liu Xing Bing Xue Za Zhi 32, 209–220.

Řeháček, J., 1965. Development of animal viruses and rickettsiae in ticks and mites. Annual Review of Entomology 10, 1–24.

Řeháček, J., Kožuch, O., 1964. Comparison of the susceptibility of pri mary tick and chick embryo cell cultures to small amounts of tick-borne encephalitis virus. Acta Virologica 8, 470–471.

Chapter 11

Skin and Arthropod-Borne Diseases: Applications to Vaccine and Diagnosis

LIVE VACCINES AGAINST PREERYTHROCYTIC MALARIA: A SKIN ISSUE?

Laura Mac-Daniel[1,2], Rogerio Amino[1] and Robert Ménard[1]
[1]1Institut Pasteur, Paris, France; [2]Loyola University Chicago, Maywood, IL, United States

INTRODUCTION

Despite the burden that malaria still imposes on many parts of the world, the situation has significantly improved in the last 15 years, including in Africa (Gething et al., 2016). This was mainly achieved by the integrated implementation of the widespread use of long-lasting insecticide-treated bed nets, indoor residual spraying of insecticide, rapid diagnostic tests, and effective artemisinin combination therapies. Between 2000 and 2013, scaling up these measures is estimated to have saved 4.3 million lives. This success has contributed to reinforce political and financial commitment and formulate new ambitious goals for the future of malaria control. In 2007, Bill and Melinda Gates issued an influential call for global malaria eradication (Roberts and Enserink, 2007). The "draft global technical strategy 2016–2030" published by the World Health Organization in 2015 aims at a reduction of global malaria mortality rates by >40% in 2020, >75% in 2025, and >90% in 2030 and malaria elimination from at least 10 countries by 2020, 20 by 2025, and 35 by 2030.

Malaria Vaccines

Malaria vaccines are expected to help near these ambitious goals. Although a highly efficient malaria vaccine has remained elusive so far, optimism remains high, largely because of the continuous development of novel experimental vaccination approaches. The updated version of the malaria vaccine technology roadmap, a strategic framework that underpins the activities of the malaria vaccine

Skin and Arthropod Vectors. https://doi.org/10.1016/B978-0-12-811436-0.00011-3

385

community, is to develop effective vaccines to enable malaria eradication. Its strategic goals are to license by 2030 (1) vaccines against *Plasmodium falciparum* and *Plasmodium vivax* with a protective efficacy of at least 75% against clinical malaria and (2) vaccines that reduce transmission of the parasite and thereby reduce the incidence of human malaria infection—to enable elimination in multiple settings (http://www.path.org/vaccineresources/details.php?i=742).

The primary difficulty in developing malaria vaccines is the complexity of the parasite and its ability to evade host immunity at many levels. The ~20–25 megabase *Plasmodium* genome encodes ~5–6000 putative proteins (Kirchner et al., 2016; http://www.plasmodb.org). Many of these proteins, particularly those interacting with the host, exhibit allelic and antigenic polymorphism—often localized to B cell or T cell epitopes—or antigenic variation (Flanagan et al., 2016; Reid, 2015). The parasite, which has coevolved with man, has adapted to cause limited but chronic infection (Mita and Jombart, 2015) and excels at escaping or manipulating host immunity at the cellular and molecular levels (Wykes et al., 2014).

As detailed in the previous chapter, the *Plasmodium* parasite life cycle runs through an *Anopheles* mosquito and consists of three phases: preerythrocytic (PE) and erythrocytic in the intermediate host and sexual/sporogonic in the mosquito. The three phases involve distinct stages of the parasite expressing stage-specific proteins; therefore, they are targeted by separate vaccine efforts. Vaccines against erythrocytic stages mainly aim at inducing antibodies directed at surface proteins of the extracellular merozoite stage to prevent their entry inside red blood cells (RBC). The goal of these vaccines is to limit blood-stage multiplication and prevent clinical malaria. Vaccines against sexual stages, called transmission blocking vaccines, aim at inducing antibodies that interfere with gamete fertilization or further zygote development in the mosquito. Their goal is to confer "herd immunity" rather than individual protection by reducing mosquito infection and transmission in the community.

Preerythrocytic Malaria Vaccines

Parasite development in the liver represents a bottleneck in parasite numbers across the life cycle, with only a few tens of individuals involved in natural conditions. Therefore, PE vaccines have the distinctive appeal to potentially block infection, i.e., prevent blood infection, along with the clinical symptoms and parasite transmission. The parasite liver stages, which develop inside hepatocytes, are also the only forms present in a nucleated cell capable of presenting parasite epitopes in association with major histocompatibility (MHC) class I molecules. PE vaccines thus aim at inducing antibodies that neutralize the extracellular motile and invasive sporozoite and cytotoxic cells, mainly CD8[+] T cells, which recognize and kill infected hepatocytes. To date, only PE vaccines have proved capable of generating sterilizing immunity, defined as clearance of challenge parasites from the vaccinated host.

PE vaccine development has been dominated during the last three decades by the development of the RTS,S vaccine, one of the rare subunit candidate that has shown efficacy in clinical trials and the sole to have undergone a phase III trial. RTS,S is a viruslike particle consisting of epitopes of the *P. falciparum* circumsporozoite protein (CS) protein, the major surface protein of the sporozoite stage, fused to the hepatitis B surface antigen. Fig. 11.1 briefly presents the main milestones in RTS,S development. Other subunit PE vaccines have been proposed based on a handful of PE parasite antigens and various delivery platforms, notably, prime–boost approaches using viral vectors (Ewer et al., 2015). So far, success of these subunit approaches has been partial, including for RTS,S (see Fig. 11.1).

In contrast to subunit preparations, attenuated "whole organisms" have long been known to be capable of inducing sterile immunity. These live attenuated parasites (LAP) are presumed to trigger broad immune responses against a large number of parasite antigens, which should help circumvent the problems posed

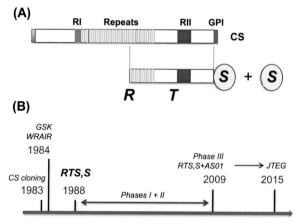

FIGURE 11.1 Development of RTS,S, the leading preerythrocytic (PE) subunit vaccine candidate. (A) Sequences of the *Plasmodium* circumsporozoite (CS) protein and RTS,S. CS is a glycosylphosphatydilinositol (GPI)-anchored protein consisting of a central asparagine-rich repeat region with N- and C-terminal motifs called region I (RI) and region II (RII) involved in host cell binding. RTS,S consists in 19 NANP repeats ("R") and the T cell epitope-containing C terminal domain ("T") of the *Plasmodium falciparum* CS from strain NF54 fused to the hepatitis B surface antigen ("S"), expressed in *Saccharomyces cerevisiae* together with free "S," yielding RTS,S viruslike particles. (B) Timeline of RTS,S development. The CS-encoding gene was first cloned in 1983 (Ellis et al., 1983). The RTS,S formulation was achieved in 1988 (Rutgers et al., 1988) by a the collaboration between GlaxoSmithKline (GSK) and the Walter Reed Army Institute of Research (WRAIR). In 2009, a multicenter phase III trial of RTS,S with the adjuvant AS01 was undertaken in 11 research centers across 7 African countries. In the two age groups of 6–12 weeks and 5–17 months, vaccine efficacy over 18 months was 26.6% and 45.7%, respectively. The joint technical expert group (JTEG) on malaria vaccines did not recommend the use of RTS,S in the younger age group and recommended for the older age group testing the efficacy of 4 doses of the vaccine in 3–5 distinct epidemiological settings in sub-Saharan Africa.

by host genetic restriction and the diversity and polymorphism of parasite targets. It is only recently, however, that LAP have been considered as vaccine for humans because of their increasingly recognized power and the suboptimal achievements of subunit vaccines.

RADIATION-ATTENUATED SPOROZOITES

Radiation-Attenuated Sporozoites in Rodents

The first success in vaccinating against PE malarial stages was achieved in the early 1940s using the avian model of *Plasmodium gallinaceum* and *Aedes albopictus*. Immunization of fowl by intravenous (IV) injections of sporozoites inactivated by ultraviolet light (Mulligan et al., 1941; Russell et al., 1941; Russell and Mohan, 1942) or killed by dehydration, formalin or freezing and thawing (Richards, 1966) induced protection against a challenge by mosquito bites. Protection was partial, however, as it decreased mortality rates but did not prevent patent, albeit low-grade, infection. The studies also revealed that vaccination by sporozoites confers resistance to a challenge by sporozoites but not blood stages of the homologous parasite. The concept was then extended to rodents, which became the leading animal model during the 1960s. Following studies on blood-stage parasites that had shown that immunization with X-irradiated blood forms of *P. gallinaceum* (Ceithaml and Evans, 1946) or *Plasmodium berghei* (Wellde and Sadun, 1967) induced partial resistance to blood-stage infection, X-radiation-attenuated sporozoites (RAS) were used as immunogen: IV injection of 75,000 *P. berghei* RAS into mice protected 2 weeks later against an IV challenge with 1000 sporozoites but not against homologous blood-stage parasites (Nussenzweig et al., 1967). Only 37% of the vaccinated animals had a detectable blood infection as compared with 90% of their controls, a first case of sterile immunity.

Since then, the RAS model in rodents has been extensively investigated to decipher the basis of protective responses against *Plasmodium* PE stages. In turn, it was hoped that RAS research would facilitate the identification of the parasite targets of protective responses and therefore guide subunit vaccine development that started in the 1980s with the advent of molecular biology. Below, we summarize some of the main lessons learned by studying RAS in rodents.

Importance of Host–Parasite Adaptation

The parasite is usually fine tuned to a single or a few closely related hosts, where it does not cause excess pathology or trigger effective responses. The natural hosts of *P. berghei* and *Plasmodium yoelii* are the tree rats *Grammomys surdaster* and *Thamnomys rutilans*, respectively, where infection tends to become chronic with persistent, low-grade parasitemia, unlike infection in mice that typically leads to death (Druilhe and Barnwell, 2007). In rodents, RAS-induced protection is inversely correlated with host susceptibility to sporozoite infection, i.e., the natural host is more difficult to protect than an experimental host (Druilhe et al., 2002). For example, using *P. berghei* RAS, a single immunization with 1000 sporozoites is enough to protect BALB/c inbred mice, three immunizations with 30,000 sporozoites are required to protect C57BL/6 mice (Ngonseu et al., 1998), and five immunizations with 30,000 RAS only protect some of the immunized *Grammomys* (Chatterjee et al., 2001). It can be assumed that parasite antigens targeted by a protective response in an unnatural host may not be targets of the protective response in the natural host.

Irradiated Liver Stage as Primary Immunogen

The irradiated sporozoite itself was initially thought to act as both inducer and target of the protective immune responses. Indeed, early emphasis was placed on antibodies to the CS protein as being sufficient to neutralize sporozoite infectivity *in vivo* (Potocnjak et al., 1980; Yoshida et al., 1980). However, it rapidly became evident that RAS-derived liver stages were playing a major role. Heat-killed sporozoites were not protective (Alger and Harant, 1976) and irradiated sporozoites normally invaded hepatocytes *in vitro* (Sigler et al., 1984). In rodents, the irradiated liver stage develops inside a parasitophorous vacuole (PV) into a young uninucleate trophozoite until the ~24th hour (Fig. 11.2); after 24 h, the wild type starts multiplying its genome, enlarges and reaches full maturity at ~60 h (when tens of thousands of merozoites are ready to egress), while the irradiated parasite remains arrested at the uninucleate stage, or displays limited nuclear divisions (Suhrbier et al., 1990; Chatterjee et al., 1996). Direct evidence that the irradiated liver stage was an important immunogen came when assessing the effect of the sporozoite irradiation on protection: over-irradiated sporozoites confer weaker or no protection both in mice and humans (Mellouk et al., 1990) and are blocked earlier during development than the optimally irradiated parasites, i.e., remain strictly uninuclear (Chatterjee et al., 1996; Silvie et al., 2002). Crucially, the irradiated liver stage arrests its development at the time when the parasite switches from expression from PE (sporozoite/liver stage-specific) to erythrocytic antigens.

Effectors of Protection

It was rapidly observed that the serum from RAS-immunized mice neutralized sporozoites (Nussenzweig et al., 1969) and blocked their motility (Vanderberg, 1974). Subsequently, a monoclonal antibody (3D11) recognizing the *P. berghei* CS protein via its repeat region was isolated and shown to abolish sporozoite infectivity (Yoshida et al., 1980), to be sufficient to fully protect animals against a challenge with 1000 *P. berghei* sporozoites (Potocnjak et al., 1980), to inhibit sporozoite invasion of cultured cells (Hollingdale et al., 1982, 1984), and to block sporozoite motility *in vitro* (Stewart et al., 1986). Moreover, intravital imaging has shown that sporozoites injected into the skin of RAS-immunized

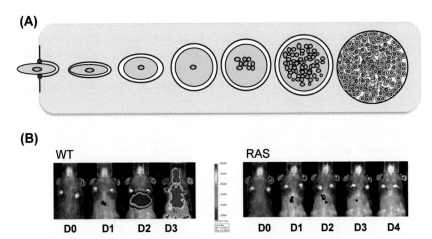

FIGURE 11.2 Radiation-attenuated sporozoites (RAS)–mediated protection. (A) The infectious sporozoite invades through a junction (*red dots*) inside a parasitophorous vacuole (PV), limited by a membrane (*blue line*), where it transforms into a spherical liver stage (LS). The LS progressively enlarges and undergoes schizogony, i.e., karyokinesis (formation of up to 30,000 nuclei in ~30 h in rodents) without cytokinesis followed by the budding off of uninucleate daughter merozoites inside the PV. Merozoites egress hepatocytes ~55–60 h after sporozoite invasion in rodents. RAS arrest their development when DNA replication starts, i.e., midway through development at ~20/24 h in rodents, although depending on the irradiation dose some nuclear division might take place. (B) Protection mediated by immunization with RAS in the most stringent *P. berghei*/C57BL6 mouse model. Naïve mice and mice immunized with injections of *P. berghei* RAS were challenged in the ear skin with *P. berghei* infectious sporozoites expressing luciferase, allowing parasite tracking in the whole body and quantification of growth. A signal of similar intensity is seen in the liver of naïve and immunized mice at day 1 (D1); however, blood-stage infection arises at D3 in naïve mice while the liver signal is abolished at D4 in RAS-immunized mice. *WT*, wild type.

mice moved normally only for about 1 min before being completely immobilized (Vanderberg and Frevert, 2004). Even more striking, anti-CS antibodies prevent the release of sporozoites from the mosquito proboscis into the skin and thus decrease the number of sporozoites ejected during the bite, on the formation of immune complexes at the tips of mosquito proboscis (Kebaier et al., 2009). Anti-CS repeats antibodies appear to be the main effectors induced by RTS,S vaccination, although their titers tend to rapidly wane with time.

CD8+ T cells were first shown to be important in RAS-mediated protection in the late 1980s with initial depletion and adoptive-transfer experiments in mice (Schofield et al., 1987; Weiss et al., 1988). CD8+ T cells were shown to recognize epitopes from the *P. berghei* CS (Romero et al., 1989) or *P. yoelii* CS (Weiss et al., 1990) and CS-specific CD8+ T cell clones were found potentially sufficient for conferring complete protection against challenge (Romero et al., 1989). To eliminate parasite liver stages, CD8+ T cells use classical effector mechanisms of cytotoxic T lymphocytes such as secretion of perforin and Fas ligand and proinflammatory cytokines, such as interferon (IFN)-γ and tumor necrosis factor (TNF)-α, which upregulate inducible nitric oxide synthase in hepatocytes (Ferreira et al., 1986). However, none of these effectors is individually essential, including IFN-γ (Tsuji et al., 1995), and their relative importance varies with the mouse strain, the method of immunization (Doolan and Hoffman, 2000), and the *Plasmodium* species (Butler et al., 2010). Other cell types are known to contribute to anti–liver-stage immunity, including cytotoxic CD4+ T cells that recognize CS (Rénia et al., 1991) or non-CS (Tsuji et al., 1990) epitopes, $\gamma\delta$ T cells (Tsuji et al., 1994), and natural killer (NK) T cells (Pied et al., 2000; Gonzalez-Aseguinolaza et al., 2000).

Since CD8+ T cells must localize and destroy a "needle in a haystack" (roughly 1 in 10^6 hepatocytes in mice and 1 in 10^9 hepatocytes in humans) in a short period (less than 48 h in rodents and 7 days in humans), the numbers of parasite-specific CD8+ T cells directly impact protection levels. A threshold of parasite-specific memory CD8+ T cells appears to be required for protection, which depends on the parasite species and the mouse genetic background (Schmidt et al., 2011). Successful RAS vaccination of mice correlates with the induction of parasite-specific CD8+ T cell numbers that range between 4% and 20% of the entire CD8+ T cell compartment (Schmidt et al., 2010), up to 100-fold the numbers of memory CD8+ T cells typically required to protect against bacterial and viral challenges (Schmidt et al., 2008).

Parasite Targets of Protective Responses

Despite intensive search, few proteins other than CS have been identified as targets of protective responses. CS appears to act as an immune-dominant antigen, in part because of its abundance in sporozoites and constant shedding during extracellular motility and cell traversal, which may hinder responses to other protective targets. The demonstration that transgenic mice rendered tolerant to CS can still develop sterile, CD8+ T cell-mediated protection after immunization with RAS (Kumar et al., 2006) shows that protective antigens other than CS do exist. Additionally, a study found that CS-specific CD8+ T cells accounted for only ~10% of the total activated T cell population following RAS immunization (Schmidt et al., 2010). To date, however, few non-CS proteins are known to provide epitopes for protective CD8+ T cells, which include TRAP (Hafalla et al., 2013) and the merozoite surface protein 1 (MSP-1) (Draper et al., 2009).

Radiation-Attenuated Sporozoites in Humans

RAS were shown to protect humans in the early 1970s. Since it was not considered ethical to inject intravenously into humans sporozoites collected from mosquito salivary glands, *P. falciparum* RAS were inoculated by the bites of hundreds of anopheline mosquitoes. In a series of studies conducted at the University of Maryland (Clyde et al., 1973, 1975; Clyde, 1990), three adult volunteers were immunized by an average of ~400 bites from *P. falciparum*-infected *Anopheles stephensi* mosquitoes over a 10-week period and challenged

4 weeks later with homologous sporozoites delivered by ~10 mosquitoes; one was protected. Another volunteer was protected against *P. vivax* (Clyde et al., 1973). A similar study was undertaken with volunteers at the Stateville Correctional Center (IL, USA), where three volunteers were immunized by exposure to 500 to 1000 X-irradiated *P. falciparum*-infected mosquitoes on six to eight occasions over 3 to 9 month-long immunization protocol; the three were protected for 2 months but not at 4 months (Rieckmann et al., 1974; Rieckmann, 1990). From 1989 to 1999, new RAS vaccine studies were performed at the Walter Reed National Military Medical Center and the University of Maryland. Fourteen volunteers were immunized by 1000–2000 bites by irradiated *P. falciparum*-infected *A. stephensi* mosquitoes, spread in ~10 feeding sessions over a ~10 month-period (Hoffman et al., 2002), and 13 were challenged 2–9 weeks after the last immunization by the bites of five *P. falciparum*-infected mosquitoes (shown to infect ~100% of humans; Hoffman, 1997); all were protected.

Therefore, RAS vaccination in humans, although building on a small number of cases during the 1973–99 period, affords a 90% rate of sterile protection a few weeks after the last immunization (Hoffman et al., 2002), albeit with highly demanding and logistically complex immunization regimens. Still, these data strengthened the notion that whole parasites are potentially powerful immunogens and, in 2003, Steve Hoffman founded the biotechnology company called Sanaria with the initial goal of manufacturing a vaccine based on cryopreserved, aseptic, purified, X-irradiated NF54 *P. falciparum* (Pf) sporozoites, called PfSPZ (Luke and Hoffman, 2003). The first doses of the PfSPZ vaccine were produced in 2010 (Hoffman et al., 2010).

Three clinical trials of the PfSPZ vaccine have been conducted in malaria naïve volunteers in the United States. In the first dose-escalation trial (Epstein et al., 2011), the PfSPZ vaccine was administered to 80 malaria-naïve adults intradermally (ID) or subcutaneously (SC). Volunteers were immunized with 4–6 doses of 7500–135,000 PfSPZ during a 3–7 month schedule and challenged 3 weeks after the last immunization by exposure to the bites of 5 *P. falciparum*-infected *A. stephensi* mosquitoes. Only two volunteers were protected, having received four injections of 30,000 PfSPZ. In the second trial (Seder et al., 2013), PfSPZ were injected intravenously. While 4–6 doses of 30,000 PfSPZ did not protect against a challenge by the bites of 5 mosquitoes 3 weeks after the last immunization, 5 doses of 135,000 PfSPZ protected all (6/6) volunteers against a similar challenge. However, of the six volunteers protected at 3 weeks, four were not protected on rechallenge at 21 weeks (Ishizuka et al., 2016). In the third trial (Ishizuka et al., 2016), using four IV injections of 270,000 PfSPZ per dose, 6/11 immunized volunteers were protected at 21 weeks after the last immunization and 5/5 of those protected at 21 weeks were still protected at 59 weeks. This constitutes the first high-level protection induced at 1 year against a *P. falciparum* homologous challenge by an injectable vaccine. However, when decreasing immunization to three IV injections of 270,000 PfSPZ, only 3/9 volunteers were protected at 3 weeks. These human trials also showed a dose-dependent increase in the frequency of PfSPZ-specific CD4$^+$, CD8$^+$, and γδ T cells in the peripheral blood after immunization and suggested that IV immunization was essential to obtain high numbers of IFN-γ-producing lymphocytes in the liver.

MORE IMMUNOGENIC THAN RAS: DAP AND GAP

In recent years, different immunization approaches based on whole parasites have been proposed, which arrest parasite development in the liver by chemotherapy (drug-arrested parasites, DAP) or as a consequence of a genetic manipulation (genetically attenuated parasites, GAP). By their versatility in allowing arrest at different steps of parasite development, DAP and GAP not only provide new vaccine tools but are also valuable probes for understanding protection against liver stages. They have now demonstrated the fact that the later the developmental block of the immunizing parasite in the liver, the stronger the induced protection.

Drug-Arrested Parasites

Many drugs that impair parasite development in the liver and beyond have been used in so-called chemoprophylaxis with sporozoites (CPS) protocols (Table 11.1). They typically confer protection against sporozoite challenge and, to various extents, against blood-stage challenge (cross-stage protection) depending on the stage of developmental arrest and the animal model. The gold-standard CPS protocol uses chloroquine (CQ-CPS) (Beaudoin et al., 1977; Golenser et al., 1977; Belnoue et al., 2004; Doll et al., 2014), which has no activity against sporozoites and liver stages but eliminates blood-stage parasites during the first cycle(s) of multiplication inside RBC. Two comparisons are particularly informative. First, CQ-CPS protects better than RAS and associates with higher and larger parasite-specific CD8$^+$ T cell responses. Importantly, protection is due only to PE immunity, not to the erythrocytic immunity that is elicited by the limited replication of blood-stage parasites afforded by CQ treatment (Pombo et al., 2002), i.e., the stronger protection conferred by CQ-CPS is only due to the full parasite development in the liver. Second, the use of azithromycin or clindamycin, antibiotics that inhibit the replication of the liver-stage apicoplast and result in the formation of nonreplicative hepatic merozoites, affords stronger and more durable protection than CQ-CPS (Friesen et al., 2010; Friesen and Matuschewski, 2011). This appears to be due to the low-grade, transient parasitemia that occurs during CQ-CPS, but not azithromycin/clindamycin-CPS, and the known suppressive effect of blood-stage multiplication on PE immunity (Orjih and Nussenzweig, 1979; Ocaña-Morgner et al., 2003). Therefore, maximal liver-stage development and minimal blood-stage development are considered to provide the strongest protection against liver stages (Nganou-Makamdop and Sauerwein, 2013).

In humans, the efficacy of CQ-CPS was investigated in a clinical trial in 2009: 10 volunteers were immunized three times by the bite of 12–15

394 Skin and Arthropod Vectors

TABLE 11.1 Drug-Arrested Parasites—Chemoprophylaxis With Sporozoites

Compound	Mode of Action	Liver Stage (LS) dev[a]	Blood Stage (BS) dev[b]	Protection	References
Antifolate					
Pyrimethamine	Inhibits DNA replication	Mid-arrest	–	+	Friesen et al. (2011)
Antibiotics					
Azithromycin[c]	Inhibits inheritance of LS apicoplast[d]	Full dev[e]	–	+++	Friesen et al. (2010)
Clindamycin[f]		Full dev	–	+++	
Quinolines					
Chloroquine	Inhibits heme detoxification	Full dev	±	++	See text
Mefloquine		Full dev	±	++	Inoue and Culleton (2011)
Piperaquine		Full dev	±	++	Pfeil et al. (2014)

[a]Liver-stage development.
[b]Blood-stage development.
[c]Macrolide.
[d]Causing the so-called delayed death phenotype, where parasite death is initiated during the first/second cycle of blood-stage parasite replication.
[e]Full development.
[f]Lincosamide.

P. falciparum-infected mosquitoes while under CQ cover (Roestenberg et al., 2009), and all 10 were protected against a challenge by five mosquito bites 8 weeks after the final immunization. Strikingly, four of six volunteers protected at 8 weeks remained fully protected when rechallenged after 2 years and displayed parasite-specific multifunctional effector memory T cells producing IFN-γ, TNFα, and IL-2 (Roestenberg et al., 2011). Protection appears to be targeted almost exclusively at the liver stage, as volunteers having undergone CQ-CPS show no evidence of cross-stage protection (Bijker et al., 2013). Similar results were obtained using mefloquine, a drug with similar activity as CQ (Bijker et al., 2014). In humans, CQ/mefloquine-CPS is estimated to be at least 20 times more efficient than RAS immunization in conferring PE protection (Bijker et al., 2014). It should be said, however, that the antimalarial CQ also enhances cross-presentation of soluble antigens to CD8+ T cells.

Genetically Attenuated Parasites

GAP are engineered in blood stages by homologous recombination and lack proteins that have no role in the blood and mosquito phases of the parasite life cycle but a vital role during liver-stage development. The first GAP was generated in *P. berghei* and lacked UIS3, a liver-stage-specific protein that associates with the membrane of the PV (PVM), where the parasite resides (Mueller et al., 2005b). Many other GAP followed that are presented in Table 11.2, divided into four groups depending on the time of developmental arrest. The groups of very early- and early-blocked GAP lack proteins that are involved in PVM formation/maintenance (P36, P52, B9, SLARP/SAP1) and function (UIS3, UIS4), respectively. They do not develop beyond the stage of RAS arrest and induce protection that is not stronger than that conferred by RAS. In contrast, the late-blocked GAP, such as mutants lacking the apicoplast-resident type II fatty acid synthesis (FAS II) pathway (Yu et al., 2008; Vaughan et al., 2009), and the very late-blocked GAP, such as the *Plasmodium*-specific apicoplast protein important for liver merozoite formation (PALM)-deficient mutant (Haussig et al., 2011), are blocked later than RAS and confer stronger protection. The late-blocked FAS II mutants are arrested before 44 h (before merozoite formation) and do not express MSP-1, an early erythrocytic stage antigen (Vaughan et al., 2009). The FABB/F− GAP induces after IV immunization superior protection in inbred and outbred mice and broader CD8+ T cell responses than RAS, as well as some degree of cross-stage and cross-species (*P. berghei*) protection (Butler et al., 2011), indicating that a larger set of antigens is recognized by FABB/F−- than RAS-mediated responses. The very late-blocked, PALM− GAP is impaired in merozoite segregation and does express MSP-1. It confers, in the most stringent *P. berghei*-C57BL/6 model, sterilizing immunity in 100% of the animals after only two doses (Haussig et al., 2011), which cannot be achieved with RAS.

The first and only phase I clinical trial of a GAP was conducted in 2008 with the *P. falciparum* P56−P32− double-mutant, very early-blocked GAP; out of the six volunteers immunized, one suffered a breakthrough infection on receiving

TABLE 11.2 Protective Genetically Attenuated Parasites

Protein	Species	Phenotype	References
Very Early			
P52[a] (P36p)	Plasmodium berghei	Increased apoptosis of infected hepatocytes after 6 h	van Dijk et al. (2005)
P36[a]+P52	Plasmodium yoelii	No PVM at 1 h—no LS[b] in vivo at 6 h	Labaied et al. (2007)
P36+P52	Plasmodium falciparum	LS rarely detected at D3 in HC-04 cell line	VanBuskirk et al. (2009)
SLARP[c]	P. berghei	LS blocked earlier than UIS—final size as WT at 12–18 h	Silvie et al. (2008)
SAP1[c]	P. yoelii	Decreased expression of UIS3/4- and P52-encoding genes	Aly et al. (2008, 2011)
B9[a]	P. berghei	No LS at 24 h—final size as WT at 5–10 h	Annoura et al. (2014)
B9	P. falciparum	No LS in PH[d] at 48 h	Annoura et al. (2014)
SLARP	P. falciparum	10 times fewer LS at 48 h in PH—LS not detected at D3	van Schaijk et al. (2014)
B9	P. falciparum	Arrest before D2 in PH	van Schaijk et al. (2014)
Early			
UIS3[e]	P. berghei	Reduced proportion of smaller LS at 24 h	Mueller et al. (2005b)
UIS4[e]	P. berghei	50% fewer LS at 24 h—no LS observed after 24 h	Mueller et al. (2005a)
Late			
FAS II[f]			
FABI	P. berghei	No nuclear division—No cytomere formation—merozoite surface protein 1 (MSP-1) (−)	Yu et al. (2008)
FABB/F	P. yoelii	Smaller LS at 44 h—no cytomere formation—MSP-1 (−)	Vaughan et al. (2009)

TABLE 11.2 Protective Genetically Attenuated Parasites—cont'd

Protein	Species	Phenotype	References
PDH[g]	*P. yoelii*	Smaller LS at 44 h—no cytomere formation—MSP-1 (−)	Pei et al. (2010)
Very Late			
PALM[h]	*P. berghei*	Normal LS staining at 48 h—merozoites not segregated—MSP-1 (+)	Haussig et al. (2011)
CGK[i]	*P. berghei*	Normal LS at 65 h—lack of merozoite egress	Falae et al. (2010)

PVM, membrane of the parasitophorous vacuole.
[a]*Member of the 6-cysteine Plasmodium-specific family of surface proteins.*
[b]*LS, liver stage.*
[c]*Posttranscriptional regulators: P. berghei SLARP (sporozoite and liver-stage asparagine-rich protein) and its ortholog in P. yoelli SAP1 (sporozoite asparagine-rich protein 1).*
[d]*PH, primary hepatocytes.*
[e]*UIS, upregulated in infective sporozoites.*
[f]*Fatty acid synthesis (FAS) II pathway of prokaryotic origin that localizes to the apicoplast.*
[g]*Pyruvate dehydrogenase provides acetyl-CoA for the FAS II pathway.*
[h]*Plasmodium-specific apicoplast protein important for liver merozoite formation.*
[i]*c-GMP-dependent protein kinase.*

~200 bites by GAP-infected mosquitoes (Spring et al., 2013), which underscores the difficulty of completely inactivating parasite infectivity with targeted genetic deletions. Another double-knockout GAP, B9⁻SLARP⁻, has been manufactured by Sanaria Inc. as an aseptic, purified, cryopreserved product and will move into clinical trials (Van Schaijk et al., 2014). A triple-knockout GAP, P36⁻P52⁻SAP1⁻, reported as sufficiently attenuated in cultured hepatocytes and mice engrafted with human hepatocytes, will likely also undergo clinical testing (Mikolajczak et al., 2014).

FOCUS ON SKIN IMMUNIZATION

So far, most vaccine studies in rodents based on whole organisms have used IV immunization to maximize protection, and in humans the PfSPZ vaccine is currently undergoing phase IIb trials using IV immunizations. As mentioned, the PfSPZ vaccine is ~10 times more protective when injected IV (Seder et al., 2013) compared with SC or ID (Epstein et al., 2011). In rodents too, immunization into the skin is less effective than IV (Spitalny and Nussenzweig, 1972; Nganou-Makamdop et al., 2012), although some studies reported similar protection levels conferred by ID and IV routes using *P. yoelii* DAP (Inoue and Culleton, 2011) and RAS (Voza et al., 2010). Importantly, when inoculated into the skin, sporozoites generate a smaller parasite load in the liver: ID inoculation was reported to yield a 10-fold (Inoue and Culleton, 2011), 30-fold (Nganou-Makamdop et al., 2012), or 50-fold (Ploemen et al., 2013) reduced parasite load in the liver at 40 h compared with IV injection, depending on the mouse/parasite system and the injection technique. These data have led to the conclusions that protective efficacy depends on the number of immunizing sporozoites reaching the liver and that protective responses need to be mounted in the liver.

Nonetheless, the immune responses to injection of live parasites into the skin have received increased attention in recent years, in parallel with studies on the behavior and fate of sporozoites inoculated into the skin detailed in the previous chapter. After mosquito delivery, ~60% of the deposited sporozoites stay in the skin, where ~10% invade and develop inside skin cells, and ~15% actively reach the proximal draining lymph node, i.e., only the remaining 25% reach the blood and the liver. Injection of sporozoites through a syringe decreases the efficiency of sporozoite egress from the skin by both blood and lymph routes, i.e., an even greater proportion of inoculated sporozoites remain in the skin.

Early Response to Sporozoite Inoculation Into the Skin

The skin immune response to the bites of anopheline mosquitoes, delivering or not *Plasmodium* sporozoites, is addressed elsewhere in this book (Chapter 4). Few studies have focused on the host response to the inoculation of large, immunizing doses of sporozoites into the skin by parenteral injection. In these conditions, sporozoites induce a massive but transient recruitment of poly-morphonuclear neutrophils (PMN) at the skin site and proximal skin-draining lymph node (sdLN), peaking at 24 h post inoculation, followed after 24 h by an infiltration of inflammatory monocytes in the skin and sdLN (Mac-Daniel et al., 2014). Flow cytometry showed that skin sporozoites rapidly associate with PMN and resident myeloid cells in the skin (Mac-Daniel et al., 2014), in line with intravital imaging that revealed an early PMN influx, starting ~20 min after inoculation, at the skin site (Amino et al., 2008). Nonetheless, neutrophil depletion prior to infection does not impair normal sporozoite fate and the establishment of a protective immune response, which likely reflects the sporozoite capacity to move swiftly in the skin and to traverse and inactivate phagocytic cells in the process.

Adaptive Immune Response in the Skin-Draining Lymph Node

Th1 Response

Among the many immune genes that are upregulated by injection of extracts from mosquito salivary glands into the skin of mice, those encoding granzyme B, CXCL9, CXCL10, and IFN-γ are specifically upregulated in the presence of *P. berghei* sporozoites (Mac-Daniel et al., 2014). The four corresponding molecules were found in increased amounts in the sdLN, composing a cyto-kine profile that indicates an early NK cell activation (granzyme B, IFN-(() and the establishment of Th1 immune responses (IFN-(, CXCL9, and CXCL10). Importantly, this Th1 immune response to immunizing doses of sporozoites differs from the response to sporozoite injection by mosquito bites. In this latter case, *Anopheles* saliva increases the levels of immunosuppressive IL-10 in the sdLN, which was shown to contribute to the downregulation of Ag (OVA)-specific T cell priming (Depinay et al., 2006).

CD8+ T Cell Priming in the Skin-Draining Lymph Node

The development of a protective CD8+ T cell response after RAS delivery into the skin has been examined using *P. yoelii* RAS and transgenic CD8+ T cells expressing a T cell receptor (TCR) specific for an MHC class I-restricted epitope of the *P. yoelii* CS protein (CS-TCR Tg CD8+ T cells; Sano et al., 2001). After RAS injection by ID microinjection or by mosquito bite into the ear skin, CS-TCR Tg CD8+ T cells were shown to be primed in the sdLN (Chakravarty et al., 2007). Two days after immunization, IFN-γ-producing CD8+ T cells were found only in the sdLN, and to a lesser extent in the spleen, but were minimally activated in the liver-dLN. Once activated, CD8+ T cells secondarily migrated to the liver and lymphoid organs, in a sphingosine-1-phosphate–dependent fashion (Chakravarty et al., 2007).

Role of Dendritic Cells and Importance of Cross-Presentation

Dendritic cells (DC) are known to play a central role in the priming of *Plasmodium*-specific CD8+ T cells, as was first shown by the inability of DC-depleted mice to mount CD8+ T cell responses to *P. yoelii* infection (Jung et al., 2002). Antigen cross-presentation by DC to naïve CD8+ T cells is required for inducing protective T cell responses (Cockburn et al., 2011). Indeed, CD8+ T cell priming is severely reduced in mice defective in endosomal function and in mice depleted of cross-presenting DC (Cockburn et al., 2011), as well as in mice lacking TAP1 (transporter associated with antigen processing 1) that transports peptides generated by the proteasome to the endoplasmic reticulum (ER). Therefore, CS is cross-presented via an endosome-to-cytosol pathway: after translocation into the DC cytosol, CS is processed into peptides that are transported by TAP into the ER, where they bind nascent MHC class I molecules. Studies using bone marrow chimeras also formally demonstrated that the effector function of CD8+ T cells no longer required antigen presentation by cells of the hematopoietic lineage; instead, CD8+ T effector cells directly recognize antigen presented on parenchymal cells, i.e., parasitized hepatocytes (Chakravarty et al., 2007; Cockburn et al., 2011). In contrast to DC, parasitized hepatocytes only present antigens that are secreted by the parasite into the cytosol across the enclosing PVM. Clearly, therefore, RAS might not induce responses to important liver-stage protective antigens and induce responses to antigens that are not expressed by infected hepatocytes.

Priming of Immune Response in the Skin-Draining Lymph Node Sufficient for Sterile Protection

The importance of T cell priming in the sdLN for protection was first shown by experiments in which the sdLN was surgically ablated prior to the transfer of CS-TCR Tg CD8 cells and subsequent immunization or when T cell egress from lymph nodes was prevented (Chakravarty et al., 2007); both conditions severely decreased the number of T cells reaching the liver and the protective capacity of CD8$^+$ T cells after a challenge 10 days later. A subsequent work demonstrated that sterile immunity could be induced by *P. yoelii* RAS vaccination into the skin in the absence of any (detectable) contribution of liver-stage immunogen (Obeid et al., 2013). This was achieved by RAS injection SC, which, unlike the ID route, did not usually cause liver and blood-stage infection in the author's conditions, or by RAS injection SC or ID into CD81-deficient mice, where hepatocytes are refractory to *P. yoelii* sporozoite invasion inside a PV and thus do not permit proper RAS maturation (Silvie et al., 2003). In both cases, immunization led to sterile protection (Obeid et al., 2013). Therefore, arrested parasites in the liver do not appear to be mandatory to achieve sterile immunity, and sterilizing immune responses can be mounted solely in the sdLN.

Which Parasite Antigens in the Skin-Draining Lymph Node?

A crucial question is the nature of the antigens presented in the sdLN. Indeed, the set of antigens acquired by antigen-presenting cells in the sdLN will determine the breadth of protection affordable after skin immunization, i.e., against sporozoites only or also against developing liver stages. As depicted in Fig. 11.3, a primary antigen source is represented by motile sporozoites reaching the sdLN via their own motility (Amino et al., 2006), where they could be phagocytosed by sdLN-resident DC. Other antigen sources are various forms of parasite material left in the skin, which may secondarily traffic to the sdLN via skin-resident migratory DC (Langerhans cells or dermal DC) that either directly present antigen or transfer it to sdLN-resident DC. Skin parasite material may be proteins shed during sporozoite motility and cell traversal, or extracellular sporozoites left dead in the skin—although killed sporozoites are known to be insufficient for triggering CD8$^+$ T cell responses. Another possible antigenic source is represented by parasites at various stages of development inside skin cells. Using both *P. berghei* and *P. yoelii*, development inside skin cells can be complete (Gueirard et al., 2010; Voza et al., 2012) and may thus, at least in theory, deliver the full spectrum of "liver-stage antigens" to the sdLN.

One study suggests that the first possibility takes place, i.e., motile sporozoites reach the sdLN by their own motility and are taken up by sdLN-resident CD8α$^+$ DC, which in turn activate CD8$^+$ T cells (Radtke et al., 2015). Dynamic and static imaging in the sdLN revealed the formation of DC-CD8$^+$ T cell clusters, which involved at early time points (8–16 h after sporozoite inoculation) CD8α$^+$ but not CD11b$^+$ DC. Furthermore, less CD8$^+$ T cell priming was

Skin

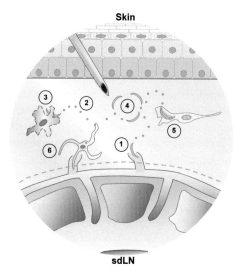

sdLN

FIGURE 11.3   **Antigenic sources after sporozoite injection into the skin.** The various types of immunogens that may get access to the skin-draining lymph node (sdLN) are indicated: sporozoites actively reaching the sdLN after invasion of a skin lymphatic vessel (1), parasite proteins released in the skin during gliding motility (2) or cell traversal (3), extracellular sporozoites left in the skin (4), and parasites having invaded/developed inside skin cells (5). Nonmotile parasite material (2, 3, 4, 5) is secondarily transferred to the sdLN via skin migratory phagocytic cells (6).

observed in mice lacking CD169+ macrophages, which appear to play an indirect role, perhaps via the capture and transfer of antigen to cross-presenting DC.

PERSPECTIVES

LAP-based vaccines against PE malaria have shown their potentially superior immunogenicity compared to available subunit vaccines, particularly when liver-stage development is allowed beyond the stage of RAS arrest. Consequently, LAP research has rapidly grown and LAP candidates are lining up for clinical testing. Nevertheless, implementation of any LAP approach still faces important hurdles, scientific and technical.

A major unanswered question is whether LAP can confer strain-transcendent immunity, not merely homologous protection against the immunizing strain. Likewise, the effect of preexisting immunity or of a subpatent blood infection on the establishment of PE immunity remains unknown and will likely be clarified by the PfSPZ phase IIb trials currently under way in adults in Africa, Europe, and the United States. An inherent limitation of LAP vaccines is parasite cryopreservation. Besides the constraints of the cold chain requiring ultralow temperature in liquid nitrogen, the current cryopreservation technique severely impairs sporozoite infectivity; cryopreserved *P. yoelii* sporozoites were reported to be ~70% (Epstein et al., 2011) and up to 86% (Ploemen et al., 2013) less infective than their freshly dissected counterparts. It is likely that the drop in infectivity caused by the cryopreservation has a greater impact on sporozoite infection from the skin, which requires extensive motility for gaining access to the blood, compared to IV, which offers a potentially rapid path to hepatocytes.

Although GAP appear as an attractive approach, in offering well-defined and homogeneous genetic manipulations without the need for a concomitant drug treatment, engineering the right GAP is not straightforward. Breakthrough infection is a primary concern; safe GAP will need to bear mutations in several genes acting in independent functional pathways and will need to undergo extensive preclinical screening in mice (Annoura et al., 2012). Furthermore, the choice of the right genes to inactivate may prove difficult. For example, the late-blocked *P. yoelii* fabb/f⁻ GAP is blocked only in the liver, but the *P. falciparum* fabb/f⁻ GAP is blocked first in the mosquito, during multiplication in the oocyst, i.e., the phenotype of murine GAP may not always accurately predict that of the homologous human GAP. Last, constructing the most protective late- or very late-blocked GAP poses a genetic problem: target genes are likely to be important for parasite multiplication in RBC, in which case the corresponding GAP cannot be created in blood-stage parasites in the first place.

Possibly the most critical issue is the route of LAP administration. The IV route can be used for therapeutic interventions in humans but is not approved for administration of any preventive vaccine in public health interventions. It is also impractical in young children and infants, the primary recipients of malaria vaccination. The IV route may thus be appropriate only for individuals in the developed world, e.g., military personnel and travelers. Studies on the skin route of immunization in mice suggest that it may be sufficient for mounting sterilizing responses. However, its efficacy will depend on the LAP ability to produce liver-stage antigens in the skin and/or the sdLN, making the occurrence of parasite development in the skin of humans a critical question. If immunization into the skin leads to the presentation in the sdLN of sporozoite antigens only, then skin injection of late- or very late-blocked GAP will not provide superior protection. If this is the case, it may still be possible to engineer GAP that express liver-stage antigens under the control of sporozoite promoters, so that the delivery of the highly protective antigens does not require GAP development.

LYME VACCINE: *BORRELIA* AND TICK AS TARGETS TO IDENTIFY VACCINE CANDIDATES AGAINST LYME BORRELIOSIS?

Ema Helezen and Nathalie Boulanger
Université de Strasbourg, Strasbourg, France

Lyme disease represents more than 90% of all vector-borne diseases in the United States and the annual count increased by 101% between 1992 and 2006 with yearly incidences in Western world >100 cases/100,000 residents (Embers and Narasimhan, 2013; Radolf et al., 2012; Schuijt et al., 2011b). This bacterial infection is transmitted by hard ticks of the *Ixodes* complex, including (by geographic region of appearance): *Ixodes scapularis* (eastern North America), *Ixodes pacificus* (western North America), *Ixodes ricinus* (Europe), and *Ixodes persulcatus* (Asia) (Steere et al., 2016). The enzootic cycle starts with uninfected tick larvae feeding on infected reservoir host, mainly rodents and birds (Humair and Gern, 2000). Larvae then molt to the nymph stage on ground, and these nymphs transmit pathogen to the animals when taking their next blood meal. *Ixodes* spp. can feed on a large variety of vertebrate species, which explains their geographic dispersion over the Northern Hemisphere. Humans are dead-end hosts, and infection of humans occurs rather incidentally and mainly by nymphs (Radolf et al., 2012).

This disease is mainly caused by three genospecies of spirochetes: *Borrelia burgdorferi* sensu stricto (ss) (USA, Euroasia), *Borrelia afzelii*, and *Borrelia garinii* (Euroasia). Clinical manifestations are different between the United States and Europe, and also between different regions in Europe; *B. burgdorferi* ss induces more articular manifestations, whereas *B. afzelii* is primarily responsible for cutaneous manifestations and *B. garinii* mainly provokes neurological manifestations (Aguero-Rosenfeld et al., 2005; Müllegger and Glatz, 2008; Steere et al., 2016). *Borrelia* has a highly complex genome. It consists of a linear chromosome with housekeeping genes of 900 kb and more than 21 different linear and circular plasmids encoding most of the outer surface lipoproteins (Fraser et al., 1997; Rosa et al., 2005). Some plasmids are instable and easily lost *in vitro*, but they are essential for the *in vivo* virulence. It explains that *in vitro* culture must be limited to maintain *Borrelia* infectivity. In addition, *Borrelia* metabolism is very limited due to its adaptation and dependence to its host.

Spirochetes possess a characteristic morphology, with inner and outer membranes surrounding periplasmic flagella giving the bacteria a particular mobility (Rosa et al., 2005). The outer membrane does not contain lipopolysaccharide. It is mainly covered by lipoproteins and glycolipids that constitute virulence factors and enable the bacteria to efficiently interact with its environment: the tick and the vertebrate host (Christodoulides et al., 2017; Kenedy et al., 2012). The synthesis of these molecules is controlled by the complex regulatory system RpoS-RpoN that allows the bacterium to adapt to variations of the enzootic cycle: pH, temperature, and bacterial density (Fikrig and Narasimhan, 2006; Radolf et al., 2012). *Borrelia* may induce persistent infections despite a host immune response. Antigenic variations and specific outer membrane proteins can explain this immune evasion. Adhesion molecules help *Borrelia* to colonize and persist in various tissues (Fikrig and Narasimhan, 2006; Kenedy et al., 2012). A certain number of these molecules have been very well characterized and identified as potential vaccine candidates. P66 binds to integrin $\alpha_{IIb}\beta_3$ on activated platelets likely facilitating bacterial dissemination (Coburn et al., 2013). BBK32 binds to fibronectin, an important glycoprotein of the extracellular matrix (Probert and Johnson, 1998); DpbA (decorin-binding protein A) binds the decorin, a proteoglycan found in different tissues, to facilitate the escape from the host immune system (Liang et al., 2004).

Despite the rapidly increasing knowledge on *Borrelia* biology and Lyme clinics, as well as diagnostic improvements, control of Lyme borreliosis seems not to be progressing (Borchers et al., 2015). In the contrary, incidences in Europe and Canada are on the rise, and various explanations have been given for this: more cases are detected thanks to better diagnostic (Aguero-Rosenfeld et al., 2005; Borchers et al., 2015), and climate changes and anthropic modifications of the environment affect the tick phenology (Dobson and Randolph, 2011; Kilpatrick and Randolph, 2012; Mead, 2015; Wenner-Moyer, 2015).

An effective vaccination would help to control the spread of the disease, but the development of a human vaccine faces several important challenges: The immunity to infection is strain-specific and decreases within 1 year after the infection (Embers and Narasimhan, 2013). Immune responses generated during infection are insufficient for long-term protection mostly due to *Borrelia's* strategy to evade antibody response—the bacteria are able to alter their antigen expression, which makes it impossible for the host immune system to establish effective memory responses (Stanek et al., 2012). Identifying suitable antigens for the induction of a persisting protective immunity is therefore a great challenge. On the other hand, as shown in the case of LYMErix, the first human vaccine against Lyme borreliosis, scientists are increasingly facing ethical concerns in the public opinion due to the fear about potential side effects of vaccines that could reveal itself very harmful for commercialization even of an effective vaccine (Abbott, 2006).

As Lyme borreliosis is first of all a zoonosis, alternative strategies to a direct human vaccination are discussed, in particular: (1) reservoir-targeted vaccines (RTVs) that could block the uptake of pathogens by ticks (Embers and Narasimhan, 2013) and (2) vector-targeted vaccines (VTVs) preventing tick feeding/pathogen colonization (de la Fuente et al., 2017a,b). This part at the end of this chapter will provide an overview of all these strategies and their limits.

HUMAN VACCINATION

LYMErix Era

The case of LYMErix is a classic example of how consumer power can override science (Abbott, 2006). In 1998, the pharmaceutical company SmithKline Beecham started to commercialize LYMErix, a recombinant OspA (outer space protein)-based vaccine (Embers and Narasimhan, 2013; Fikrig et al., 1990; Poland, 2011; Zhong et al., 1997). OspA, a lipoprotein that is abundant in culture and when *Borrelia* is in the tick midgut (Schwan et al., 1995), is quite immunogenic (de Silva et al., 1996; Pal et al., 2000; Schaible et al., 1993). The immunization with OspA leads to the elimination of *Borrelia* from ticks by serum antibodies during tick feeding (Luft et al., 2002; Schaible et al., 1993). The vaccine contained lipidated OspA adsorbed onto aluminum hydroxide adjuvant in phosphate buffered saline. It was produced using recombinant DNA technology where OspA gene from *B. burgdorferi* ss ZS7 strain is inserted and grown in *Escherichia coli*. OspA was produced as a single polypeptide chain of 257 amino acids. The lipid moiety, which is essential for immunogenicity, was covalently bound to its N-terminal part after translation (Luft et al., 2002; Steere et al., 1998). Each vaccine dose contained 30 µg of OspA that was administrated intramuscularly in three injections (Embers and Narasimhan, 2013).

LYMErix was withdrawn from market in 2002 due to poor sales. Hundreds of vaccine recipients were reporting side effects including autoimmunity (arthritis) (Poland, 2011; Scheckelhoff et al., 2006). Subsequently, a new hypothesis arose that *Borrelia* escaped the vaccine effect by molecular mimicry, a strategy in which part of a bacterial protein is structurally similar to part of a patient's protein (Abbott, 2006; Embers and Narasimhan, 2013). A portion of OspA$_{165-173}$ shares a very similar amino acid sequence with human protein LFA-1 (lymphocyte function-associated antigen) (Steere et al., 2001). However, this cross-reactivity has never been shown *in vivo* (Abbott, 2006). Moreover, a number of other proteins were able to slightly activate OspA-specific immune cells, some of them even without showing much similarity to OspA. Therefore, molecular mimicry as escape strategy appears highly unlikely. Furthermore, subjects developed arthritis in both the placebo and the active vaccine groups in phase III trials with LYMErix (Steere et al., 2001).

Post-LYMErix Era

Surprisingly, another OspA vaccine was successfully tested later by the US company Baxter in a phase I/II trial (Wressnigg et al., 2013), based on the conclusion that an effective vaccine should include multiple OspA from different species to achieve a protective immunity (Gern et al., 1997). This vaccine was composed of three recombinant antigens: (1) One antigen containing epitopes OspA-1 (*B. burgdorferi* ss) and OspA-2 (*B. afzelii*); (2) one antigen containing epitopes OspA-3 and OspA-5 (*B. garinii*); and (3) one antigen containing epitopes OspA-6 (*B. garinii*) and OspA-4 (*Borrelia bavariensis*). This combination should warrant protection against all *Borrelia* species pathogenic for humans in the United States and in the Europe. The hypothetical risk of cross-reactivity with LFA-1 was eliminated by replacing the putative cross-reactive OspA-1 epitope with the corresponding OspA-2 sequence (Wressnigg et al., 2013). The vaccine proved to be safe, well tolerated (only mild local and systemic reactions), and immunogenic in healthy adults. Moreover, the induced OspA-specific antibodies were borreliacidal, killing *B. burgdorferi* ss, *B. afzelii*, *B. garinii*, and *B. bavariensis*. However, further development has been stopped in the meantime and the vaccine abandoned by Baxter Laboratories.

Since 2016, Valneva, a spin-off company of Baxter, is developing a novel multivalent subunit vaccine in patients in a Phase I trial in the United States and Europe (Poljak et al., 2012; Wressnigg et al., 2013). Trials are in progress (source: Valneva website). In fact, OspA seems not to be an ideal antigen for a vaccine against borreliosis since its expression profile during *Borrelia* development is mainly present in the tick (Earnhart and Marconi, 2007). OspA is downregulated as soon as a tick begins its blood meal (Ohnishi et al., 2001; Schwan et al., 1995). The majority of spirochetes do not express OspA on their surface when they enter a host. So the protection of OspA-based vaccine relies on high OspA-specific antibody levels in vertebrate host, which are ingested by the feeding tick, blocking *Borrelia* in its migration toward the salivary glands. It is a typical "transmission blocking vaccine." It blocks the pathogen transmission by the arthropod vector.

OspC could in theory be a vaccine candidate. Other than OspA, OspC is upregulated during tick feeding, and it is expressed during spirochetes transmission from tick to mammal and during the first weeks of mammalian infection (Fig. 11.4) (Antonara et al., 2010; Önder et al., 2012). However, OspC is highly variable among *Borrelia* species (Ohnishi et al., 2001), and vaccination with recombinant OspC is only protective against spirochetes carrying OspC of identical or very similar sequence. The production of polyvalent chimeric molecules may enhance its potential (Earnhart and Marconi, 2007; Embers and Narasimhan, 2013; Schuijt et al., 2011b).

A first tetravalent OspC vaccine including types A, B, K, and D was tested some years ago (Earnhart and Marconi, 2007). These four types were selected from patients having an invasive infection: A, B, C, D, H, I, K, and N (Wormser et al., 2008). In patients, hematogenous dissemination is mainly associated with types A, B, I, and H among the 16 actual OspC genotypes (Wormser et al., 2008). These genetic markers are currently the best to define specific clinical isolates of *B. burgdorferi* and define their virulence (Jones et al., 2006). These *Borrelia* serotypes induced antibodies specific to each type in mice. These antibodies were borreliacidal by complement-dependent mechanism, but a vaccine for Europe and North America has to include more types. Therefore, an octavalent OspC vaccine was developed. The vaccine included OspC types A, B, K, D, E, N, I, and C and a conserved C-terminal motif of OspC. The resulting 41.3 kDa recombinant protein had 384 amino acids. It had advantages over

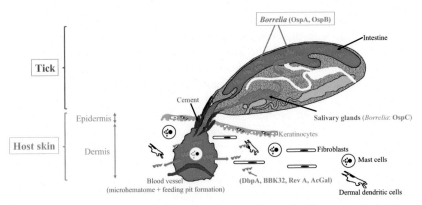

FIGURE 11.4 Lyme vaccine candidates targeting *Borrelia* proteins (in green and in bold).

using a polyvalent mixture of full length OspCs since the mixture was proven to be inefficient, causing a misdirection of antibody response to irrelevant regions of OspC. The problem was solved by creating epitope-based chimeric vaccines, which allowed better control of the response to specific epitopes (Earnhart and Marconi, 2007). Although OspC is essential in *Borrelia* transmission, its high variability did not make it a reliable vaccine candidate.

Other Borrelia Protein Targets

OspB plays an important role for the colonization of *B. burgdorferi* in the midgut of *I. scapularis* (Ohnishi et al., 2001) (Fig. 11.4). OspB vaccine induced protective response and borreliacidal antibodies that function in a complement-independent way. These antibodies disrupt the outer membrane of *B. burgdorferi* leading to an osmotic killing of the bacteria (LaRocca et al., 2009; Schuijt et al., 2011b). OspB is a vaccine candidate that should be more investigated.

B. burgdorferi upregulates several proteins in the mammalian host (Radolf et al., 2012). These proteins such as **BBK32**, a 47 kDa protein (Fikrig and Narasimhan, 2006; Li et al., 2006; Probert and Johnson, 1998), or **Rev A**, a surface exposed 17 kDa outer membrane protein (Brissette et al., 2010), both fibronectin-binding proteins, could be potential candidates for a vaccine, but no vaccination study has been done yet. **DpbA** is another *Borrelia* protein interacting with the extracellular matrix. Immunization of mice with this protein protected animals only if they were syringe inoculated with *Borrelia* but not when spirochetes were transmitted through infected nymphs (Hagman et al., 2000).

It was shown that a mixture of multiple antigens can potentiate the protective effect. Combinations of OspA and DbpA or OspC, BBK32, and DbpA showed a stronger protective effect in mice than a single immunogenic vaccine candidate. Moreover, it was possible to generate long-term immunity this way, as it was shown that patients with late symptoms are less susceptible to reinfection than patients with early symptoms, possibly due to a greater exposure to borrelial antigens (Schuijt et al., 2011b).

Borrelia glycolipids have recently been proposed as another vaccine candidate. Eighty percent of patients in late stages of the disease develop higher levels of antibodies directed against ACGal (6-O-acylated β-D-galactopyranoside), which is the most abundant glycolipid, representing 45% of all *Borrelia* glycolipids (Schröder et al., 2008). ACGal is conserved among all three species *B. burgdorferi* ss, *B. afzelii*, and *B. garinii*. However, no vaccination study has been designed yet.

Nevertheless, there is agreement that the expansion of Lyme borreliosis into new geographical areas cannot be controlled only by a direct human vaccination. Humans are an accidental host and their vaccination will not reduce *B. burgdorferi* sensu lato (sl) burden in its enzootic cycle. Therefore, vaccines targeting the reservoir or the tick vector have been undertaken.

RESERVOIR-TARGETED VACCINES

As Lyme borreliosis is a zoonosis, immunization of reservoir hosts might constitute an approach to prevent the disease in humans and in domestic animals (Embers and Narasimhan, 2013; Gomes-Solecki, 2014). A variety of reservoirs in small mammals and birds exists in North America (mouse, shrew, squirrel, striped skunk, ground birds) and in Euroasia (mouse, voles, squirrel, ground birds) (Humair and Gern, 2000; Kurtenbach et al., 2006; Ogden and Tsao, 2009). Thus, eliminating *B. burgdorferi* sl from nature is unrealistic, but diminishing their threat to humans is an achievable goal. One possible strategy to achieve this goal could be the development of effective RTVs.

The principle of RTVs is to reduce the infective agents in both, the reservoir hosts and in the vector feeding on reservoir hosts without eradicating the hosts or the vectors from the ecosystem. This strategy is viable for zoonotic infections with geographical hot spots and it was shown already effective for the prevention of rabies virus and plague in animal population health (Embers and Narasimhan, 2013; Gomes-Solecki, 2014). To translate this to the borrelial epidemiology, *B. burgdorferi* sl should be reduced in the various reservoir hosts but also concurrently in the tick. Most human infections are transmitted by ticks in the nymph stage. To interrupt this transmission, it would be necessary to block the bacterial colonization at the larval stage.

For the development of RTV, the selection of the antigen type, the route of delivery, the delivery system, and the implementation protocol are key criteria. OspA is emerging as the most promising vaccine in animals. It is able to block the transmission and has a helpful advantage of already having been developed in laboratory mice (Bensaci et al., 2012).

Initial studies were based on "catch-and-release strategy" of wild rodents and the OspA-based vaccination showed an impact on the percentage of infected ticks the following year (Embers and Narasimhan, 2013). The reduction correlated positively with the density of the tick and mice populations. Targeting mouse-dense areas had a significant impact on the borrelial carriage. But one of the main limitations of the study was the contribution of nonmouse species, which had not been taken in consideration. Subsequent studies developed more practical strategies for large-scale vaccination, in particular the baited oral vaccination strategy (Embers and Narasimhan, 2013; Gomes-Solecki, 2014; Scheckelhoff et al., 2006). This strategy is noninvasive and very effective. The ideal vector for a vaccine should be selected using several important criteria: (1) it should be able to produce protection after one single dose because the uptake by wild animals is unpredictable; (2) it should be stable under a variety of environmental conditions; and (3) it should be nontoxic to the targeted and to the nontargeted wildlife species (Scheckelhoff et al., 2006). Thus, two main vaccination strategies were developed: (1) an oral vaccination based on *E. coli* expressing recombinant OspA (Embers and Narasimhan, 2013) and (2) an oral vaccination based on vaccinia virus (VV) expressing OspA (Bensaci et al., 2012). Both strategies are equally powerful to produce protective levels of OspA-specific antibodies in *Mus musculus* and *Peromyscus leucopus* and are equally effective in decreasing *B. burgdorferi* from infected ticks when feeding on vaccinated mice (Gomes-Solecki, 2014). Further options will be discussed below and in section c).

Escherichia coli Expressing OspA

This RTV was developed by transformation of *E. coli* with a pET9c plasmid containing cloned full-length OspA. This RTV based on *E. coli* expressing OspA protected mice in about 89% and it caused also a reduction of *Borrelia* in ticks. Detectable antibody titers were peaking 64 days after inoculation. This RTV was effective without having dangerous side effects, and the induced immunity did not decline over time (Embers and Narasimhan, 2013). However, a higher number of vaccine doses were required to induce protective immunity, which burdened distribution plans. In one study a 23% reduction of the infection rate was achieved in nymphs 1 year after the vaccine deployment and a reduction by 76% after 5 years (Gomes-Solecki, 2014).

Vaccinia Virus Expressing OspA

The VV was extensively studied as a vaccine for smallpox and as a vector for vaccination against viral (HIV, rabies) and parasitic (malaria) diseases. RTV based on VV expressing OspA was tested by oral gavage on C3H mice. Detectable antibody titers were peaking 42 days after inoculation. Only 17% of ticks remained positive for *Borrelia* infection after feeding on vaccinated mice (Embers and Narasimhan, 2013; Scheckelhoff et al., 2006). The vaccine was not tested in the field but an empiric model predicted a reduction of the infection prevalence in nymphs by at least 56% (Gomes-Solecki, 2014).

This RTV has several advantages: it requires only a low number of doses to induce robust humoral and cellular responses (one vaccination is normally sufficient); VV has a broad range of hosts and it is stable under digestive conditions; VV is able to accept large fragments of foreign DNA without affecting its infectivity; proteins are expressed at high levels; and the ingestion of virus does not cause a disease and it is not transmissible amongst infected animals (Embers and Narasimhan, 2013; Scheckelhoff et al., 2006). The main concern remains its potential to transmit the virus to unwanted recipients including immunocompromised humans or persons suffering from eczema (Embers and Narasimhan, 2013; Gomes-Solecki, 2014). Additionally, VV formulation has been only tested in laboratory conditions, mainly in C3H mice, but not on *P. leucopus* or other natural reservoir hosts. This vaccine could also include additional antigens that target other tick-transmitted pathogens such as *Anaplasma phagocytophilum* and *Babesia microti*, a bacterium and a parasite respectively, transmitted by *Ixodes* ticks (Scheckelhoff et al., 2006).

To reduce the potential for human infection, another novel RTV based on a VV encapsulated with pH-sensitive polymers Eudragit was designed aiming at inactivating the virus until it gets in contact with stomach fluids. Eudragit is cationic, nontoxic, and from a biocompatible family of polymers. These polymethylacrylate-based copolymers are designed to dissociate at a specific pH (pH < 5). They have not been used to be encapsulated in living viruses due to the permanent inactivation of the virus by solvents used in the coating process, but in this study the solvents were optimized and thus a safer RTV based on VV could be developed (Kern et al., 2016).

Other Reservoir-Targeted Vaccines

Subolesin. An alternative to pathogen proteins is the inclusion of tick proteins such as subolesin in the recombinant VV vector (Bensaci et al., 2012; Embers and Narasimhan, 2013). Subolesin is a regulatory protein involved in the control of multiple cellular pathways. In the evolution it has been conserved in invertebrates and vertebrates and presents sequence homology to akirins, a recently renamed group of proteins that were proposed to function as transcription factors in *Drosophila* and mice (Galindo et al., 2009). It is possibly involved in regulation of NFκB-dependent and -independent gene expression (Bensaci et al., 2012). The recombinant subolesin protein is used in immunization trials to protect vertebrate hosts against tick infestations. This vaccination reduced tick survival and reproduction and caused degeneration of tick gut, salivary gland, reproductive tissues, and embryos (Kocan et al., 2007). Moreover, subolesin is common in ticks, mosquitoes, and sandflies. Such a vaccination could target therefore multiple zoonosis transmitted by multiple vectors (Schuijt et al., 2011b). The goal of the subolesin-based RTV would be to prevent tick feeding and the uptake of pathogens from reservoir. Laboratory experiments showed a reduction of tick feeding by 52% and a reduction of *Borrelia* transmission to vaccinated mice by 40%. One of the major advantages of this vaccine is that it has a potential to prevent also other infections transmitted by ticks (*Babesia bigemina, A. phagocytophilum*) or coinfections (Bensaci et al., 2012). Due to its incomplete protection against tick feeding and acquisition of *B. burgdorferi*, this RTV cannot be used alone. It would be interesting to test a combination of subolesin together with other *Borrelia* proteins in future vaccination projects (Bensaci et al., 2012; Embers and Narasimhan, 2013).

Doxycycline. Another alternate approach is the doxycycline rodent bait formulation. An injectable formulation of doxycycline was effective in preventing *B. burgdorferi* and *A. phagocytophilum* infection in a murine model (Dolan et al., 2008). In this study, 500 mg of doxycycline prevented *B. burgdorferi* transmission to vertebrates and it cured infections in mice. A major concern in this approach is that the widespread distribution of antibiotics in humans could induce the development of resistance in target and nontarget pathogens. It would be interesting to identify and eventually to develop antibiotics with similar antispirochetal activity, which are not used as first-line treatment of zoonoses, to minimize the risk of resistance.

Osps. The major strength of RTVs is that they bypass human immunization and therefore avoid potential vaccine failures and side effects. They can reduce the pathogen in its ecosystem while leaving all other living components of the enzootic cycle undisturbed. Several limitations, however, exist: RTVs based on OspA neutralize spirochetes only in the tick midgut. A vaccine based on OspA and OspC would be more effective because it would target the host and the vector in the same time. Indeed, OspA is expressed on *Borrelia* in the tick midgut and OspC is upregulated in tick salivary glands and is essential in the early transmission to the vertebrate host (Grimm et al., 2004; Ohnishi et al., 2001). Up to now, most studies targeted mice, but the number of different reservoir species for Lyme borreliosis is important. Therefore, other small vertebrates should be included. In addition, OspA-based RTVs are not designed to reduce tick density, which is an important ecoepidemiological parameter of Lyme disease risk. An RTV approach supplemented by acaricide treatment and human vaccination is the most effective route to reduce drastically the incidence of Lyme borreliosis (Gomes-Solecki, 2014).

VECTOR-TARGETED VACCINES

In arthropod-borne diseases (ABDs), most approaches to develop a vaccine have used live attenuated pathogens or selected antigens of them. Then, the important role of the arthropod saliva in pathogen transmission became increasingly obvious in certain diseases (Fontaine et al., 2011). This topic has been particularly well investigated in Lyme borreliosis (Brossard and Wikel, 1997; Wikel, 1999, 2013) and Leishmaniasis. Subsequently, targeting of the vector has strongly evolved as a new vaccine approach in ABD control. To illustrate this approach, we chose in this book *Ixodes* ticks, which represents the most advanced model of vector-based vaccines, together with the sand fly, the vector of Leishmaniasis (see Chapter 4). The process of the hard tick bite and its long-lasting blood meal of 3–10 days depending on the tick stage has been particularly well studied (Wikel, 1999). Tick saliva modulates the host pharmacology and immunity to facilitate blood uptake and tick attachment. Tick salivary compounds involved in this process have been thoroughly investigated and characterized (Francischetti et al., 2010; Kazimírová and Štibrániová, 2013; Kotal et al., 2015; Nuttall and Labuda, 2004).

Acquired Resistance to Ticks

Observations of tick blood meal in certain animal models revealed the existence of an acquired immunity to repeated tick bites. Rabbits infested with *Dermacentor* ticks develop robust immune responses against tick components that result in a rapid rejection of ticks after successive blood meals (Embers and

Narasimhan, 2013). *I. scapularis* feeding on nonreservoir hosts results in rejection but the feeding is successful on reservoir hosts (Wikel, 1999). The hallmark of tick resistance is swelling and redness at the bite site probably due to cutaneous basophil hypersensitivity or due to rapid recruitment of basophils mediated by concerted activation of humoral and cellular responses. The mechanism of recruitment of basophils and the subsequent degranulation, which causes the end of tick feeding and promotes tick mortality, is unknown. Some proteins secreted into the tick bite site provoke immune responses in the host and the subsequent recruitment of basophils. When nymphs infected with *B. burgdorferi* were feeding on guinea pigs, a nonreservoir species, it was observed that the transmission of bacteria was impaired too (Embers and Narasimhan, 2013). In humans leaving in endemic areas for Lyme borreliosis, it was found that people regularly bitten by *Ixodes* ticks had less risk to acquire Lyme disease. The presence of antibodies to tick proteins and a hypersensitivity to tick bite (itch) was identified as responsible of this process (Burke et al., 2005). Exploring the acquired tick resistance will lead to identification of tick salivary proteins that are natural targets of host immunity. Tick salivary proteins could then be selected as candidates for a vaccine.

Tick Salivary Proteins

Larvae are rarely infected as transovarian transmission of pathogens is low. Nymphs are the most involved stage in pathogen transmission since they are small enough to be undetected and stay long time attached (3–5 days). Infected female adults are also able to transmit pathogens but their larger size makes them detectable and easily removed. Therefore, they have much less opportunity for transmission. Therefore, the general consensus is to target the nymphs (Embers and Narasimhan, 2013). On the bite, tick salivary glands produce pharmacologically and immunologically active molecules that disturb the host immune defense including complement, procoagulants, proteases, histamine-binding proteins, and immune cells, thus preparing and facilitating pathogen transmission, also named "saliva-activated transmission" (Kazimírová and Štibrániová, 2013) (see Chapter 5).

When searching for the VTV, the feeding process of the tick was focused because it is central for tick survival and pathogen transmission. A successful tick feeding depends on its ability to surmount host defense responses. Up to now, a number of salivary proteins were tested and some of them provided impaired feeding, but none of them was able to block totally the feeding process.

Then, the effect of tick saliva on host pharmacology and immunity was investigated. An array of molecules inhibiting host immune responses including histamine-binding proteins, anticoagulant proteins, anticomplement proteins, peroxidases, and protease inhibitors were identified:

Tick salivary proteins **ISAC** (*I. scapularis* anticomplement protein), **Salp15**, **Salp20**, and **TSLPI** (tick salivary pathway inhibitor) (Fig. 11.5) inhibit different arms of host complement pathways.

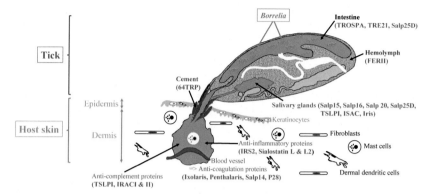

FIGURE 11.5   Tick vaccine candidates targeting tick proteins.

Anticoagulant proteins **Ixolaris**, **Penthalaris**, **Salp14**, and **P28** inhibit different components of the coagulation cascade, which is one of host defenses because the activation of coagulation results in an inflammation (Kazimírová and Štibrániová, 2013; Liu and Bonnet, 2014); to obtain an efficient vaccine, these proteins should be targeted in combination.

A novel serpin, **IRS-2**, inhibits host inflammation and thrombin-induced platelet aggregation. Homologs of IRS-2 were assembled in a cocktail to target several anticoagulants. A genome and transcriptome analysis identified several putative secreted proteins that could also be potential vaccine targets (Liu and Bonnet, 2014).

64TRP (tick recombinant protein) is a secreted cement protein 64P of the hard tick *Rhipicephalus appendiculatus* (Trimnell et al., 2005) (Fig. 11.5). Tick cement is a protein material secreted by salivary glands, which spreads over the host skin and into the dermis to form a wedge-shaped anchor that seals tick mouthparts to the host skin (Sonenshine and Anderson, 2014). 64TRP was tested as vaccine candidate in guinea pigs, hamsters, rabbits, and cattle. It impaired adult tick feeding by 23%–25% and resulted in a 45% reduction of the tick's egg mass. 64TRP-specific antibodies react with multiple epitopes in the tick midgut and cause its rupture. Several 64TRPs (full-length 64TRP5, 6 and overlapping N-terminal fragment 64TRP2) present structural homology and cross-reactivity with proteins found in more than one tick species (*I. ricinus* and *I. scapularis*), so they have a potential to become a broad spectrum vaccine and to inhibit transmission of multiple tick-transmitted pathogens (Embers and Narasimhan, 2013; Trimnell et al., 2005). In an experiment performed in mice during the transmission of tick-borne encephalitis by *I. ricinus*, antibodies directed against 64TRP protected mice from virus infection (Labuda et al., 2006).

Two salivary cysteine protease inhibitors of *I. scapularis*, **Sialostatin L** and **Sialostatin L2**, promote *B. burgdorferi* survival and infection in the skin. Sialostatin L inhibits cathepsin S, which results in decrease of inflammation thus facilitating feeding. Sialostatin L2 modulates inflammation responses, tissue remodeling, and angiogenesis by inhibiting intra- and extracellular cathepsins of the innate immune system (Kotsyfakis et al., 2006). This makes them also candidates for vaccine blocking transmission (Embers and Narasimhan, 2013).

I. scapularis **TSLPI** is an anticomplement protein that inhibits the lectin pathway of the complement cascade by interacting with mannose-binding lectin, which prevents its ability to activate downstream signaling. It confers protection to *B. burgdorferi* against the complement. Immunization against TSLPI did not succeed to block tick feeding but immunization against TSLPI and RNA interference impaired bacterial transmission (Schuijt et al., 2011a).

IRAC (*I. Ricinus* anticomplement protein) I and IRAC II are anticomplement proteins that inhibit the alternative pathway. They present structural and functional homology to *I. scapularis* ISAC. A cocktail of IRACs and ISAC could become a broad spectrum vaccine against these two tick species (Schroeder et al., 2007).

I. ricinus **FER2** is a secreted protein present in tick hemolymph. It supplies ovary tissues and salivary glands with iron derived from the blood meal. It was shown that silencing of FER2 impaired tick feeding. An immunization of rabbits and cattle was effective (Schuijt et al., 2011b).

Iris, a salivary protein expressed in nymph and adult stages, inhibits inflammatory cytokines. A vaccination with Iris impaired tick feeding in rabbits. A neutralization of Iris effect could lead to more effective immune responses against *B. burgdorferi* (Schuijt et al., 2011b). Various antiinflammatory proteins from *Ixodes* species could be vaccine candidates but they are yet to be tested (Embers and Narasimhan, 2013).

Targeting Tick Gut

Tick midgut is an essential organ for tick-borne pathogens. It is a tick–host–pathogen interface comprising proteases, anticomplement proteins, antibacterial peptides, and anticoagulants to protect the epithelial barrier from pathogens and host-mediated damage and to maintain blood meal fluid during long feeding period. Targeting tick gut components, which are essential for the tick, is also an approach to block pathogen transmission (de la Fuente et al., 2017a,b).

Salp25D is a peroxiredoxin expressed in the salivary glands and in the gut. It is critical for *B. burgdorferi* acquisition because it neutralizes reactive oxygen species at the vector–host interface and therefore facilitates viability of *B. burgdorferi* in the tick gut (Embers and Narasimhan, 2013; Schuijt et al., 2011b). Targeting Salp25D and the later described TROSPA (tick receptor for OspA) would help blocking *B. burgdorferi* acquisition from murine hosts. It is applicable to host and could impact infection prevalence (Embers and Narasimhan, 2013).

Bm86 and Bm95 are glycoproteins present on *Boophilus (Rhipicephalus) microplus* midgut epithelium. These glycoproteins are used to control *Rhipicephalus* ticks on cattle. Two vaccines are commercialized—TickGARD and Gavac (Carreon et al., 2012; de la Fuente et al., 1999; Garcia-Garcia et al., 2000).

These VTVs offer partial protection not only against tick bites (85.2%–99.6% depending on *Rhipicephalus* species) but also protection against *Babesia* infection (Schuijt et al., 2011b). A cross-protection against *Ixodes* ticks has not been studied yet.

The tick midgut is a physical and immunocompetent barrier for tick-borne pathogens (Hajdusek et al., 2013; Rudenko et al., 2005). The mechanisms by which *B. burgdorferi* escapes this barrier are unknown. The discovery of a tick microbiota has increased the complexity of the interaction vector–pathogen and renders our understanding of *B. burgdorferi* acquisition and transmission more difficult (Narasimhan et al., 2014).

Targeting Specific Borrelia-Vector Interactions

TROSPA. Due to its large impact in human health, Lyme vaccine research is particularly well developed (Mead, 2015). TROSPA, a tick gut protein helping to tether *B. burgdorferi* by binding OspA, facilitates spirochete colonization by directly interacting with OspA (Pal et al., 2000). A decrease in TROSPA expression impairs *B. burgdorferi* acquisition. Moreover, if *B. burgdorferi* does not express OspA, it is unable to colonize the tick midgut. Therefore, this tick–bacteria interaction is specific and could be targeted by a vaccine (Embers and Narasimhan, 2013).

Salp15 is a multifunctional tick salivary protein, which is specifically upregulated in ticks infected with *B. burgdorferi* (Ramamoorthi et al., 2005). It inhibits the activation of CD4+ T cells, complement activity, and DC functions (Anguita et al., 2002; Garg et al., 2006; Hovius et al., 2008). Salp15 binds to OspC during *B. burgdorferi* exit from salivary glands. It is also critical for early infection because it hides *B. burgdorferi* from the host immune system. Mice immunized to recombinant Salp15 and challenged with infected nymphs were partially protected from infection. Antibodies directed against Salp15 sequester Salp15, which results in OspC exposed to the immune system or Salp15-specific antibodies binding to Salp15-coated *B. burgdorferi* and delivering them to phagocytes, enhancing therefore bacterial clearance (Dai et al., 2009).

Salp16 is important also for acquisition of *B. burgdorferi* from host to tick. It could be useful for wildlife vaccination (Schuijt et al., 2011b).

B. burgdorferi's outer surface protein BB31 binds TRE31, a tick gut protein. It enhances *B. burgdorferi* egress from the gut by unknown mechanisms. Compromising this interaction would inhibit spirochete exit from the gut (Embers and Narasimhan, 2013).

Perspectives

Despite increasing efforts, no human vaccine has emerged to the market since the commercial failure of LYMErix in 2002. Based on the same concept of transmission-blocking vaccine, a new vaccine based on OspA is currently under clinical development in humans. Lyme vaccines have been developed in domestic animals, especially in dogs, but providing only partial protection. They are based not only on OspA molecules but also on *Borrelia* lysates (Krupka and Straubinger, 2010; Töpfer and Straubinger, 2007). There is increasing need for a vaccine, considering the spread of ticks due to environmental, ecological and climatic changes, acaricide resistance, and increase in international trade (Kilpatrick and Randolph, 2012; Kuleš et al., 2016; Randolph, 2010). Since Lyme borreliosis is first of all a zoonosis widespread in wild animals, a series of RTVs based mainly on OspA are investigated and tested in the field (Piesman and Eisen, 2008; Richer et al., 2014). The importance of tick saliva in pathogen transmission has also opened new avenues of research with tick-targeted vaccines (de la Fuente et al., 2017a,b). Only combined approaches based on effective methods to contain *B. burgdorferi* within its enzootic cycle and a direct vaccination of humans will lead to synergistic effect in human and animal health.

CONCLUSIONS

Vector-borne diseases are a major threat to human and animal health. Few vaccines have been successfully commercialized, targeting mainly virus diseases such as yellow fever, tick-borne encephalitis, and Japanese encephalitis. New technologies including genomics, transcriptomics, and proteomics should help to identify appropriate vaccine candidates (Kuleš et al., 2016). However, although the list of vaccine candidates covering pathogen and arthropod saliva molecules, which are essential in the transmission, is important, most ABD vaccines have failed so far. The skin constitutes a key interface in all these diseases since this is the site where pathogens and arthropod saliva are inoculated (Bernard et al., 2015, 2014; Wikel, 2013). The complexity of skin immunity is more and more unraveled (Nestle et al., 2009; Pasparakis et al., 2014). The discovery of skin microbiota and their role in inflammation should also be investigated (Belkaid and Segre, 2014; Gallo and Hooper, 2012; Grice and Segre, 2011; Lai et al., 2009) (Chapter 1). Its involvement and importance has been clearly demonstrated in various ABDs as a site of multiplication and persistence such as in malaria (Chapter 7), trypanosomiasis (Chapter 6), and Lyme (Chapter 9), among others. The development of proteomics on infected skin in the context of ABDs should help to identify new vaccine candidates in this specific cutaneous environment (Kuleš et al., 2016; Schnell et al., 2015). Once pathogen candidates will be clearly identified, the development of appropriate formulations and adjuvants for a skin inoculation will be a key requisite to successfully induce a protective immunity (Chen et al., 2015; Ita, 2016; Kaurav et al., 2016).

REFERENCES

Abbott, A., 2006. Lyme disease: uphill struggle. Nature 439, 524–525.

Aguero-Rosenfeld, M., Wang, G., Schwartz, I., Wormser, G., 2005. Diagnosis of lyme borreliosis. Clinical Microbiology Reviews 18, 484–509.

Alger, N.E., Harant, J., 1976. *Plasmodium berghei*: heat-treated sporozoite vaccination of mice. Experimental Parasitology 40, 261–268.

Aly, A.S., Mikolajczak, S.A., Rivera, H.S., Camargo, N., Jacobs-Lorena, V., Labaied, M., Coppens, I., Kappe, S.H., 2008. Targeted deletion of SAP1 abolishes the expression of infectivity factors necessary for successful malaria parasite liver infection. Molecular Microbiology 69, 152–163.

Aly, A.S.I., Lindner, S.E., MacKellar, D.C., Peng, X., Kappe, S.H.I., 2011. SAP1 is a critical post-transcriptional regulator of infectivity in malaria parasite sporozoite stages. Molecular Microbiology 79, 929–939.

Amino, R., Thiberge, S., Martin, B., Celli, S., Shorte, S., Frischknecht, F., Ménard, R., 2006. Quantitative imaging of *Plasmodium* transmission from mosquito to mammal. Nature Medicine 12, 220–224.

Amino, R., Giovannini, D., Thiberge, S., Gueirard, P., Boisson, B., Dubremetz, J.-F., Prévost, M.-C., Ishino, T., Yuda, M., Ménard, R., 2008. Host cell traversal is important for progression of the malaria parasite through the dermis to the liver. Cell Host and Microbe 3, 88–96.

Anguita, J., Ramamoorthi, N., Hovius, J.W.R., Das, S., Thomas, V., Persinski, R., Conze, D., Askenase, P.W., Rincón, M., Kantor, F.S., Fikrig, E., 2002. Salp15, an *Ixodes scapularis* salivary protein, inhibits CD4+ T cell activation. Immunity 16, 849–859.

Annoura, T., Ploemen, I.H., van Schaijk, B.C., Sajid, M., Vos, M.W., van Gemert, G.J., Chevalley-Maurel, S., Franke-Fayard, B.M., Hermsen, C.C., Gego, A., Franetich, J.F., Mazier, D., Hoffman, S.L., Janse, C.J., Sauerwein, R.W., Khan, S.M., 2012. Assessing the adequacy of attenuation of genetically modified malaria parasite vaccine candidates. Vaccine 30, 2662–2670.

Annoura, T., van Schaijk, B.C.L., Ploemen, I.H.J., Sajid, M., Lin, J.-W., Vos, M.W., Dinmohamed, A.G., Inaoka, D.K., Rijpma, S.R., van Gemert, G.-J., Chevalley-Maurel, S., Kielbasa, S.M., Scheltinga, F., Franke-Fayard, B., Klop, O., Hermsen, C.C., Kita, K., Gego, A., Franetich, J.F., Mazier, D., Hoffman, S.L., Janse, C.J., Sauerwein, R.W., Khan, S.M., 2014. Two *Plasmodium* 6-cys family-related proteins have distinct and critical roles in liver-stage development. The FASEB Journal 28, 2158–2170.

Antonara, S., Ristow, L., Mccarthy, J., Coburn, J., 2010. Effect of *Borrelia burgdorferi* OspC at the site of inoculation in mouse skin. Infection and Immunity 78, 4723–4733. https://doi.org/10.1128/IAI.00464-10.

Beaudoin, R.L., Strome, C.P., Mitchell, F., Tubergen, T.A., 1977. *Plasmodium berghei*: immunization of mice against the ANKA strain using the unaltered sporozoite as an antigen. Experimental Parasitology 42, 1–5.

Belkaid, Y., Segre, J., 2014. Dialogue between skin microbiota and immunity. Science 346, 954–959.

Belnoue, E., Costa, F.T., Frankenberg, T., Vigário, A.M., Voza, T., Leroy, N., Rodrigues, M.M., Landau, I., Snounou, G., Rénia, L., 2004. Protective T cell immunity against malaria liver stage after vaccination with live sporozoites under chloroquine treatment. The Journal of Immunology 172, 2487–2495.

Bensaci, M., Bhattacharya, D., Clark, R., Hu, L.T., 2012. Oral vaccination with vaccinia virus expressing the tick antigen subolesin inhibits tick feeding and transmission of *Borrelia burgdorferi*. Vaccine 30, 6040–6046.

Bernard, Q., Jaulhac, B., Boulanger, N., 2014. Smuggling across the border: how arthropod-borne pathogens evade and exploit the host defense system of the skin. The Journal of Investigative Dermatology 134, 1211–1219.

Bernard, Q., Jaulhac, B., Boulanger, N., 2015. Skin and arthropods: an effective interaction used by pathogens in vector-borne diseases. European Journal of Dermatology 25 (Suppl. 1), 18–22.

Bijker, E.M., Bastiaens, G.J., Teirlinck, A.C., van Gemert, G.J., Graumans, W., van de Vegte-Bolmer, M., Siebelink-Stoter, R., Arens, T., Teelen, K., Nahrendorf, W., Remarque, E.J., Roeffen, W., Jansens, A., Zimmerman, D., Vos, M., van Schaijk, B.C., Wiersma, J., van der Ven, A.J., de Mast, Q., van Lieshout, L., Verweij, J.J., Hermsen, C.C., Scholzen, A., Sauerwein, R.W., 2013. Protection against malaria after immunization by chloroquine prophylaxis and sporozoites is mediated by preerythrocytic immunity. Proceedings of the National Academy of Sciences of the United States of America 110, 7862–7867.

Bijker, E.M., Schats, R., Obiero, J.M., Behet, M.C., van Gemert, G.J., van de Vegte-Bolmer, M., Graumans, W., van Lieshout, L., Bastiaens, G.J., Teelen, K., Hermsen, C.C., Scholzen, A., Visser, L.G., Sauerwein, R.W., 2014. Sporozoite immunization of human volunteers under mefloquine prophylaxis is safe, immunogenic and protective: a double-blind randomized controlled clinical trial. PLoS One 9, e112910.

Borchers, A., Keen, C., Huntley, A., Gershwin, M., 2015. Lyme disease: a rigorous review of diagnostic criteria and treatment. Journal of Autoimmunity 57, 82–115.

Brissette, C.A., Rossmann, E., Bowman, A., Cooley, A.E., Riley, S.P., Hunfeld, K.-P., Bechtel, M., Kraiczy, P., Stevenson, B., 2010. The borrelial fibronectin-binding protein RevA is an early antigen of human Lyme disease. Clinical and Vaccine Immunology 17, 274–280.

Brossard, M., Wikel, S.K., 1997. Immunology of interactions between ticks and hosts. Medical and Veterinary Entomology 11, 270–276.

Burke, G., Wikel, S.K., Spielman, A., Telford, S.R., McKay, K., Krause, P.J., 2005. Hypersensitivity to ticks and Lyme disease risk. Emerging Infectious Diseases 11, 36–41.

Butler, N.S., Schmidt, N.W., Harty, J.T., 2010. Differential effector pathways regulate memory CD8 T cell immunity against *Plasmodium berghei* versus *P. yoelii* sporozoites. The Journal of Immunology 184, 2528–2538.

Butler, N.S., Schmidt, N.W., Vaughan, A.M., Aly, A.S., Kappe, S.H.I., Harty, J.T., 2011. Superior antimalarial immunity after vaccination with late liver stage-arresting genetically attenuated parasites. Cell Host and Microbe 9, 451–462.

Carreon, D., de la Lastra, J.M., Almazan, C., Canales, M., Ruiz-Fons, F., Boadella, M., Moreno-Cid, J.A., Villar, M., Gortazar, C., Reglero, M., Villarreal, R., de la Fuente, J., 2012. Vaccination with BM86, subolesin and akirin protective antigens for the control of tick infestations in white tailed deer and red deer. Vaccine 30, 273–279.

Ceithaml, J., Evans, E.A., 1946. The biochemistry of the malaria parasite; the *in vitro* effects of x-rays upon *Plasmodium gallinaceum*. The Journal of Infectious Diseases 78, 190–197.

Chakravarty, S., Cockburn, I.A., Kuk, S., Overstreet, M.G., Sacci, J.B., Zavala, F., 2007. CD8+ T lymphocytes protective against malaria liver stages are primed in skin-draining lymph nodes. Nature Medicine 13, 1035–1041.

Chatterjee, S., François, G., Duilhe, P., Timperman, G., Wéry, M., 1996. Immunity to *Plasmodium berghei* exoerythrocytic forms derived from irradiated sporozoites. Parasitology Research 82, 297–303.

Chatterjee, S., Ngonseu, E., van Overmeir, C., Correwyn, A., Druilhe, P., Wéry, M., 2001. Rodent malaria in the natural host – irradiated sporozoites of *Plasmodium berghei* induce liver-stage specific immune responses in the natural host *Grammomys surdaster* and protect immunized *Grammomys* against *P. berghei* sporozoite challenge. African Journal of Medicine and Medical Sciences 30 (Suppl.), 25–33.

Chen, D., Bowersock, T., Weeratna, R., Yeoh, T., 2015. Current opportunities and challenges in intradermal vaccination. Therapeutic Delivery 6, 1101–1108.

Christodoulides, A., Boyadjian, A., Kelesidis, T., 2017. Spirochetal lipoproteins and immune evasion. Frontiers in Immunology 8, 364.

Clyde, D.F., Most, H., McCarthy, V.C., Vanderberg, J.P., 1973. Immunization of man against sporozoite-induced falciparum malaria. The American Journal of the Medical Sciences 266, 169–177.

Clyde, D.F., McCarthy, V.C., Miller, R.M., Woodward, W.E., 1975. Immunization of man against falciparum and vivax malaria by use of attenuated sporozoites. The American Journal of Tropical Medicine and Hygiene 24, 397–401.

Clyde, D.F., 1990. Immunity to falciparum and vivax malaria induced by irradiated sporozoites: a review of the University of Maryland studies, 1971-75. Bulletin of the World Health Organization 68 (Suppl.), 9–12.

Coburn, J., Leong, J., Chaconas, G., 2013. Illuminating the roles of the *Borrelia burgdorferi* adhesins. Trends in Microbiology 21, 372–379.

Cockburn, I.A., Tse, S.-W., Radtke, A.J., Srinivasan, P., Chen, Y.-C., Sinnis, P., Zavala, F., 2011. Dendritic cells and hepatocytes use distinct pathways to process protective antigen from *Plasmodium in vivo*. PLoS Pathogens 7, e1001318.

Dai, J., Wang, P., Adusumilli, S., Booth, C.J., Narasimhan, S., Anguita, J., Fikrig, E., 2009. Antibodies against a tick protein, Salp15, protect mice from the Lyme disease agent. Cell Host and Microbe 6, 482–492.

de la Fuente, J., Rodriguez, M., Montero, C., Redondo, M., Garcia-Garcia, J.C., Mendez, L., Serrano, E., Valdes, M., Enriquez, A., Canales, M., Ramos, E., Boue, O., Machado, H., Lleonart, R., 1999. Vaccination against ticks (Boophilus spp.): the experience with the Bm86-based vaccine Gavac. Genetics Analysis 15, 143–148.

de la Fuente, J., Antunes, S., Bonnet, S., Cabezas-Cruz, A., Domingos, A., Estrada-Peña, A., Johnson, N., Kocan, K., Mansfield, K., Nijhof, A., Papa, A., Rudenko, N., Villar, M., Alberdi, P., Torina, A., Ayllón, N., Vancova, M., Golovchenko, M., Grubhoffer, L., Caracappa, S., Fooks, A., Gortazar, C., Rego, R., 2017a. Tick-pathogen interactions and vector competence: identification of molecular drivers for tick-borne diseases. Frontiers in Cellular and Infection Microbiology 7, 114.

de la Fuente, J., Contreras, M., Estrada-Peña, A., Cabezas-Cruz, A., 2017b. Targeting a global health problem: vaccine design and challenges for the control of tick-borne diseases. Vaccine 35, 5089–5094, S0264–410X(17)31023–X.

de Silva, A., Telford 3rd, S., Brunet, L., Barthold, S., Fikrig, E., 1996. *Borrelia burgdorferi* OspA is an arthropod-specific transmission-blocking Lyme disease vaccine. The Journal of Experimental Medicine 183, 271–275.

Depinay, N., Hacini, F., Beghdadi, W., Peronet, R., Mécheri, S., 2006. Mast cell-dependent down-regulation of antigen-specific immune responses by mosquito bites. The Journal of Immunology 176, 4141–4146.

Dobson, A.D.M., Randolph, S.E., 2011. Modelling the effects of recent changes in climate, host density and acaricide treatments on population dynamics of Ixodes ricinus in the UK. Journal of Applied Ecology 48, 1029–1037.

Dolan, M., Zeidner, N., Gabitzsch, E., Dietrich, G., Borchert, J., Poché, R., Piesman, J., 2008. A doxycycline hyclate rodent bait formulation for prophylaxis and treatment of tick-transmitted *Borrelia burgdorferi*. The American Journal of Tropical Medicine and Hygiene 803–805.

Doll, K.L., Butler, N.S., Harty, J.T., 2014. CD8 T cell independent immunity after single dose infection-treatment-vaccination (ITV) against *Plasmodium yoelii*. Vaccine 32, 483–491.

Doolan, D.L., Hoffman, S.L., 2000. The complexity of protective immunity against liver-stage malaria. The Journal of Immunology 165, 1453–1462.

Draper, S.J., Goodman, A.L., Biswas, S., Forbes, E.K., Moore, A.C., Gilbert, S.C., Hill, A.V., 2009. Recombinant viral vaccines expressing merozoite surface protein-1 induce antibody- and T cell-mediated multistage protection against malaria. Cell Host and Microbe 5, 95–105.

Druilhe, P., Barnwell, J.W., 2007. Pre-erythrocytic stage malaria vaccines: time for a change in path. Current Opinion in Microbiology 10, 371–378.

Druilhe, P., Hagan, P., Rook, G.A., 2002. The importance of models of infection in the study of disease resistance. Trends in Microbiology 10 (Suppl.), S38–S46.

Earnhart, C., Marconi, R., 2007. An octavalent lyme disease vaccine induces antibodies that recognize all incorporated OspC type-specific sequences. Human Vaccines 3, 281–289.

Ellis, J., Ozaki, L.S., Gwadz, R.W., Cochrane, A.H., Nussenzweig, V., Nussenzweig, R.S., Godson, G.N., 1983. Cloning and expression in *E. coli* of the malarial sporozoite surface antigen gene from *Plasmodium knowlesi*. Nature 302, 536–538.

Embers, M., Narasimhan, 2013. Vaccination against Lyme disease: past, present, and future. Frontiers in Cellular and Infection Microbiology 3.

Epstein, J.E., Tewari, K., Lyke, K.E., Sim, B.K.L., Billingsley, P.F., Laurens, M.B., Gunasekera, A., Chakravarty, S., James, E.R., Sedegah, M., Richman, A., Velmurugan, S., Reyes, S., Li, M., Tucker, K., Ahumada, A., Ruben, A.J., Li, T., Stafford, R., Eappen, A.G., Tamminga, C., Bennett, J.W., Ockenhouse, C.F., Murphy, J.R., Komisar, J., Thomas, N., Loyevsky, M., Birkett, A., Plowe, C.V., Loucq, C., Edelman, R., Richie, T.L., Seder, R.A., Hoffman, S.L., 2011. Live attenuated malaria vaccine designed to protect through hepatic CD8$^+$ T cell immunity. Science 334, 475–480.

Ewer, K.J., Sierra-Davidson, K., Salman, A.M., Illingworth, J.J., Draper, S.J., Biswas, S., Hill, A.V., 2015. Progress with viral vectored malaria vaccines: a multi-stage approach involving 'unnatural immunity'. Vaccine 33, 7444–7451.

Falae, A., Combe, A., Amaladoss, A., Carvalho, T., Ménard, R., Bhanot, P., 2010. Role of *Plasmodium berghei* cGMP-dependent protein kinase in late liver stage development. The Journal of Biological Chemistry 285, 3282–3288.

Ferreira, A., Schofield, L., Enea, V., Schellekens, H., van der Meide, P., Collins, W.E., Nussenzweig, R.S., Nussenzweig, V., 1986. Inhibition of development of exoerythrocytic forms of malaria parasites by gamma-interferon. Science 232, 881–884.

Fikrig, E., Narasimhan, S., 2006. *Borrelia burgdorferi*–traveling incognito? Microbes and Infection 8, 1390–1399.

Fikrig, E., Barthold, S., Kantor, F., Flavell, R., 1990. Protection of mice against the Lyme disease agent by immunizing with recombinant OspA. Science 250, 553–556.

Flanagan, K.L., Wilson, K.L., Plebanski, M., 2016. Polymorphism in liver-stage malaria vaccine candidate proteins: immune evasion and implications for vaccine design. Expert Review of Vaccines 15, 389–399.

Fontaine, A., Diouf, I., Bakkali, N., Missé, D., Pagès, F., Fusai, T., Rogier, C., Almeras, L., 2011. Implication of haematophagous arthropod salivary proteins in host-vector interactions. Parasites and Vectors 28, 187.

Francischetti, I., Sa-Nunes, A., Mans, B., Santos, I., Ribeiro, J., 2010. The role of saliva in tick feeding. Frontiers in Bioscience 14, 2051–2088.

Fraser, C.M., Casjens, S., Huang, W.M., Sutton, G.G., Clayton, R., Lathigra, R., White, O., Ketchum, K.A., Dodson, R., Hickey, E.K., Gwinn, M., Dougherty, B., Tomb, J.F., Fleischmann, R.D., Richardson, D., Peterson, J., Kerlavage, A.R., Quackenbush, J., Salzberg, S., Hanson, M., van Vugt, R., Palmer, N., Adams, M.D., Gocayne, J., Weidman, J., Utterback, T., Watthey, L., McDonald, L., Artiach, P., Bowman, C., Garland, S., Fuji, C., Cotton, M.D., Horst, K., Roberts, K., Hatch, B., Smith, H.O., Venter, J.C., 1997. Genomic sequence of a Lyme disease spirochaete, *Borrelia burgdorferi*. Nature 390, 580–586.

Friesen, J., Matuschewski, K., 2011. Comparative efficacy of pre-erythrocytic whole organism vaccine strategies against the malaria parasite. Vaccine 29, 7002–7008.

Friesen, J., Silvie, O., Putrianti, E.D., Hafalla, J.C., Matuschewski, K., Borrmann, S., 2010. Natural immunization against malaria: causal prophylaxis with antibiotics. Science Translational Medicine 2, 40ra49.

Friesen, J., Borrmann, S., Matuschewski, K., 2011. Induction of antimalaria immunity by pyrimethamine prophylaxis during exposure to sporozoites is curtailed by parasite resistance. Antimicrobial Agents and Chemotherapy 55, 2760–2767.

Galindo, R., Doncel-Pérez, E., Zivkovic, Z., Naranjo, V., Gortazar, C., Mangold, A., 2009. Tick subolesin is an ortholog of the akirins described in insects and vertebrates. Developmental and Comparative Immunology 33, 612–617.

Gallo, R., Hooper, L., 2012. Epithelial antimicrobial defence of the skin and intestine. Nature Reviews Immunology 12, 503–516.

Garcia-Garcia, J.C., Montero, C., Redondo, M., Vargas, M., Canales, M., Boue, O., Rodriguez, M., Joglar, M., Machado, H., Gonzalez, I.L., Valdes, M., Mendez, L., de la Fuente, J., 2000. Control of ticks resistant to immunization with Bm86 in cattle vaccinated with the recombinant antigen Bm95 isolated from the cattle tick, Boophilus microplus. Vaccine 18, 2275–2287.

Garg, R., Juncadella, I.J., Ramamoorthi, N., Ashish, Ananthanarayanan, S.K., Thomas, V., Rincón, M., Krueger, J.K., Fikrig, E., Yengo, C.M., Anguita, J., 2006. Cutting edge: CD4 is the receptor for the tick saliva immunosuppression, Salp15. The Journal of Immunology 177, 6579–6583.

Gern, L., Hu, C., Voet, P., Hauser, P., Lobet, Y., 1997. Immunization with a polyvalent OspA vaccine protects mice against Ixodes ricinus tick bites infected by Borrelia burgdorferi ss, Borrelia garinii and Borrelia afzelii. Vaccine 15, 1551–1557.

Gething, P.W., Casey, D.C., Weiss, D.J., Bisanzio, D., Bhatt, S., Cameron, E., Battle, K.E., Dalrymple, U., Rozier, J., Rao, P.C., Kutz, M.J., Barber, R.M., Huynh, C., Shackelford, K.A., Coates, M.M., Nguyen, G., Fraser, M.S., Kulikoff, R., Wang, H., Naghavi, M., Smith, D.L., Murray, C.J., Hay, S.I., Lim, S.S., 2016. Mapping Plasmodium falciparum mortality in Africa between 1990 and 2015. The New England Journal of Medicine 375, 2435–2445.

Golenser, J., Heeren, J., Verhave, J.P., Kaay, H.J., Meuwissen, J.H., 1977. Crossreactivity with sporozoites, exoerythrocytic forms and blood schizonts of Plasmodium berghei in indirect fluorescent antibody tests with sera of rats immunized with sporozoites or infected blood. Clinical and Experimental Immunology 29, 43–51.

Gomes-Solecki, M., 2014. Blocking pathogen transmission at the source: reservoir targeted OspA-based vaccines against Borrelia burgdorferi. Frontiers in Cellular and Infection Microbiology 4, 136.

Gonzalez-Aseguinolaza, G., de Oliveira, C., Tomaska, M., Hong, S., Bruna-Romero, O., Nakayama, T., Taniguchi, M., Bendelac, A., Van Kaer, L., Koezuka, Y., Tsuji, 2000. Alpha-galactosylceramide-activated V alpha 14 natural killer T cells mediate protection against murine malaria. Proceedings of the National Academy of Sciences of the United States of America 97, 8461–8466.

Grice, E.A., Segre, J.A., 2011. The skin microbiome. Nature Reviews Microbiology 9, 244–253.

Grimm, D., Tilly, K., Byram, R., Stewart, P., Krum, J.G., Bueschel, D., Schwan, T., Policastro, P., Elias, A.F., Rosa, P., 2004. Outer-surface protein C of the Lyme disease spirochete: a protein induced in ticks for infection of mammals. Proceedings of the National Academy of Sciences of the United States of America 101, 3142–3147.

Gueirard, P., Tavares, J., Thiberge, S., Bernex, F., Ishino, T., Milon, G., Franke-Fayard, B., Janse, C.J., Ménard, R., Amino, R., 2010. Development of the malaria parasite in the skin of the mammalian host. Proceedings of the National Academy of Sciences of the United States of America 107, 18640–18645.

Hafalla, J.C., Bauza, K., Friesen, J., Gonzalez-Aseguinolaza, G., Hill, A.V., Matuschewski, K., 2013. Identification of targets of CD8+ T cell responses to malaria liver stages by genome-wide epitope profiling. PLoS Pathogens 9, e1003303.

Hagman, K., Yang, X., Wikel, S., Schoeler, G., Caimano, M., Radolf, J., Norgard, M., 2000. Decorin-binding protein A (DbpA) of *Borrelia burgdorferi* is not protective when immunized mice are challenged via tick infestation and correlates with the lack of DbpA expression by B. burgdorferi in ticks. Infection and Immunity 68, 4759–4764.

Hajdusek, O., Sima, R., Ayllon, N., Jalovecka, M., Perner, J., de la Fuente, J., Kopacek, P., 2013. Interaction of the tick immune system with transmitted pathogens. Frontiers in Cellular and Infection Microbiology 3, 26.

Haussig, J.M., Matuschewski, K., Kooij, T.W.A., 2011. Inactivation of a *Plasmodium* apicoplast protein attenuates formation of liver merozoites. Molecular Microbiology 81, 1511–1525.

Hoffman, S.L., Goh, L.M., Luke, T.C., Schneider, I., Le, T.P., Doolan, D.L., Sacci, J., de la Vega, P., Dowler, M., Paul, C., Gordon, D.M., Stoute, J.A., Church, L.W., Sedegah, M., Heppner, D.G., Ballou, W.R., Richie, T.L., 2002. Protection of humans against malaria by immunization with radiation-attenuated *Plasmodium falciparum* sporozoites. The Journal of Infectious Diseases 185, 1155–1164.

Hoffman, S.L., Billingsley, P.F., James, E., Richman, A., Loyevsky, M., Li, T., Chakravarty, S., Gunasekera, A., Chattopadhyay, R., Li, M., Stafford, R., Ahumada, A., Epstein, J.E., Sedegah, M., Reyes, S., Richie, T.L., Lyke, K.E., Edelman, R., Laurens, M.B., Plowe, C.V., Sim, B.K., 2010. Development of a metabolically active, non-replicating sporozoite vaccine to prevent *Plasmodium falciparum* malaria. Human Vaccines 6, 97–106.

Hoffman, S.L., 1997. Experimental challenge of volunteers with malaria. Annals of Internal Medicine 127, 233–235.

Hollingdale, M.R., Zavala, F., Nussenzweig, R.S., Nussenzweig, V., 1982. Antibodies to the protective antigen of *Plasmodium berghei* sporozoites prevent entry into cultured cells. The Journal of Immunology 128, 1929–1930.

Hollingdale, M.R., Nardin, E.H., Tharavanij, S., Schwartz, A.L., Nussenzweig, R.S., 1984. Inhibition of entry of *Plasmodium falciparum* and *P. vivax* sporozoites into cultured cells; an in vitro assay of protective antibodies. The Journal of Immunology 132, 909–913.

Hovius, J.R., De Jong, M.A.W.P., Dunnen, Jd., Litjens, M., Fikrig, E., Van Der Poll, T., Gringhuis, S.I., Geijtenbeek, T.B.H., 2008. Salp15 binding to DC-SIGN inhibits cytokine expression by impairing both nucleosome remodeling and mRNA stabilization. PLoS Pathogens 4.

Humair, P., Gern, L., 2000. The wild hidden face of Lyme borreliosis in Europe. Microbes and Infection 2, 915–922.

Inoue, M., Culleton, R.L., 2011. The intradermal route for inoculation of sporozoites of rodent malaria parasites for immunological studies. Parasite Immunology 33, 137–142.

Ishizuka, A.S., Lyke, K.E., DeZure, A., Berry, A.A., Richie, T.L., Mendoza, F.H., Enama, M.E., Gordon, I.J., Chang, L.J., Sarwar, U.N., Zephir, K.L., Holman, L.A., James, E.R., Billingsley, P.F., Gunasekera, A., Chakravarty, S., Manoj, A., Li, M., Ruben, A.J., Li, T., Eappen, A.G., Stafford, R.E., C, N.K., Murshedkar, T., DeCederfelt, H., Plummer, S.H., Hendel, C.S., Novik, L., Costner, P.J., Saunders, J.G., Laurens, M.B., Plowe, C.V., Flynn, B., Whalen, W.R., Todd, J.P., Noor, J., Rao, S., Sierra-Davidson, K., Lynn, G.M., Epstein, J.E., Kemp, M.A., Fahle, G.A., Mikolajczak, S.A., Fishbaugher, M., Sack, B.K., Kappe, S.H., Davidson, S.A., Garver, L.S., Björkström, N.K., Nason, M.C., Graham, B.S., Roederer, M., Sim, B.K., Hoffman, S.L., Ledgerwood, J.E., Seder, R.A., 2016. Protection against malaria at 1 year and immune correlates following PfSPZ vaccination. Nature Medicine 22, 614–623.

Ita, K., 2016. Transdermal delivery of vaccines – recent progress and critical issues. Biomedicine Pharmacotherapy 83, 1080–1088.

Jones, K., Glickstein, L., Damle, N., Sikand, V., McHugh, G., Steere, A., 2006. *Borrelia burgdorferi* genetic markers and disseminated disease in patients with early lyme disease. Journal of Clinical Microbiology 44, 4407–4413.

Jung, S., Unutmaz, D., Wong, P., Sano, G., De los Santos, K., Sparwasser, T., Wu, S., Vuthoori, S., Ko, K., Zavala, F., Pamer, E.G., Littman, D.R., Lang, R.A., 2002. *In vivo* depletion of CD11c⁺ dendritic cells abrogates priming of CD8⁺ T cells by exogenous cell-associated antigens. Immunity 17, 211–220.

Kaurav, M., Minz, S., Sahu, K., Kumar, M., Madan, J., Pandey, R.N., 2016. Nanoparticulate mediated transcutaneous immunization: myth or reality. Nanomedicine 12, 1063–1081.

Kazimírová, M., Štibrániová, I., 2013. Tick salivary compounds: their role in modulation of host defences and pathogen transmission. Frontiers in Cellular and Infection Microbiology 3, 43.

Kebaier, C., Voza, T., Vanderberg, J.P., 2009. Kinetics of mosquito-injected *Plasmodium* sporozoites in mice: fewer sporozoites are injected into sporozoite-immunized mice. PLoS Pathogens 5, e1000399.

Kenedy, M.R., Lenhart, T.R., Akins, D.R., 2012. The role of *Borrelia burgdorferi* outer surface proteins. FEMS Immunology and Medical Microbiology 66, 1–19.

Kern, A., Zhou, C., Jia, F., Xu, Q., Hu, L., 2016. Live-vaccinia virus encapsulation in pH-sensitive polymer increases safety of a reservoir-targeted Lyme disease vaccine by targeting gastrointestinal release. Vaccine 34, 4507–4513.

Kilpatrick, A., Randolph, S., 2012. Drivers, dynamics, and control of emerging vector-borne zoonotic diseases. Lancet 380, 1946–1955.

Kirchner, S., Power, B.J., Waters, A.P., 2016. Recent advances in malaria genomics and epigenomics. Genome Medicine 8, 92.

Kocan, K., Manzano-Roman, R., de la Fuente, J., 2007. Transovarial silencing of the subolesin gene in three-host ixodid tick species after injection of replete females with subolesin dsRNA. Parasitology Research 100, 1411–1415.

Kotal, J., Langhansova, H., Lieskovska, J., Andersen, J.F., Francischetti, I.M., Chavakis, T., Kopecky, J., Pedra, J.H., Kotsyfakis, M., Chmelar, J., 2015. Modulation of host immunity by tick saliva. Journal of Proteomics 128, 58–68.

Kotsyfakis, M., Sá-Nunes, A., Francischetti, I.M.B., Mather, T.N., Andersen, J.F., Ribeiro, J.M.C., 2006. Antiinflammatory and immunosuppressive activity of Sialostatin L, a salivary cystatin from the tick *Ixodes scapularis*. Journal of Biological Chemistry 281, 26298–26307.

Krupka, I., Straubinger, R., 2010. Lyme borreliosis in dogs and cats: background, diagnosis, treatment and prevention of infections with *Borrelia burgdorferi* sensu stricto. Veterinary Clinics of North America: Small Animal Practice 40, 1103–1119.

Kuleš, J., Horvatić, A., Guillemin, N., Galan, A., Mrljak, V., Bhide, M., 2016. New approaches and omics tools for mining of vaccine candidates against vector-borne diseases. Molecular Biosystems 12, 2680–2694.

Kumar, K.A., Sano, G., Boscardin, S., Nussenzweig, R.S., Nussenzweig, M.C., Zavala, F., Nussenzweig, V., 2006. The circumsporozoite protein is an immunodominant protective antigen in irradiated sporozoites. Nature 444, 937–940.

Kurtenbach, K., Hanincová, K., Tsao, J., Margos, G., Fish, D., Ogden, N., 2006. Fundamental processes in the evolutionary ecology of Lyme borreliosis. Nature Reviews Microbiology 4, 660–669.

Labaied, M., Harupa, A., Dumpit, R.F., Coppens, I., Mikolajczak, S.A., Kappe, S.H.I., 2007. *Plasmodium yoelii* sporozoites with simultaneous deletion of P52 and P36 are completely attenuated and confer sterile immunity against infection. Infection and Immunity 75, 3758–3768.

Labuda, M., Trimnell, A.R., Licková, M., Kazimírová, M., Davies, G.M., Lissina, O., Hails, R.S., Nuttall, P. a, 2006. An antivector vaccine protects against a lethal vector-borne pathogen. PLoS Pathogens 2, e27.

Lai, Y., Di Nardo, A., Nakatsuji, T., Leichtle, A., Yang, Y., Cogen, A.L., Wu, Z.-R., Hooper, L.V., Schmidt, R.R., von Aulock, S., Radek, K.A., Huang, C.-M., Ryan, A.F., Gallo, R.L., 2009. Commensal bacteria regulate Toll-like receptor 3-dependent inflammation after skin injury. Nature Medicine 15, 1377–1382.

LaRocca, T., Holthausen, D., Hsieh, C., Renken, C., Mannella, C., Benach, J., 2009. The bactericidal effect of a complement-independent antibody is osmolytic and specific to Borrelia. Proceedings of the National Academy of Sciences of the United States of America 106, 10752–10757.

Li, X., Liu, X., Beck, D.S., Kantor, F.S., Fikrig, E., 2006. *Borrelia burgdorferi* lacking BBK32, a fibronectin-binding protein, retains full pathogenicity. Infection and Immunity 74, 3305–3313.

Liang, F.T., Brown, E.L., Wang, T., Iozzo, R.V., Fikrig, E., 2004. Protective niche for *Borrelia burgdorferi* to evade humoral immunity. The American Journal of Pathology 165, 977–985. https://doi.org/10.1016/S0002-9440(10)63359-7.

Liu, X.Y., Bonnet, S.I., 2014. Hard tick factors implicated in pathogen transmission. PLoS Neglected Tropical Diseases 8, e2566.

Luft, B., Dunn, J., Lawson, C., 2002. Approaches toward the directed design of a vaccine against *Borrelia burgdorferi*. The Journal of Infectious Diseases 185, S46–S51.

Luke, T.C., Hoffman, S.L., 2003. Rationale and plans for developing a non-replicating, metabolically active, radiation-attenuated *Plasmodium falciparum* sporozoite vaccine. The Journal of Experimental Biology 206, 3803–3808.

Mac-Daniel, L., Buckwalter, M.R., Berthet, M., Virk, Y., Yui, K., Albert, M.L., Gueirard, P., Ménard, R., 2014. Local immune response to injection of *Plasmodium* sporozoites into the skin. The Journal of Immunology 193, 1246–1257.

Mead, P., 2015. Epidemiology of lyme disease. Infectious Disease Clinics of North America 29, 187–210.

Mellouk, S., Lunel, F., Sedegah, M., Beaudoin, R.L., Druilhe, P., 1990. Protection against malaria induced by irradiated sporozoites. Lancet 335, 721.

Mikolajczak, S.A., Lakshmanan, V., Fishbaugher, M., Camargo, N., Harupa, A., Kaushansky, A., Douglass, A.N., Baldwin, M., Healer, J., O'Neill, M., Phuong, T., Cowman, A., Kappe, S.H., 2014. A next-generation genetically attenuated *Plasmodium falciparum* parasite created by triple gene deletion. Molecular Therapy 22, 1707–1715.

Mita, T., Jombart, T., 2015. Patterns and dynamics of genetic diversity in *Plasmodium falciparum*: what past human migrations tell us about malaria. Parasitology International 64, 238–243.

Mueller, A.-K., Camargo, N., Kaiser, K., Andorfer, C., Frevert, U., Matuschewski, K., Kappe, S.H.I., 2005a. *Plasmodium* liver stage developmental arrest by depletion of a protein at the parasite-host interface. Proceedings of the National Academy of Sciences of the United States of America 102, 3022–3027.

Mueller, A.-K., Labaeid, M., Kappe, S.H.I., Maruschewski, K., 2005b. Genetically modified *Plasmodium* parasites as a protective experimental malaria vaccine. Nature 433, 164–167.

Müllegger, R., Glatz, M., 2008. Skin manifestations of lyme borreliosis: diagnosis and management. American Journal of Clinical Dermatology 9, 355–368.

Mulligan, H.W., Russell, P.F., Mohan, B.N., 1941. Active immunization of fowls against *Plasmodium gallinaceum* by injections of killed homologous sporozoites. Journal of the Malaria Institute of India 4, 25–34.

Narasimhan, S., Rajeevan, N., Liu, L., Zhao, Y.O., Heisig, J., Pan, J., Eppler-Epstein, R., Deponte, K., Fish, D., Fikrig, E., 2014. Gut microbiota of the tick vector Ixodes scapularis modulate colonization of the Lyme disease spirochete. Cell Host and Microbe 15, 58–71.

Nestle, F.O., Di Meglio, P., Qin, J.-Z., Nickoloff, B.J., 2009. Skin immune sentinels in health and disease. Nature Reviews Immunology 9, 679–691.

Nganou-Makamdop, K., Sauerwein, R.W., 2013. Liver or blood-stage arrest during malaria sporozoite immunization: the later the better? Trends in Parasitology 29, 304–310.

Nganou-Makamdop, K., Ploemen, I., Behet, M., van Gemert, G.-J., Hermsen, C., Roestenberg, M., Sauerwein, R.W., 2012. Reduced *Plasmodium berghei* sporozoite liver load associates with low protective efficacy after intradermal immunization. Parasite Immunology 34, 562–569.

Ngonseu, E., Chatterjee, S., Wéry, M., 1998. Blocked hepatic-stage parasites and decreased susceptibility to *Plasmodium berghei* infections in BALB/c mice. Parasitology 117, 419–423.

Nussenzweig, R.S., Vanderberg, J.P., Most, H., Orton, C., 1967. Protective immunity produced by the injection of X-irradiated sporozoites of *Plasmodium berghei*. Nature 216, 161–163.

Nussenzweig, R.S., Vanderberg, J., Most, H., 1969. Protective immunity produced by the injection of X-irradiated sporozoites of Plasmodium berghei. IV. Dose response, specificity and humoral immunity. Military Medicine 134 (Suppl.), 1176–1182.

Nuttall, P., Labuda, M., 2004. Tick-host interactions: saliva-activated transmission. Parasitology 129 (Suppl.), S177–S189.

Obeid, M., Franetich, J.-F., Lorthiois, A., Gego, A., Grüner, A.-C., Tefit, M., Boucheix, C., Snounou, G., Mazier, D., 2013. Skin-draining lymph node priming is sufficient to induce sterile immunity against pre-erythrocytic malaria. EMBO Molecular Medicine 5, 250–263.

Ocaña-Morgner, C., Mota, M.M., Rodriguez, A., 2003. Malaria blood stage suppression of liver stage immunity by dendritic cells. The Journal of Experimental Medicine 197, 143–151.

Ogden, N.H., Tsao, J.I., 2009. Biodiversity and Lyme disease: dilution or amplification? Epidemics 1, 196–206.

Ohnishi, J., Piesman, J., de Silva, A., 2001. Antigenic and genetic heterogeneity of *Borrelia burgdorferi* populations transmitted by ticks. Proceedings of the National Academy of Sciences of the United States of America 98, 670–675.

Önder, Ö., Humphrey, P.T., McOmber, B., Korobova, F., Francella, N., Greenbaum, D.C., Brisson, D., 2012. OspC is potent plasminogen receptor on surface of *Borrelia burgdorferi*. Journal of Biological Chemistry 287, 16860–16868.

Orjih, A.U., Nussenzweig, R.S., 1979. *Plasmodium berghei*: suppression of antibody response to sporozoite stage by acute blood stage infection. Clinical and Experimental Immunology 38, 1–8.

Pal, U., de Silva, A., Montgomery, R., Fish, D., Anguita, J., Anderson, J., Lobet, Y., Fikrig, E., 2000. Attachment of *Borrelia burgdorferi* within *Ixodes scapularis* mediated by outer surface protein A. Journal of Clinical Investigation 106, 561–569.

Pasparakis, M., Haase, I., Nestle, O., 2014. Mechanisms regulating skin immunity and inflammation. Nature Reviews Immunology 14, 289–301.

Pei, Y., Tarun, A.S., Vaughan, A.M., Herman, R.W., Soliman, J.M., Erickson-Wayman, A., Kappe, S.H., 2010. *Plasmodium* pyruvate dehydrogenase activity is only essential for the parasite's progression from liver infection to blood infection. Molecular Microbiology 75, 957–971.

Pfeil, J., Sepp, K.J., Heiss, K., Meister, M., Mueller, A.K., Borrmann, S., 2014. Protection against malaria by immunization with non-attenuated sporozoites under single-dose piperaquine-tetraphosphate chemoprophylaxis. Vaccine 32, 6005–6011.

Pied, S., Roland, J., Louise, A., Voegtle, D., Soulard, V., Mazier, D., Cazenave, P.A., 2000. Liver CD4−CD8−NK1.1+ TCR alpha beta intermediate cells increase during experimental malaria infection and are able to exhibit inhibitory activity against the parasite liver stage in vitro. The Journal of Immunology 164, 1463–1469.

Piesman, J., Eisen, L., 2008. Prevention of tick-borne diseases. Annual Review of Entomology 53, 323–343.

Ploemen, I.H., Chakravarty, S., van Gemert, G.-J., Annoura, T., Khan, S.M., Janse, C.J., Hermsen, C.C., Hoffman, S.L., Sauerwein, R.W., 2013. *Plasmodium* liver load following parenteral sporozoite administration in rodents. Vaccine 31, 3410–3416.

Poland, G., 2011. Vaccines against Lyme disease: what happened and what lessons can we learn? Clinical Infectious Diseases 52, s253–s258.

Poljak, A., Comstedt, P., Hanner, M., Schüler, W., Meinke, A., Wizel, B., Lundberg, U., 2012. Identification and characterization of Borrelia antigens as potential vaccine candidates against Lyme borreliosis. Vaccine 30, 4398–4406.

Pombo, D.J., Lawrence, G., Hirunpetcharat, C., Rzepczyk, C., Bryden, M., Cloonan, N., Anderson, K., Mahakunkijcharoen, Y., Martin, L.B., Wilson, D., Elliott, S., Elliott, S., Eisen, D.P., Weinberg, J.B., Saul, A., Good, M.F., 2002. Immunity to malaria after administration of ultralow doses of red cells infected with Plasmodium *falciparum*. The Lancet 360, 610–617.

Potocnjak, P., Yoshida, N., Nussenzweig, R.S., Nussenzweig, V., 1980. Monovalent fragments (Fab) of monoclonal antibodies to a sporozoite surface antigen (Pb44) protect mice against malarial infection. The Journal of Experimental Medicine 151, 1504–1513.

Probert, W., Johnson, B., 1998. Identification of a 47 kDa fibronectin-binding protein expressed by *Borrelia burgdorferi* isolate B31. Molecular Microbiology 30, 1003–1015.

Radolf, J.D., Caimano, M.J., Stevenson, B., Hu, L.T., 2012. Of ticks, mice and men: understanding the dual-host lifestyle of Lyme disease spirochaetes. Nature Reviews Microbiology 10, 87–99.

Radtke, A.J., Kastenmüller, W., Espinosa, D.A., Gerner, M.Y., Tse, S.-W., Sinnis, P., Germain, R.N., Zavala, F., Cockburn, I.A., 2015. Lymph-node resident CD8α+ dendritic cells capture antigens from migratory malaria sporozoites and induce CD8+ T cell responses. PLoS Pathogens 11, e1004637.

Ramamoorthi, N., Narasimhan, S., Pal, U., Bao, F.K., Yang, X.F.F., Fish, D., Anguita, J., Norgard, M.V., Kantor, F.S., Anderson, J.F., Koski, R.A., Fikrig, E., 2005. The Lyme disease agent exploits a tick protein to infect the mammalian host. Nature 436, 573–577.

Randolph, S.E., 2010. To what extent has climate change contributed to the recent epidemiology of tick-borne diseases? Veterinary Parasitology 167, 92–94.

Reid, A., 2015. Large, rapidly evolving gene families are at the forefront of host-parasite interactions in Apicomplexa. Parasitology 142 (Suppl. 1), S57–S70.

Rénia, L., Marussig, M.S., Grillot, D., Pied, S., Corradin, G., Miltgen, F., Del Giudice, G., Mazier, D., 1991. In vitro activity of CD4+ and CD8+ T lymphocytes from mice immunized with a synthetic malaria peptide. Proceedings of the National Academy of Sciences of the United States of America 88, 7963–7967.

Richards, W.H.G., 1966. Immunology – active immunization of chicks against *Plasmodium gallinaceum* by inactivated homologous sporozoites and erythrocytic parasites. Nature 212, 1492–1494.

Richer, L., Brisson, D., Melo, R., Ostfeld, R., Zeidner, N., Gomes-Solecki, M., 2014. Reservoir targeted vaccine against *Borrelia burgdorferi*: a new strategy to prevent Lyme disease transmission. The Journal of Infectious Diseases 209, 1972–1980.

Rieckmann, K.H., Carson, P.E., Beaudoin, R.L., Cassells, J.S., Sell, K.W., 1974. Sporozoite induced immunity in man against an Ethiopian strain of *Plasmodium falciparum*. Transactions of the Royal Society of Tropical Medicine and Hygiene 68, 258–259.

Rieckmann, K.H., 1990. Human immunization with attenuated sporozoites. Bulletin of the World Health Organization 68 (Suppl.), 13–16.

Roberts, L., Enserink, M., 2007. Malaria. Did they really say … eradication? Science 318, 1544–1545.

Roestenberg, M., McCall, M., Hopman, J., Wiersma, J., Luty, A.J., van Gemert, G.J., van de Vegte-Bolmer, M., van Schaijk, B., Teelen, K., Arens, T., Spaarman, L., de Mast, Q., Roeffen, W., Snounou, G., Rénia, L., van der Ven, A., Hermsen, C.C., Sauerwein, R., 2009. Protection against a malaria challenge by sporozoite inoculation. The New England Journal of Medicine 361, 468–477.

Roestenberg, M., Teirlinck, A.C., McCall, M.B., Teelen, K., Makamdop, K.N., Wiersma, J., Arens, T., Beckers, P., van Gemert, G., van de Vegte-Bolmer, M., van der Ven, A.J., Luty, A.J., Hermsen, C.C., Sauerwein, R.W., 2011. Long-term protection against malaria after experimental sporozoite inoculation: an open-label follow-up study. The Lancet 377, 1770–1776.

Romero, P., Maryanski, J.L., Corradin, G., Nussenzweig, R.S., Nussenzweig, V., Zavala, F., 1989. Cloned cytotoxic T cells recognize an epitope in the circumsporozoite protein and protect against malaria. Nature 341, 323–326.

Rosa, P., Tilly, K., Stewart, P., 2005. The burgeoning molecular genetics of the Lyme disease spirochaete. Nature Reviews Microbiology 3, 129–143.

Rudenko, N., Golovchenko, M., Edwards, M., Grubhoffer, L., January 2005. Differential expression of Ixodes ricinus tick genes induced by blood feeding or Borrelia burgdorferi infection. Journal of Medical Entomol. 42 (1), 36–41.

Russell, P.F., Mohan, B.N., 1942. The immunization of fowls against mosquito-borne Plasmodium gallinaceum by injections of serum and of inactivated homologous sporozoites. The Journal of Experimental Medicine 76, 477–495.

Russell, P.F., Mulligan, H.W., Mohan, B.N., 1941. Inactivation of Plasmodium gallinaceum sporozoites infectivity. Journal of the Malaria Institute of India 4, 15–24.

Rutgers, T., Gordon, D., Gathoye, A.M., Hollingdale, M., Hockmeyer, W., Rosenberg, M., De Wilde, M., 1988. Hepatitis B surface antigen as carrier matrix for the repetitive epitope of the circumsporozoite protein of Plasmodium falciparum. Nature Biotechnology 6, 1065–1070.

Sano, G., Hafalla, J.C., Morrot, A., Abe, R., Lafaille, J.J., Zavala, F., 2001. Swift development of protective effector functions in naïve CD8(+) T cells against malaria liver stages. The Journal of Experimental Medicine 194, 173–180.

Schaible, U., Wallich, R., Kramer, M., Gern, L., Anderson, J.F., Museteanu, C., Simon, M., 1993. Immune sera to individual Borrelia burgdorferi isolates or recombinant OspA thereof protect SCID mice against infection with homologous strains but only partially or not at all against those of different OspA/OspB genotype. Vaccine 11, 1049–1054.

Scheckelhoff, M., Telford, S., Hu, L., 2006. Protective efficacy of an oral vaccine to reduce carriage of Borrelia burgdorferi (strain N40) in mouse and tick reservoirs. Vaccine 24, 1949–1957.

Schmidt, N.W., Podyminogin, R.L., Butler, N.S., Badovinac, V.P., Tucker, B.J., Bahjat, K.S., Lauer, P., Reyes-Sandoval, A., Hutchings, C.L., Moore, A.C., Gilbert, S.C., Hill, A.V., Bartholomay, L.C., Harty, J.T., 2008. Memory CD8 T cell responses exceeding a large but definable threshold provide long-term immunity to malaria. Proceedings of the National Academy of Sciences of the United States of America 105, 14017–14022.

Schmidt, N.W., Butler, N.S., Badovinac, V.P., Harty, J.T., 2010. Extreme CD8 T cell requirements for anti-malarial liver-stage immunity following immunization with radiation attenuated sporozoites. PLoS Pathogens 15, e1000998.

Schmidt, N.W., Butler, N.S., Harty, J.T., 2011. Plasmodium-host interactions directly influence the threshold of memory CD8+ T cells required for protective immunity. The Journal of Immunology 186, 5873–5884.

Schnell, G., Boeuf, A., Westermann, B., Jaulhac, B., Carapito, C., Boulanger, N., Ehret-Sabatier, L., 2015. Discovery and targeted proteomics on cutaneous biopsies: a promising work toward an early diagnosis of Lyme disease. Molecular and Cellular Proteomics 14, 1254–1264.

Schofield, L., Villaquiran, J., Ferreira, A., Schellekens, H., Nussenzweig, R., Nussenzweig, V., 1987. Gamma interferon, CD8+ T cells and antibodies required for immunity to malaria sporozoites. Nature 330, 664–666.

Schröder, N.W.J., Eckert, J., Stübs, G., Schumann, R.R., 2008. Immune responses induced by spirochetal outer membrane lipoproteins and glycolipids. Immunobiology 213, 329–340.

Schroeder, H., Daix, V., Gillet, L., Renauld, J.-C., Vanderplasschen, A., 2007. The paralogous salivary anti-complement proteins IRAC I and IRAC II encoded by Ixodes ricinus ticks have broad and complementary inhibitory activities against the complement of different host species. Microbes and Infection 9, 247–250.

Schuijt, T.J., Coumou, J., Narasimhan, S., Dai, J., Deponte, K., Wouters, D., Brouwer, M., Oei, A., Roelofs, J.J., van Dam, A.P., van der Poll, T., Van't Veer, C., Hovius, J.W., Fikrig, E., 2011a. A tick mannose-binding lectin inhibitor interferes with the vertebrate complement cascade to enhance transmission of the lyme disease agent. Cell Host and Microbe 10, 136–146.

Schuijt, T.J., Hovius, J.W., van der Poll, T., van Dam, A.P., Fikrig, E., 2011b. Lyme borreliosis vaccination: the facts, the challenge, the future. Trends in Parasitology 27, 40–47.

Schwan, T.G., Piesman, J., Golde, W.T., Dolan, M.C., Rosa, P.A., 1995. Induction of an outer surface protein on *Borrelia burgdorferi* during tick feeding. Proceedings of the National Academy of Sciences of the United States of America 92, 2909–2913.

Seder, R.A., Chang, L.-J., Enama, M.E., Zephir, K.L., Sarwar, U.N., Gordon, I.J., Holman, L.A., James, E.R., Billingsley, P.F., Gunasekera, A., Richman, A., Chakravarty, S., Manoj, A., Velmurugan, S., Li, M., Ruben, A.J., Li, T., Eappen, A.G., Stafford, R.E., Plummer, S.H., Hendel, C.S., Novik, L., Costner, P.J.M., Mendoza, F.H., Saunders, J.G., Nason, M.C., Richardson, J.H., Murphy, J., Davidson, S.A., Richie, T.L., Sedegah, M., Sutamihardja, A., Fahle, G.A., Lyke, K.E., Laurens, M.B., Roederer, M., Tewari, K., Epstein, J.E., Kim Lee Sim, B., Ledgerwood, J.E., Graham, B.S., Hoffman, S.L., the VRC 312 Study Team, 2013. Protection against malaria by intravenous immunization with a nonreplicating sporozoite vaccine. Science 341, 1359–1365.

Sigler, C.I., Leland, P., Hollingdale, M.R., 1984. *In vitro* infectivity of irradiated *Plasmodium berghei* sporozoites to cultured hepatoma cells. The American Journal of Tropical Medicine and Hygiene 33, 544–547.

Silvie, O., Semblat, J.P., Franetich, J.F., Hannoun, L., Eling, W., Mazier, D., 2002. Effects of irradiation on *Plasmodium falciparum* sporozoite hepatic development: implications for the design of pre-erythrocytic malaria vaccines. Parasite Immunology 24, 221–223.

Silvie, O., Rubinstein, E., Franetich, J.-F., Prenant, M., Belnoue, E., Rénia, L., Hannoun, L., Eling, W., Levy, S., Boucheix, C., Mazier, D., 2003. Hepatocyte CD81 is required for *Plasmodium falciparum* and *Plasmodium yoelii* sporozoite infectivity. Nature Medicine 9, 93–96.

Silvie, O., Goetz, K., Matuschewski, K., 2008. A sporozoite asparagine-rich protein controls initiation of Plasmodium liver stage development. PLoS Pathogens 4, e1000086.

Sonenshine, D.E., Anderson, J.M., 2014. Mouthparts and digestive system. In: Sonenshine, D.E., Michael Rose, R. (Eds.), Biology of Ticks. Oxford University Press, New York, pp. 122–162.

Spitalny, G.L., Nussenzweig, R.S., 1972. Effects of various routes of immunization and methods of parasite attenuation on the development of protection against sporozoite-induced rodent malaria. Military Medicine 39, 506–514.

Spring, M., Murphy, J., Nielsen, R., Dowler, M., Bennett, J.W., Zarling, S., Williams, J., de la Vega, P., Ware, L., Komisar, J., Polhemus, M., Richie, T.L., Epstein, J., Tamminga, C., Chuang, I., Richie, N., O'Neil, M., Heppner, D.G., Healer, J., O'Neill, M., Smithers, H., Finney, O.C., Mikolajczak, S.A., Wang, R., Cowman, A., Ockenhouse, C., Krzych, U., Kappe, S.H., 2013. First-in-human evaluation of genetically attenuated *Plasmodium falciparum* sporozoites administered by bite of *Anopheles* mosquitoes to adult volunteers. Vaccine 31, 4975–4983.

Stanek, G., Wormser, G., Gray, J., Strle, F., 2012. Lyme borreliosis. Lancet 379, 461–473.

Steere, A., Sikand, V., Meurice, F., Parenti, D., Fikrig, E., Schoen, R., Nowakowski, J., Schmid, C., Laukamp, S., Buscarino, C., Krause, D., 1998. Vaccination against lyme disease with recombinant *Borrelia burgdorferi* outer-surface lipoprotein A with adjuvant. Lyme disease vaccine study group. The New England Journal of Medicine 339, 209–215.

Steere, A.C., Gross, D., Meyer, A.L., Huber, B.T., 2001. Autoimmune mechanisms in antibiotic treatment-resistant lyme arthritis. Journal of Autoimmunity 16, 263–268.

Steere, A., Strle, F., Wormser, G., Hu, L., Branda, J., Hovius, J., Li, X., Mead, P., 2016. Lyme borreliosis. Nature Reviews Disease Primers 2, 1–13.

Stewart, M.J., Nawrot, R.J., Schulman, S., Vanderberg, J.P., 1986. *Plasmodium berghei* sporozoite invasion is blocked *in vitro* by sporozoite-immobilizing antibodies. Infection and Immunity 51, 859–864.

Suhrbier, A., Winger, L.A., Castellano, E., Sinden, R.E., 1990. Survival and antigenic profile of irradiated malarial sporozoites in infected liver cells. Infection and Immunity 58, 2834–2839.

Töpfer, K., Straubinger, R., 2007. Characterization of the humoral immune response in dogs after vaccination against the Lyme borreliosis agent A study with five commercial vaccines using two different vaccination schedules. Vaccine 25, 314–326.

Trimnell, A., Davies, G., Lissina, O., Hails, R., Nuttall, P., 2005. A cross-reactive tick cement antigen is a candidate broad-spectrum tick vaccine. Vaccine 23, 4329–4341.

Tsuji, M., Romero, P., Nussenzweig, R.S., Zavala, F., 1990. CD4+ cytolytic T cell clone confers protection against murine malaria. The Journal of Experimental Medicine 172, 1353–1357.

Tsuji, M., Mombaerts, P., Lefrançois, L., Nussenzweig, R.S., Zavala, F., Tonegawa, S., 1994. Gamma delta T cells contribute to immunity against the liver stages of malaria in alpha beta T-cell-deficient mice. Proceedings of the National Academy of Sciences of the United States of America 91, 345–349.

Tsuji, M., Miyahira, Y., Nussenzweig, R.S., Aguet, M., Reichel, M., Zavala, F., 1995. Development of antimalaria immunity in mice lacking IFN-gamma receptor. The Journal of Immunology 154, 5338–5344.

van Dijk, M.R., Douradinha, B., Franke-Fayard, B., Heussler, V., van Dooren, M.W., van Schaijk, B., van Gemert, G.-J., Sauerwein, R.W., Mota, M.M., Waters, A.P., Janse, C.J., 2005. Genetically attenuated, P36p-deficient malarial sporozoites induce protective immunity and apoptosis of infected liver cells. Proceedings of the National Academy of Sciences of the United States of America 102, 12194–12199.

van Schaijk, B.C.L., Ploemen, I.H.J., Annoura, T., Vos, M.W., Foquet, L., van Gemert, G.-J., Chevalley-Maurel, S., van de Vegte-Bolmer, M., Sajid, M., Franetich, J.-F., Lorthiois, A., Leroux-Roels, G., Meuleman, P., Hermsen, C.C., Mazier, D., Hoffman, S.L., Janse, C.J., Khan, S.M., Sauerwein, R.W., 2014. A genetically attenuated malaria vaccine candidate based on *P. falciparum* b9/slarp gene-deficient sporozoites. eLife 3, e03582.

VanBuskirk, K.M., O'Neill, M.T., De La Vega, P., Maier, A.G., Krzych, U., Williams, J., Dowler, M.G., Sacci, J.B., Kangwanrangsan, N., Tsuboi, T., Kneteman, N.M., Heppner, D.G., Murdock, B.A., Mikolajczak, S.A., Aly, A.S.I., Cowman, A.F., Kappe, S.H., 2009. Preerythrocytic, live-attenuated *Plasmodium falciparum* vaccine candidates by design. Proceedings of the National Academy of Sciences of the United States of America 106, 13004–13009.

Vanderberg, J.P., Frevert, U., 2004. Intravital microscopy demonstrating antibody-mediated immobilisation of *Plasmodium berghei* sporozoites injected into skin by mosquitoes. International Journal of Parasitology 34, 991–996.

Vanderberg, J.P., 1974. Studies on the motility of *Plasmodium* sporozoites. Journal of Protozoology 21, 527–537.

Vaughan, A.M., O'Neill, M.T., Tarun, A.S., Camargo, N., Phuong, T.M., Aly, A.S.I., Cowman, A.F., Kappe, S.H., 2009. Type II fatty acid synthesis is essential only for malaria parasite late liver stage development. Cellular Microbiology 11, 506–520.

Voza, T., Kebaier, C., Vanderberg, J.P., 2010. Intradermal immunization of mice with radiation-attenuated sporozoites of *Plasmodium yoelii* induces effective protective immunity. Malaria Journal 9, 362.

Voza, T., Miller, J.L., Kappe, S.H., Sinnis, P., 2012. Extrahepatic exoerythrocytic forms of rodent malaria parasites at the site of inoculation: clearance after immunization, susceptibility to primaquine, and contribution to blood-stage infection. Infection and Immunity 80, 2158–2164.

Weiss, W.R., Sedegah, M., Beaudoin, R.L., Miller, L.H., Good, M.F., 1988. CD8$^+$ T cells (cytotoxic/suppressors) are required for protection in mice immunized with malaria sporozoites. Proceedings of the National Academy of Sciences of the United States of America 85, 573–576.

Weiss, W.R., Mellouk, S., Houghten, R.A., Sedegah, M., Kumar, S., Good, M.F., Berzofsky, J.A., Miller, L.H., Hoffman, S.L., 1990. Cytotoxic T cells recognize a peptide from the circumsporozoite protein on malaria-infected hepatocytes. The Journal of Experimental Medicine 171, 763–773.

Wellde, B.T., Sadun, E.H., 1967. Resistance produced in rats and mice by exposure to irradiated *Plasmodium berghei*. Experimental Parasitology 21, 310–324.

Wenner-Moyer, M., 2015. Tick trouble. Nature 524, 406–408.

Wikel, S.K., 1999. Tick modulation of host immunity: an important factor in pathogen transmission. International Journal of Parasitology 29, 851–859.

Wikel, S.K., 2013. Ticks and tick-borne pathogens at the cutaneous interface: host defenses, tick countermeasures, and a suitable environment for pathogen establishment. Frontiers in Microbiology 4, 337.

Wormser, G., Brisson, D., Liveris, D., Hanincová, K., Sandigursky, S., Nowakowski, J., Nadelman, R., Ludin, S., Schwartz, I., 2008. *Borrelia burgdorferi* genotype predicts the capacity for hematogenous dissemination during early Lyme disease. The Journal of Infectious Diseases 198, 1358–1364.

Wressnigg, N., Pöllabauer, E., Aichinger, G., Portsmouth, D., Löw-Baselli, A., Fritsch, S., Livey, I., Crowe, B., Schwendinger, M., Brühl, P., Pilz, A., Dvorak, T., Singer, J., Firth, C., Luft, B., Schmitt, B., Zeitlinger, M., Müller, M., Kollaritsch, H., Paulke-Korinek, M., Esen, M., Kremsner, P., Ehrlich, H., Barrett, P., 2013. Safety and immunogenicity of a novel multivalent OspA vaccine against Lyme borreliosis in healthy adults: a double-blind, randomised, dose-escalation phase 1/2 trial. The Lancet Infectious Diseases 13, 680–689.

Wykes, M.N., Horne-Debets, J.M., Leow, C.Y., Karunarathne, D.S., 2014. Malaria drives T cells to exhaustion. Frontiers in Microbiology 5, 249.

Yoshida, N., Nussenzweig, R.S., Potocnjak, P., Nussenzweig, V., Aikawa, M., 1980. Hybridoma produces protective antibodies directed against the sporozoite stage of malaria parasite. Science 207, 71–73.

Yu, M., Kumar, T.R., Nkrumah, L.J., Coppi, A., Retzlaff, S., Li, C.D., Kelly, B.J., Moura, P.A., Lakshmanan, V., Freundlich, J.S., Valderramos, J.C., Vilcheze, C., Siedner, M., Tsai, J.H., Falkard, B., Sidhu, A.B., Purcell, L.A., Gratraud, P., Kremer, L., Waters, A.P., Schiehser, G., Jacobus, D.P., Janse, C.J., Ager, A., Jacobs, W.R., Sacchettini, J.C., Heussler, V., Sinnis, P., Fidock, D.A., 2008. The fatty acid biosynthesis enzyme FabI plays a key role in the development of liver-stage malarial parasites. Cell Host and Microbe 4, 567–578.

Zhong, W., Stehle, T., Museteanu, C., Siebers, A., Gern, L., Kramer, M., Wallich, R., Simon, M.M., 1997. Therapeutic passive vaccination against chronic Lyme disease in mice. Proceedings of the National Academy of Sciences of the United States of America 94, 12533–12538.

Chapter 12

Tools to Decipher Vector-Borne Pathogen and Host Interactions in the Skin

Pauline Formaglio[1], Joppe W. Hovius[2], Chetan Aditya[3], Joana Tavares[4], Lauren M.K. Mason[2], Robert Ménard[3], Nathalie Boulanger[5], Rogerio Amino[3]

[1]*Otto-von-Guericke University, Magdeburg, Germany;* [2]*University of Amsterdam, Amsterdam, The Netherlands;* [3]*Institut Pasteur, Paris, France;* [4]*Universidade do Porto, Porto, Portugal;* [5]*Université de Strasbourg, Strasbourg, France*

VISUALIZING A SKIN PHASE IN THE LIFE HISTORY OF VECTOR-BORNE PATHOGENS

Introduction

At the end of the 19th century, a sequence of seminal reports was published showing that hematophagous arthropods can harbor (Manson, 1878) and actively transmit parasites (Bruce, 1895; Grassi, 1898; Ross, 1898; Smith and Kilborne, 1893), bacteria (Simond, 1898), and viruses (Reed et al., 1900) to vertebrate hosts during blood feeding. These results opened a new perspective regarding the way pathogens, and thus the disease they cause, are transmitted from invertebrate vectors to vertebrate hosts, leading ultimately to the concept of vector-borne diseases (VBDs). VBDs are responsible for more than 1 billion cases, causing more than 1 million deaths every year, malaria being the most important contributor to the VBDs' deadly toll (http://www.who.int/mediacentre/factsheets/fs387/en/).

Strikingly, although the transmission mode of malaria parasites was described rather soon after the identification of its etiological agent, the characterization of the early events following the mosquito bite, and leading to the hepatic and subsequent blood infection, has been more challenging. In particular, the precise site of *Plasmodium* sporozoites inoculation, either in the vessels or in the extravascular parts of the skin, and their mode of dissemination, lymphatic or hematogenous, have remained uncertain for over a century. Here, we first review how successive findings, related to *Plasmodium* and other

Skin and Arthropod Vectors. https://doi.org/10.1016/B978-0-12-811436-0.00012-5

arthropod-borne pathogens, have led to the identification of a skin phase in the life cycle of these microorganisms and forged our current understanding of their systemic dissemination. The instrumental role of intravital imaging techniques and their continuous improvement is highlighted in the end of this first section.

Despite their informative power, *in vivo* imaging approaches are mostly restricted to small laboratory animal models, frequently rodents, which precludes the study of human pathogens that are not adapted to infect these hosts. The second section of this chapter reviews alternative methods to dissect the behavior of human-infecting pathogens using primary human cells and skin tissue cultures *in vitro*. The advances in the generation of humanized mice harboring human skin explants and human immune system emerge as a solution to study the dynamic interactions of pathogens and human cells *in vivo* using imaging and are discussed in the end of the second section.

Who's in: Intravascular or Extravascular Transmission of Pathogens?

Whether infected blood-feeding vectors inject pathogens directly into the blood circulation or extravascularly in the skin tissue has long raised controversy in the community. The first suggested evidence of an intravascular inoculation of malaria parasites during transmission came from an experiment performed by Boyd and Stratman-Thomas in 1934. In this experiment, a patient was first exposed to the bites of an infected mosquito during 6 min and the inoculation site was then excised within the next 7 min. Thirteen days later, the patient developed symptoms of malaria, leading the authors to conclude that *"sporozoites were evidently introduced directly in the circulation"* (Boyd and Stratman-Thomas, 1934). That same year, Alberto Missiroli conducted an experiment where he introduced cannulae in the wings of birds, which enabled the extravascular delivery of sporozoites. Following injection, he excised the inoculation site after 1, 5, 10, or 20 min and checked for infection. Intriguingly, all animals became infected, except for those excised 1 min after inoculation. The author concluded that *"sporozoites leave within 1 to 5 minutes the point of inoculation through the lymphatics"* (Missiroli, 1934). In 1934, it was therefore clear that sporozoites deposited extravascularly in the skin could rapidly leave the inoculation site and infect the host within the first minutes following their transmission. This fast kinetics of exit challenged the interpretation of Boyd and Stratman, suggesting that the 13 min preceding skin excision in their experiment might have been enough for extravascular sporozoites to leave the bite site and infect the host, possibly using the lymphatics as claimed by Missiroli.

Missiroli's discovery cast a doubt on the way, intravascular or extravascular, infected mosquitoes inject sporozoites in the skin of the host during natural transmission. Consequently, his study also questioned the route used by sporozoites to get access to the blood. Of note, at this time, active motility of sporozoites had not been reported yet and only a passive transport, via the lymph or blood circulation, was considered as a reasonable option for parasite exit from the inoculation site. Further evidence supporting the notion that "extra-vascular" malaria sporozoites are infectious to the host include experiments showing that mosquitoes biting avascular blisters can infect humans (Boyd and Kitchen, 1939), or more impressively, that sporozoites, but not infected red blood cells, can infect rodents when administered orally (Yoeli and Most, 1971). Yet, in spite of these findings, uncertainty regarding the site of sporozoite delivery persisted. The classic manuscript published by Fairley et al. in 1947 illustrates clearly how unsettled the dispute about the mode of sporozoite transmission remained. In this report, human donors were exposed for 5–8 min to mosquitoes infected with the two most important species causing human malaria, *Plasmodium falciparum* and *Plasmodium vivax*. During the bite session and up to 120 min after, 150–500 mL of the donor's blood were withdrawn and transferred to nonimmune recipients. To assess the infectivity of the subinoculated volumes, recipients were then monitored for the appearance of malaria symptoms and the presence of infected erythrocytes in their blood. As observed in Fig. 12.1, the percentage of infected recipients was inversely proportional to the time the blood was withdrawn from bitten donors, and according to the authors, "… *viable sporozoites may be demonstrated in the circulating blood for short periods (1/2 to 1 hour) after inoculation of sporozoites by anopheline mosquitoes into the tissues or directly into the blood vessels*" (Fairley, 1947).

Studies on the blood-feeding behavior of mosquitoes and the role of saliva in this process next provided further insights into the scenario of sporozoite

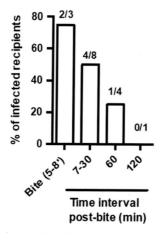

FIGURE 12.1   **Most sporozoites reach the blood circulation in the first hour postinoculation.** Donors were bitten by *Plasmodium vivax*- and *Plasmodium falciparum*-infected mosquitoes. After the time indicated in the graph, 150–500 mL of blood were withdrawn from bitten donors and transferred to naive recipients. Bars show the percentage of infected recipients following transfusion. *(Data adapted from Fairley, N.H., 1947. Sidelights on malaria in man obtained by subinoculation experiments. Transactions of the Royal Society of Tropical Medicine and Hygiene 40 (5), 621–676.)*

transmission. According to Griffiths and Gordon, blood feeding can be divided in two phases (Griffiths and Gordon, 1952). In the first exploratory phase, called probing phase, mosquitoes insert their specialized mouthparts into the host skin while salivating copiously during their search for blood vessels, which only represent a small fraction of the skin volume. After rupturing capillaries or puncturing a larger vessel, they start imbibing blood from hemorrhagic pools or blood vessels, respectively, which constitutes the second phase of the feeding process (Gordon and Lumsden, 1939). Notably, lack of saliva does not impair blood uptake (Hudson et al., 1960), but mosquitoes take longer to obtain their blood meal, with a great increase in their probing time (Ribeiro et al., 1984). These results indicate that saliva is important for finding a blood vessel but does not play a direct role in blood ingestion (Ribeiro, 1987). Accordingly, saliva, and thus the pathogens it contains, should mostly be injected in the extravascular regions of the skin and not inside blood vessels.

This hypothesis could indeed be verified for the extracellular and motile parasites *Trypanosoma brucei rhodesiense*, *Trypanosoma brucei brucei*, and *Trypanosoma congolense*, which are transmitted to mammals by tsetse flies. Following an infective bite, a local inflammatory reaction (chancre) develops and analysis of skin biopsies reveals the presence of parasites in the skin. Local growth at the site of infection was found to precede the ensuing detection of parasites, first, in the lymphatic system and, a few days later, in the blood (Akol and Murray, 1986; Barry and Emergy, 1984; Caljon et al., 2016; Dwinger et al., 1990; Fairbairn and Godfrey, 1957; Roberts et al., 1969) (Fig. 12.2).

In the 1990s, the group of Andrew Spielman consolidated the notion that pathogens transmitted during the bite of hematophagous vectors are mostly,

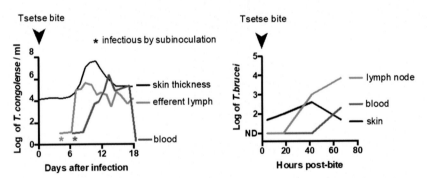

FIGURE 12.2 Skin inoculation and lymphohematogenous dissemination of African trypanosomes following natural transmission. The graphs show the kinetics of dissemination of *Trypanosoma congolense* and *Trypanosoma brucei* delivered in the host skin by tsetse flies. *ND*, not detected. *(Data adapted from Dwinger, R.H., Murray, M., Moloo, S.K., 1990. Parasite kinetics and cellular responses in goats infected and superinfected with Trypanosoma congolense transmitted by* Glossina morsitans *centralis. Acta Tropica 47, 23–33; Caljon, G., Van Reet, N., De Trez, C., Vermeersch, M., Perez-Morga, D., Van Den Abbeele, J., 2016. The dermis as a delivery site of Trypanosoma brucei for tsetse flies. PLoS Pathogens 12, e1005744.)*

if not all, deposited in the extravascular regions of the host skin, from where they disseminate throughout the body of the host with different kinetics. To reach this conclusion, they allowed infected vectors to feed on naive recipients. After a few minutes to several days, the bite site was excised and systemic infection evaluated. For the tick-borne, Lyme disease-causing spirochete *Borrelia burgdorferi* sensu lato, while all control animals became infected after the vector bite, no infection was detected if the inoculation site was excised up to 2 days following tick detachment, showing that infected ticks do not inoculate bacteria directly into the blood circulation (Shih et al., 1992).

The same occurred when mosquitoes infected with Rift Valley fever virus were allowed to bite the tail of rodents. Excision of the inoculation site up to 5 min after exposure to the vector led to ~10% of lethality, while ~80% of nonexcised controls died after 50 h, showing again that most of viruses were inoculated extravascularly and exhibited a delayed migration toward others tissues. Notably, there was no significant association between mouse survival and blood acquisition during the infective bite, supporting the notion that viruses are not directly inoculated into the vasculature during blood ingestion (Turell and Spielman, 1992).

In 1997, Sidjanski and Vanderberg conducted an experiment similar to Missiroli's but using *Plasmodium yoelii* sporozoites naturally transmitted to rodents by mosquito bites. The authors found that most animals excised either immediately or 5 min after a 2-min bite session were free of infection, whereas a great percentage of the mice became infected when the bite site was removed 15 min postbite or left untouched (Fig. 12.3). These results strikingly paralleled the phenotype of delayed migration out of the inoculation site reported for the first time by Missiroli in 1934 after extravascular injection of sporozoites. The fact that a small proportion of the animals excised at the earliest time point still developed malaria suggests that extravascular sporozoites could rapidly migrate from the site of inoculation during the 2-min bite session and infect the host, such as demonstrated by Missiroli. Alternatively, it is also possible that mosquitoes sporadically inoculate sporozoites directly into the blood circulation. In both cases, it is, however, evident that, in most animals, infection is caused by sporozoites inoculated outside blood vessels.

All these experiments, using a variety of vectors relying on distinct blood-feeding strategies and microorganisms from different kingdoms, strongly suggest that pathogens transmitted by the bite of infected vectors share a common skin phase characterized by their extravascular inoculation followed, later on, by their entry into the blood circulation according to extremely variable kinetics. Despite possessing active motility, trypanosomes and spirochetes, whose vectors—tsetse flies and ticks—exclusively feed on hemorrhagic pools (Lavoipierre, 1965), both require extensive amounts of time, in the range of days, to disseminate out of the skin. This long-delayed migration is likely a consequence of their initial cutaneous replication. Indeed, these extracellular pathogens initially stay at the bite site where they multiply causing a skin

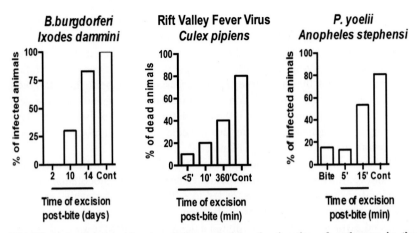

FIGURE 12.3 Extravascular inoculation and delayed migration of pathogens in the skin. Anesthetized mice were bitten by *Ixodes* tick, *Culex*, and *Anopheles* mosquitoes, infected by *B. burgdorferi* sensu lato (Lyme disease), Rift Valley fever virus, and *Plasmodium yoelii* parasite (malaria), respectively. Following bite, the inoculation site was excised at the indicated time points and the percentage of infected animals determined. Controls (Cont) are bitten, nonexcised, or contralateral excised animals. *(Adapted from Shih, C.M., Pollack, R.J., Telford 3rd, S.R., Spielman, A., 1992. Delayed dissemination of Lyme disease spirochetes from the site of deposition in the skin of mice. The Journal of Infectious Diseases 166, 827–831; Turell, M.J., Spielman, A., 1992. Nonvascular delivery of Rift Valley fever virus by infected mosquitoes. The American Journal of Tropical Medicine and Hygiene 47, 190–194; Sidjanski, S., Vanderberg, J.P., 1997. Delayed migration of Plasmodium sporozoites from the mosquito bite site to the blood. The American Journal of Tropical Medicine and Hygiene 57, 426–429.)*

manifestation in the form of a chancre (trypanosomes) or an expanding erythema migrans (*B. burgdorferi* sl). Their kinetics of entry into the systemic circulation should therefore be influenced by their doubling time and by their ability to cross the lymphatics or, alternatively, to directly invade a blood vessel.

On the other hand, viruses and sporozoites transmitted by mosquitoes, which combine pool and vessel-feeding behaviors (O'Rourke, 1956), exhibit a faster dissemination out of the bite site. Since viruses are nonmotile, their relatively rapid translocation is either occurring by the means of an infected host cell moving out of the bite site or by passive transport via the lymph. As for trypanosomes and spirochetes, replication of the Rift Valley fever virus at the inoculation site has been proposed to determine the kinetics of virus dissemination (Turell and Spielman, 1992). In stark comparison, the rapid exit of malaria sporozoites from the skin, which occurs within minutes to 1 h (Figs. 12.1 and 12.3), is incompatible with parasite replication at the bite site, indicating that sporozoites must have evolved a different strategy to get access to the blood circulation during the initial step of malaria infection.

Can You Show Us the Way: Blood or Lymph Dissemination Route?

Once it was established that arthropod vectors mostly inoculate pathogens in the extravascular parts of the skin, figuring out how microorganisms can propagate from the bite site to the blood was the next logical question to answer. Two main routes can be considered for pathogens to disseminate: progression through the lymphatic system, which eventually drains into the subclavian veins, or a direct entry into the blood vessels. While analysis of clinical symptoms and tissue biopsies has provided compelling arguments for many infectious agents, the way *Plasmodium* sporozoites gain access to the bloodstream has puzzled scientists until the beginning of the 21st century.

After being inoculated in the host skin tissue, a great number of motile and nonmotile, extracellular, and intracellular vector-borne pathogens use the lymph or host cells circulating through the lymphatics to reach the blood circulation. Arboviruses, African trypanosomes, *Leishmania* spp., filarial worms, *B. burgdorferi* sl, *Rickettsia* spp., and *Yersinia pestis* seem to preferentially adopt this lymphohematogenous dissemination route to infect the host as evidenced by two facts. The first is their detection in the lymph or lymph nodes draining the bite site, where their development can often result in a lymphadenopathy (e.g., bubonic plague, bubonic leishmaniasis, Winterbottom's sign). The second is that their presence in the lymphatic system precedes their detection in the blood circulation. From there, they infect other organs causing distinctive pathology (encephalitis, pneumonia, septicemia) and are eventually transmitted back to a vector. Fig. 12.2 shows the typical profile of a lymphohematogenous dissemination (in this case African trypanosomes), which appears to be the propagation pathway selected by many vector-borne pathogens. Not surprisingly, several nonmotile pathogens using this route infect host cells, such as dermal dendritic cells (DDCs) (Lozach et al., 2011), which can circulate through the lymphatics. In addition, this dissemination path seems to be facilitated by the discontinuous nature of the button-like interendothelial junctions of initial lymphatic capillaries (Baluk et al., 2007), which favors the entry of liquids and cells into lymph vessels, and by the ability of motile pathogens to swim at high speeds in solution. Nonetheless, a direct hematogenous dissemination of extravascularly inoculated pathogens from the cutaneous site of infection and multiplication (chancre, erythema, tache noire) cannot be formally excluded. To do so, pathogens must, however, necessarily have the ability to cross the perivascular and endothelial barriers that constitute cutaneous blood vessels.

In the case of malaria parasites, the existence of suggestive evidence pleading in favor of both lymphohematogenous and hematogenous dissemination has long left the question of sporozoite propagation to the liver in a *statu quo*. In 1939, Boyd and Kitchen indeed observed *P. vivax* sporozoites in the lymph node draining the site of parasite inoculation at 24 h postbite. The absence of signs of degradation, invasion, and intracellular development suggested that these parasites had "neither reached their destination nor achieved their destiny" (Boyd and Kitchen, 1939) and were possibly en route to the liver. However, according to subinoculation (Fig. 12.1) and excision (Fig. 12.3) experiments, most sporozoites get access to the blood before 1-h postbite. The discrepancy with the rapid kinetics of sporozoite entry in the circulation thus suggested that these "late" lymph node sporozoites, if they can reach the bloodstream, are unlikely to significantly contribute to the primary liver infection. While it became undeniable that sporozoites can use the lymphatic route to exit the skin, the contribution of this lymphatic parasites to the hepatic infection remained uncertain and their exact fate in the draining lymph node (DLN) unclear. On the other hand, many sporozoites of the phylum Apicomplexa, which *Plasmodium* belongs to, are invasive cells capable of establishing a primary intracellular infection in organs that are far away from the initial site of transmission. This peculiarity can be observed in bite-transmitted *Plasmodium*, *Hepatocystis*, and *Leucocytozoon* sporozoites released in the host skin and which establish a primary infection in the liver of the host. In some other species, sporozoites are released in the digestive tube of the host following oral contamination. After crossing the intestinal epithelial barrier, these sporozoites somehow infect distal organs, such as the seminal vesicles of earthworms (*Monocystis*) and the liver of vertebrate hosts (*Hepatozoon* and *Karyolysus*) (Olsen, 1986). Because invertebrates completely lack a lymphatic system, the example of *Monocystis* supported the hypothesis of sporozoite dissemination via intravasation into blood vessels (Olsen, 1986), a behavior that could be conserved throughout the phylum. In spite of these circumstantial elements, the direct evidence demonstrating the nature, either hematogenous or lymphohematogenous, of sporozoite dissemination was lacking and was only provided in the beginning of the 21st century when direct *in vivo* observation of fluorescent sporozoites following natural transmission identified blood vessel invasion as the only productive mode of dissemination.

In Vivo Imaging of Malaria Parasites in the Skin

The Long Road Leading to the Fluorescent Microscopic World

The observation and analysis of the phenomenal world are the basis of natural history. However, the spatial resolution of our naked eyes impedes the observation of what is below a certain size limit. Extraordinarily, despite the inaccessibility of this microscopic world to our direct perception, the existence of "minute animals (animalcules), invisible to the eye" and responsible for causing diseases, was hypothesized as early as c.30 BCE (Varro and Storr-Best, 1912), i.e., seventeen centuries before the first observation of an animalcule.

The invention of the microscope at the end of the 16th century was the first revolutionary step leading to a new vision and understanding of this imperceptible microscopic world (Van Helden, 2010). In 1674, Antony van Leeuwenhoek stunned the world with the groundbreaking discovery of microorganisms, observed with his unique homemade microscopes capable of magnifying more than

200 times the observed specimen (Dobell and Leeuwenhoek, 1932). Microscopes were also fundamental for establishing a causal link between minute living creatures and human disease as shown for itch mites and scabies in 1687 by Giovan Cosimo Bonomo and Diacinto Cestoni (Ramos-e-Silva, 1998) or for plasmodial parasites and malaria in 1880 by Alphonse Laveran (1881). Van Leeuwenhoek is not only one of the precursors of microbiology but also the pioneer of *in vivo* imaging microscopy describing with precision in 1688, the blood circulation in the translucent tail of live aquatic specimens using his eel-viewer microscope (Harris, 1921). Yet, pathogenic microorganisms usually live inside invertebrate and vertebrate host tissues, which are often neither translucent nor accessible to *in vivo* observation. In addition, the lack of contrast between pathogens and host cells impedes their discrimination without the use of distinctive staining, hindering the direct observation of their dynamic behavior inside their hosts.

The solution to these problems started to emerge in 1911, when Oskar Heimstädt using ultraviolet (UV) darkfield illumination imaged autofluorescent bacteria with the prototype of a fluorescence microscope. In 1929, based on this UV illumination, Phillip Ellinger and August Hirt built the first epifluorescence "intravital microscope" and imaged the kidney and liver of rodents injected with fluorochromes. The presence of a filter between the objective and the ocular blocked the reflected UV excitation light, but not the fluorescent light emitted by fluorescein, used to label body fluids or by a nuclear stain, trypaflavine, used to reveal the tissue architecture (Ellinger and Hirt, 1929). If the conceptual basis of fluorescence microscopes was defined at this time, the efforts to establish specific, stable, and innocuous fluorescent staining of live cells were only beginning. Fluorochrome-coupled antibodies or fluorescent vital stains could be used to specifically image labeled cells *in vivo*, but the great revolution started in 1962 with the discovery of the green fluorescent protein (GFP) by Shimomura et al. (1962). More than 30 years later, Chalfie et al. (1994) used this fluorescent reporter to visualize the pattern of β-tubulin gene expression in the worm *Caenorhabditis elegans*, revolutionizing life sciences by allowing the direct observation and quantification of dynamic events in living organisms in a wide spatiotemporal and color resolution (Tsien, 1998). In 1996, the publication of stable gene integration in the *Plasmodium* genome (van Dijk et al., 1996; Wu et al., 1996) opened the possibility for the generation of the first malaria parasite stably expressing GFP as a genetic marker in 2001 (Natarajan et al., 2001). Therefore, thanks to the development of fluorescence microscopes and GFP-expressing parasites, the behavior of pathogens transmitted by arthropod bite could finally be observed in the skin of the host.

The Triple Hallmark of a Novel Skin Phase of Malarial Infection: Invasion, Invasion, and Invasion

In 2003, the first movie of a fluorescent sporozoite gliding in a mammalian host was presented at the Molecular Parasitology Meeting in Woods Hole, where Friedrich Frischknecht showed a *Plasmodium berghei* sporozoite gliding in the dermis of a mouse and, surprisingly, actively invading a cutaneous blood capillary (Amino et al., 2006). In the following year, Jerome Vanderberg and Ute Frevert independently confirmed this invasive behavior and showed, for the first time, infected mosquitoes biting the skin of a mouse and inoculating fluorescent sporozoites in the extravascular regions of the tissue (Vanderberg and Frevert, 2004). Quantitative analysis revealed that ~25% of the parasites inoculated by mosquitoes in the skin of anesthetized mice eventually invade blood vessels and thus reach the circulation (Amino et al., 2006). This population is, in addition, responsible for establishing the liver infection.

Sporozoites delivered in the skin by mosquitoes not only intravasate into the blood circulation but can also invade lymph vessels. The invasion of lymphatics is an active process (Fig. 12.4), and accordingly, heat-treated sporozoites barely reach the DLN as opposed to motile parasites. Following lymphatic vessel invasion, sporozoites are passively transported with the lymph to the DLN (Amino et al., 2006), where their journey normally ends. Indeed, most parasites are trapped and subsequently killed by dendritic cells (DCs) (Amino et al., 2006; Radtke et al., 2015). This population of lymphatic sporozoites represents ~15%–20% of the parasites inoculated in the skin (Amino et al., 2006; Yamauchi et al., 2007) and is important to prime specific CD8[+] T cells in the DLN, and thus, mount a protective immune response against the parasite (Chakravarty et al., 2007; Radtke et al., 2015).

Intriguingly, 1 h after infection, most of the sporozoites still remain at the inoculation site (Amino et al., 2006; Kebaier et al., 2009; Yamauchi et al., 2007) where ~10% of the bite inoculum invades skin cells and develops into cutaneous exoerythrocytic forms. These parasites exhibit a slower development and do not grow as much as their hepatic counterparts (Gueirard et al., 2010; Voza et al., 2012) but, nonetheless, give rise to infectious red blood cell-infecting merozoites (Gueirard et al., 2010). Surprisingly, cutaneous parasites can be detected weeks after sporozoite inoculation and are usually associated with hair follicles, which are known immune-privileged sites of the skin. Whether these forms represent a dormant or slow-growing cutaneous reservoir of parasites remains to be elucidated. Evolutionarily, the development of exoerythrocytic forms outside of the liver has been observed in many species of avian malaria (Huff, 1957) but if primate-infecting sporozoites have the capacity to develop in the skin or outside the liver is still unknown.

Following their extravascular inoculation in the skin, *P. berghei* sporozoites can therefore encounter three major fates depending on their invasion target. They can invade a blood vessel to gain access to and develop in the liver; a lymph vessel to reach the DLN and act as source of antigen to elicit an immune response, or a skin cell to eventually develop into merozoites. Whether this invasive fate is predetermined by different degrees of transcriptional and translational maturity of sporozoites remains to be tested. The impact and consequences of the discovery of a skin phase in malaria infection is discussed below in further details.

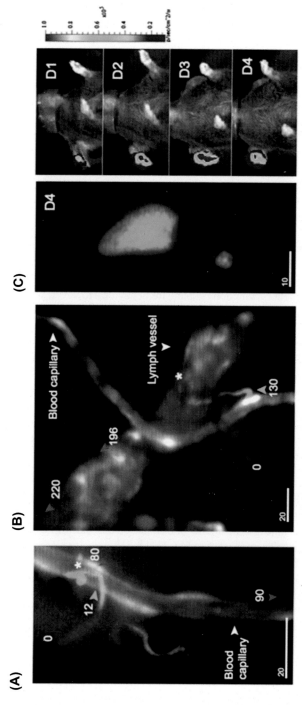

FIGURE 12.4 **Invasive fate of malaria sporozoites in the skin.** (A) Sporozoite gliding and invading a blood vessel in the skin. Blood capillary is depicted in white (flk1-gfp transgenic mouse). Blue–green–red projection shows the trajectory of the invasive sporozoite during 90 s of movement (blue: 0–12 s, green: 12–80 s, red: 80–90 s). *Arrowheads* point the direction of the movement. *Asterisk* indicates the point of invasion. (B) A sporozoite gliding and invading a lymph vessel in the skin. Vessels are depicted in white (flk1-gfp transgenic mouse). Lymph and blood vessels are as indicated. Blue–green–red projection shows the trajectory of the invasive sporozoite during 220 s of movement (blue: 0–130 s, green: 130–196 s, red: 196–220 s). *Arrowheads* point the direction of the movement. *Asterisk* indicates the point of invasion. Notice the lateral drifting described by Amino et al. (2006), as the sporozoite is passively transported inside the lymphatics. (C) The left panel shows a skin cell infected in situ by GFP-expressing sporozoites, containing red blood cell-infective merozoites at day 4 postinfection. The right panel shows the development of luciferase-expressing skin parasites as assessed by the increasing emission of light from the cutaneous site of infection. Animals were treated with 25 mg/kg of primaquine by i.p. injection at days 0 and 1 postinfection to eliminate hepatic parasites. *GFP*, green fluorescent protein.

Functional Imaging of the Skin Phase

The finding that mosquitoes inoculate sporozoites in the extravascular regions of the skin rather than directly into the blood vessels has constituted a milestone in the study of *Plasmodium* biology. Prompting the community to investigate this newly described cutaneous phase, it has enabled the identification of new steps in the parasite life cycle and unraveled unsuspected requirements sporozoites must fulfill to establish an infection. Indeed, for parasites to reach the bloodstream and the liver, dissemination from the site of inoculation into the constrained environment of the dermis, localization of a blood vessel, and passage across its wall are absolute prerequisites. Gliding motility, which enables sporozoites to migrate in host tissues at an average speed of ~1–2 µm/s (Amino et al., 2006), and cell traversal activity, which endows sporozoites with the capacity to transmigrate through host cells by breaching their plasma membrane (Mota et al., 2001), have both been found crucial for the successful completion of these steps.

Intravital microscopy has notably been instrumental to dissect and understand how these two cellular activities underlie efficient parasite progression in the host. On one hand, imaging of mutant sporozoites has allowed assessing the impact of loss or impairment of function on parasite behavior. However, when a protein is crucial for successive steps of the life cycle, constitutive genetic deletion only enables the study of the most upstream events. Knockout parasites may, in addition, rely on compensatory but nonphysiological mechanisms to overcome protein loss. Functional assays that enable the detection of parasite and host cellular activities *in vivo* and in real time have therefore been developed and applied to study the behavior of wild-type sporozoites. These two complementary approaches constitute together what we refer to as "functional imaging." In this section of the chapter, we review how it has substantially contributed to our understanding of *Plasmodium* sporozoite hematogenous dissemination and can be applied to others pathogens.

Moving and Traversing Cells to Progress in the Skin

Using a widefield microscope, Vanderberg and Frevert published in 2004 the first direct evidence that, during a mosquito bite, most of the sporozoites are inoculated in the extravascular parts of the skin. Following parasite release in the dermis, induction of motility occurs after a short delay (Vanderberg and Frevert, 2004), and sporozoites subsequently exhibit a robust forward motility, moving along tortuous trajectories (Amino et al., 2006; Vanderberg and Frevert, 2004). Strikingly, transition from the mosquito salivary system to the mammalian host is associated with a strong increase in gliding speed (Frischknecht et al., 2004). This switch from the sluggish motility observed in the *Anopheles* to a rapid migratory behavior in the mouse dermis seems essential for successful parasite exit from the skin. Indeed, mutant sporozoites that lack the N-terminal region of the major surface molecule, the circumsporozoite protein (CSP), glide in the dermis at significantly lower speed than their wild-type counterpart (Hopp et al., 2015). As a consequence, they do not disseminate as well in the tissue and this altered exploratory behavior might underlie their reported inability to gain access to the bloodstream (Coppi et al., 2011). The thrombospondin related anonymous protein (TRAP)-like mutants, which also exhibit a defect in exit from the skin, similarly, display a lower gliding speed associated with decreased dispersal in an *in vitro* pillar array mimicking obstacles in the skin (Hellmann et al., 2011). Longitudinal tracking of wild-type parasite migration in the skin shows, in addition, that sporozoites' speed decreases over the first hour postinoculation while their trajectories concomitantly become more constrained. At 2 h, only about 30% of the parasites still exhibit motility (Hopp et al., 2015). Over time, sporozoites therefore probe their environment less efficiently, which likely contributes to the gradual decrease in blood vessel invasion rates observed by Amino et al. (2006). Whether this progressive decline of sporozoite migratory capacities reflects the exhaustion of the gliding machinery or is associated with a shift toward a more cell-invasive behavior of the parasite is unclear. It is, however, in line with *in vitro* experiments demonstrating loss of sporozoite motility and infectivity on prolonged incubation at 37°C (Vanderberg, 1974). While ~10% of the inoculated parasites eventually invade skin cells (Gueirard et al., 2010), the fate of the remaining sporozoites that linger in the dermis after 1–2 h is still an open question. Do they simply die in the tissue? Or do some of them still manage to exit the dermis? If so, are they still able to infect the liver? Collectively, all these observations suggest that time is a critical feature during the "search" for a blood vessel and the exploration "strategies" developed by sporozoites likely reflect the optimization of this parameter.

In addition to efficient motility, cell traversal activity is also important for sporozoites to progress in the dermis. Indeed, imaging of the cell traversal-deficient SPECT (sporozoite protein essential for cell traversal) and SPECT2 (sporozoite protein essential for cell traversal 2) mutants revealed that, following intradermal inoculation, these parasites are rapidly arrested in the skin, presumably by cells they cannot traverse. Furthermore, these parasites are also quickly found in association with CD11b[+] phagocytes, thus suggesting that cell traversal enables sporozoites to disseminate in the tissue and escape phagocytosis (Amino et al., 2008). A functional assay, based on the incorporation of a cell-impermeant nuclear dye by wounded cells, was, in addition, developed to allow the dynamic detection of cell traversal activity *in vivo* (Formaglio et al., 2014). It enabled the visualization of stained nuclei appearing on the path followed by wild-type sporozoites as they migrate in the skin, therefore confirming that sporozoites do indeed traverse cells as they progress in the dermis. Furthermore, quantitative analysis revealed that the number of wounded cells in the tissue increases with time and as a function of the number of parasites in the imaged field. Complementation of this assay with the application of a second cell-impermeant dye at the end of the imaging session further allowed to assess the fate of traversed cells. It showed that most wounded cells are unable to reseal their breached membrane, suggesting they are likely dead and thus release intracellular compounds and danger signals in the tissue. Whether this has an impact on sporozoite motility and underlies the decline of speed and trajectory straightness reported by Hopp et al. is unknown. However, increasing numbers of dying cells in the tissue might over time promote the recruitment of immune cells to the site of infection and thus indirectly hinder the successful escape of sporozoites from the dermis. This stresses again the importance of the time factor for blood vessel invasion. Interestingly, another type of transcellular migration involving the formation of transient vacuoles has been observed *in vitro* (Rankin et al., 2010; Risco-Castillo et al., 2015). Although it has never been reported *in vivo* so far, this mechanism might also participate to efficient sporozoite progression in the host.

Searching and Finding a Blood Vessel

As the vasculature represents only ~5% of the skin volume, finding a blood vessel constitutes a challenging task, which ~25% of the sporozoites inoculated during a bite, however, successfully complete. Although the exact time window during which parasites can remain in the skin before they enter the circulation is still a matter of debate, blood vessel invasion could be observed within minutes after parasite inoculation (Amino et al., 2006; Hopp et al., 2015), indicating that their migration patterns in the dermis enable sporozoites to rapidly encounter blood vessels. At first sight, parasite trajectories in the skin appear tortuous and seemingly random (Amino et al., 2006). Detailed analysis of sporozoite tracks shows, in addition, that the migration pattern of sporozoites cannot be explained by a mere diffusion process, such as Brownian motion (Hellmann et al., 2011; Hopp et al., 2015). This is in good agreement with conclusions drawn from the study of stochastic processes. It has, indeed, been suggested that standard random walks, which possess strong oversampling properties and therefore allow for low exploration of the environment, are not optimal search strategies to find rare targets (Viswanathan et al., 1999). Sporozoite trajectories might therefore be best described by other variations of the random walk. Alternatively, sporozoite "search" for blood vessels might be a directed process, biased by gradient-based or discontinuous cues that attract parasites toward the vasculature. So far, chemotaxis in response to soluble factors has never been demonstrated for salivary gland sporozoites. However, one study has provided evidence that structural cues can influence sporozoite trajectories (Hellmann et al., 2011). Indeed, when injected either in the tail or in the ear of a mouse, two areas of the skin exhibiting a very different organization, sporozoites display strikingly distinct migration patterns moving along rather linear or meandering paths, respectively. Using obstacle arrays with variable pillar-to-pillar distances, these two behaviors could be reproduced *in vitro*, thus suggesting that sporozoite trajectories are governed by successive deflections on structural elements. Whether this environmental guidance has a positive impact on the localization of blood vessels remains unclear and the potential existence of additional stimuli cannot be excluded.

Crossing the Cutaneous Endothelial Barrier

Once a sporozoite encounters a blood vessel, it does not necessarily penetrate immediately in the circulation but rather often interacts with vascular segments for variable amounts of time. Contact with the vasculature is associated with a significant decrease in migration speed as parasites glide, around or parallel to the vessels, or even pause before they cross the endothelial barrier (Amino et al., 2006; Hopp et al., 2015). Interestingly, all interactions between sporozoites and blood vessels do not lead to a productive invasion. Instead, sporozoites sometimes move back into the nonvascular parts of the tissue after contacting a vessel (Hopp et al., 2015), thus suggesting that all the segments of the vasculature might not equally support parasite entry in the circulation. Alternatively, parasite-intrinsic differences might affect the ability of sporozoites to invade blood vessels and underlie the heterogeneity of their fates in the skin. Indeed, sporozoite development in the mosquito is a highly asynchronous process and salivary glands may therefore harbor parasites at different stages of maturation. The observation of a bimodal distribution of gliding speed as early as 3 min after the bite (Amino et al., 2006) is consistent with this hypothesis and suggests that a given inoculum might contain "fast" and "slow" subpopulations of sporozoites. It cannot be excluded, however, that this variability reflects in fact the influence of discrete structural cues in the tissue.

As the vast majority of SPECT- and SPECT2-deficient mutants are rapidly arrested in the dermis after their inoculation, their entry in blood or lymph vessels could never be directly observed. Intravital microscopy therefore has never provided evidence for a role of cell traversal in the passage of endothelial barriers in the skin. However, following intradermal injection, these mutants can be found in the lymph nodes draining the site of infection and can infect clodronate-treated rats (Amino et al., 2008). Clodronate enables the depletion of macrophages in the liver but not in the skin and therefore makes it possible for mutant parasites that have reached the bloodstream to infect the liver as efficiently as wild-type sporozoites (Ishino et al., 2004, 2005). Altogether, these results thus indicate that cell traversal is not necessary to cross the walls of either lymph or blood vessels. Yet, it cannot be excluded that wild-type sporozoites rely on several strategies to enter the circulation and favor cell traversal-dependent mechanisms to migrate through cutaneous endothelial barriers as has been demonstrated for the passage across the wall of hepatic sinusoids (Tavares et al., 2013).

Functional imaging of the skin phase has substantially increased our understanding of *Plasmodium* biology, which constitutes a first step toward the development of a long-awaited malaria vaccine. So far, only a few studies have directly looked at sporozoite transmission, progression, and development in the dermis of mice immunized either with irradiated parasites or by passive transfer of anti-CSP antibodies (Kebaier et al., 2009; Vanderberg and Frevert, 2004; Voza et al., 2012). Mosquitoes were found to inject less sporozoites in the skin of immunized animals, a finding that was correlated to the formation of immune complexes at the tip of mosquito proboscis probing the dermis of passively immunized mice (Kebaier et al., 2009). A second study showed that sporozoites transmitted to immunized mice are swiftly immobilized in the dermis and arrested before they can reach and invade a blood vessel (Vanderberg and Frevert, 2004). Animals immunized with irradiated sporozoites also showed a decrease in the number of infected skin cells after natural transmission of sporozoites (Voza et al., 2012). *In vivo* imaging in immunized mice has therefore enabled the identification of the skin phase as an attractive target for therapeutic intervention. In the future, it should, in addition, help us dissect the mechanisms of protection and support the ongoing effort to design efficient vaccines against VBDs.

Imaging at Your Convenience Now and Beyond

The advent of intravital fluorescence microscopy (IFM) has been capital to reveal vistas hitherto unknown to us. However, as all techniques, IFM has its own limitations and drawbacks. Due to practical and ethical constraints, most of IFM studies are restricted to mouse models, which sometimes do not recapitulate what happens in humans. The second general issue faced while imaging a living tissue is the phototoxicity, i.e., the damage caused to the sample by the energy transferred during light irradiation. This is especially problematic at shorter wavelengths, such as used in the first UV fluorescence microscopes. Moreover, under constant illumination, fluorophores can bleach, which causes a gradual decrease of fluorescence and thus makes long-lasting observations challenging. These two deleterious effects, phototoxicity and photobleaching, are important impeding factors that need to be kept minimal during image acquisition (Helmchen and Denk, 2005). Illumination with the lowest amount of power allowing a discriminative signal-to-noise ratio combined with the use of highly sensitive detectors should thus be aimed at to overcome these limitations. In addition, in the skin of live animals, autofluorescence background, light scattering, and the absorptive properties of the tissue severely affect signal-to-noise ratio and light penetration (So et al., 2000). Microscopic observation, which is per se restricted to the sole visualization of fluorescently labeled populations, is therefore with most approaches further limited to the skin surface. Altogether, these limitations render sometimes cumbersome the analysis and interpretation of the partial information imaging techniques can provide us with. Finally, experimental conditions should be carefully controlled as they can also give rise to artifacts. For instance, the necessity to immobilize animals for intravital imaging usually implies the use of liquid or gaseous anesthesia. These chemicals generally affect the physiology of the animal and can potentially have a direct impact on the observed phenomenon. A typical consequence of general anesthesia is hypothermia, which should be compensated for by the use of heat pads or heat chambers. Similarly, attention should be paid to the positioning of the animal as it can compromise the lymphatic circulation.

In spite of the difficulties and limitations associated with the implementation of intravital imaging, widefield, confocal, and two-photon microscopy (TPM), the three main modalities of IFM, have largely contributed to our understanding of parasitic infections of the skin. Widefield fluorescence microscopy, the most basic setup for IFM, is simple, relatively inexpensive, and flexible and thus remains a popular and widespread approach. It has notably enabled the first visualization of sporozoite release in the mouse skin during a mosquito bite (Vanderberg and Frevert, 2004). However, in this approach, excitation and emission light are, respectively, delivered through and collected from the whole volume of the sample. Due to the thickness of most biological tissues, a considerable amount of out-of-focus light is therefore detected, which decreases optical resolution and limits the scope of widefield fluorescence microscopy for *in vivo* imaging (Diaspro, 2002).

Fortunately, significant strides have been made to push the boundaries of our observational prowess (Huang et al., 2009). The implementation of new techniques, such as confocal and TPM, together with the development of highly sensitive detectors has increased the spatial and temporal resolution, reduced photodamage, and improved the signal-to-noise ratio (Diaspro, 2002; So et al., 2000). Laser scanning confocal microscopy can notably yield high-resolution images. The addition of an adjustable pinhole on the light path, which blocks out-of-focus fluorescence, indeed enables optical sectioning and thus increases resolution, although at the cost of brightness (Diaspro, 2002). In this setup, however, a focused laser beam must be scanned over samples point by point to produce images. As a consequence, the photodamage caused by the intense irradiation and out-of-focus illumination is important and acquisition speed is limited. For intravital microscopy, spinning disk confocal microscopy is therefore usually preferred (Graf et al., 2005; Ichihara et al., 1996). In this configuration, the single pinhole is replaced by rotating disks perforated with thousands of pinholes. These spinning disks rapidly project the light over the sample and block out-of-focus fluorescence while a highly sensitive camera collects the signal from the whole imaged area. Although the spinning disk configuration does not achieve optical sectioning as well as its laser scanning counterpart, it produces significantly less photodamage as the laser power is spread over many pinholes and allows for faster acquisition than the point-detection approach. It has thus been frequently used to record the dynamic behavior of *Plasmodium* sporozoites in the mouse skin (Amino et al., 2006, 2008, 2007; Hopp et al., 2015) and to capture relatively short-lived events such as skin cell traversal (Formaglio et al., 2014) and the crossing of liver sinusoids (Sturm et al., 2006; Tavares et al., 2013). Spinning disk confocal microscopy was also successfully used to image Lyme borreliosis pathogen. *B. burgdorferi* motility in the skin and bacteria translocation from the vasculature into the skin were clearly visualized (Moriarty et al., 2008). Similarly, this technique enabled the visualization of highly motile

African trypanosomes in the mouse dermis at a location distinct from the bite site, indicating that these parasites can reinvade the skin following their hematogenous dissemination (Capewell et al., 2016). Last but not least of the main modalities of IFM, TPM exploits the fact that the simultaneous absorption of two photons of lower energy and longer wavelength can excite electrons and thus produce fluorescence emission, provided their combined energy matches the electronic transition. Excitation at longer wavelength is associated with less scattering in living samples and thus better light penetration, allowing to image deeper in the tissues. In addition, because a high density of photons is required for the two-photon effect to occur, excitation of the fluorophores is achieved only at the precise point where the laser is focused, which limits photodamage outside of the imaged plane, enables optical sectioning in the absence of a pinhole, and improves signal-to-noise ratio (Diaspro, 2002). As for laser scanning confocal microscopy, point detection and the necessity to thus scan over the sample limits acquisition speed. Yet, this technique still offers sufficient temporal resolution to investigate dynamic interactions between immune cells and pathogens in the dermis. The migratory behavior of DCs (Ng et al., 2008), the recruitment of neutrophils (Peters et al., 2008), and the heterogeneous sampling of T lymphocytes (Filipe-Santos et al., 2009) in *Leishmania*-infected skin could indeed be visualized with TPM. Using this same technique, the process of *Borrelia* transmission to mice during the tick bite has been visualized, as well as the acquisition of pathogens from infected mice (Bockenstedt et al., 2014). Intravital imaging of cutaneous tissues with TPM should, however, always be performed with caution, especially in pigmented skin. Indeed, melanin-containing cells readily absorb the infrared light of the TPM laser, which can cause the tissue to burn and lead to inflammatory immune response as evidenced by neutrophil recruitment at sites of irradiation-induced speckling (Li et al., 2012). On the other hand, the benefits of TPM to image deeper tissues remain undisputed. Accordingly, this technique was used to identify extravasated *T. b. brucei* and *Toxoplasma gondii* in the brain parenchyma (Frevert et al., 2012; Konradt et al., 2016) and *Plasmodium* sporozoite uptake by DCs in the lymph node (Radtke et al., 2015).

Thanks to the many improvements microscopic instrumentation has undergone over the past decades, single-cell resolution can nowadays be routinely achieved *in vivo*. However, traditional imaging techniques are physically limited by light diffraction and their resolution typically does not exceed 200 nm laterally and 500 nm axially (Huang et al., 2009), which is insufficient to characterize subcellular organelles or molecular interactions. To overcome these limitations, a group of techniques, collectively referred to as super-resolution microscopy, has therefore been developed. As it requires light intensities up to six orders of magnitude lower than other super-resolution approaches (Grotjohann et al., 2012; Klar et al., 2000), RESOLFT (reversible saturable optical linear fluorescence transitions) appears as the most promising candidate for *in vivo* imaging. This technique has built on the development of photoswitchable fluorescent proteins that possess the ability to reversibly alternate between an "on" (fluorescent) and "off" (dark) state under appropriate illumination (Hell and Wichmann, 1994; Hofmann et al., 2005; Huang et al., 2009). Practically, a donut-shaped depletion laser is first used to switch off all the fluorescent molecules in the vicinity of a region of subdiffractive dimensions, which then enables imaging the center of the donut with a nanoscopic resolution of about 50 nm (Hofmann et al., 2005; Huang et al., 2009). The recent development of fast-switching proteins has further enhanced the temporal performance of the technique (Grotjohann et al., 2012) as recently demonstrated by Schnorrenberg et al. who captured *in vivo* the dynamics of *Drosophila melanogaster* cytoskeleton (Schnorrenberg et al., 2016). In this study, adaptation of the depletion laser for 3D imaging and combination with confocal microscopy enabled, in addition, the acquisition of stacks with a fourfold increase in axial resolution. However, time lapse and 3D nanoscopy were both performed on very limited volumes that would not be suitable to track molecular events in cells moving in a tissue. Further improvements are therefore required to achieve imaging of larger volumes with sufficient temporal resolution *in vivo*. To this end, strategies supporting parallelized acquisition are under investigation. For instance, RESOLFT nanoscopy generating more than 100,000 donuts simultaneously has successfully been implemented. Live cells could be imaged in the range of seconds over fields of 100 μm × 120 μm, acquisition being only limited by the kinetics of fluorophore switching and camera frame rate (Chmyrov et al., 2013). Similarly, multifocal plane microscopy, which enables the acquisition of up to 25 focal planes in one single step (Abrahamsson et al., 2013), has been coupled to super-resolutive photo-activated localization microscopy and stochastic optical reconstruction microscopy to achieve faster multicolor 3D imaging of whole fixed cells with lateral and axial resolutions of ~20 and ~50 nm, respectively (Hajj et al., 2014). Altogether, these promising developments suggest super-resolution techniques might in the future become part of the *in vivo* imaging toolbox. Interestingly, the discovery of light-responsive fluorescent proteins has not only benefited the development of super-resolution microscopy techniques but also provided scientists with new tools to assess physiological parameters within living pathogens. Photoconvertible proteins, which undergo an irreversible shift in fluorescence emission upon photoactivation, have, for instance, been used to follow protein redistribution during *Toxoplasma* cell division (Ouologuem and Roos, 2014) or measure the metabolic activity of *Leishmania major* parasites *in vivo* (Müller et al., 2013).

Conclusion and Perspectives

The skin is the largest organ of the body and the entry site of innumerous vector-borne pathogens. Nevertheless, the study of host–pathogen interactions in this organ remains largely neglected. This disregard is probably due to the common and misleading view that pathogens transmitted by the bite of hematophagous vectors are injected directly into the blood circulation, skin thus being a mere fleeting site of transit. On the contrary, strong evidence indicates that during the infective bite, most vector-borne pathogens are mainly deposited extravascularly in the skin. From there, they gain access to the blood and other distant tissues with different kinetics of dissemination but always with a "delay" as compared to pathogens inoculated intravascularly. This delayed migration from the bite site is a distinctive characteristic of this unappreciated skin phase of infection and encompasses the development of pathogens in the skin as well as their dissemination through the host body by intravasation into blood or lymph vessels. In both cases, this implies that pathogens are able to survive the host innate and adaptive immune response in the skin. Consequently, this cutaneous phase has an important impact not only regarding pathogenesis (Sebbane et al., 2006) and responsiveness to drug and vaccine administration (Gueirard et al., 2010; Vanderberg and Frevert, 2004; Voza et al., 2012) but also affects pathogen dissemination (Kubba et al., 1987; Sebbane et al., 2006; Shih et al., 1992; Sidjanski and Vanderberg, 1997; Turell and Spielman, 1992), persistence (Belkaid et al., 2002; Capewell et al., 2016; Gueirard et al., 2010), and transmission back to the hematophagous vector (Bockenstedt et al., 2014; Caljon et al., 2016; Capewell et al., 2016).

Its accessibility to IFM makes skin the ideal organ to perform longitudinal and noninvasive studies of pathogens inoculated focally in the tissue, either by bite or by microinjection. Thanks to the development of functional imaging assays and the large availability of knockout and transgenic mice displaying cell type-specific expression of fluorescent reporters, IFM has thus become an invaluable tool to directly observe and dissect the interactions between pathogens, vector saliva, and host cells in the skin, both in a qualitative and quantitative manner and at a molecular and cellular level. In addition, through the comparison of pathogen behavior within susceptible and immune hosts, assessing the determinants and mechanisms of protection against pathogens in the skin is now within our reach. We have come a long way since the invention of fluorescence microscopy and the future of imaging seems more promising than ever. With the advance of technology, we hope functional imaging *in vivo* will soon allow us to scrutinize, with nanometric resolution and in multiple color dimensions, the dynamic molecular interactions at play during pathogen and host cell struggle for survival in the skin.

CULTURE OF SKIN CELLS

Despite all advantages of functional *in vivo* imaging to scrutinize the intricacies of pathogen–host interactions in the skin of rodents, unfortunately, some vector-borne pathogens, such as, e.g., Dengue virus (DENV), *P. falciparum*, and human filarial *Brugia malayi*, are unadapted to infect immune-competent rodents. This species-specific barrier can be a consequence of the human-infecting pathogen elimination by the mouse immune system (Perry et al., 2011) or the necessity of specific human host cell surface molecules for invasion (Ploss et al., 2009).

Therefore, in this second part of the chapter we will discuss how host skin cells, with special emphasis on human skin cells and explant cultures, can assist to further dissect pathogen–host interactions at a cellular and molecular level.

In Vitro Culture of Skin Cells

The skin is a complex organ made of resident skin cells (keratinocytes and fibroblasts) and immune cells (DCs, neutrophils, mast cells, CD8[+] T cells, and CD4[+] T cells) (Di Meglio et al., 2011; Nestle et al., 2009) (see Chapter 1). Immune cells are the easiest to culture since generally they are isolated from the blood, from the spleen, and from bone marrow (mast cells, for example). Immune cells have been the first one to be isolated since the medium required to culture them were more or less easy (Roswell Park Memorial Institute medium (RPMI) or Dulbecco's Modified Eagle Medium, supplemented with fetal calf serum and antibiotics). Resident cell cultures require more complex medium, especially when the work is accomplished on primary culture cells.

Alternatively, cell lines can be used for keratinocytes, fibroblasts, and monocytes, but they generally behave differently to the different stimuli and are far from the reality. For example, Hacat cell line behaves differently than primary human keratinocytes when stimulated by different T helper (Th) cytokines. Their gene transcriptional profile of cornified envelope-associated proteins, such as filaggrin, loricrin, involucrin, and keratin 10, differs significantly (Seo et al., 2012). Similarly, THP1 cell lines, used in experiments as antigen-presenting cells to replace DCs, do not prime efficiently naïve T cells (Chanput et al., 2014). Therefore, we do not recommend their use and we present data mainly relying on primary cells.

Culture of Immune Cells

Dendritic Cells

On entry of the host's skin, one of the first immune cells that pathogens encounter is DCs. These cells patrol the skin for invading pathogens, which they phagocytose and process to present to T cells when migrated to DLNs, thereby bridging innate and adaptive immune responses (Banchereau and Steinman, 1998). Vector–host–pathogen interactions at the skin interface are of paramount importance in the (early) immunopathogenesis of VBDs. DCs also play an important role in the pathogenesis of many VBDs, among which leishmaniasis (Bagirova et al., 2016; Kautz-Neu et al., 2011), malaria (Yu et al., 2016), arboviral (Bowen et al., 2017; Olagnier et al., 2014; Sprokholt et al., 2017) and bunyavirus infections (Akinci et al., 2013). The same also holds true for Lyme borreliosis, caused by spirochetes belonging to the *B. burgdorferi* sl group, which are transmitted by *Ixodes* ticks (Hovius et al., 2007). DDCs and Langerhans cells (LCs) have been shown to be able to phagocytose *B. burgdorferi* sl (Filgueira et al., 1996; Suhonen et al., 2003) and recognize *B. burgdorferi* sl lipoproteins, for example, by Toll-like receptor-2 (TLR2) (Hirschfeld et al., 1999). This in turn leads to activation and maturation of DCs, transcription of various immune genes, and upregulation of the costimulatory molecules, such as CD80, CD86, CD40, CD83, and MHC II (Moore et al., 2007; Petzke et al., 2009; Suhonen et al., 2003). In the lymph nodes, following maturation, antigen presentation, and cytokine production, DCs prime T cells to become effector T_H cells, such as T_H1, T_H2, T_H17, or regulatory T cells (T_{regs}).

Tick saliva is able to modulate diverse aspects of DC function, as reviewed in Mason et al. (2014), including inhibition of phagocytosis, expression of costimulatory molecules, secretion of proinflammatory cytokines, and antigen presentation to T cells, which impairs T cell proliferation. Inhibition of DC responses by tick saliva could be beneficial for tick survival, by dampening the host immune response enabling the tick to stay attached during its blood meal, and survival of the pathogens they transmit, by making the tick–host–pathogen interface a less hostile environment for invading microorganisms. Several specific compounds responsible for the observed inhibitory effects have been identified in *Ixodes* tick saliva, for example, prostaglandin E2 (Lieskovska and Kopecky, 2012; Sa-Nunes et al., 2007) and protein kinase A. Also sialostatin L, a cysteine protease inhibitor in *Ixodes* tick saliva, was shown to inhibit the production of interleukin (IL)-12 and tumor necrosis factor (TNF)-α, upregulation of CD80 and CD86 by DCs, and interfered with antigen presentation (Sa-Nunes et al., 2009). Finally, an *Ixodes* saliva protein of 15 kDa, Salp15, has been shown to bind to the C-type lectin receptor DC-SIGN (dendritic cell-specific intercellular adhesion molecule-3-grabbing nonintegrin) (Hovius et al., 2008). DC-SIGN is involved in pathogen recognition and is expressed by DDCs (Geijtenbeek et al., 2000). The interaction of Salp15 with DC-SIGN triggered a Raf-1 MEK1/MEK2-dependent signaling cascade, inhibiting IL-12, IL-6, and TNF-α production by immature DCs activated with TLR-ligands or viable *B. burgdorferi* sl. Finally, the presence of Salp15 and the presence of *Ixodes* tick saliva attenuated DC-induced T cell activation (Hovius et al., 2008). The latter study has been performed with human monocyte-derived DCs (moDCs), as it is difficult to identify and isolate sufficient numbers of DCs from human blood. Indeed, human moDCs can be used as a convenient model to study vector-human DC–pathogen interactions *in vitro*. The protocol to generate and stimulate immature moDC for this purpose is straightforward and is outlined below. The protocol is largely based on Romani et al. (1994) and has been previously described in detail (Hovius et al., 2008, 2009; Mason et al., 2015, 2016). Briefly, peripheral blood mononuclear cells from heparinized blood are isolated by density gradient using Ficoll paque, and monocytes are identified as adherent cells at 37°C after isolation using Percoll. Monocytes are subsequently cultured in RPMI with fetal calf serum and L-glutamine supplemented with IL-4, which suppresses differentiation into macrophages, and granulocyte macrophage colony-stimulating factor, which maintains the viability of the cells, at 37°C (5% CO_2) for 1 week to become immature moDCs. Next, moDCs, usually approximately 1×10^5 cells depending on the experiment and volume, can be stimulated with various stimuli, for example, TLR agonists, such as lipopolysaccharide or Pam3CSK4, heat-killed or viable vector-borne pathogens, or (recombinant) derivatives thereof. By performing these stimulations in the presence of vector salivary gland extract, saliva, or recombinant salivary gland proteins, the effect of these substances on DC activation can be determined. Routine readouts include cytokine production (e.g., proinflammatory, antiinflammatory, and type I interferons measured by ELISA), expression of maturation markers (CD80, CD83, CD86, and HLA-DR) measured by flow cytometry, quantitative reverse transcription polymerase chain reaction or microarray using RNA/cDNA, or mixed leukocyte reaction or T cell differentiation assays (cocultures with peripheral blood lymphocytes) to determine subsequent adaptive immune responses. Using this model, the influence of the DC microenvironment is overlooked, and DC migration and the effect of the vector substances under study on LCs cannot be determined.

The role of DCs has been also particularly investigated in DENV. Dengue is an acute infectious disease caused by DENV that affects approximately 400 million people annually, being one of the most prevalent human arthropod-borne diseases. When an *Aedes* mosquito transmits DENV, the virus is inoculated into the skin, where DCs become infected and it leads to production of proinflammatory cytokines and chemokines (Schneider et al., 2012; Sprokholt et al., 2017). Several viral proteins, particularly NS1, which is a secreted protein, modulate activation of human DCs. This protein enhances viral replication and proinflammatory cytokine production in human DCs (Alayli and Scholle, 2016). DCs play a dual role as both targets of DENV replication and mediators of innate and adaptive immunity. DENV develops immune evasion strategies and impairs the function of infected DCs (Schmid et al., 2014).

Neutrophils

Neutrophils are essential components of the innate immune system. They are involved in secretion of antimicrobial molecules, in tissue repair, and they migrate rapidly to the site of inflammation (Nathan, 2006). There, they release reactive oxygen radicals and the content of their intracellular granules, and they produce DNA traps in the extracellular space. Their half-life in the bloodstream of adult mice is 12 h. Their role is then essential at the cutaneous interface where they are rapidly recruited on injury. Different chemoattractants operate their migration from the blood. The recognition of danger-associated molecular patterns triggers the activation of skin cells: tissue macrophages and resident skin cells secreting IL-1β, TNF-α, and CXCL2; T cells secreting IL-17A (Jain et al., 2016). As described in previous chapters, their role has been particularly emphasized in flagellate parasites such as leishmaniasis (Laskay et al., 2003; Peters et al., 2008; Ribeiro-Gomes et al., 2012) and trypanosomiasis (Caljon et al., 2016; Stijlemans et al., 2014). In leishmaniasis, flagellate parasites target the host macrophages where they escape the immune response as amastigote forms. *L. major* promastigotes first infect and induce apoptosis of neutrophils and are secondarily captured by DDCs as they engulf apoptotic, infected neutrophils; this suppresses DC ability to stimulate adaptive CD4$^+$ T cell-mediated responses and the early development of antiparasitic immunity. It has been recently shown that once inoculated into the skin they are first phagocytosed by neutrophils and then by macrophages (Peters et al., 2008; Ribeiro-Gomes et al., 2012). In malaria, the depletion of neutrophils in mice before intradermal inoculation of sporozoites, the infectious stage transmitted by mosquito, did not modify their distribution and dissemination (Mac-Daniel et al., 2014). The role of neutrophils in arthropod-borne diseases is generally studied directly in situ in mouse ear using fluorescent pathogens, and specific neutrophil markers such as Ly6G or CD11 are used. Certain pathogens use neutrophils as host cells in the vertebrate host. *Anaplasma phagocytophilum*, a Gram-negative bacterium transmitted by *Ixodes* tick, targets the host neutrophils (Dumler et al., 2005). Within the host cells, *Anaplasma* inhibits the fusion of lysosomes with the phagosome, escaping the lytic effect and inhibiting apoptosis (Carlyon and Fikrig, 2003; Choi et al., 2005). It also modulates IL-8 secretion, a potent chemokine attracting more neutrophils to the inflammation site (Akkoyunlu et al., 2001). The same phenomenon occurs with *Y. pestis* responsible of the plague. The bacteria use neutrophils at the inoculation site to escape the host immune response (Shannon et al., 2015; Spinner et al., 2014).

T Cells, CD4, and CD8

The diversity of T cells discovered these last years has complicated the understanding of skin immunity, and it is clear that their role is essential due to the amount of these cells present in this organ. Twenty billion of T cells are present in the entire skin, twice the number of T cells in the blood circulation. Most of T cells are skin homing cells expressing CD45RO, CCR4, and cutaneous lymphocyte antigen (Nomura and Shinohara, 2016). CD4 memory T cells reside in the dermis, while CD8 T cells are present in the epidermis. CD4 T cells have evolved into Th cells and immune suppressive T_{regs}. Cytokine environment is going to shape the diversity of Th cells.

T cells have been studied *in vitro* in the context of tick-borne diseases. In T cell proliferation assays, tick saliva has been shown to inhibit their proliferation (Wikel and Bergman, 1997). In *Ixodes ricinus*, the European vector of Lyme disease, a salivary gland protein, has been shown to suppress T lymphocyte proliferation and to induce a Th2 cytokine profile (Leboulle et al., 2002). In *Ixodes scapularis*, the American vector of Lyme disease, the tick saliva protein Salp15, has been described to specifically bind to the CD4 receptor and to inhibit T cell proliferation (Juncadella et al., 2007). More generally, tick saliva tends to orientate the T cell response toward a Th2 response (Skallová et al., 2008; Wikel, 2013). The role of T cells has been also extensively studied in leishmaniasis, also showing a polarization toward a Th2 response. These studies have been mainly conducted in animal models (da Silva Santos and Brodskyn, 2014).

(A)   **(B)**

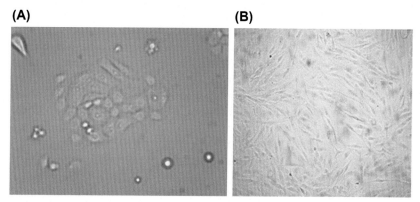

FIGURE 12.5 Primary culture of human skin resident cells: (A) Keratinocytes; (B) fibroblasts.

Culture of Resident Cells

Keratinocytes

The role of keratinocytes has been emphasized by the discovery of TLRs and antimicrobial peptides, making them key cells at the skin interface, environment, and host (Schauber and Gallo, 2009). TLR molecules are present on keratinocytes surface that allow the recognition of pathogen-associated molecular patterns (Pivarcsi et al., 2005). Keratinocytes can be isolated from abdominal skin obtained by abdominoplasty, but their isolation from human foreskin is less tedious (Fig. 12.5A). They can be studied either at confluence or at 70%–80% confluence. Alternatively, they are commercially available. Their role has been studied in few VBDs and is likely underestimated. They secrete antimicrobial peptides, cytokines, and chemokines on the interaction with *B. burgdorferi* sl (Marchal et al., 2009; Marchal et al., 2011) or DENV (Surasombatpattana et al., 2012).

Fibroblasts

As the main resident cells of the dermis, fibroblasts play a major role by secreting extracellular matrix and they interact tightly with the keratinocytes (Werner et al., 2007). They are cultured easily as primary cells (Bernard et al., 2017) (Fig. 12.5B). Their role is also likely underestimated in arthropod-borne diseases, although most of pathogens inoculated by arthropods are injected into the dermis. Few studies have shown that they react to pathogens such as *Borrelia* (Marchal et al., 2009; Schramm et al., 2012) by secreting inflammatory molecules (chemokines and antimicrobial peptides).

Ex Vivo Culture of Skin Explants

Skin explants offer a better approach to understand the role of the whole skin in the immune response to arthropod-borne pathogens, especially the role of DCs.

The microenvironment of DCs is crucially important and affects the function of DCs. Also, in human skin various subsets of DC are present. These factors are ignored during *in vitro* stimulation experiments using moDCs (as discussed above). These factors can, however, be studied in a human skin explant model in which pathogens can interact with human skin DCs, both DDCs and LCs, in a more physiological setting. Indeed, full-thickness skin explant models using human skin surplus to plastic surgical procedures have been used to study various aspects of DC biology (de Gruijl et al., 2006; Flacher et al., 2010; Schneider et al., 2012). In early studies, epidermal skin explant derived from skin blister has been tested to measure the role of LCs in the infection induced by *B. malayi*, one of the lymphatic filariasis in humans. The parasites were clearly shown to inhibit the activation of the LCs (Semnani et al., 2004). Such models are closer to the *in vivo* situation in that they comprise all cell types present in the skin. The ex vivo skin explant model has been developed to study *B. burgdorferi* sl–DC interactions (Mason et al., 2016) (Fig. 12.6). Briefly, surplus human skin is cleansed with 70% alcohol and sterile water and injected intradermally with an inoculum of viable *B. burgdorferi* sl in phosphate buffered saline (PBS) or other positive (e.g., TLR agonists) and negative

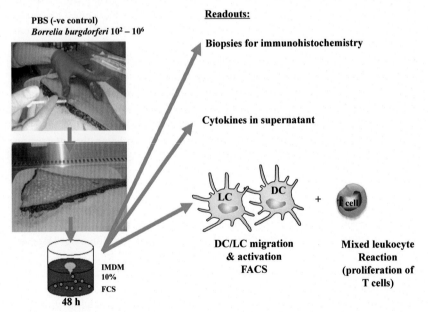

FIGURE 12.6 Schematic overview of the ex vivo skin explant model studying the effect of injection of the tick-borne pathogen *Borrelia burgdorferi*. Readouts include histopathological examination of skin sections, cytokine production in the supernatant, dendritic cells (DC) and Langerhans cell (LC) migration, or DC/LC–T cell interactions, for instance, by a mixed leukocyte reaction. *(This schematic overview is modified from Mason, L., Wagemakers, A., van't Veer, C., Oei, A., van der Pot, W., Ahmed, K., van der Poll, T., Geijtenbeek, T.B., Hovius, J., 2016. Borrelia burgdorferi induces TLR2-mediated migration of activated dendritic cells in an ex vivo human skin model. PLoS One 11, e0164040.)*

(vehicle control) stimuli. Next, using (multiple) 8 mm biopsy punch, a biopsy is taken immediately from around the inoculated skin and after washing is transferred to culture medium with the epidermis facing upward in a culture plate and incubated at 37°C (5% CO_2) for 24–72 h. By performing these experiments in the presence of vector salivary gland extract, saliva, or recombinant salivary gland proteins, the effect of these substances on DDC and/or LC activation can be determined. Standard readouts include counting of the DC population (HLA-DR$^+$/CD11$^+$) migrated from the skin explants into the culture medium and characterization of the DC population by staining for CD1a/langerin to distinguish between LCs and DDCs, and other markers, such as CD14, and determination of maturation (CD80, CD83, CD86, and HLA-DR expression by flow cytometry) and cytokine production (ELISA). The collected DCs may also be cocultured with T lymphocytes (mixed leukocyte reaction) to study adaptive immune responses. Biopsies can either be discarded, snap-frozen to make cryosections for histopathology, or saved in liquid nitrogen for (later) RNA isolation. Apart from being able to study the influence of the DC micro-environment and effects on various subsets of DCs, the model also allows to determine the migration potential of DCs and how this is affected by vector proteins or substances. It should be realized that all DCs that have migrated—also in response to a negative control (PBS)—have a mature and activated phenotype. This model could be extrapolated to other vector-borne pathogens and it provides an alternative to the use of animal models.

In Vivo Culture of Human Skin Cells and Explants

Despite all advantages offered by the *in vitro* skin cell and ex vivo explant cultures, the *in vivo* environment is unique in offering a complex and dynamic scenario for the interplay of pathogens and human cells in the skin. However, ethical and technical impediments hamper the direct analysis of pathogen behavior in humans. The discovery that immunodeficient mice can permanently sustain skin allografts and xenografts, including those of full-thickness human skin explants (Manning et al., 1973), opened the way for the study of pathogens within the human skin engrafted in living rodents. This humanized mouse model has been used to visualize by IFM the dynamic interactions of human-infecting pathogens, such as *Neisseria meningitidis* (Melican et al., 2013) or *P. falciparum*-infected human red blood cells (Ho et al., 2000), with the microvasculature of the human skin. In addition, this type of approach can be supplemented by the introduction of human immune cells (Centlivre et al., 2017; Wrone-Smith and Nickoloff, 1996), bioengineered skin equivalents (Guerrero-Aspizua et al., 2010), and other human cell types (Fujiwara, 2017).

Conclusions and Perspectives

In vitro culture of skin cells helps to decipher the complex interactions of pathogens with the skin at the molecular and cellular level. It allows to better appreciate the role of the different cell types, especially keratinocytes, in the transmission of vector-borne pathogens. The utility of *in vitro* skin cell cultures has been shown in studies on arthropod-borne viral infections (Briant et al., 2014), underlining the specific role of the innate immune response in the development of these infections. Studies using *in vitro* skin cell cultures also emphasize the crucial role of arthropod saliva, containing numerous molecules and proteins that interfere with host defense mechanisms and facilitate pathogen transmission (Bernard et al., 2015; Wikel, 2013). Coculture of skin cells and ex vivo culture of human skin explants further demonstrate the complex signaling between different skin cells and the role of the skin environment during pathogen transmission by arthropods (Mason et al., 2016). All the tools discussed in this chapter, i.e., *in vivo* imaging, *in vitro* skin cell cultures, and ex vivo culture of human skin explants, are necessary to better define the role of the skin as a homing site for vector-borne pathogens and will allow for the development of efficient strategies, including vaccines, to control arthropod-borne diseases.

REFERENCES

Abrahamsson, S., Chen, J., Hajj, B., Stallinga, S., Katsov, A.Y., Wisniewski, J., Mizuguchi, G., Soule, P., Mueller, F., Dugast Darzacq, C., Darzacq, X., Wu, C., Bargmann, C.I., Agard, D.A., Dahan, M., Gustafsson, M.G., 2013. Fast multicolor 3D imaging using aberration-corrected multifocus microscopy. Nature Methods 10, 60–63.

Akinci, E., Bodur, H., Leblebicioglu, H., 2013. Pathogenesis of Crimean-Congo hemorrhagic fever. Vector Borne and Zoonotic Diseases 13, 429–437.

Akkoyunlu, M., Malawista, S., Anguita, J., Fikrig, E., 2001. Exploitation of interleukin-8-induced neutrophil chemotaxis by the agent of human granulocytic ehrlichiosis. Infection and Immunity 69, 5577–5588.

Akol, G.W., Murray, M., 1986. Parasite kinetics and immune responses in efferent prefemoral lymph draining skin reactions induced by tsetse-transmitted *Trypanosoma congolense*. Veterinary Parasitology 19, 281–293.

Alayli, F., Scholle, F., 2016. Dengue virus NS1 enhances viral replication and pro-inflammatory cytokine production in human dendritic cells. Virology 496, 227–236.

Amino, R., Giovannini, D., Thiberge, S., Gueirard, P., Boisson, B., Dubremetz, J.F., Prevost, M.C., Ishino, T., Yuda, M., Menard, R., 2008. Host cell traversal is important for progression of the malaria parasite through the dermis to the liver. Cell Host and Microbe 3, 88–96.

Amino, R., Thiberge, S., Blazquez, S., Baldacci, P., Renaud, O., Shorte, S., Menard, R., 2007. Imaging malaria sporozoites in the dermis of the mammalian host. Nature Protocols 2, 1705–1712.

Amino, R., Thiberge, S., Martin, B., Celli, S., Shorte, S., Frischknecht, F., Menard, R., 2006. Quantitative imaging of *Plasmodium* transmission from mosquito to mammal. Nature Medicine 12, 220–224.

Bagirova, M., Allahverdiyev, A.M., Abamor, E.S., Ullah, I., Cosar, G., Aydogdu, M., Senturk, H., Ergenoglu, B., 2016. Overview of dendritic cell-based vaccine development for leishmaniasis. Parasite Immunology 38, 651–662.

Baluk, P., Fuxe, J., Hashizume, H., Romano, T., Lashnits, E., Butz, S., Vestweber, D., Corada, M., Molendini, C., Dejana, E., McDonald, D.M., 2007. Functionally specialized junctions between endothelial cells of lymphatic vessels. The Journal of Experimental Medicine 204, 2349–2362.

Banchereau, J., Steinman, R., 1998. Dendritic cells and the control of immunity. Nature 392, 245–252.

Barry, J.D., Emergy, D.L., 1984. Parasite development and host responses during the establishment of *Trypanosoma brucei* infection transmitted by tsetse fly. Parasitology 88 (Pt 1), 67–84.

Belkaid, Y., Piccirillo, C.A., Mendez, S., Shevach, E.M., Sacks, D.L., 2002. CD4+CD25+ regulatory T cells control Leishmania major persistence and immunity. Nature 420, 502–507.

Bernard, Q., Jaulhac, B., Boulanger, N., 2015. Skin and arthropods: an effective interaction used by pathogens in vector-borne diseases. European Journal of Dermatology 25 (Suppl. 1), 18–22.

Bernard, Q., Jaulhac, B., Boulanger, N., 2017. Chapter 24: *In vitro* models of cutaneous inflammation. In: Pal, U., Buyuktanir, O. (Ed.), *Borrelia burgdorferi*: Methods and Protocols. Springer.

Bockenstedt, L.K., Gonzalez, D., Mao, J., Li, M., Belperron, A.A., Haberman, A., 2014. What ticks do under your skin: two-photon intravital imaging of *Ixodes scapularis* feeding in the presence of the lyme disease spirochete. The Yale Journal of Biology and Medicine 87, 3–13.

Bowen, J.R., Quicke, K.M., Maddur, M.S., O'Neal, J.T., McDonald, C.E., Fedorova, N.B., Puri, V., Shabman, R.S., Pulendran, B., Suthar, M.S., 2017. Zika virus antagonizes type I interferon responses during infection of human dendritic cells. PLoS Pathogens 13, e1006164.

Boyd, M.F., Kitchen, S.F., 1939. The demonstration of sporozoites in human tissues. The American Journal of Tropical Medicine and Hygiene s1–19, 27–31.

Boyd, M.F., Stratman-Thomas, W.K., 1934. Studies on benign tertian malaria. 7. Some observations on inoculation and onset. American Journal of Epidemiology 20, 488–495.

Briant, L., Desprès, P., Choumet, V., Missé, D., 2014. Role of skin immune cells on the host susceptibility to mosquito-borne viruses. Virology 464–465, 26–32.

Bruce, D., 1895. Preliminary Report on the Tsetse Fly Disease or Nagana, in Zululand. Bennett & Davis, Durban.

Caljon, G., Van Reet, N., De Trez, C., Vermeersch, M., Perez-Morga, D., Van Den Abbeele, J., 2016. The dermis as a delivery site of *Trypanosoma brucei* for tsetse flies. PLoS Pathogens 12, e1005744.

Capewell, P., Cren-Travaille, C., Marchesi, F., Johnston, P., Clucas, C., Benson, R.A., Gorman, T.A., Calvo-Alvarez, E., Crouzols, A., Jouvion, G., Jamonneau, V., Weir, W., Stevenson, M.L., O'Neill, K., Cooper, A., Swar, N.K., Bucheton, B., Ngoyi, D.M., Garside, P., Rotureau, B., MacLeod, A., 2016. The skin is a significant but overlooked anatomical reservoir for vector-borne African trypanosomes. Elife 5.

Carlyon, J.A., Fikrig, E., 2003. Invasion and survival strategies of *Anaplasma phagocytophilum*. Cellular Microbiology 5, 743–754.

Centlivre, M., Petit, M., Hutton, A.J., Dufossee, M., Boccara, D., Mimoun, M., Soria, A., Combadiere, B., 2017. Analysis of the skin of mice humanized for the immune system. Experimental Dermatology 26 (10), 963–966.

Chakravarty, S., Cockburn, I.A., Kuk, S., Overstreet, M.G., Sacci, J.B., Zavala, F., 2007. CD8+ T lymphocytes protective against malaria liver stages are primed in skin-draining lymph nodes. Nature Medicine 13, 1035–1041.

Chalfie, M., Tu, Y., Euskirchen, G., Ward, W.W., Prasher, D.C., 1994. Green fluorescent protein as a marker for gene expression. Science 263, 802–805.

Chanput, W., Mes, J., Wichers, H., November 2014. THP-1 cell line: an in vitro cell model for immune modulation approach. International Immunopharmacology 23 (1), 37–45.

Chmyrov, A., Keller, J., Grotjohann, T., Ratz, M., d'Este, E., Jakobs, S., Eggeling, C., Hell, S.W., 2013. Nanoscopy with more than 100,000 'doughnuts'. Nature Methods 10, 737–740.

Choi, K., Park, J., Dumler, J., 2005. *Anaplasma phagocytophilum* delay of neutrophil apoptosis through the p38 mitogen-activated protein kinase signal pathway. Infection and Immunity 73, 8209–8218.

Coppi, A., Natarajan, R., Pradel, G., Bennett, B.L., James, E.R., Roggero, M.A., Corradin, G., Persson, C., Tewari, R., Sinnis, P., 2011. The malaria circumsporozoite protein has two functional domains, each with distinct roles as sporozoites journey from mosquito to mammalian host. The Journal of Experimental Medicine 208, 341–356.

da Silva Santos, C., Brodskyn, C., 2014. The role of CD4 and CD8 T cells in human cutaneous leishmaniasis. Frontiers in Public Health 2, 165.

de Gruijl, T., Sombroek, C., Lougheed, S.M., Oosterhoff, D., Buter, J., Van Eertwegh, A., Scheper, R.J., Pinedo, H., 2006. A postmigrational switch among skin-derived dendritic cells to a macrophage-like phenotype is predetermined by the intracutaneous cytokine balance. The Journal of Immunology 176, 7232–7242.

Di Meglio, P., Perera, G.K., Nestle, F.O., 2011. The multitasking organ: recent insights into skin immune function. Immunity 35, 857–869.

Diaspro, A., 2002. Confocal and Two-Photon Microscopy: Foundations, Applications, and Advances. Wiley-Liss, New York.

Dobell, C., Leeuwenhoek, A.V., 1932. Antony van Leeuwenhoek and his "Little Animals"; Being Some Account of the Father of Protozoology and Bacteriology and his Multifarious Discoveries in These Disciplines. Harcourt, Brace and Company, New York.

Dumler, J., Choi, K., Garcia-Garcia, J., Barat, N., Scorpio, D., Garyu, J., Grab, D., Bakken, J., 2005. Human granulocytic anaplasmosis and *Anaplasma phagocytophilum*. Emerging Infectious Diseases 11, 1828–1834.

Dwinger, R.H., Murray, M., Moloo, S.K., 1990. Parasite kinetics and cellular responses in goats infected and superinfected with *Trypanosoma congolense* transmitted by *Glossina morsitans* centralis. Acta Tropica 47, 23–33.

Ellinger, P., Hirt, A., 1929. Mikroskopische Beobachtungen an lebenden Organen mit Demonstrationen (Intravitalmikroskopie). Naunyn-Schmiedebergs Archiv Fur Experimentelle Pathologie Und Pharmakologie 147, 63.

Fairbairn, H., Godfrey, D.G., 1957. The local reaction in man at the site of infection with *Trypanosoma rhodesiense*. Annals of Tropical Medicine and Parasitology 51, 464–470.

Fairley, N.H., 1947. Sidelights on malaria in man obtained by subinoculation experiments. Transactions of the Royal Society of Tropical Medicine and Hygiene 40, 621–676.

Filgueira, L., Nestle, F., Rittig, M., Joller, H., Groscurth, P., 1996. Human dendritic cells phagocytose and process *Borrelia burgdorferi*. The Journal of Immunology 157, 2998–3005.

Filipe-Santos, O., Pescher, P., Breart, B., Lippuner, C., Aebischer, T., Glaichenhaus, N., Spath, G.F., Bousso, P., 2009. A dynamic map of antigen recognition by CD4 T cells at the site of *Leishmania* major infection. Cell Host and Microbe 6, 23–33.

Flacher, V., Tripp, C., Stoitzner, P., Haid, B., Ebner, S., Del Frari, B., Koch, F., Park, C., Steinman, R., Idoyaga, J., Romani, N., 2010. Epidermal Langerhans cells rapidly capture and present antigens from C-type lectin-targeting antibodies deposited in the dermis. The Journal of Investigative Dermatology 130, 755–762.

Formaglio, P., Tavares, J., Menard, R., Amino, R., 2014. Loss of host cell plasma membrane integrity following cell traversal by *Plasmodium* sporozoites in the skin. Parasitology International 63, 237–244.

Frevert, U., Movila, A., Nikolskaia, O.V., Raper, J., Mackey, Z.B., Abdulla, M., McKerrow, J., Grab, D.J., 2012. Early invasion of brain parenchyma by African trypanosomes. PLoS One 7, e43913.

Frischknecht, F., Baldacci, P., Martin, B., Zimmer, C., Thiberge, S., Olivo-Marin, J.C., Shorte, S.L., Menard, R., 2004. Imaging movement of malaria parasites during transmission by Anopheles mosquitoes. Cellular Microbiology 6, 687–694.

Fujiwara, S., 2017. Humanized mice: a brief overview on their diverse applications in biomedical research. Journal of Cellular Physiology.

Geijtenbeek, T., Torensma, R., van Vliet, S., van Duijnhoven, G., Adema, G.J., Van Kooyk, Y., Figdor, C., 2000. Identification of DC-SIGN, a novel dendritic cell-specific ICAM-3 receptor that supports primary immune responses. Cell 100, 575–585.

Gordon, R.M., Lumsden, W.H.R., 1939. A study of the behaviour of the mouth-parts of mosquitoes when taking up blood from living tissue; together with some observations on the ingestion of microfilariae. Annals of Tropical Medicine and Parasitology 33, 259–278.

Graf, R., Rietdorf, J., Zimmermann, T., 2005. Live cell spinning disk microscopy. Advances in Biochemical Engineering/Biotechnology 95, 57–75.

Grassi, B., 1898. La malaria propagata per mezzo di peculiari insetti. Atti della Reale Accademia dei Lincei. Rendiconti. Classe di scienze fisiche matematiche e naturali. Serie 5 7, 234–240.

Griffiths, R.B., Gordon, R.M., 1952. An apparatus which enables the process of feeding by mosquitoes to be observed in the tissues of a live rodent; together with an account of the ejection of saliva and its significance in malaria. Annals of Tropical Medicine and Parasitology 46, 311–319.

Grotjohann, T., Testa, I., Reuss, M., Brakemann, T., Eggeling, C., Hell, S.W., Jakobs, S., 2012. rsEGFP2 enables fast RESOLFT nanoscopy of living cells. Elife 1, e00248.

Gueirard, P., Tavares, J., Thiberge, S., Bernex, F., Ishino, T., Milon, G., Franke-Fayard, B., Janse, C.J., Menard, R., Amino, R., 2010. Development of the malaria parasite in the skin of the mammalian host. Proceedings of the National Academy of Sciences of the United States of America 107, 18640–18645.

Guerrero-Aspizua, S., Garcia, M., Murillas, R., Retamosa, L., Illera, N., Duarte, B., Holguin, A., Puig, S., Hernandez, M.I., Meana, A., Jorcano, J.L., Larcher, F., Carretero, M., Del Rio, M., 2010. Development of a bioengineered skin-humanized mouse model for psoriasis: dissecting epidermal-lymphocyte interacting pathways. The American Journal of Pathology 177, 3112–3124.

Hajj, B., Wisniewski, J., El Beheiry, M., Chen, J., Revyakin, A., Wu, C., Dahan, M., 2014. Whole-cell, multicolor superresolution imaging using volumetric multifocus microscopy. Proceedings of the National Academy of Sciences of the United States of America 111, 17480–17485.

Harris, D.F., 1921. Anthony Van Leeuwenhoek the first bacteriologist. The Scientific Monthly 12, 150–160.

Heimstädt, O., 1911. Das Fluoreszenzmikroskop. Zeitschrift Fur Wissenschaftliche Mikroskopie Und Mikroskopische Technik 28, 330–337.

Hell, S.W., Wichmann, J., 1994. Breaking the diffraction resolution limit by stimulated emission: stimulated-emission-depletion fluorescence microscopy. Optics Letters 19, 780–782.

Hellmann, J.K., Munter, S., Kudryashev, M., Schulz, S., Heiss, K., Muller, A.K., Matuschewski, K., Spatz, J.P., Schwarz, U.S., Frischknecht, F., 2011. Environmental constraints guide migration of malaria parasites during transmission. PLoS Pathogens 7, e1002080.

Helmchen, F., Denk, W., 2005. Deep tissue two-photon microscopy. Nature Methods 2, 932–940.

Hirschfeld, M., Kirschning, C.J., Schwandner, R., Wesche, H., Weis, J.H., Wooten, R.M., Weis, J.J., 1999. Cutting edge: inflammatory signaling by *Borrelia burgdorferi* lipoproteins is mediated by toll-like receptor 2. The Journal of Immunology 163, 2382–2386.

Ho, M., Hickey, M.J., Murray, A.G., Andonegui, G., Kubes, P., 2000. Visualization of *Plasmodium falciparum*-endothelium interactions in human microvasculature: mimicry of leukocyte recruitment. The Journal of Experimental Medicine 192, 1205–1211.

Hofmann, M., Eggeling, C., Jakobs, S., Hell, S.W., 2005. Breaking the diffraction barrier in fluorescence microscopy at low light intensities by using reversibly photoswitchable proteins. Proceedings of the National Academy of Sciences of the United States of America 102, 17565–17569.

Hopp, C.S., Chiou, K., Ragheb, D.R., Salman, A.M., Khan, S.M., Liu, A.J., Sinnis, P., 2015. Longitudinal analysis of *Plasmodium* sporozoite motility in the dermis reveals component of blood vessel recognition. Elife 4.

Hovius, J.W., Bijlsma, M.F., van der Windt, G.J., Wiersinga, W.J., Boukens, B.J., Coumou, J., Oei, A., de Beer, R., de Vos, A.F., van't Veer, C., van Dam, A.P., Wang, P., Fikrig, E., Levi, M.M., Roelofs, J.J., van der Poll, T., 2009. The urokinase receptor (uPAR) facilitates clearance of *Borrelia burgdorferi*. PLoS Pathogens 5, e1000447.

Hovius, J.W.R., De Jong, M.A.W.P., Dunnen, J.D., Litjens, M., Fikrig, E., Van Der Poll, T., Gringhuis, S.I., Geijtenbeek, T.B.H., 2008. Salp15 binding to DC-SIGN inhibits cytokine expression by impairing both nucleosome remodeling and mRNA stabilization. PLoS Pathogens 4.

Hovius, J.W.R., van Dam, A.P., Fikrig, E., 2007. Tick-host-pathogen interactions in Lyme borreliosis. Trends in Parasitology 23, 434–438.

Huang, B., Bates, M., Zhuang, X., 2009. Super-resolution fluorescence microscopy. Annual Review of Biochemistry 78, 993–1016.

Hudson, A., Bowman, L., Orr, C.W., 1960. Effects of absence of saliva on blood feeding by mosquitoes. Science 131, 1730–1731.

Huff, C.G., 1957. Organ and tissue distribution of the exoerythrocytic stages of various avian malarial parasites. Experimental Parasitology 6, 143–162.

Ichihara, A., Tanaami, T., Isozaki, K., Sugiyama, Y., Kosugi, Y., Mikuriya, K., Abe, M., Uemura, I., 1996. High-speed confocal fluorescence microscopy using a Nipkow scanner with microlenses for 3-D imaging of single fluorescent molecule in real time. Bioimages 4, 57–62.

Ishino, T., Chinzei, Y., Yuda, M., 2005. A *Plasmodium* sporozoite protein with a membrane attack complex domain is required for breaching the liver sinusoidal cell layer prior to hepatocyte infection. Cellular Microbiology 7, 199–208.

Ishino, T., Yano, K., Chinzei, Y., Yuda, M., 2004. Cell-passage activity is required for the malarial parasite to cross the liver sinusoidal cell layer. PLoS Biology 2, E4.

Jain, R., Mitchell, A., Tay, S., Roediger, B., Weninger, W., 2016. Neutrophils. In: Kabashima, K. (Ed.), Immunology of the Skin: Basic and Clinical Sciences in Skin Immune Responses. Spinger, p. 510.

Juncadella, I.J., Garg, R., Ananthnarayanan, S.K., Yengo, C.M., Anguita, J., 2007. T-cell signaling pathways inhibited by the tick saliva immunosuppressor, Salp15. FEMS Immunology and Medical Microbiology 49, 433–438.

Kautz-Neu, K., Noordegraaf, M., Dinges, S., Bennett, C.L., John, D., Clausen, B.E., von Stebut, E., 2011. Langerhans cells are negative regulators of the anti-*Leishmania* response. The Journal of Experimental Medicine 208, 885–891.

Kebaier, C., Voza, T., Vanderberg, J., 2009. Kinetics of mosquito-injected *Plasmodium* sporozoites in mice: fewer sporozoites are injected into sporozoite-immunized mice. PLoS Pathogens 5, e1000399.

Klar, T.A., Jakobs, S., Dyba, M., Egner, A., Hell, S.W., 2000. Fluorescence microscopy with diffraction resolution barrier broken by stimulated emission. Proceedings of the National Academy of Sciences of the United States of America 97, 8206–8210.

Konradt, C., Ueno, N., Christian, D.A., Delong, J.H., Pritchard, G.H., Herz, J., Bzik, D.J., Koshy, A.A., McGavern, D.B., Lodoen, M.B., Hunter, C.A., 2016. Endothelial cells are a replicative niche for entry of Toxoplasma gondii to the central nervous system. Nature Microbiology 1, 16001.

Kubba, R., El-Hassan, A.M., Al-Gindan, Y., Omer, A.H., Kutty, M.K., Saeed, M.B., 1987. Dissemination in cutaneous leishmaniasis. I. Subcutaneous nodules. International Journal of Dermatology 26, 300–304.

Laskay, T., van Zandbergen, G., Solbach, W., 2003. Neutrophil granulocytes–Trojan horses for *Leishmania* major and other intracellular microbes? Trends in Microbiology 11, 210–214.

Laveran, A., 1881. Nature parasitaire des accidents de l'impaludisme:description d'un nouveau parasite trouvé dans le sang des malades atteints de fièvre palustre. J.-B. Baillière, Paris.

Lavoipierre, M.M., 1965. Feeding mechanism of blood-sucking arthropods. Nature 208, 302–303.

Leboulle, G., Crippa, M., Decrem, Y., Mejri, N., Brossard, M., Bollen, A., Godfroid, E., 2002. Characterization of a novel salivary immunosuppressive protein from Ixodes ricinus ticks. The Journal of Biological Chemistry 277, 10083–10089.

Li, J.L., Goh, C.C., Keeble, J.L., Qin, J.S., Roediger, B., Jain, R., Wang, Y., Chew, W.K., Weninger, W., Ng, L.G., 2012. Intravital multiphoton imaging of immune responses in the mouse ear skin. Nature Protocols 7, 221–234.

Lieskovska, J., Kopecky, J., 2012. Effect of tick saliva on signalling pathways activated by TLR-2 ligand and *Borrelia afzelii* in dendritic cells. Parasite Immunology 34, 421–429.

Lozach, P.Y., Kuhbacher, A., Meier, R., Mancini, R., Bitto, D., Bouloy, M., Helenius, A., 2011. DC-SIGN as a receptor for phleboviruses. Cell Host and Microbe 10, 75–88.

Mac-Daniel, L., Buckwalter, M., Berthet, M., Virk, Y., Yui, K., Albert, M., Gueirard, P., Ménard, R., 2014. Local immune response to injection of *Plasmodium* sporozoites into the skin. The Journal of Immunology 193, 1246–1257.

Manning, D.D., Reed, N.D., Shaffer, C.F., 1973. Maintenance of skin xenografts of widely divergent phylogenetic origin of congenitally athymic (nude) mice. The Journal of Experimental Medicine 138, 488–494.

Manson, P., 1878. On the development of Filaria sanguinis hominis, and on the mosquito considered as a nurse. Journal of the Linnean Society of London, Zoology 14, 304–311.

Marchal, C., Luft, B., Yang, X., Sibilia, J., Jaulhac, B., Boulanger, N., 2009. Defensin is suppressed by tick salivary gland extract during the in vitro interaction of resident skin cells with *Borrelia burgdorferi*. The Journal of Investigative Dermatology 129, 2515–2517.

Marchal, C., Schramm, F., Kern, A., Luft, B.J., Yang, X., Schuijt, T.J., Hovius, J., Jaulhac, B., Boulanger, N., 2011. Antialarmin effect of tick saliva during the transmission of Lyme disease. Infection and Immunity 79, 774–785.

Mason, L., Herkes, E., Krupna-Gaylord, M., Oei, A., van der Poll, T., Wormser, G., Schwartz, I., Petzke, M., Hovius, J., 2015. *Borrelia burgdorferi* clinical isolates induce human innate immune responses that are not dependent on genotype. Immunobiology 220, 1141–1150.

Mason, L., Wagemakers, A., van't Veer, C., Oei, A., van der Pot, W., Ahmed, K., van der Poll, T., Geijtenbeek, T.B., Hovius, J., 2016. *Borrelia burgdorferi* induces TLR2-mediated migration of activated dendritic cells in an ex vivo human skin model. PLoS One 11, e0164040.

Mason, L.M.K., Veerman, C.C., Geijtenbeek, T.B.H., Hovius, J.W.R., 2014. Ménage à trois: Borrelia, dendritic cells, and tick saliva interactions. Trends in Parasitology 30, 95–103.

Melican, K., Michea Veloso, P., Martin, T., Bruneval, P., Dumenil, G., 2013. Adhesion of *Neisseria meningitidis* to dermal vessels leads to local vascular damage and purpura in a humanized mouse model. PLoS Pathogens 9, e1003139.

Missiroli, A., 1934. Sullo sviluppo dei parassiti malarici. Rivista di Malariologia 13, 539–552.

Moore, M., Cruz, A., LaVake, C., Marzo, A., Eggers, C., Salazar, J., Radolf, J., 2007. Phagocytosis of *Borrelia burgdorferi* and *Treponema pallidum* potentiates innate immune activation and induces gamma interferon production. Infection and Immunity 75, 2046–2062.

Moriarty, T.J., Norman, M.U., Colarusso, P., Bankhead, T., Kubes, P., Chaconas, G., 2008. Real-time high resolution 3D imaging of the lyme disease spirochete adhering to and escaping from the vasculature of a living host. PLoS Pathogens 4, e1000090.

Mota, M.M., Pradel, G., Vanderberg, J.P., Hafalla, J.C., Frevert, U., Nussenzweig, R.S., Nussenzweig, V., Rodriguez, A., 2001. Migration of *Plasmodium* sporozoites through cells before infection. Science 291, 141–144.

Müller, A.J., Aeschlimann, S., Olekhnovitch, R., Dacher, M., Spath, G.F., Bousso, P., 2013. Photoconvertible pathogen labeling reveals nitric oxide control of *Leishmania* major infection in vivo via dampening of parasite metabolism. Cell Host and Microbe 14, 460–467.

Natarajan, R., Thathy, V., Mota, M.M., Hafalla, J.C., Menard, R., Vernick, K.D., 2001. Fluorescent *Plasmodium berghei* sporozoites and pre-erythrocytic stages: a new tool to study mosquito and mammalian host interactions with malaria parasites. Cellular Microbiology 3, 371–379.

Nathan, C., 2006. Neutrophils and immunity: challenges and opportunities. Nature Reviews. Immunology 6, 173–182.

Nestle, F.O., Di Meglio, P., Qin, J.-Z., Nickoloff, B.J., 2009. Skin immune sentinels in health and disease. Nature Reviews. Immunology 9, 679–691.

Ng, L.G., Hsu, A., Mandell, M.A., Roediger, B., Hoeller, C., Mrass, P., Iparraguirre, A., Cavanagh, L.L., Triccas, J.A., Beverley, S.M., Scott, P., Weninger, W., 2008. Migratory dermal dendritic cells act as rapid sensors of protozoan parasites. PLoS Pathogens 4, e1000222.

Nomura, T., Shinohara, A., 2016. T cells. In: Kabashima, K. (Ed.), Immunology of the Skin: Basic and Clinical Sciences in Skin Immune Responses. Springer, Japan, Tokyo, pp. 57–94.

O'Rourke, F.J., 1956. Observations on pool and capillary feeding in *Aedes aegypti* (L.). Nature 177, 1087–1088.

Olagnier, D., Peri, S., Steel, C., van Montfoort, N., Chiang, C., Beljanski, V., Slifker, M., He, Z., Nichols, C.N., Lin, R., Balachandran, S., Hiscott, J., 2014. Cellular oxidative stress response controls the antiviral and apoptotic programs in dengue virus-infected dendritic cells. PLoS Pathogens 10, e1004566.

Olsen, O.W., 1986. Animal Parasites: Their Life Cycles and Ecology. Dover, New York.

Ouologuem, D.T., Roos, D.S., 2014. Dynamics of the Toxoplasma gondii inner membrane complex. Journal of Cell Science 127, 3320–3330.

Perry, S.T., Buck, M.D., Lada, S.M., Schindler, C., Shresta, S., 2011. STAT2 mediates innate immunity to Dengue virus in the absence of STAT1 via the type I interferon receptor. PLoS Pathogens 7, e1001297.

Peters, N.C., Egen, J.G., Secundino, N., Debrabant, A., Kimblin, N., Kamhawi, S., Lawyer, P., Fay, M.P., Germain, R.N., Sacks, D., 2008. In vivo imaging reveals an essential role for neutrophils in leishmaniasis transmitted by sand flies. Science 321, 970–974.

Petzke, M.M., Brooks, A., Krupna, M.a., Mordue, D., Schwartz, I., 2009. Recognition of *Borrelia burgdorferi*, the Lyme disease spirochete, by TLR7 and TLR9 induces a type I IFN response by human immune cells. Journal of Immunology (Baltimore, Md.: 1950) 183, 5279–5292.

Pivarcsi, A., Nagy, I., Kemeny, L., 2005. Innate immunity in the skin: how keratinocytes fight against pathogens. Current Immunology Reviews 1, 29–42.

Ploss, A., Evans, M.J., Gaysinskaya, V.A., Panis, M., You, H., de Jong, Y.P., Rice, C.M., 2009. Human occludin is a hepatitis C virus entry factor required for infection of mouse cells. Nature 457, 882–886.

Radtke, A.J., Kastenmuller, W., Espinosa, D.A., Gerner, M.Y., Tse, S.W., Sinnis, P., Germain, R.N., Zavala, F.P., Cockburn, I.A., 2015. Lymph-node resident CD8alpha⁺ dendritic cells capture antigens from migratory malaria sporozoites and induce CD8⁺ T cell responses. PLoS Pathogens 11, e1004637.

Ramos-e-Silva, M., 1998. Giovan Cosimo Bonomo (1663–1696): discoverer of the etiology of scabies. International Journal of Dermatology 37, 625–630.

Rankin, K.E., Tavares, J., Struck, N.S., Menard, R., Heussler, V.T., Amino, R., 2010. A novel type of *Plasmodium* sporozoite transcellular migration. Malaria Journal 9, P41.

Reed, W., Carroll, J., Agramonte, A., Lazear, J.W., 1900. The etiology of yellow fever-a preliminary note. Public Health Papers and Reports 26, 37–53.

Ribeiro, J.M., 1987. Role of saliva in blood-feeding by arthropods. Annual Review of Entomology 32, 463–478.

Ribeiro, J.M., Rossignol, P.A., Spielman, A., 1984. Role of mosquito saliva in blood vessel location. The Journal of Experimental Biology 108, 1–7.

Ribeiro-Gomes, F.L., Peters, N.C., Debrabant, A., Sacks, D.L., 2012. Efficient capture of infected neutrophils by dendritic cells in the skin inhibits the early anti-leishmania response. PLoS Pathogens 8, e1002536.

Risco-Castillo, V., Topcu, S., Marinach, C., Manzoni, G., Bigorgne, A.E., Briquet, S., Baudin, X., Lebrun, M., Dubremetz, J.F., Silvie, O., 2015. Malaria sporozoites traverse host cells within transient vacuoles. Cell Host and Microbe 18, 593–603.

Roberts, C.J., Gray, M.A., Gray, A.R., 1969. Local skin reactions in cattle at the site of infection with *Trypanosoma congolense* by *Glossina morsitans* and *G. tachinoides*. Transactions of the Royal Society of Tropical Medicine and Hygiene 63, 620–624.

Romani, N., Gruner, S., Brang, D., Kämpgen, E., Lenz, A., Trockenbacher, B., Konwalinka, G., Fritsch, P., Steinman, R., Schuler, G., 1994. Proliferating dendritic cell progenitors in human blood. The Journal of Experimental Medicine 180, 83–93.

Ross, R., 1898. The role of mosquito in the evolution of the malarial parasite. The Lancet 152, 488–490.

Sa-Nunes, A., Bafica, A., Antonelli, L.R., Choi, E.Y., Francischetti, I.M., Andersen, J.F., Shi, G.P., Chavakis, T., Ribeiro, J.M., Kotsyfakis, M., 2009. The immunomodulatory action of sialo-statin L on dendritic cells reveals its potential to interfere with autoimmunity. The Journal of Immunology 182, 7422–7429.

Sa-Nunes, A., Bafica, A., Lucas, D.A., Conrads, T.P., Veenstra, T.D., Andersen, J.F., Mather, T.N., Ribeiro, J.M., Francischetti, I.M., 2007. Prostaglandin E2 is a major inhibitor of dendritic cell maturation and function in *Ixodes scapularis* saliva. The Journal of Immunology 179, 1497–1505.

Schauber, J., Gallo, R., 2009. Antimicrobial peptides and the skin immune defense system. The Journal of Allergy and Clinical Immunology 124, R13–R18.

Schmid, M., Diamond, M., Harris, E., 2014. Dendritic cells in dengue virus infection: targets of virus replication and mediators of immunity. Frontiers in Immunology 5, 647.

Schneider, L., Schoonderwoerd, A.J., Moutaftsi, M., Howard, R., Reed, S., de Jong, E., Teunissen, M., 2012. Intradermally administered TLR4 agonist GLA-SE enhances the capacity of human skin DCs to activate T cells and promotes emigration of Langerhans cells. Vaccine 30, 4216–4224.

Schnorrenberg, S., Grotjohann, T., Vorbruggen, G., Herzig, A., Hell, S.W., Jakobs, S., 2016. In vivo super-resolution RESOLFT microscopy of *Drosophila melanogaster*. Elife 5.

Schramm, F., Kern, A., Barthel, C., Nadaud, S., Meyer, N., Jaulhac, B., Boulanger, N., 2012. Microarray analyses of inflammation response of human dermal fibroblasts to different strains of *Borrelia burgdorferi* sensu stricto. PLoS One 7, e40046.

Sebbane, F., Jarrett, C.O., Gardner, D., Long, D., Hinnebusch, B.J., 2006. Role of the *Yersinia pestis* plasminogen activator in the incidence of distinct septicemic and bubonic forms of flea-borne plague. Proceedings of the National Academy of Sciences of the United States of America 103, 5526–5530.

Semnani, R., Law, M., Kubofcik, J., Nutman, T., May 15, 2004. Filaria-induced immune evasion: suppression by the infective stage of *Brugia malayi* at the earliest host-parasite interface. The Journal of Immunology 172 (10), 6229–6238.

Seo, M., Kang, T., Lee, C., Lee, A., Noh, M., 2012. HaCaT keratinocytes and primary epidermal keratinocytes have different transcriptional profiles of cornified envelope-associated genes to T helper cell cytokines. Biomolecules and Therapeutics (Seoul) 20, 171–176.

Shannon, J.G., Bosio, C.F., Hinnebusch, B.J., 2015. Dermal neutrophil, macrophage and dendritic cell responses to *Yersinia pestis* transmitted by fleas. PLoS Pathogens 11, e1004734.

Shih, C.M., Pollack, R.J., Telford 3rd, S.R., Spielman, A., 1992. Delayed dissemination of Lyme disease spirochetes from the site of deposition in the skin of mice. The Journal of Infectious Diseases 166, 827–831.

Shimomura, O., Johnson, F.H., Saiga, Y., 1962. Extraction, purification and properties of aequorin, a bioluminescent protein from the luminous hydromedusan, Aequorea. Journal of Cellular and Comparative Physiology 59, 223–239.

Sidjanski, S., Vanderberg, J.P., 1997. Delayed migration of *Plasmodium* sporozoites from the mosquito bite site to the blood. The American Journal of Tropical Medicine and Hygiene 57, 426–429.

Simond, P.-L., 1898. La propagation de la peste. Annales de l'Institut Pasteur, pp. 625–687.

Skallová, A., Iezzi, G., Ampenberger, F., Kopf, M., Kopecky, J., 2008. Tick saliva inhibits dendritic cell migration, maturation, and function while promoting development of Th2 responses. Journal of Immunology (Baltimore, Md.: 1950) 180, 6186–6192.

Smith, T., Kilborne, F.L., 1893. Investigations into the nature, causation, and prevention of Texas or southern cattle fever. Bureau of Animal Industry 1, 301–324.

So, P.T., Dong, C.Y., Masters, B.R., Berland, K.M., 2000. Two-photon excitation fluorescence microscopy. Annual Review of Biomedical Engineering 2, 399–429.

Spinner, J., Winfree, S., Starr, T., Shannon, J., Nair, V., Steele-Mortimer, O., Hinnebusch, B., 2014. *Yersinia pestis* survival and replication within human neutrophil phagosomes and uptake of infected neutrophils by macrophages. Journal of Leukocyte Biology 95, 389–398.

Sprokholt, J.K., Kaptein, T.M., van Hamme, J.L., Overmars, R.J., Gringhuis, S.I., Geijtenbeek, T.B.H., 2017. RIG-I-like receptor triggering by dengue virus drives dendritic cell immune activation and TH1 differentiation. The Journal of Immunology 198, 4764–4771.

Stijlemans, B., Leng, L., Brys, L., Sparkes, A., Vansintjan, L., Caljon, G., Raes, G., Van Den Abbeele, J., Van Ginderachter, J., Beschin, A., Bucala, R., De Baetselier, P., 2014. MIF contributes to *Trypanosoma brucei* associated immunopathogenicity development. PLoS Pathogens 10, e1004414.

Sturm, A., Amino, R., van de Sand, C., Regen, T., Retzlaff, S., Rennenberg, A., Krueger, A., Pollok, J.M., Menard, R., Heussler, V.T., 2006. Manipulation of host hepatocytes by the malaria parasite for delivery into liver sinusoids. Science 313, 1287–1290.

Suhonen, J., Komi, J., Soukka, J., Lassila, O., Viljanen, M., 2003. Interaction between *Borrelia burgdorferi* and immature human dendritic cells. Scandinavian Journal of Immunology 58, 67–75.

Surasombatpattana, P., Patramool, S., Luplertlop, N., Yssel, H., Misse, D., 2012. *Aedes aegypti* saliva enhances dengue virus infection of human keratinocytes by suppressing innate immune responses. The Journal of Investigative Dermatology 132, 2103–2105.

Tavares, J., Formaglio, P., Thiberge, S., Mordelet, E., Van Rooijen, N., Medvinsky, A., Menard, R., Amino, R., 2013. Role of host cell traversal by the malaria sporozoite during liver infection. The Journal of Experimental Medicine 210, 905–915.

Tsien, R.Y., 1998. The green fluorescent protein. Annual Review of Biochemistry 67, 509–544.

Turell, M.J., Spielman, A., 1992. Nonvascular delivery of Rift Valley fever virus by infected mosquitoes. The American Journal of Tropical Medicine and Hygiene 47, 190–194.

van Dijk, M.R., Janse, C.J., Waters, A.P., 1996. Expression of a *Plasmodium* gene introduced into subtelomeric regions of *Plasmodium berghei* chromosomes. Science 271, 662–665.

Van Helden, A., 2010. The Origins of the Telescope. KNAW Press, Amsterdam.

Vanderberg, J.P., 1974. Studies on the motility of *Plasmodium* sporozoites. The Journal of Protozoology 21, 527–537.

Vanderberg, J.P., Frevert, U., 2004. Intravital microscopy demonstrating antibody-mediated immobilisation of *Plasmodium berghei* sporozoites injected into skin by mosquitoes. International Journal for Parasitology 34, 991–996.

Varro, M.T., Storr-Best, L., 1912. Varro on Farming. M. Terenti Varronis Rerum Rusticarum Libri Tres. G. Bell and Sons, Ltd., London.

Viswanathan, G.M., Buldyrev, S.V., Havlin, S., da Luz, M.G., Raposo, E.P., Stanley, H.E., 1999. Optimizing the success of random searches. Nature 401, 911–914.

Voza, T., Miller, J.L., Kappe, S.H., Sinnis, P., 2012. Extrahepatic exoerythrocytic forms of rodent malaria parasites at the site of inoculation: clearance after immunization, susceptibility to primaquine, and contribution to blood-stage infection. Infection and Immunity 80, 2158–2164.

Werner, S., Krieg, T., Smola, H., 2007. Keratinocyte-fibroblast interactions in wound healing. The Journal of Investigative Dermatology 127, 998–1008.

Wikel, S.K., 2013. Ticks and tick-borne pathogens at the cutaneous interface: host defenses, tick countermeasures, and a suitable environment for pathogen establishment. Frontiers in Microbiology 4, 337.

Wikel, S.K., Bergman, D., 1997. Tick-host immunology: significant advances and challenging opportunities. Parasitology Today (Personal Edition) 13, 383–389.

Wrone-Smith, T., Nickoloff, B.J., 1996. Dermal injection of immunocytes induces psoriasis. The Journal of Clinical Investigation 98, 1878–1887.

Wu, Y., Kirkman, L.A., Wellems, T.E., 1996. Transformation of *Plasmodium falciparum* malaria parasites by homologous integration of plasmids that confer resistance to pyrimethamine. Proceedings of the National Academy of Sciences of the United States of America 93, 1130–1134.

Yamauchi, L.M., Coppi, A., Snounou, G., Sinnis, P., 2007. *Plasmodium* sporozoites trickle out of the injection site. Cellular Microbiology 9, 1215–1222.

Yoeli, M., Most, H., 1971. Sporozoite-induced infections of *Plasmodium berghei* administered by the oral route. Science 173, 1031–1032.

Yu, X., Cai, B., Wang, M., Tan, P., Ding, X., Wu, J., Li, J., Li, Q., Liu, P., Xing, C., Wang, H.Y., Su, X.Z., Wang, R.F., 2016. Cross-regulation of two type I interferon signaling pathways in plasmacytoid dendritic cells controls anti-malaria immunity and host mortality. Immunity 45, 1093–1107.

Index

Printed in the United States
By Bookmasters